KU-439-861

Behavioral Adaptation to Intertidal Life

Edited by

Guido Chelazzi

and

Marco Vannini

University of Florence
Florence, Italy

Plenum Press
New York and London
Published in cooperation with NATO Scientific Affairs Division

Proceedings of a NATO Advanced Research Workshop on
Behavioral Adaptation to Intertidal Life,
held May 19-24, 1987,
in Castiglioncello, Italy

Library of Congress Cataloging in Publication Data

NATO Advanced Research Workshop on Behavioral Adaptation to Intertidal Life (1987: Castiglioncello, Italy)
Behavioral adaptation to intertidal life.

(NATO ASI series. Series A, Life sciences; v. 151)
"Proceedings of a NATO Advanced Research Workshop on Behavioral Adaptation to Intertidal Life, held May 19-24, 1987, in Castiglioncello, Italy" —T.p. verso.
"Published in cooperation with NATO Scientific Affairs Division."
Includes bibliographical references and index.
1. Intertidal fauna—Behavior—Congresses. I. Chelazzi, Guido. II. Vannini, Marco. III. North Atlantic Treaty Organization. Scientific Affairs Division. IV. Title. V. Series.
QL121.N27 1987 591.5′2638 88-17956
ISBN 0-306-42930-6

A Division of Plenum Publishing Corporation
233 Spring Street, New York, N.Y. 10013

Printed in the United States of America

Preface

The NATO Advanced Research Workshop on "Behavioural Adaptation to Intertidal Life" held in Castiglioncello, Italy (May, 1987) was attended by 50 participants, most of whom presented requested lectures. It was perhaps the first time that specialists of various animal groups, from cnidarians to birds, were able to meet and discuss the importance of behavioural adaptation to this peculiar, sometimes very harsh environment. But the taxonomic barrier is not the only one which the meeting attemped to overcome. Lately, the research on intertidal biology has spread from pure taxonomy and static analysis of community structure to such dynamic aspects as intra- and interspecific relationships, and physiological mechanisms aimed at avoiding stress and exploitation of limited-resources. This increasing interest stems not only from an inclination for this particular ecological system and some of its typical inhabitants, but also from the realization that rocky and sandy shore communities are suitable models for testing and improving some global theories of evolutionary biology, behavioural ecology and sociobiology.

The number of eco-physiological and eco-ethological problems emerging from the study of intertidal animals is fascinatingly large and a complete understanding of this environment cannot be reached using a strictly "reductionistic" or a pure "holistic" approach. It was our conviction that discussion between experts on such differening topics as orientation and communication mechanisms and community ecology was probably the only way to link the microscopic to the macroscopic phenomena, the behaviour of individuals to species zonation, geographical distribution and temporal fluctuation of intertidal communities. Among the key topics were: i) identification of the main behavioural strategies in the intertidal fauna, despite its great taxonomic diversity; ii) the analysis of behaviourally mediated relationships between individual species and their physical environment, with special reference to the complexity and predictability of the temporal and spatial structure of the latter; iii) assessment of the importance of coevolution in shaping behaviour, a factor regulating intra- and interspecific relationships.

Another goal of the meeting was the solution of methodological problems, held to be of paramount importance in behavioural ecology research. Rather than a discussion on strictly instrumental virtuosisms in data collection and analysis, we requested critical reconsideration of study strategy and experimental design in the analysis of such problems as the determinism of rhythmic activity, interlocking of different orientation mechanisms, and behavioural regulation of population density. Our expectations were not disappointed: these topics were touched in many papers. This attempt to compare different areas and experiences was not exempt from costs and a surface look at this volume may produce the impression that merging was not always accomplished. This is true not only for marginal or detailed topics, but also in major problems such as the degree of intertidal environment predictability which dominated many evening discussions, and which remains open for further debate. In general, the physio-ethologists felt more "optimistic" concerning predictability, because the behaviour they analyze (rhythms, orientation, communication) becomes an important adaptive tool when and where ecologically significant signals emerge above the ground noise. On the contrary, most "large scale" eco-ethologists are much impressed by the complexity and unpredictability of the intertidal environment. The impression is that these two positions are not reconcilable since both have their roots in complementary aspects of this environment. A second problem which reveals some asynchrony between experts is the concept of limiting factor. This may depend on differences in the study cases chosen to test the different hypotheses; nevertheless, a clear distinction between proximate and ultimate factors may improve the discussion on this problem.

We hope that, despite some "physiological" heterogeneity, this volume will help in putting the picture of the intertidal environment into focus. Today, this effort for a non applicative field could appear inappropriate, considering the increasingly serious threat to coastal integrity. Despite their interest in the theoretical aspects of intertidal biology, the participants to the Castiglioncello NATO ARW were definitely not indifferent to the problems of natural conservation and took the opportunity of petitioning the competent Authorities to keep a closer vigilance on preserving the coastal environment, particularly in the Mediterranean sea.

We take this opportunity to thank the NATO Scientific Affairs Division for having encouraged and supported the B.A.I.L. Workshop. The contribution of the Universita' degli Studi di Firenze and of the Centro di Studio per la Faunistica ed Ecologia Tropicali del C.N.R. is also acknowledged, together with that of the Accademia dei Lincei. We wish to thank the Municipality of Rosignano Marittima for having put the beautiful Castello Pasquini at our disposal, and Alitalia and Banca Toscana for their consistent help.
We are pleased to acknowledge the important organizational work done by Sarah Whitman. We are also personally grateful to Stefano Focardi and Alessia Mascherini who assisted with the preparation of the work for the press. Silvia Guidi, Fiorenza Micheli, Federica Tarducci, Giovanni Checcucci and Paolo Della Santina were of relevant assistance prior and during the meeting; this Workshop would not have been accomplished without their help and cooperation.

Guido Chelazzi

Marco Vannini

Contents

RHYTHMIC BEHAVIOUR

SPATIAL BEHAVIOUR

Clock-Controlled Behaviour in Intertidal Animals

Ernest Naylor

University College of North Wales
Bangor, U.K.

INTRODUCTION

The behavioural patterns of sessile and motile intertidal animals often consist of rhythmic sequences of movements which are correlated with environmental variables of daily or tidal periodicity. Many such rhythms are driven solely by environmental variables and hence are responses, but others free-run in constant conditions under the control of internal physiological pacemakers. The latter are often expressed at periodicities which approximate to those of geophysical variables defined, for example, as circatidal, circadian and circasemilunar rhythms (see Naylor 1985). Such endogenous rhythmicity appears to be a general feature amongst eukaryotes (Brady 1982) and most evidence suggests that it is innate and therefore a true genetic adaptation (Naylor 1987).

In the context of the general topic of behavioural adaptations to intertidal life it is appropriate to consider a number of aspects of clock-controlled rhythmic behaviour. These include the adaptive value of tidal and diel rhythms of behaviour in relation to tidal rise and fall, the relative significance of responsive compared with endogenous patterns of activity, the role of behavioural rhythms in the maintenance of intertidal zonation, and orientational aspects of diel and tidal migrational rhythms. In the same context it is also important to compare the behaviour of coastal animals from areas of high tidal amplitudes, as on Atlantic European coasts, with that of related animals from areas of low tidal amplitude, as in the Mediterranean.

BEHAVIOURAL RHYTHMS AND TIDAL OSCILLATIONS

Casual observations on marine shores with extensive tidal oscillations suggest that many animals there are active at one phase of the tidal cycle and inactive at the antiphase. For example, the European shore crab *Carcinus maenas* which remains quiescent beneath stones at low tide has been observed by divers to forage actively between tidemarks at high tide (Naylor 1958). In contrast, the North American fiddler crab *Uca pugilator* is active mainly at low tide and retires to a burrow at high tide (Barnwell 1968). Such observations have been confirmed by laboratory studies of these two crabs which indicate that their respective tidally-rhythmic behaviour patterns are biological clock-controlled (Naylor 1958;

Barnwell 1968; Atkinson & Naylor 1973). Reliance on observational information rather than field experiments when seeking to correlate field and laboratory behaviour patterns has, however, been questioned by Williams, Naylor and Chatterton (1985). For the New Zealand mud crab *Helice crassa* casual observations suggest that animals on the shore are low tide active, yet recordings in constant conditions in the laboratory show that crabs exhibit peak locomotor activity at times of expected high tides. Prompted by this finding field experiments were carried out in which a grid of pitfall traps was sampled at 3 hour intervals over several tidal cycles (Williams et al. 1985). In these traps greater numbers of crabs were taken at high tide than at low tide, confirming the prediction from the laboratory experiments that *Helice crassa* exhibits a circatidal rhythm of locomotor activity with peaks at high tide (Fig.1). Similarly, laboratory studies on the sub-tidal prawn *Nephrops norvegicus* indicated nocturnally-phased locomotor activity (Atkinson & Naylor 1976; Hammond & Naylor 1977) which was unexpected on the basis of catches by commercial fishermen (Moller & Naylor 1980) and from direct observations in the field (Chapman & Howard 1979).

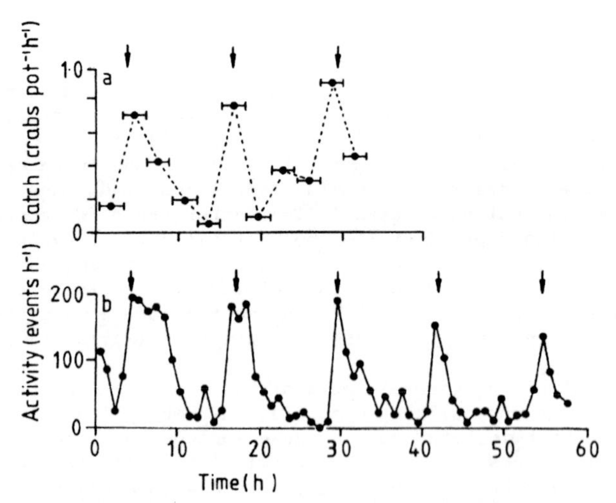

Fig. 1. Field catches and endogenous locomotor rhythms of the New Zealand mud crab *Helice crassa*: a - number of crabs collected in a grid of pitfall traps emptied at 3h intervals (horizontal bars) over three consecutive tides (arrows - times of high tide); b - total hourly walking activity of five crabs in infra-red beam aktographs during 60h in continuous dim light at constant 15°C (arrows - times of 'expected' high tide). (After Williams, Naylor & Chatterton 1985).

Gibson (1978) has pointed out that the activity patterns of coastal fishes recorded in the laboratory are not necessarily identical to those exhibited in the field. Indeed it would not be surprising if only part of an animal's behavioural repertoire was under the control of an endogenous timer (Williams et al. 1985). It is not unreasonable, therefore, to propose that there is a need for comparative field and laboratory experi-

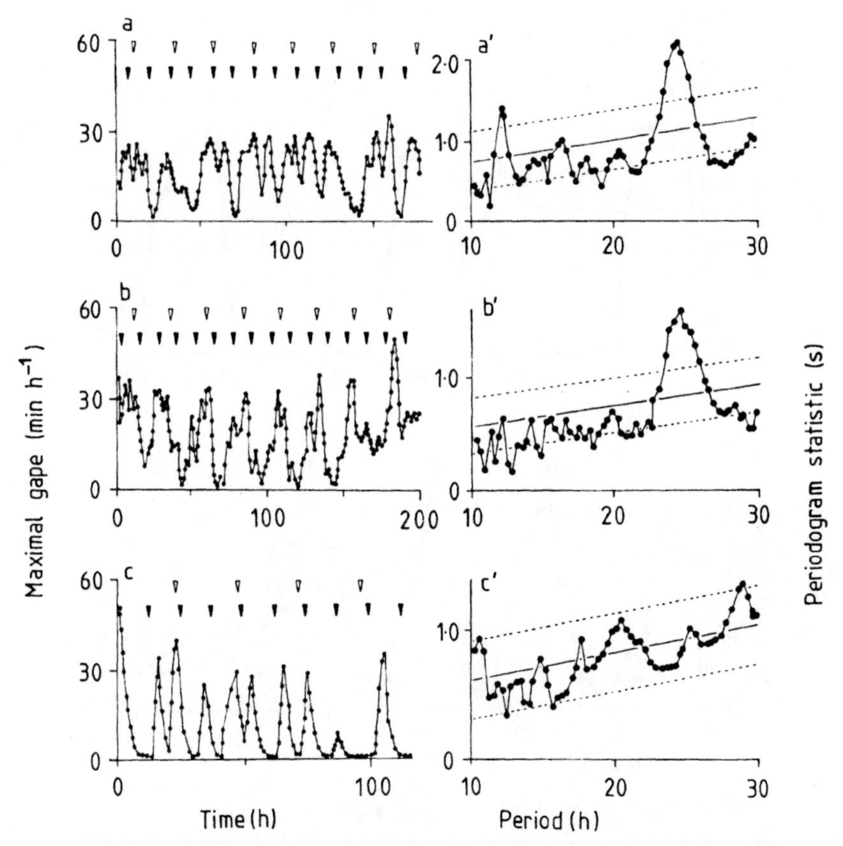

Fig. 2. Shell gape patterns of three unfed Mytilus edulis maintained continuously immersed for 120-200h in constant light at 15°C. Vertical axes in a, b and c plot the time (in min.) during successive hourly intervals when the mussels showed maximal shell gape. Open arrows - 'expected' midnight; closed arrows - 'expected' high tide. a', b' and c' are periodograms of the corresponding data sets. (After Akumfi & Naylor 1987a).

ments in establishing the descriptive framework of responsive and clock-controlled behavioural rhythms of tidal and diel periodicity.

The simplest hypothesis to explain the rhythmic behaviour in intertidal animals is that they primarily respond to tidal and daily variables, and this appears to be the case for some sessile species such as barnacles (Sommer 1972) and mussels (Akumfi & Naylor 1987a). In Mytilus edulis kept continuously immersed in seawater in constant conditions in the laboratory there are spontaneous changes in shell gape, but these changes are not circatidal and appear to be largely random with only a weak circadian component. Three examples of such patterns are illustrated in Fig.2 which, out of ten mussels studied, represent the only example of possible weak circatidal periodicity (Fig.2a, a'), the best example of circadian rhythmicity (Fig.2b, b') and a typical random pattern (Fig.2c, c').

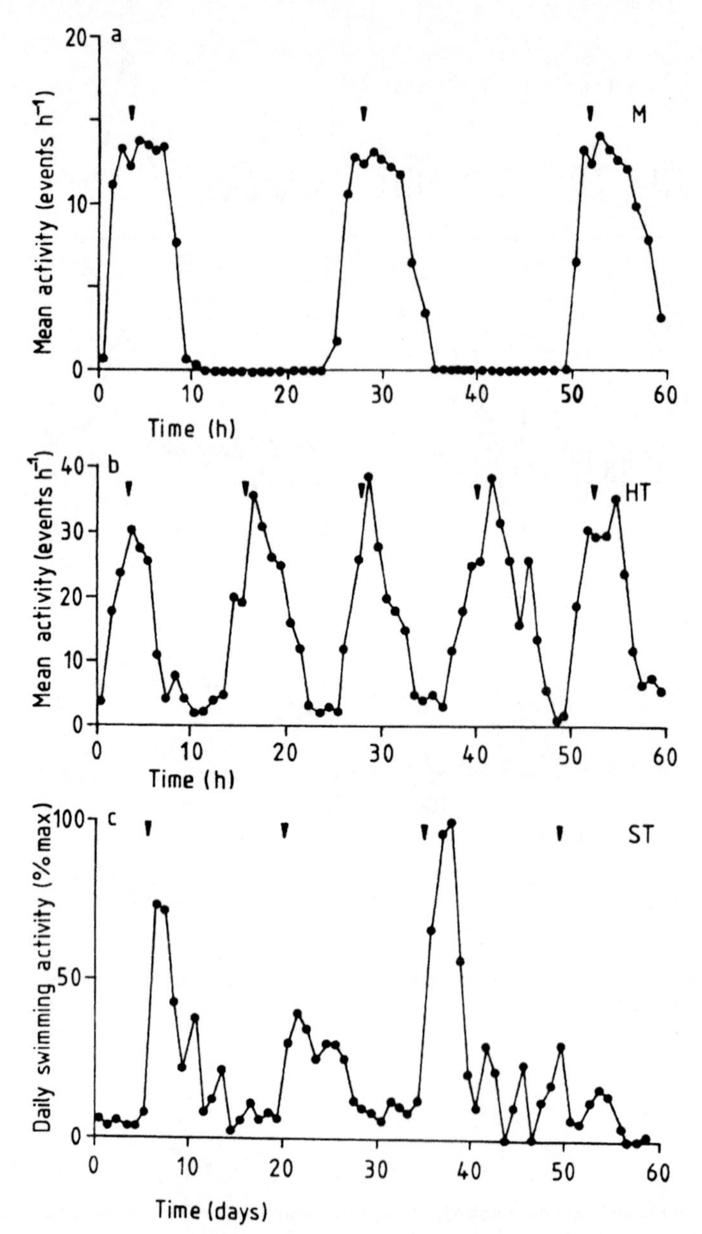

Fig. 3. Free-running locomotor activity rhythms of circadian, circatidal and circasemilunar periodicities in coastal crustaceans kept in constant conditions in the laboratory: a - circadian walking rhythm of 32 Talitrus saltator recorded for 60h (arrows are times of 'expected' midnight - M) (After Bregazzi & Naylor 1972); b - circatidal swimming rhythm of 5 Eurydice pulchra recorded for 60h (arrows are times of 'expected' high tide - HT) (After Jones & Naylor 1970); c - circasemilunar swimming rhythm of 20 Eurydice pulchra recorded for 60 days (arrows are times of 'expected' spring tides - ST) (After Reid & Naylor 1985).

If mussels on the shore show tidal patterns of shell gape, such patterns would seem more likely to reflect exogenous responses to tidal rise and fall than the presence of an internal 'tidal clock' (Akumfi & Naylor 1987a).

In contrast, many mobile species exhibit tidally, daily or lunar-phased locomotor activity patterns which are often highly persistent without reinforcement in constant conditions in the laboratory (Fig.3). An ecological advantage of diel or tidal timekeeping is that it provides a mechanism whereby nocturnal or high tide active intertidal animals could seek shelter beneath stones or burrow in sand in anticipation of dawn or tidal ebb. There is evidence that this is the case in the nocturnally active amphipod _Talitrus saltator_ (Bregazzi & Naylor 1972; Williams 1980), and in the high tide active crab _Carcinus maenas_ (Naylor 1958) and sand-

Table 1. Measures of the persistence (column 1) and precision (columns 2 and 3) of the free-running locomotor rhythms of two burrowing supralittoral amphipods (_Talitrus saltator_ and _Talorchestia deshaysi_) and two surface living forms, the amphipod _Orchestia gammarella_ and the isopod _Ligia oceanica_ which are cryptozoic but non-burrowing (After Williams 1983).

Species	Persistence of distinct circadian rhythm (days) in DD	Period length of circadian rhythm (h ± s.d.)	Circadian periodogram statistic (arbitrary units)
Talitrus saltator	35	24.56 ± 0.43	53
Talorchestia deshaysi	20	24.64 ± 0.54	32
Orchestia gammarella	12	24.51 ± 1.0	22
Ligia oceanica	12	24.42 ± 0.88	20

beach isopod _Eurydice pulchra_ (Jones & Naylor 1970). Furthermore the hypothesis has been proposed (Atkinson & Naylor 1973; Enright 1975; Williams 1980) and supported (Williams 1983) that among mobile intertidal species, burrowing forms exhibit more precise and persistent endogenous rhythmicity than epifaunal forms. In the burrowing amphipods _Talitrus saltator_ and _Talorchestia deshaysi_ rhythms are more persistent and more precise than in the epifaunal amphipod _Orchestia gammarella_ and the isopod _Ligia oceanica_ which are cryptozoic but non-burrowing (Table 1). In populations in which not all individuals emerge from burrows to forage during every 'active' period, a strong endogenous component of rhythmicity would permit animals to remain burrowed for several nights whilst also allowing those individuals to emerge more or less in synchrony with other members of the population on later nights (Naylor 1976; Williams 1983).

Alternatively, a strong endogenous component of rhythmicity in burrowing species may be regarded as a mechanism whereby animals are induced by internal timing processes to move to the sand surface where they would be exposed to environmental reinforcing factors by a form of self-entrainment. However, such hypotheses remain to be tested. Interestingly among bivalve molluscs, Chione stutchburyi alone so far has been shown unequivocally to exhibit pronounced endogenous circatidal rhythmicity (Beentjes & Williams 1986), and this species is a burrowing form.

Finally, comparative studies of the rhythmic behaviour of a coastal species in tidal and non-tidal conditions have been carried out using the shore crab Carcinus. C. maenas from a non-tidal harbour basin adjacent to a tidally influenced coastline exhibited rhythmicity which was superficially circadian in pattern but which under experimental conditions was shown to be circatidal like that of open coast forms, indicative of the deep-seated and probable genetic basis of such behaviour (Naylor 1960). In contrast, C. mediterraneus, a species which has been isolated from extensive tidal conditions for geological periods of time, showed purely circadian locomotor rhythmicity (Naylor 1961).

RESPONSIVE AND CLOCK-DRIVEN MODES OF RHYTHMIC BEHAVIOUR IN INDIVIDUAL ANIMALS

Earlier it was suggested that despite the fact that there may be strong selective advantage for clock-controlled behaviour in mobile intertidal animals, nevertheless there may be both clock-driven and responsive rhythmic behaviour in the repertoire of a single species. This occurs in the shore crab Carcinus maenas and is apparent from comparative studies of the entrainment of locomotor activity rhythms in that crab, using simulated tidal cycles of environmental variables. Exposure to 6:6 h. cycles of temperature (Williams & Naylor 1969) and hydrostatic pressure (Naylor & Atkinson 1972) entrains an endogenous locomotor rhythm of circatidal periodicity with peaks at times of low temperature and high pressure, equivalent to high tide conditions. In contrast, entrainment of C. maenas is quite different in simulated tidal cycles of salinity change, whether these are of square wave (Taylor & Naylor 1977) or sine wave (Bolt & Naylor 1985, 1986) form. Exposure to the episodes of salinity of 34 p.p.t. in such cycles resulted in increased locomotor activity, in a pattern which persisted as a spontaneous rhythm of circatidal periodicity in constant conditions after treatment. In contrast, exposure to the episodes of low salinity during treatment also induced locomotor activity, but these responses did not continue at constant salinity after treatment; they were evidently not clock-coupled.

Exposure to 34 p.p.t. is a repeated occurrence at high tide for crabs in the intertidal zone of Atlantic European coasts, and it is not unreasonable to hypothesize that responses to such conditions have become coupled to the biological clock. In contrast, crabs would in general find themselves in conditions of low salinity only intermittently, for example, at low tide in estuaries or where freshwater flows down the shore. Under such conditions it seems unlikely that temporal adaptation would take place; avoidance of unusual salinities appears to occur partly by increased unorientated locomotor activity purely as responsive behaviour, defined as halokinesis (Taylor & Naylor 1977; Thomas, Lasiak & Naylor 1981).

In further studies of the avoidance of persistent unfavourable salinities by Carcinus maenas, Akumfi and Naylor (1987b) have shown that this may in fact be achieved by a combination of responsive halokinesis and spontaneous locomotor activity. This was demonstrated in experiments in which movements of crabs were monitored continuously when offered pairs of salinities in a two-choice chamber apparatus. When offered pairs of

Table 2. Carcinus maenas. Behavioural responses of crabs offered pairs of salinities in 2-choice chamber experiments. Mean percentage times in each chamber were obtained using batches of 4 crabs in 5 replicate experiments, each over a 12 h period, for each pair of salinities. Mean transits from one chamber to the other were obtained for individual crabs recorded over 24 h, and 20 crabs were tested in each pair of salinities. (* significantly different from 50:50 or from control at 5% level, ** significant at 1% level or less) (After Akumfi and Naylor 1987b).

Salinity Choice (p.p.t.)		Behavioural response		Comment
Chamber A	Chamber B	Mean % time in Chamber A. $12h^{-1}$	Mean transits between Chambers A and B. $24h^{-1}$	
34	34	58.2	26.5	Control
5	9	8.5**	69.8*	Avoidance of low salinity: significant high halokinesis
9	13	9.0**	52.7*	
13	17	5.9**	54.9*	
17	21	23.7**	31.4	Avoidance of low salinity: no significant halokinesis
25	29	24.0*	27.7	
29	34	28.7**	24.9	
17	34	28.7**	15.5	
9	34	1.7**	9.2*	Avoidance of low salinity: significant low halokinesis
40	34	47.2	40.8	cf. Control

salinities of 34 p.p.t. or less, with differentials of as little as 4 p.p.t., Carcinus maenas always showed statistically significant avoidance of the lower of the two salinities (Table 2) (Akumfi & Naylor 1987b). In pairs of salinities in which the greater was 17 p.p.t. or less the total number of transits from each chamber to the other was also significantly greater than in controls, indicating sustained high halokinesis. In nature, therefore, crabs which are suddenly exposed to salinities of 17 p.p.t. or lower in estuaries or in rock pools may avoid such conditions by halokinesis. In contrast, in pairs of salinities in which the lowest value was 17 p.p.t. or more, though significant avoidance of the lower of the salinities still occurred, there was no evidence of sustained halokinesis. Presumably this is partially explained by the fact that any initial high halokinesis by the crabs in the lower salinity of a pair was compensated by low halokinesis when they had moved to a preferred salinity, as suggested by the extreme case of crabs given the choice of 9 and 34 p.p.t. However, control crabs given 34 p.p.t. in each half of the choice chamber, also showed spontaneous locomotor activity, which in freshly collected crabs varies in a tidally rhythmic manner (Table 2). It seems likely, therefore, that endogenous locomotor activity patterns may in some circumstances play a role in achieving avoidance of unfavourable environmental conditions.

LOCOMOTOR RHYTHMS AND MAINTENANCE OF ZONATION

A number of studies have suggested that several coastal animals use internal clock-controlled locomotor rhythms in maintaining their 'preferred' distribution patterns on the shore, notably in sand-beach crustaceans which burrow at low tide and swim at high tide. For example, spontaneous swimming at the times of the ebb tide appears to be of ecological advantage in reducing the risk of being stranded high on the shore at low tide, as has been demonstrated for the amphipods Synchelidium sp (Enright 1963) and Corophium volutator (Morgan 1965). Similarly the isopod Eurydice pulchra which has a 'preferred' distribution above MTL (Fig.4a), shows spontaneous

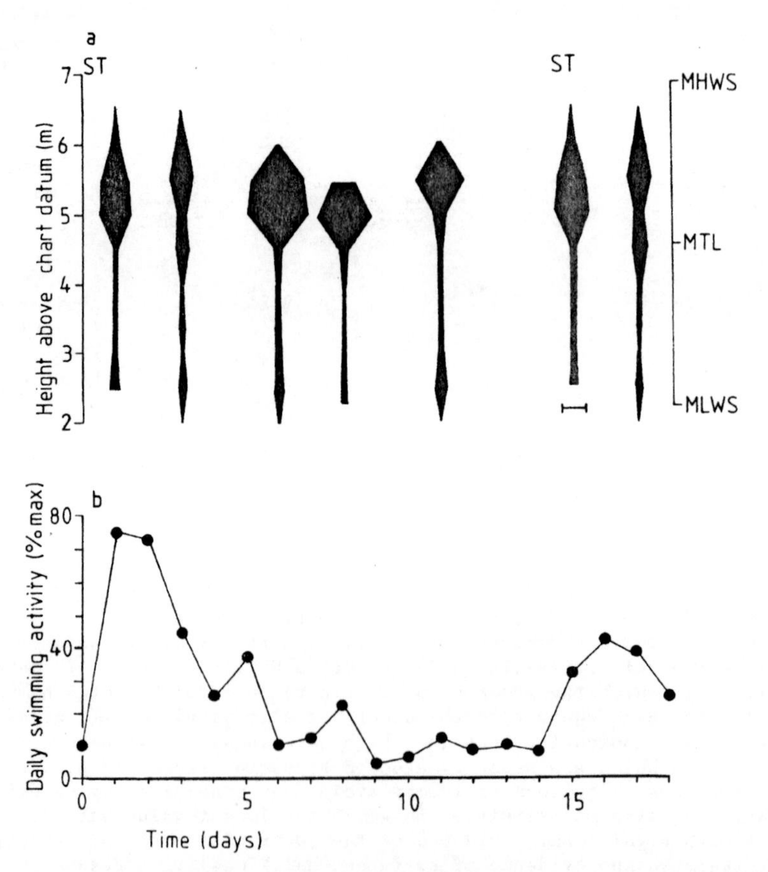

Fig. 4. a: distribution of Eurydice pulchra in transects down a tidally influenced sandy shore on various days throughout a neaps/springs cycle (ST = times of spring tides; scale bar indicates 5000 isopods m^{-2}) (After Hastings 1980). b: total daily swimming activity of Eurydice pulchra on each day throughout the neaps/springs cycle (data extracted from Fig.3) (After Reid & Naylor 1985).

cessation of swimming 2-3 h after high tide (Fig.3b). This was proposed as a behavioural characteristic which would ensure that the animals reburrow in the sand and avoid being carried too far downshore on the ebbing tide (Jones & Naylor 1970).

That hypothesis concerning E. pulchra has been tested in relation to neaps/springs changes in its zonation pattern (Alheit & Naylor 1976) and in relation to observed differences in the high tide swimming behaviour between juveniles and adults (Hastings & Naylor 1980). E. pulchra, when dug from sand at low tide, show a rise and fall in the upper limit of their zone of occurrence on the shore which varies with the springs/neaps cycle (Fig.4a). In adult E. pulchra such changes are correlated with changes in the extent of spontaneous tidal swimming; the circatidal swimming rhythm of specimens kept in tanks with sand is more strongly expressed at the times of 'expected' falling spring tides than at other times of the springs/neaps cycle (Fig.3c, 4b) (Alheit & Naylor 1976; Reid & Naylor 1985). The endogenous semilunar pattern of swimming by adults appears to be correlated with the fact that they, unlike juveniles, do not swim at every high tide. Clock-controlled emergence and swimming on falling spring tides appears to provide a mechanism whereby adult Eurydice pulchra avoid being stranded above HWN at neap tides (Hastings & Naylor 1980). In contrast, juveniles which do not show a semilunar rhythm of swimming, swim abundantly on every tide, which provides a mechanism whereby they, like the adults, also avoid stranding high on the shore at neap tides (Hastings & Naylor 1980).

Recently the semilunar rhythm of swimming in Eurydice pulchra has been shown to free-run for 60 days in constant conditions (Fig.3c) (Reid & Naylor 1985). It is, therefore, a truly physiological circasemilunar rhythm and it does not arise from the 'beat' effect of circatidal and circadian rhythms (Reid & Naylor 1985). Preliminary evidence, for one locality, suggested that the circasemilunar rhythm is entrained by artificial tidal cycles of simulated wave action applied at times of the 24 h cycle when high spring tides occur (Hastings 1981). Subsequent studies show that this is so irrespective of the locality from which the isopods are collected (Reid & Naylor 1986), indicating that there are regionally specific differences in circadian sensitivity to the tidal synchronizing factor in the semilunar entraining mechanism. Endogenous circatidal and circasemilunar rhythms of behaviour, therefore, provide a mechanism whereby dynamical changes occur in the zonation of Eurydice pulchra on strongly tidal shores.

ENDOGENOUS RHYTHMS OF BEHAVIOURAL RESPONSIVENESS

Some mobile intertidal animals are distributed more widely during one phase of the tidal or diel cycle than at the antiphase. Such excursions are partly explained by the occurrence of endogenous locomotor activity rhythms, as in the upshore and downshore tidal movements of the amphipod Synchelidium (Enright 1963) and the brown shrimp Crangon crangon (Al-Adhub & Naylor 1975). However, other factors may also be involved, concerning tidal or daily changes in the responsiveness of such animals to environmental variables. In Synchelidium, for example, Forward (1980) demonstrated tidal changes in the sign of phototaxis during the swimming phase of this amphipod; animals on rising tides were more negatively phototaxic and less sensitive to light than on falling tides. The rhythmic changes in light responsiveness were functionally related to the up- and downshore transport of the amphipods and Forward (1980) demonstrated that the rhythm of light responsiveness persisted for 40 h in constant conditions, suggestive of the fact that it was clock-controlled.

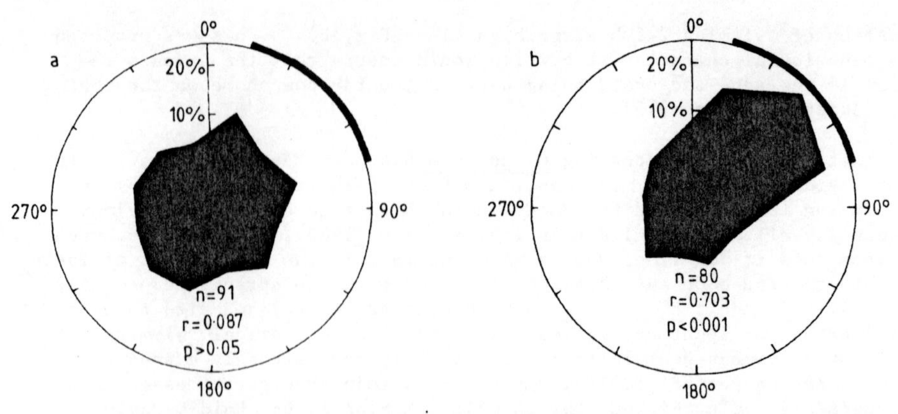

Fig. 5. Escape directions shown by Talitrus saltator released individually at the centre of a 70cm diameter arena in experiments carried out (a) at 0400h, and (b) at 1200h. In each case a vertical rectangle of black cardboard extending over a 60° arc was centred at 45° on the 25cm high white perimeter wall. n - number of amphipods tested, r - an estimate of the non-uniformity of the circular distributions given as mean vector length by the Rayleigh test. (After Edwards & Naylor 1987).

Unequivocal evidence of clock-controlled changes in the sign of an orientational response has recently been obtained for the semi-terrestrial sand-beach amphipod Talitrus saltator (Edwards & Naylor 1987). This migratory species on N. European shores burrows in sand above HWM by day and emerges at night to forage on the sand surface between EHWS and MHWN (Bregazzi & Naylor 1972; Williams 1983), and sometimes down to MTL (Williamson 1951). In constant conditions in the laboratory they show a persistent circadian rhythm with an initial peak which throughout the year maintains a constant phase-relationship, about 3 h in advance of the time of expected dawn (Williams 1980). Navigational orientation upshore to the 'preferred' burrowing zone has been shown to occur as a visual response to the dune/sky boundary, which can be mimicked by dark/light shapes presented under laboratory conditions (Williamson 1951).

The time-base of navigational orientation, in relation to the previously unaddressed question concerning the mechanism of downshore migration has been studied by Edwards and Naylor (1987). Talitrus saltator released at the centre of a circular arena by day moved towards a black rectangle covering a 30° arc on an otherwise white arena wall (Fig.5b). In contrast, when the same experiment was carried out at night, under identical conditions of illumination, the amphipods dispersed randomly from the release point (Fig.5a). In the absence of a blackened arc on the arena wall dispersion was random at all times, suggesting that there was no component of orientation which could be attributed to geophysical magnetic fields as suggested by Arendse (1978) and Arendse and Kruswijk (1981). The visual orientation response was in fact apparent from just before dawn until sunset and was absent from 1800 to 0600 h (Fig.6). Moreover the daily rhythm of change of the orientational response was shown to be endogenous, persisting in animals kept in continuous dim light and being appropriately phase-shifted in amphipods kept in perturbed light/dark regimes (Edwards & Naylor 1987). The time of onset of orientational behaviour in T. saltator, 1-2 h before dawn, coincided with the last part of each

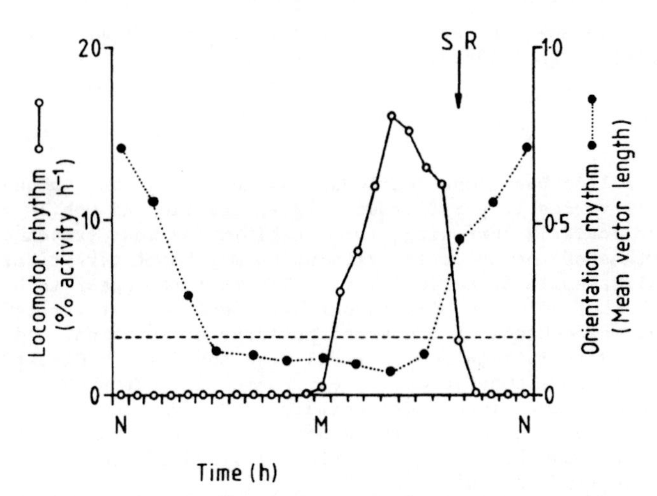

Fig. 6. Circadian rhythms of locomotor activity and of visual orientation in the high shore amphipod Talitrus saltator during the first 24h in constant conditions. Locomotor rhythm data after Williams (1980); orientation rhythm data after Edwards & Naylor (1987) with mean vector lengths (r) derived as in Fig.5. in experiments carried out every two hours throughout the 24h cycle. Horizontal dashed line indicates upper 95% confidence limit of random orientation. Both sets of data refer to times of the year when sunrise (SR) occurred around 0800h.

episode of activity which characterizes the circadian locomotor rhythm (Williams 1980). This would ensure that amphipods which forage at night would spontaneously migrate upshore towards the dune/sky boundary before sunrise. The orientational response, which continued throughout the day, is exhibited when the amphipods are on wet sand but is lost when they reach dry sand, into which they burrow (Williamson 1951), thus ensuring that the amphipods return to their 'preferred' burrowing zone just above HWM.

The spontaneous change from orientated movement to random movement occurred just before dusk, before the amphipods emerge on to the sand surface at night. Thus when Talitrus saltator emerges to forage at about 2200-0100 h, depending upon the time of year (Williams 1980), they are indifferent to the dune sky boundary and might, therefore, be expected to disperse randomly. However, extensive migrations inland at night which occur in Mediterranean T. saltator (Gepetti & Tongiorgi 1967) have not been observed on tidally influenced shores in Britain (Williams 1983). There the species forages intertidally, particularly in strand-line algae, but often moving downshore as far as MTL (Williamson 1951; Williams 1983) presumably by avoiding dry sand.

Talitrus saltator on European Atlantic shores possess a clock-based navigational system which is evidently different from that of Mediterranean forms, which use a celestial compass system to return to HWM after foraging above that tidal level (Papi & Pardi 1953; Pardi 1960). The same species on Mediterranean coasts has recently been shown to use landscape patterns in orientation, but so far there is no evidence of clock-control of those responses, which in any case are considered to be only of minor importance (Ugolini, Scapini & Pardi 1986).

CONCLUSIONS

Clock-controlled behaviour is widespread in eukaryotes; amongst intertidal animals it varies in persistence between sessile and mobile species. Mobile, and particularly burrowing, forms exhibit the most pronounced endogenous rhythms of locomotor activity which may be of circatidal, circadian or circasemilunar periodicity. The rhythms appear to be correlated with the avoidance of unfavourable conditions at particular states of tide, maintenance of preferred patterns of zonation, and anticipatory activity in relation to tidal rise and fall. Circatidal and circadian rhythms of responsiveness to environmental factors have also been demonstrated in migratory intertidal animals. Differences in rhythmic behaviour patterns between conspecifics and congenerics from Mediterranean and Atlantic coasts can be related functionally to differences in tidal amplitudes on the two types of shore.

ACKNOWLEDGEMENTS

The author is most grateful to N.E.R.C. for financial support towards this work.

REFERENCES

Akumfi, C.A. and Naylor, E., 1987, Temporal patterns of shell gape in Mytilus edulis. Mar. Biol. (In press).

Akumfi, C.A. and Naylor, E., 1987b, Spontaneous and induced components of salinity preference behaviour in Carcinus maenas. Mar. Ecol. Prog. Ser. (In press).

Al-Adhub, A.H.Y. and Naylor, E., 1975, Emergence rhythms and tidal migrations in the brown shrimp, Crangon crangon (L.). J. mar. biol. Assoc. U.K. 55: 801-810.

Alheit, J. and Naylor, E., 1976, Behavioural basis of intertidal zonation in Eurydice pulchra Leach. J. exp. mar. Biol. Ecol. 23: 135-144.

Arendse, M.C., 1978, Magnetic field detection is distinct from light detection in the invertebrates Tenebrio and Talitrus. Nature, 274: 358-362.

Arendse, M.C. and Kruyswijk, C.J., 1981, Orientation of Talitrus saltator to magnetic fields. Neth. J. Sea Res. 15: 23-32.

Atkinson, R.J.A. and Naylor, E., 1973, Activity rhythms in some burrowing decapods. Helgolander wiss. meeresunters. 24: 192-201.

Atkinson, R.J.A. and Naylor, E., 1976, An endogenous activity rhythm and the rhythmicity of catches of Nephrops norvegicus (L.). J. exp. mar. biol. ecol. 25: 95-108.

Barnwell, F.H., 1968, The role of rhythmic systems in the adaptation of fiddler crabs to the intertidal zone. Am. Zool. 8: 569-583.

Beentjes, M.P. and Williams, B.G., 1986, Endogenous circatidal rhythmicity in the New Zealand cockle Chione stutchburyi (Bivalvia, Veneridae). Mar. Behav. Physiol. 12: 171-180.

Bolt, S.R.L. and Naylor, E., 1985, Interaction of endogenous and exogenous factors controlling locomotor activity rhythms in Carcinus exposed to tidal salinity cycles. J. exp. mar. Biol. Ecol. 85: 47-56.

Bolt, S.R.L. and Naylor, E., 1986, Entrainability by salinity cycles of rhythmic locomotor activity in normal and eyestalk-ablated Carcinus maenas (L). Mar. Beh. Phys. 12: 257-267.

Brady, J., 1982, Introduction to biological timekeeping, in: Biological Timekeeping, J. Brady, ed., Cambridge C.U.P. pp.33-48.

Bregazzi, P.K. and Naylor, E., 1972, The locomotor activity rhythm of Talitrus saltator (Montagu) (Crustacea, Amphipoda). J. Exp. Biol. 57: 375-391.

Chapman, C.J. and Howard, F.G., 1979, Field observations on the emergence rhythm of the Norway lobster Nephrops norvegicus, using different methods. Mar. Biol. 51: 157-165.

Edwards, J.M. and Naylor, E., 1987, Endogenous circadian changes in orientational behaviour of Talitrus saltator. J. mar. biol. Ass. U.K. 67: 17-26.

Enright, J.T., 1963, The tidal rhythm of activity of a sand-beach amphipod. Zeit. f. vergl. Physiol. 47: 276-313.

Enright, J.T., 1975, Orientation in time: endogenous clocks, in: "Marine Ecology 2(2) Physiological Mechanisms" O. Kinne, ed., pp.917-944, Wiley-Interscience.

Forward, R.B., 1980, Phototaxis of a sand-beach amphipod: physiology and tidal rhythms. J. Comp. Physiol. 135: 243-250.

Gepetti, L. and Tongiorgi, L., 1967, Nocturnal migrations of Talitrus saltator (Montagu) (Crustacea, Amphipoda). Monit. Zool. Ital. NS 1: 37-40.

Gibson, R.N., 1978, Lunar and tidal rhythms in fish, in: "Rhythmic activity of fishes" J.E. Thorpe, ed., Academic Press, New York, pp.201-213.

Hammond, R.D. and Naylor, E., 1977, Effects of dusk and dawn on locomotor activity rhythms of the Norway Lobster Nephrops norvegicus. Mar. Biol. 38: 253-260.

Hastings, M.H., 1980, Aspects of the ecology of sandy-shore Crustacea. Ph.D. Thesis, University of Liverpool, 122pp.

Hastings, M.H., 1981, The entraining effect of turbulence on the circatidal activity rhythm and its semilunar modulation in Eurydice pulchra. J. mar. Biol. Ass. U.K. 61: 151-160.

Hastings, M.H. and Naylor, E., 1980, Ontogeny of an endogenous rhythm in Eurydice pulchra. J. exp. mar. Biol. Ecol. 46: 137-145.

Jones, D.A. and Naylor, E., 1970, The swimming rhythm of the sand beach isopod Eurydice pulchra. J. exp. Mar. Biol. Ecol. 4: 188-199.

Moller, T.H. and Naylor, E., 1980, Environmental influence on locomotor activity in Nephrops norvegicus (Crustacea: Decapoda). J. mar. biol. Ass. U.K. 60: 103-113.

Morgan, E., 1965, The activity rhythm of the amphipod Corophium volutator (Pallas) and its possible relationship to changes in hydrostatic pressure associated with tides. J. Anim. Ecol. 34: 731-746.

Naylor, E., 1958, Tidal and diurnal rhythms of locomotory activity in Carcinus maenas (L.). J. exp. Biol. 35: 602-610.

Naylor, E., 1960, Locomotory rhythms in Carcinus maenas (L.) from non-tidal conditions. J. Exp. Biol. 37: 481-488.

Naylor, E., 1961, Spontaneous locomotor rhythm in Mediterranean Carcinus. Pubbl. Staz. Zool. Napoli. 32: 58-63.

Naylor, E., 1976, Rhythmic behaviour and reproduction in marine animals, in: "Adaptation to environment: essays om the physiology of marine animals" 393-429, R.C. Newell, ed., Butterworths, London.

Naylor, E., 1985, Tidally rhythmic behaviour of marine animals. Symp. Soc. Exp. Biol. 39: 63-93.

Naylor, E., 1987, Temporal aspects of adaptation in the behavioural physiology of marine animals, in: "21st Eur. Mar. Biol. Symp., Gdansk" (In press).

Naylor, E. and Atkinson, R.J.A., 1972, Pressure and the rhythmic behaviour of inshore animals. Symp. Soc. Exp. Biol. 26: 395-415.

Papi, F. and Pardi, L., 1953, Ricerche sull' orientamento di Talitrus saltator. II sui fattori che regolana la variazone dell' angolo di orientamento nel carso del giorno. Zeit. f. vergl. Physiol. 35: 490-518.

Pardi, L., 1960, Innate components in the solar orientation of littoral amphipods. Cold Spring Harb. Symp. Quant. Biol. 25: 395-401.

Reid, D.G. and Naylor, E., 1985, Free-running, endogenous semilunar rhythmicity in a marine isopod crustacean. J. mar. biol. Assoc. U.K. 65: 85-91.

Reid, D.G. and Naylor, E., 1986, An entrainment model for semilunar rhythmic swimming behaviour in the marine isopod Eurydice pulchra (Leach). J. exp. Mar. Biol. Ecol. 100: 25-35.

Sommer, H.H., 1972, Endogene und exogene Periodik in der Aktivitat eines mederen Krebses (Balanus balanus L..). Zeit. f. vergl. Physiol. 76: 177-192.

Taylor, A.C. and Naylor, E., 1977, Entrainment of the locomotor rhythm of Carcinus by cycles of salinity change. J. mar. biol. Ass. U.K. 57: 273-277.

Thomas, N.J., Lasiak, T.A., and Naylor, E., 1981, Salinity preference behaviour in Carcinus. Mar. Behav. Physiol. 7: 277-283.

Ugolini, A., Scapini, F., and Pardi, L., 1986, Interaction between solar orientation and landscape visibility in Talitrus saltator (Crustacea: Amphipoda). Mar. Biol. 90: 449-460.

Williams, J.A., 1980, The light response rhythm and seasonal entrainment of the endogenous circadian locomotor rhythm of Talitrus saltator (Crustacea: Amphipoda). J. mar. biol. Ass. U.K. 60: 773-785.

Williams, J.A., 1983, The endogenous locomotor activity rhythm of four supralittoral peracarid crustaceans. J. mar. biol. Ass. U.K. 63: 481-492.

Williams, B.G. and Naylor, E., 1969, Synchronization of the locomotor tidal rhythm of Carcinus. J. Exp. Biol. 51: 715-725.

Williams, B.G., Naylor, E., and Chatterton, T.D., 1985, The activity patterns of New Zealand mud crabs under field and laboratory conditions. J. exp. mar. biol. Ecol. 89: 269-282.

Williamson, D.I., 1951, Studies on the biology of Talitridae (Crustacea, Amphipoda): visual orientation in Talitrus saltator. J. mar. biol. Ass. U.K. 30: 91-99.

Migration or Shelter? Behavioural Options for Deposit Feeding Crabs on Tropical Sandy Shores

Alan D. Ansell

Dunstaffnage Marine
Research Laboratory
Argyll, Scotland, U.K.

INTRODUCTION

Species which live in the intertidal zone are subject to a wide range of environmental fluctuation arising from tidal and diel changes. Temperature, salinity, humidity, light intensity, wave action and other factors all show changes which affect both sedimentary and hard substratum shores.

Some of these changes, such as tidal immersion and emmersion and daily changes in light intensity, are predictable, cyclical phenomena. Others, like those dependent on the weather or interactions with other organisms, are little if at all predictable.

Behavioural responses similarly fall into two categories. Some are the necessary outcome of morphological and physiological adaptations which enable the organism to exploit the intertidal environment, and help define a particular environment niche; these relate mainly to the more predictable environment changes. Others provide the flexibility to survive the more extreme and unpredictable fluctuations which affect that particular niche. The former include e.g., the endogenous rhythms which control the tidal and diel activities of many beach organisms and the main components of those activities, particularly feeding. The latter include responses which fine tune such rhythms and allow the organism flexibility in response to short term changes. The scope of such responses may be no more than the ability to withdraw into shelter until the stress is gone, or it may comprise a suite of responses to different disturbances, which allow activity to continue with a minimum of interruption.

This paper will examine the behaviour of a group of mainly tropical and subtropical surface deposit feeding crabs with particular regard to their normal patterns of behaviour and to flexibility in those patterns shown in response to disturbance of their intertidal habitat.

The crabs concerned belong to two families of Brachyura, the Mictyridae and the Ocypodidae. Within the Ocypodidae only the sub-family Scopimerinae will be discussed in detail thus limiting the scope of the paper to strictly intertidal inhabitants of mainly sandy shores. The Ocypodidae (Ocypode and Uca spp., both of which have received much greater attention elsewhere, will be referred to only very briefly, as will the third sub-family of the Ocypodidae, the Macrophthalminae, whose biology and behaviour have received scant attention although their taxonomy and distribution have been very thoroughly reviewed (see Barnes, 1977 for references).

Geographically the Scopimerinae and Mictyridae are restricted to the Indo-Pacific region, unlike the Ocypodinae whose genera have a circum-tropical distribution.

MORPHOLOGICAL AND PHYSIOLOGICAL ADAPTATIONS

Scopimerine and mictyrid crabs all show a similar range of morphological and physiological adaptations to intertidal life, in two main respects; firstly, to a semi-terrestrial habit, and secondly, to feeding on surface deposits left by the receding tide.

The adaptations for a semi-terrestrial habit include: a) the development of special structures for aerial gas exchange. In most Scopimerinae, these take the form of tympani, or gas windows (Maitland, 1986) on the meral segments of the legs, and in some cases also on the sternal plates of the thorax. In the Mictyridae, epibranchial lungs are developed, a feature evolved here in parallel to similar structures in the Ocypodinae. b) The development of special structures for water uptake and conservation in all species, including tufts or rows of setae which aid extraction of water from moist sand surfaces, and c) the development of physiological mechanisms which aid thermal regulation by evaporative cooling.

Their adaptations for feeding on surface deposits include particularly the development of sorting mechanisms involving the mouth parts, a particular characteristic of the sand dwelling species being the development of 'spoon shaped' setae on the 2nd maxillipeds (Bigalke, 1921; Ono, 1965; Vogel, 1984).

These adaptations set strict intertidal limits to scopimerine and mictyrid distribution. As surface deposit feeders they need an area regularly covered by the tide so that their surface food supply is renewed; as 'air breathers' they need an area regularly uncovered to provide sufficient time for feeding during the low tide period. Subsidiary requirements, for sediments of a type (particle size, organic content, water content, etc.,) suitable for both burrowing and feeding; for shelter from wave action during the immersed period; and for other needs in individual species, lead to considerable habitat separation, by sediment type, tidal zonation, type of beach exposure, etc. (Ono, 1965; Hartnoll, 1975).

TIDAL AND DIEL RHYTHMS OF ACTIVITY

All scopimerine and mictyrid crabs which have been studied are active during low tide periods during the day and remain inactive in a burrow in the sediment during darkness and during periods of immersion by the tide during the day. There is little doubt that these rhythms have an endogenous component, but no studies of this aspect of behaviour have been made in either group. During the period spend in the burrow, the crabs are inactive. The burrow is sealed before the tide reaches it so that it traps air, and the crabs remain at the air/water interface allowing aerial gas exchange to take place.

PATTERNS OF ACTIVITY IN THE SCOPIMERINAE

In most scopimerine crabs which have been observed, the main activities are usually centred around an individual burrow within the main area of distribution of the population. The burrow may be defended, and marks the centre of the feeding range, as well as serving as a shelter from disturbance, predation and physical extremes, and as a source of replenishment of water. In most species also, however, such burrow-orientated activity is abandoned under certain conditions in favour of wandering, or migration, generally beyond the confines of the normal zone.

Burrow-Orientated Activities

Burrow structure and construction. The burrows of scopimerine crabs are generally simple slightly drop-shaped or straight-sided, vertical or slightly sloping, and with often a slight spiral form. The shaft may be expanded into a chamber (Silas and Sankarankutty, 1967). In some species, the burrow extends to the water table (Fielder, 1970), in others it may stop short, but in all it provides a refuge and a source for the renewal of moisture loss. The burrow diameter generally reflects closely the size of the occupant (Fishelson, 1983).

Burrows are generally occupied, and defended, by a single crab, but in *Scopimera proxima* (personal observation) burrows are frequently occupied by pairs of crabs, while in *Ilyoplax pusillus*, small males often cohabit with large males (Wada, 1981a; 1983c).

In *Scopimera globosa*, when burrowing in moist sand, the crab lies on the one side and progresses by scooping sand from below with the chela and walking legs, passing it below the thorax as a rough ball to the upper side, where the chela and walking legs fix it to the roof above the body. The body rotates about 360° between each transfer so producing the spiral form and the crab moves down together with an air bubble. Similar movements are made in burrow construction, but the sand is removed above ground in large pellets and deposited outside the burrow. In *S. globosa* the sand is carried between the walking legs, but in *S. inflata*, the sand pellets are pushed in front (Fielder, 1970). Burrowing is usually followed by cleaning for which the mouthparts are used.

Emergence, and emergence patterns. Individual crabs emerge some time after the burrow is uncovered by the tide, breaking through the plug to form a round hole in the sand. This may be followed by a period of cautious exploration (Fielder, 1970) before the crab finally emerges and begins feeding. Emergence may also be followed by a period of burrow clearing or smoothing of the burrow margin (Fielder, 1970), or feeding may begin immediately.

Scopimerine crabs emerge onto the sand surface at varying times after the area has been uncovered. Counts of the total number emerged follow a roughly sigmoid curve. In *D. wichmanni* in Hong Kong, the first crabs emerged within one hour of being uncovered, and 50% of those which emerged had appeared within 3 h.(personal observation). In *Scopimera intermedia*, 50% emergence occurred 1.5 h after uncovering (Tweedie, 1950); in *S. inflata* (Fielder, 1970) and *S. pilula* (Silas and Sankarankutty, 1967) after 3 h. *Dotilla fenestrata* (Hartnoll,1973) reached 50% emergence after 1.1 to 1.5 h at different tidal levels on spring tides, and 0.7 to 1.6 h on neap tides. *D. sulcata* in the Red Sea, however, began to emerge 3-4 min after the water receded with the most massive emergence occurring 7-8 min later, and passing the peak at 15-16 mins of exposure (Fishelson, 1983). *D. mictyroides* also emerges close behind the receding tide (MacIntyre, 1968; Tweedie, 1950). Clearly there are great differences in the pattern of emergence, in part related to the level of zonation (Hartnoll, 1973) although other factors such as the physiological state of the individuals concerned play a part.

Emergence may be completely inhibited when weather conditions are completely unsuitable for feeding (Wada, 1983a).

Burrow-orientated feeding. In most scopimerine crabs, feeding normally occurs around the burrow entrance, although individual species differ in the degree of attachment to the burrow and the length of time the same burrow is occupied. Essentially similar modes of burrow-orientated feeding have been described for *Dotilla fenestrata* (Hartnoll, 1973) *D. mictyrioides* (Altevogt, 1957; MacIntyre, 1968), *D. sulcata* (Fishelson, 1983), *D. wichmanni* (personal observation), *Scopimera inflata* (Fielder, 1970) *S. pilula* (Silas and Sankarankutty, 1967) *S. intermedia* (Tweedie, 1950), *S. globosa* (Ono, 1965), and *S. proxima* (personal observation).

In the most structured types of burrow-orientated feeding, seen in *Scopimera* and *Dotilla* species, the crab moves sideways in a straight line along a feeding trench radiating from the burrow. As it progresses, the chelae are used alternatively to collect sand and pass it to the mouthparts at the base of the third maxillipeds. Within the mouthparts, the organic rich fine particles are scrubbed off and sorted from the larger 'inorganic' particles by a similar flotation process to that described for *Uca* (Miller, 1961). The sand particles collect together and are moulded into a pellet which emerges at the apex of the third maxillipeds. From here the pellet is removed by the chelae and rejected. At some distance from the burrow entrance, related to its size, the crab moves forward slightly and begins to feed back towards the burrow, so that it gradually excavates an increasing segment of a circle surrounding the burrow entrance as the tidal period progresses. Other species, like *Iloplax pusillus* (Wada, 1984), feed within a home range around the burrow but in a more random fashion.

Although the main elements of the feeding process are similar in all scopimerine crabs, there is much variation in detail between species, e.g., in the mode of rejection of pseudofaecal pellets. In *Scopimera globosa*, the right and left chelae are used indescriminantly to remove sand pellets, which are placed to the side of the crab between the body and the periopods, which pass them posteriorly, the rear leg being used finally to kick the pellet out behind the crab. The feeding trench in use is kept completely clear of pellets which accumulate behind the crab often several deep obscuring previously used trenches. Somewhat similar behaviour has been reported for *D. fenestrata* (Bigalke, 1921; Hartnoll, 1973) and *D. blandfordi* (Altevogt, 1957). *S. inflata* uses both chelae working together to remove pellets (Fielder, 1970), and in *S. proxima* they are knocked off by a sharp movement of the right chelae (Silas and Sankarankutty, 1967). In *Dotilla wichmanni* , the pellet is always removed by the chela on the trailing side, i.e., the side of the crab away from the direction in which it is feeding. The pellet is then dropped by the chela to the side and is not passed back, so that the feeding trenches with their rows of rejected pellets remain clearly defined.

Although so much is known of the method of feeding of scopimerine crabs, there is relatively little information in the literature on the rate of feeding, surprisingly in view of the fact that in many situations the process is clearly recorded as patterns of feeding trenches and rejected pellets. Fishelson (1983) calculated the time available for feeding and other activities by *D. sulcata* at different tidal levels, but did not record the actual time spent by individuals. Hartnoll (1979) also calculated the time available for feeding at different tidal levels for *Dotilla fenestrata*, and states that cessation of feeding and closure of the burrows is a well synchronized activity with all crabs which emerge remaining active until a few minutes before flooding. Wada (1986) has recently published data for *Scopimera globosa* which show that the duration of surface activity is very variable in all size groups with most crabs active considerably less than the whole tidal exposure.

In Hong Kong, individuals of *D. wichmanni* showed great individual variation in the length of time spent feeding, and in the timing of this activity. Most individuals began feeding without any initial period of burrow maintenance, but in some, burrowing activity took place at the cessation of feeding. The period spent actually feeding ranged from < 1 h to < 4 h, although the entire tidal period was not monitored. Most burrows remained open when observations were discontinued, and it is possible that there was renewed activity before the burrows were finally plugged. A few individuals plugged the burrow temporarily during the period of observation.

The rate of production of feeding pellets varies between species (Hartnoll, 1973). The relationship between the number of chela movements and the number of feeding pellets produced is variable but generally greater than the 1 : 1 ratio which is seen during deposit feeding in *Ocypode* species (references in Eshky, 1985).

The changing area excavated by individual scopimerine crabs provides a convenient estimate of feeding rate but has rarely been measured. For *Dotilla wichmanni* preliminary estimates of area, based on the mean length of the feeding trench and the circular angle of the segment excavated (personal observation) show that some individuals maintain a steady rate of feeding, but that there is much individual variation in this aspect of feeding behaviour. The mean rate of uninterrupted feeding measured as the increase in circular angle cleared was 90°(±7.4°) h.

In *D. wichmanni*, there is a fairly close relationship between the length of the feeding trench and the size of the crab, such that, assuming similar efficiencies of sorting and assimilation of organics from the surface sand, the metabolic requirements of different sized individuals (measured as rates of oxygen consumption under standard conditions) would be met by similar angular excavation rates. During a single tidal period for *D. wichmanni*, the mean angle of segment excavated was 350° for 21 crabs (33°), with a range in individuals from 70° to 540°, suggesting that control of the path length enables each crab to satisfy its metabolic requirement by means of one complete 360° excavation per day.

If this relationship is not just the result of chance then it suggests that there may be behavioural limits on population density, determined by the size of feeding area. Any such control is not exact however, since individual crabs show angular clearances of > 360°, i.e. they rework some areas already excavated earlier in the tidal cycle. Also the feeding areas of adjacent crabs may overlap, i.e., a crab may rework an area excavated earlier in the tidal period by a neighbour. At low population densities, *D. wichmanni* avoids neighbours' feeding areas, but this avoidance breaks down at higher densities. For *Scopimera globosa*, Sugiyama (1961) suggested three ways in which adjustments are made at higher population densities; by excavating deeper; by alternating periods of emergence; and by migration. Ono (1965) distinguished between species in which the feeding area is reduced at high densities (e.g., *S. globosa*) and others which maintain fixed areas (e.g., *Ilyoplax pusillus*). Large *I. pusillus* show the unique behaviour of 'barricade building' against smaller neighbours (Wada, 1984).

Water uptake and temperature regulation. All scopimerine crabs, in common with ocypodids and *Mictryis*, possess groups of special setae arranged in tufts or lines at the termination of channels leading into the gill chamber. These act to help draw water from moist surfaces by capillary attraction into the gill chamber. Eshky (1985) has shown that the process is aided in *Ocypode saratan* by reversed pumping of the scaphognathite. When conditions are suitable the crabs pause for short periods during surface feeding to replenish the water in the gill chambers, lowering the body to bring the tufts of setae into contact with the surface water film (Fielder, 1970). In species which live in very dry substrata, however, the burrow may provide the only available source and the crabs must then return to it at frequent intervals to restore losses from feeding activity and by evaporation (Fishelson, 1983).

At high temperatures water loss is increased by the use of evaporative cooling for temperature regulation. Dotilla species have deep grooves in the carapace along which water is pumped for this purpose (Fishelson, 1983). In Scopimera globosa exposed to high temperatures, water is pumped out of the anterior opening from the gill chamber and spread by the chelae to form a film along the sides of the carapace and over the gas windows on the legs (personal observation). It seems possible that this serves to help control body temperature, although there is no experimental evidence for this suggestion.

Escape/flight responses. During burrow-orientated feeding the normal escape response of scopimerine crabs to the approach of predators or to other disturbances is to return rapidly to the burrow along the feeding trench. This response may be elicited by any rapid movement, including that of the observer and e.g., by sea-birds flying overhead (Fielder, 1970). There must be a balance between the advantage of a shorter escape route, and a need for a longer feeding path if metabolic requirements are high. There is some indication that this balance changes as the tidal period progresses since Fielder (1970) records that S. inflata is more sensitive to disturbance early in the tidal cycle, and in D. wichmanni there is a tendency in many feeding tracks for the length to increase later in the tidal period (personal observation).

A second escape response, noted by Fielder (1970), for S. inflata involves cessation of feeding only, followed by the crab flattening itself onto the sand surface and remaining immobile for up to several minutes, making use of cryptic colouration for concealment.

Aggressive behaviour. Aggressive behaviour in scopimerine crabs takes the form of adoption of the threat posture. The walking legs are straightened to a nearly vertical position, so that the body is raised to its maximum extent. Simultaneously the chelaepeds are raised to their full extension above the body. This position is held very briefly (1-2 secs) before the chelae and body are dropped (Fielder, 1970). The threat posture is similar in all scopimerine crabs although the vertical extension may be increased by the crabs lifting the first walking legs off the ground (personal observation). Threats of this kind are used in defence of the burrow, or of the feeding trench, and usually result in one crab retiring. Prolonged confrontation may, however, be followed by sparing or wrestling of the two crabs with the chelae interlocked, as described for Dotilla myctiroides by Altevoght (1957).

Successful combatants may make further threat gestures or a 'triumphal dance' after their opponent has retreated (Tweedie, 1950; Altevoght, 1957; Fielder, 1970).

In some species the threat posture may be exhibited regularly in the apparent absence of any intruder. Individuals of Tmethypocoelis ceratophora on Hong Kong beaches threaten at frequent intervals during surface activity, this frequency perhaps reflecting its wandering habits. T. ceratophora also exhibits an unusual response to disturbance by running to the nearest burrow entrance. This may be the burrow of Scopimera globosa, of Uca lactea or Uca vocans with all of which its distribution overlaps, or it may be even the siphon tube of a bivalve or the tip of a brachiopod hole.

During the tidal cycle, aggressive behaviour occurs most frequently in some species during bouts of so-called aggressive wandering. This generally involves the larger males which wander through the area of distribution, often attacking other crabs and making threat gestures (Tweedie, 1950; Fielder, 1970).

These aggressive encounters between wandering crabs and crabs with burrows may result in ejection of one crab by another. In Scopimera globosa in Japan (Wada, 1986), burrow owners were displaced frequently during the low tide period, and burrow usurpation accounted for about half of the cases where a crab left its burrow. In most instances the crab usurping a burrow was larger than its opponent, irrespective of sex. Wada suggests that burrow usurpation is a factor accounting for the different zonation of adults and young on the beach, a phenomenon common in the distribution of several other species of scopimerine crabs also (Fielder, 1971; Hails and Yaziz, 1982).

Cessation of activity. Reburial. After the cessation of feeding the crabs retire to their burrows prior to inundation by the incoming tide. This process may take place more-or-less simultaneously in a population, a very short time before the tide reaches the crabs (Hartnoll, 1973) or may be more extended (Wada, 1986). In some species it may be preceded by a period of wandering and increased aggression by some individuals.

Final retirement to the burrow always involves sealing its mouth. This may be done by the crab excavating a plug of sand of the correct size from near the burrow entrance, then entering the burrow and pulling the plug in behind. In S. globosa in Hong Kong this process took place only seconds before the arrival of the tide in some cases and was extremely quick. Alternatively retreat may involve the erection of an igloo-like structure over the burrow entrance, built from large pellets of sand excavated from the burrow or nearby (Fielder, 1970).

Migration/wandering

Populations of scopimerine crabs are dynamic and move in response to such factors as differences in tidal height over the lunar cycle, to wave action, and changes in slope of the beach (Fielder, 1971). In Scopimera globosa and Ilyoplax pusillus the burrow location is frequently moved over short distances, only some patterns of movement being dependent on the tidal phase (Wada, 1983a, b).

Other movements involve large numbers of individuals moving away from the normal distribution zone to feed. Feeding away from the burrow in this way apparently takes place in response to a number of environmental disturbances which disrupt burrow-orientated feeding. In the Red Sea, changes in sea level occur seasonally which result in the upper levels of the shore being out of reach of the highest tides in the summer/autumn period; during these months Dotilla sulcata from these high levels make migrations to the water line to feed, many returned to the same burrows as the tide rises (Fishelson, 1983). In Dotilla fenestrata (Hartnoll, 1973), crabs not uncovered by extreme neap tides emerge and feed from the undrained sand. In Dotilla wichmanni, in Hong Kong, many crabs were observed to move away from the burrow zone to feed on a lower zone of the beach where the sand surface remained waterlogged at low tide. There individual crabs maintained much of the same pattern of feeding activity seen in burrow-orientated feeding, moving in a straight line and occasionally reversing direction either spontaneously or when approaching another individual. As a result they remained very evenly spaced over the surface. This behaviour was observed only following a period of heavy rainfall when the upper levels of the beach were greatly affected by surface runoff, so that much of the normal burrow zone failed to drain at low tide. The lack of drainage, perhaps combined with low salinity resulting from seepage, seems to have been the stimulus for migration to the lower level. Later in the tidal period, individual D. wichmanni moved back up the beach and could be seen burrowing into the sand, to produce sealed dome-like structures (igloos) in the zone in which burrow-orientated feeding was observed at other times (personal observation). Wada (1981b) suggested that waterlogging of the normal burrowing zone was a factor in causing wandering of Scopimera globosa in Japan.

High population density or increased metabolic demand by the crabs may also be factors leading to wandering or migration (Yamaguchi and Tanaka, 1974; Iwata et al., 1983).

During wandering phases away from the burrow the normal response to disturbance is immediate burial. Under these circumstances, burial of scopimerine crabs occurs by the very characteristic corkscrewing motion (Tweedie, 1950). Burial in this way is dependent on the sand being fairly wet, and this may be the determining factor in the choice of location for feeding away from the burrow.

Scopimerine crabs share with species of Uca and Ocypode, and with Mictyris, (Kraus and Tautz, 1981) an acute visual awareness of their surroundings. They rapidly melt into the sediment at some distance from a moving observer or other disturbance. Like other crabs though they have little defence against human predators who can remember the position at which they disappeared and dig them up. Fortunately few natural predators seem to have developed this technique.

Reproductive Behaviour

Only very few observations of the reproductive behaviour of scopimerine crabs have been published and no clear pattern of reproductive behaviour emerges. In Scopimera globosa, males may chase females for up to a metre before copulation takes place on the surface, or a male may capture a female, carry her to the burrow, push her in and follow, plugging the burrow behind (Yamaguchi, et al., 1979; Wada, 1981a; 1983a). In S. proxima males wander in search of receptive females (Silas and Sankarankutty, 1967), and copulation takes place outside the burrow. In Ilyoplax pusillus, copulation always takes place in the burrow. The male waves to attract a female which then follows him into the burrow; after copulation the male emerges and plugs the burrow.

Fielder (1971) suggested that the berried females of S. inflata remain in the burrow until just before release of the larvae, as occurs in many Uca species (Salmon, 1965; Crane, 1975; Christy, 1978). This is not generally the case in scopimerine crabs, however; in Dotilla wichmanni in Hong Kong, numerous berried females were observed feeding on the surface.

PATTERNS OF ACTIVITY IN THE MICTYRIDAE

In contrast to most scopimerine crabs, the common mode of feeding in Mictyris longicarpus is away from the burrow (Cameron, 1966; Quinn and Fielder, 1978; Quinn, 1986). After emergence, the crabs gradually come together into larger feeding aggregations, or armies, which move down the beach to feed in areas where the surface of the sand remains wet. Mictyris differs from the scopimerine crabs in moving forwards whilst feeding (Sleinis and Silvey, 1980); also sand is lifted by the chelae to the apex of the third maxillipeds and passes downwards during sorting to emerge at the base of the third maxillipeds from where it drops off or is wiped away by one of the chelae (McNeil, 1926; Lazarus, 1945). In this respect Mictyris is similar to Ocypode and Uca. It seems probable that this difference in direction of particle movement is related to the use by all these species of air filled epibranchial chambers for aerial gas exchange and the resulting conflict between the needs for water pumping for feeding and air ventilation, both of which involve the scaphognathite (Eshky, 1985).

During their feeding migrations, Mictyris respond to disturbance by burrowing rapidly into the moist sand by a similar spiral movement to that of scopimerine crabs (McNeil, 1926). Static objects which present a distinct visual contrast with their surroundings are avoided (Kraus and Tautz, 1981). Water supplies for flotation feeding are replaced by crabs pausing, lowering the fringe of water absorbing hairs along the posterior margin of the carapace into the surface film, and drawing water into the gill chambers by forward pumping of the scaphognathite (Quinn, 1980).

Like scopimerine crabs, Mictyris also shows an alternative feeding mode, in this case burrow-orientated, but differing basically from the scopimerine pattern. Here it takes the form of hummocking (Cameron, 1966; Quinn, 1986). Sand excavated by crabs close to the sand surface in their burrows, is compacted into pellets and used to build a small round mound one pellet thick. Further sand excavated is passed to the maxillipeds and sorted, and the rejected pellet is lifted by the chela and placed above and in front of the crab. This results in the formation of long hummocks consisting of a feeding trench covered over by a curving roof of feeding pellets. In Hong Kong, hummocking was the only activity of Mictyris longicarpus when conditions during the low tide period were overcast or raining, although some hummocking always occured before emergence of Mictyris, even on days when gregarious feeding followed. The number of Mictyris which emerge is dependent on environmental conditions, including temperature (Kelemic, 1979). Quinn (1980), Quinn and Fielder (1978) and Quinn (1986) have described various aspects of both surface feeding and hummocking in M. longicarpus from Australian beaches, including quantitative aspects of selection and utilization of sediment.

Mictyris shows a very similar form of threat posture to that of scopimerine crabs (Cameron, 1966). This is most apparent during an aggressive wandering phase which precedes burrowing at the end of surface activity. During this phase, pairs of males frequently display, and these threat displays may be joined by a third, fourth, and sometimes fifth individual (Cameron, 1966).

DISCUSSION

The scopimerine and mictyrid crabs, although superficially similar, have evolved the semi-terrestrial, surface deposit feeding habit independently, and show contrasting adaptations of behaviour. Both, however, exhibit similar combinations of stereotyped behaviour and flexible response, which allow both the efficient exploitation of tidally-replenished food resources under good conditions and the maintenance of some activity when conditions are less favourable. Thus most scopimerine and mictyrid crabs which have been studied exhibit two different feeding modes, one orientated to a semi-permanent burrow, the other away from any permanent burrow, each accompanied by appropriate escape responses and other behaviours. In individual species one mode is usually preferred, the other a response to environmental conditions which are unfavourable to the preferred mode.

In scopimerine crabs, a structured, burrow-orientated feeding pattern is followed under most circumstances, with escape responses also orientated to the shelter of the burrow; migratory feeding occurs in response to conditions which disrupt this pattern. In *Mictyris*, in contrast, gregarious migratory feeding is the norm, and burrow orientated feeding the response to less ideal conditions. In both, the ability to 'switch' modes of activity allows response to both the predictable and the unpredictable elements of the intertidal environment.

From this brief, and selective, review of the behavior of scopimerine and mictyrid crabs it is not possible to draw general conclusions, indeed many of the observations made are capable of more than one interpretation. The behaviour of both groups appears to lack some of the complexity shown by both *Uca* and *Ocypode* species, but this may reflect merely the less developed state of knowledge of these geographically more restricted groups. Certainly, many critical studies remain to be done. The coexistence of many species of scopimerine and mictyrid crabs together with *Uca* and *Ocypode* species and macrophthalmids, on Indo-Pacific shores, provides an opportunity for comparative studies so far sadly underexploited.

ACKNOWLEDGEMENTS

The data from personal observations referred to here were collected during study visits to Hong Kong during 1983 and 1986, and will be reported in detail elsewhere. I am grateful to the Jeffries Association Ltd for financial support, to Professor Brian Morton and members of the staff of the Department of Biology, University of Hong Kong for provision of essential help and facilities during these visits, and to the Council of the Scottish Marine Biological Association for allowing me leave of absence.

REFERENCES

Altevoght, R., 1957, Beitrage zur Biologie und Ethologie von *Dotilla blandfordii* Alcock und *Dotilla myctiroides* (Milne-Edwards) (Crustacea Decapoda). *Z. Morph. u Okol. Tiere*, 46: 369.

Barnes, R.S.K., 1967, The Macrophthalminae of Australia: with a review of the evolution in morphological diversity of the type genus *Macrophthalmus* (Crustacea, Brachyura). *Trans. Zool. Soc. Lond.*, 31: 195.

Bigalke, R., 1921, On the habits of the crab *Dotilla fenestrata* Hilgendorf, with special reference to the mode of feeding. *S. Afr. J. nat. Hist.*, 3: 205.

Cameron, A.M., 1966, Some aspects of the behaviour of the soldier crab *Mictyris longicarpus*. *Pacif. Sci.*, 20: 224.

Christy, J.H., 1978, Adaptive significance of reproductive cycles in the fiddler crab *Uca pugilator*: A hypothesis. *Science*, 199: 453.

Crane, J., 1975, *Fiddler Crabs of the World (Ocypodidae. genus Uca)*. Princeton University Press, Princeton, New Jersey. 736 pp.

Eshky, A.A., 1985, Aspects of the ecology, behaviour and physiology of the ghost crab *Ocypode saratan* (Forskal). Ph.D. Thesis, University of Glasgow. 250 pp.

Fielder, D.R., 1970, The feeding behaviour of the sand crab, *Scopimera inflata* (Decapoda, Ocypodidae). *J. Zool., Lond.*, 160: 35.

Fielder, D.R., 1971, Some aspects of distribution and population structure in the sand bubbler crab *Scopimera inflata* Milne Edwards 1873. (Decapoda Ocypodidae). *Aust. J. Mar. Freshwat. Res.*, 22: 41.

Fishelson, L., 1983, Population ecology and biology of Dotilla sulcata (Crustacea, Ocypodidae) typical for sand beaches of the Red Sea, in: Sandy beaches as Ecosystems, A. McLachlan and T. Erasmus, ed., Dr W. Junk, Publishers, The Hague/Boston/Lancaster, 643-654.

Hails, A.J. and Yaziz, S., 1982, Abundance, breeding and growth of the Ocypodid crab Dotilla myctiroides (Milne-Edwards) on a west Malaysian beach, Est. Coastal Shelf Sci., 15: 229.

Hartnoll, R.G., 1973, Factors affecting the distribution and behaviour of the crab, Dotilla fenestrata on East African shores, Estuar. cstl. mar. Sci., 1: 137.

Hartnoll, R.G., 1975, The Grapsidae and Ocypodidae (Decapoda: Brachyura) of Tanzania, J. Zool., Lond., 177: 305.

Iwata, K., Arita, T., and Ino, K., 1983, Seasonal changes in energy reserves of Scopimera globosa (Crustacea: Ocypodidae) with special reference to its wandering behaviour, Zool. Mag. Tokyo., 92: 306.

Kelemic, J.A., 1979, Effects of temperature on the emergence from burrows of the soldier crab, Mictyris longicarpus (Latreille), Aust. J. Mar. Freshwat. Res., 30: 463.

Kraus, H-J. and Tautz, J., 1981, Visual distance keeping in the soldier crab, Mictyris platycheles Latreille (Grapsoidea: Mictyridae) A field study, Mar. Behav. Physiol., 8: 123.

Lazarus, M., 1945, Mictyrus, The soldier crab, M.Sc., Thesis, University of Sydney, N.S.W. Australia, 62 pp. (quoted by Cameron, 1966).

MacIntyre, A.D., 1968, The meiofauna and macrofauna of some tropical beaches. J. Zool., Lond., 156: 377.

Maitland, D.P., 1986, Crabs that breath air with their legs - Scopimera and Dotilla, Nature, Lond., 319: 493.

McNeil, F.A., 1926, Studies in Australian carcinology. II.A revision the family Mictyridae. Rec. Aust. Mus., 15: 100.

Miller, D.C., 1961, The feeding mechanism of fiddler crabs, with ecological considerations of feeding adaptations. Zoologica, N.Y., 46: 89.

Ono, Y., 1965, On the ecological distribution of ocypodid crabs in the estuary. Mem. Fac. Sci. Kyushu Univ., Ser. E (Biol), 4: 1.

Quinn, R.H., 1980, Mechanisms for obtaining water for flotation feeding in the soldier crab, Mictyris longicarpus Latreille, 1806 (Decapoda, Mictyridae). J. exp. mar. Biol. Ecol., 43: 49.

Quinn, R.H., 1986, Experimental studies of food ingestion and assimilation of the soldier crab, Mictyris longicarpus Latreille (Decapoda, Mictyridae) J. exp. mar. Biol. Ecol., 102: 167.

Quinn, R.H., and Fielder, D.R., 1978, A laboratory bench system for prolonged maintenance of the sand crabs, Mictyris longicarpus Latreille, 1806 and Scopimera globosa de Haan, 1833 (Decapoda, Brachyura), Crustaceana, 34: 310.

Salmon, M., 1965, Waving display and sound production in the courtship behaviour of Uca pugilator, with comparisons to U. minax and U. pugnax, Zoologica, N.Y., 50: 123.

Silas, E.G. and Sankarankutty, C., 1967, Field investigations on the shore crabs of the gulf of Manaar and Palk Bay, with special reference to the ecology and behaviour of the pellet crab, Scopimera proxima Kemp, Proc. Symposium on Crustacea. Mar. Biol. Ass. India, Symposium Ser. 2, pt III: 1008.

Sleinis, S. and Silvey, G.E., 1980, Locomotion in a forward walking crab, Mictyris platycheles. J. Comp. Physiol., A, 136: 301.

Sugiyama, Y., 1961, The social structure of a sand crab, Scopimera globosa De Haan, with special reference to its population. Physiol. Ecol., Japan. 10: 10.

Tweedie, M.W.F., 1950, Notes on grapsoid crabs from Raffles Museum, II, On the habits of three ocypodid crabs, Bull. Raffles Mus., 23: 317.

Vogel, F., 1984, Comparative and functional morphology of the spoon-tipped setae on the second maxillipeds in Dotilla Stimpson, 1858 (Decapoda, Brachyra, Ocypodidae), Crustaceana, 47: 225.

Wada, K., 1981a, Growth, breeding and recruitment in Scopimera globosa and Ilyoplax pusillus (Crustacea, Ocypodidae) in the estuary of Waka River, middle Japan, Publ. Seto mar. biol. Lab., 26: 243.

Wada, K., 1981b, Wandering in Scopimera globosa (Crustacea: Ocypodidae). Publ. Seto mar. biol. Lab., 26: 447.

Wada, K., 1983a, Movement of burrow location in Scopimera globosa and Ilyoplax pusillus (Decapoda Ocypodidae), Physiol. Ecol. Japan, 20:1.

Wada, K., 1983b, Temporal changes of spatial distributions of Scopimera globosa and Ilyoplax pusillus (Decapoda: Ocypodidae) at co-occuring areas. Jap. J. Ecol., 33: 1.

Wada, K., 1983c, Spatial distributions and population structures in Scopimera globosa and Ilyplax pusillus (Decapoda: Ocypodidae). Publ. Seto mar. biol. Lab., 27: 281.

Wada, K., 1984, Barricade building in Ilyoplax pusillus (Crustacea: Brachyura). J. exp. mar. Biol. Ecol., 83: 73.

Wada, K., 1986, Burrow usurpation and duration of surface activity in Scopimera globosa (Crustacea: Brachura: Ocypodidae), Publ. Seto mar. biol. Lab., 31: 327.

Yamaguchi, T., Noguchi, Y, and Ogawara, N., 1979, Studies of the courtship behaviour and copulation of the sand bubbler crab, Scopimera globosa. Publ. Amakusa mar. biol. Lab., 5: 31.

Yamaguchi, T. and Tanaka, M., 1974, Studies on the ecology of a sand bubbler crab, Scopimera globosa de Haan (Decapoda, Ocypodidae) 1. Seasonal variation of population structure, Jap. J. Ecol.; 24: 165.

Activity Rhythms in *Siphonaria thersites*

George M. Branch

University of Cape Town
South Africa

INTRODUCTION

The existence of both exogenous and endogenous activity rhythms is now beyond doubt, and their nature has been extensively analysed in the laboratory (Naylor, 1988). The underlying causes of activity rhythms and their adaptive value in the field are, however, often more difficult to explore. The major thrust of this paper is to describe the patterns of activity diplayed by a siphonariid limpet, Siphonaria thersites Carpenter, and to explain them in terms of its biology. To achieve this, the scope of the paper is broader than the primary aim, because behavioural patterns need to be viewed in the context of an animal's morphology, physical environment and biotic interactions.

Homing and activity rhythms are characteristic of many limpets, including most, if not all, species of Siphonaria (e.g. Cook, 1971, 1976; Thomas, 1973; Cook and Cook, 1978, 1981; Bertness, Garrity and Levings, 1981; Hulings, 1985). The adaptive value of homing and/or activity rhythms in Siphonaria species has been related to avoidance of potentially lethal osmotic stress (McAlister and Fisher, 1968; Branch and Cherry, 1985), desiccation (Verderber, Cook and Cook, 1983), wave action (Branch and Cherry, 1985), and predation (Garrity and Levings, 1983). Changes in the frequency of homing have been suggested as a means of controlling local densities of Cellana tramoserica, a patellid limpet, in relation to the availability of food (MacKay and Underwood, 1977). More rigorously controlled experiments have, however, shown that this interpretation was probably based on an experimental artifact (Underwood, 1988). In any event, homing and activity rhythms clearly have many potential functions.

Siphonaria thersites is unusual in a number of respects. Being distributed from the Aleutian Islands to Washington, it is the only representative of the genus on the north-west coast of North America. Yonge (1960) noted that it is closely associated with macroalgal beds, and described it as "a siphonariid with specialised and restricted distribution" and "a specialized inhabitant of the upper shore". In view of this high-shore zonation, it is curious that S. thersites has a reduced shell, which is small and fragile and cannot cover the large, flexible foot. Given these peculiar features, and others that emerge below, the central question posed in this paper is how the activity rhythms displayed by S. thersites relate to its ecology, including its responses to desiccation and wave action, its vulnerability to predation, and its interaction with its foodplant.

METHODS

Study Sites, Densities and Zonation Patterns

Most of the research was undertaken at Tatoosh and Waadah Islands, off the Olympic Peninsula of the Washington coast of North America. Additional observations were made at ShiShi on the Olympic Peninsula and at Bamfield and Botanical Beach, Vancouver Island.

Densities were determined at Waadah and Tatoosh on eight 1m-wide transects running from the upper to the lower limits of zonation of S. thersites. During periods when the limpets were active, counts were made of all S. thersites visible in 100 randomly located $324cm^2$ quadrats on each transect. This technique was supplemented by destructive removal of all algae within 25 of these quadrats, allowing more accurate determination of the total numbers of S. thersites. This revealed that the actual densities were 8.2 times higher than those obtained by counting visible animals. Using this information, the numbers of active S. thersites could be converted to densities.

Activity Rhythms

Activity was monitored in fixed quadrats, at 30 minute intervals throughout the tidal cycle, by day and night. Observations were initially made during high tide by snorkelling, but were discontinued because it soon became apparent that the animals were never active when covered by water. Five sites were monitored at each locality, with four replicate quadrats of $324cm^2$ at each. Observations were made on May 3-6, August 15-18, and October 14-16 at Waadah, and May 6-9, September 13-16 and November 9-13 at Tatoosh. In testing for correlations between activity and the predicted height of the low tide, the data on activity were expressed as percentages of the maximum numbers active at each site, thus allowing direct comparison between sites with different densities.

Desiccation

The rate and lethal limits of water loss experienced by S. thersites were measured in a wind tunnel, at constant but realistic conditions simulating those experienced in the field on hot, dry, calm days, i.e. 29°C, 50%RH and an air speed of $2.5m.s^{-1}$. Rates of water loss of 19 animals (with wet body masses ranging from 0.02 to 0.25g) were measured every 15 mins for up to 120 mins. At the end of the experiment the animals were oven-dried to constant weight at 30°C and their original water content determined. Rate of loss was non-linear, so the results are presented as percentage water loss after periods of 15, 60 and 120 mins.

In the field, survival of desiccation was determined when the animals were: a) allowed to home to crevices after feeding; b) left in beds of the alga Iridaea cornucopiae, in sites where there were no crevices; or c) placed on bare rock where they had access to neither crevices nor foliar algae. Survival was monitored at 15 min intervals. Two replicate lots of 30 animals (shell lengths 7-9mm) were used for each treatment. Over bare sections of rock air temperatures spanned 16-18°C and humidities 65-72%.

Effects of wave action

Cotton threads were attached to the apexes of shells and tenacity determined by measuringt the vertical force required for detachment of limpets when they were either stationary on rocks (n=24) or actively moving on algae (n=20). After dislodgement the limpets were placed on a glass sheet, allowed to reattach, and their pedal surface areas measured.

The survival of batches of marked animals (n=20) over the high-tide period was determined when the limpets were either allowed to home to crevices, or interfered with in various ways to prevent homing. Details are outlined in the text. The abundance of S. thersites in relation to wave action was quantified at North Island, Tatoosh, on a shore where there is a clear gradient of wave action from moderate shelter to extreme exposure. Ten random samples of $400cm^2$ were taken within beds of I. cornucopiae at each of seven sites along the gradient.

Responses of Predators

Preliminary observations were made on the responses of predators to S. thersites, firstly by direct observation of oystercatchers and other avian predators and, secondly, by feeding S. thersites to two types of tidepool predators/scavengers, the sculpin Oligocottus maculatus, and a common hermit crab, Pagurus hirsutiusculus. Individual S. thersites were removed from algal beds, and then either left intact or prodded or damaged to induce release of mucus before being introduced to these predators in the field. The responses of the predators were categorised according to whether they ignored, mouthed/handled, or ingested the limpets. As a comparison, similar sized individuals of a second species of limpet, Collisella pelta, were also introduced to the predators, on the rationale that this species is not known to have any chemical defense against predators. The behaviour of the predators towards this limpet thus acted as a form of control. Two trials (n=10 limpets in each) were run for each species and treatment.

Interactions between S. thersites and its foodplant

Patterns of grazing on Iridaea cornucopiae were first established in the field. Following this, laboratory experiments were designed to test whether S thersites preferred different phases of the algal life cycle or different portions of the alga. Duplicate batches of 6 or 10 limpets were offered ca.1g wet weight of the alga, and the rate of consumption (loss of blotted wet mass) recorded after periods of up to 96h. In the initial experiments the limpets were offered a choice between whole blades of infertile and tetraporic plants. Subsequently the blades were cut into two equal portions (distal and proximal) and batches of limpets separately provided with basal or apical portions of the blades of infertile, tetrasporic or carposporic plants. As in other species of Iridaea, male plants could not be readily distinguished from infertile plants (Hansen and Doyle, 1976; Hannach and Santelices, 1985) and may have been included amongst the latter, although sections cut of these plants failed to reveal any mature male plants. Control batches of algae were maintained without limpets.

As an index of the toughness of algal blades, thin sections were cut and the thicknesses of the cuticle, medulla and the whole blade measured for infertile, tetrasporic and carposporic plants. More direct measurements of toughness were made using a penetrometer attached to a strain gauge. This enabled the force and work required to achieve penetration of blades to be measured. These data are, however, not used in the paper, partly because the action of a blunt penetration probe is unlikely to mimic the effects of a Siphonaria radula, and also because opposite trends were obtained when probes of 0.6 and 1.2mm were used (blunter probes stretching the blade and thus measuring flexibility rather than penetrability).

RESULTS AND DISCUSSION

Zonation Pattern

Figure 1 summarises the zonation of S. thersites at Waadah and Tatoosh. Three features emerge. Firstly, the limpets occur very high on the shore, at heights ranging from 1.0 to 2.3m above low low water during spring tides (when the tidal amplitude is about 3.8m, and ranges from -0.8m to +3.0m). No other species of Siphonaria occurs exclusively in the high-shore, and the restriction of S. thersites to this zone is curious considering its small fragile shell and its unprotected foot: features likely to make it both susceptible to water loss and vulnerable to predators.

Secondly, S. thersites is closely associated with beds of the alga Iridaea cornucopiae, itself a high-shore occupant of open coasts, and grazed extensively on it. It also feeds on Halosaccion glandiforme, and has the potential to affect this alga substantially, because it consumes the entire upright portion of the plant. It was, however, never found sheltering in Halosaccion beds, and only ate Halosaccion abutting on the lower limits of beds of I. cornucopiae. Small numbers were seen feeding on Ulva when it grew within beds of I. cornucopiae. Cladophora, tufts of which grew commonly near the lower limits of I. cornucopiae, was never grazed by the limpets. S. thersites was found in beds of Fucus at Tatoosh but, although it grazed on the Fucus, it was scarce there, with a mean density of 29m^{-2}, compared with 944m^{-2} on I. cornucopiae. Furthermore, the average and maximum sizes of the limpets on Fucus were smaller than those on I. cornucopiae (x=3.4mm, max.=4.5mm on Fucus; x=6.2mm, max.=9.6mm on I. cornucopiae). In the areas investigated, all on the open coast, there was no doubt that I. cornucopiae was the major foodplant. Further north, in Alaska, and in a tiny isolated population at San Juan Island, the limpets do, however, occur predominantly in beds of Fucus (Yonge, 1960). One speculative possibility explaining this geographic difference is that Fucus is an adequate host in quieter waters while the more dense, bushy I. cornucopiae provides better protection on exposed shores.

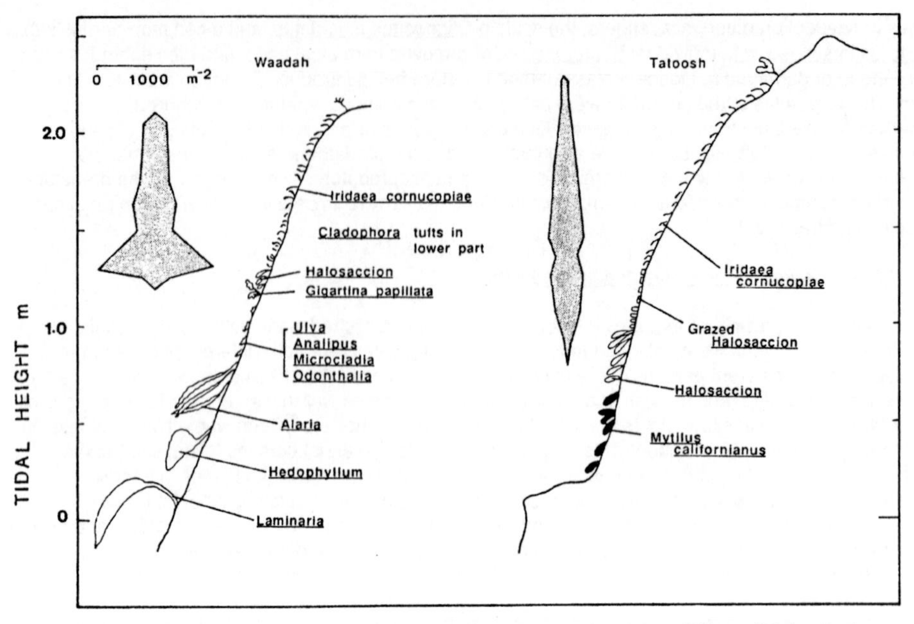

Fig. 1. The density and zonation of Siphonaria thersites at Waadah and Tatoosh Islands, in relation to the vertical distribution of dominant bed-forming algae. Tidal height is expressed as height above low low water.

Thirdly, the limpets were patchily distributed, being absent or very rare in certain beds (e.g. at ShiShi and at Bamfield). Where they were found, they usually occurred at extremely high densities, regularly exceeding $500m^{-2}$ and often attaining $2500m^{-2}$ in beds of I. cornucopiae. The only other organisms occupying these beds in any numbers were amphipods and two littorines, Littorina scutulata and L. sitkana. The latter was extraordinarily abundant, reaching densities an order of magnitude higher than those of S. thersites.

Activity rhythms

Siphonaria thersites was only active during low-tide, when it was visible feeding on the tips of the blades of foliar algae, notably Iridaea cornucopiae. Activity began shortly after the limpets were fully exposed by the receeding tide, and lasted for about 4h, after which the limpets converged on holes and crevices, disappearing from sight. Activity ceased well in advance of immersion by the rising tide. The limpets aggregated in particular shelters, individuals returning faithfully to the same shelter after each period of activity, although within the crevices their positions varied from one period of activity to the next. At sites where holes and crevices were absent, the limpets sheltered between mussels or at the bases of algae, hidden by the canopy.

The daily pattern of activity is summarised in Figure 2. In summer (May to August), the limpets were active only during the early morning low tide, ceasing activity well before mid-day (Fig. 2A-C). This period of the day was relatively cool, and coincided with the lower of the two daily low tides. As the year progressed, tidal amplitudes altered, and there was a brief period during September when the two low tides were almost equal. At that stage there were two periods of activity per day, one in the early morning and the other during the evening low tide (Fig. 2D). By October the tidal pattern had switched to one in which the lowest low tides occurred by night, and the period of activity was then confined to the night-time low tide. This suggested that the height of the low tide is critical in determining whether the limpets are active. To test this, observations were undertaken over two neap tides, when tidal ranges were dampened, leading up to a period of spring tides when the nocturnal low tides were lower than during the neaps. Fig. 2E shows that very few S. thersites were active during the neap tides, but increasing numbers became active as the low tides dropped progressively.

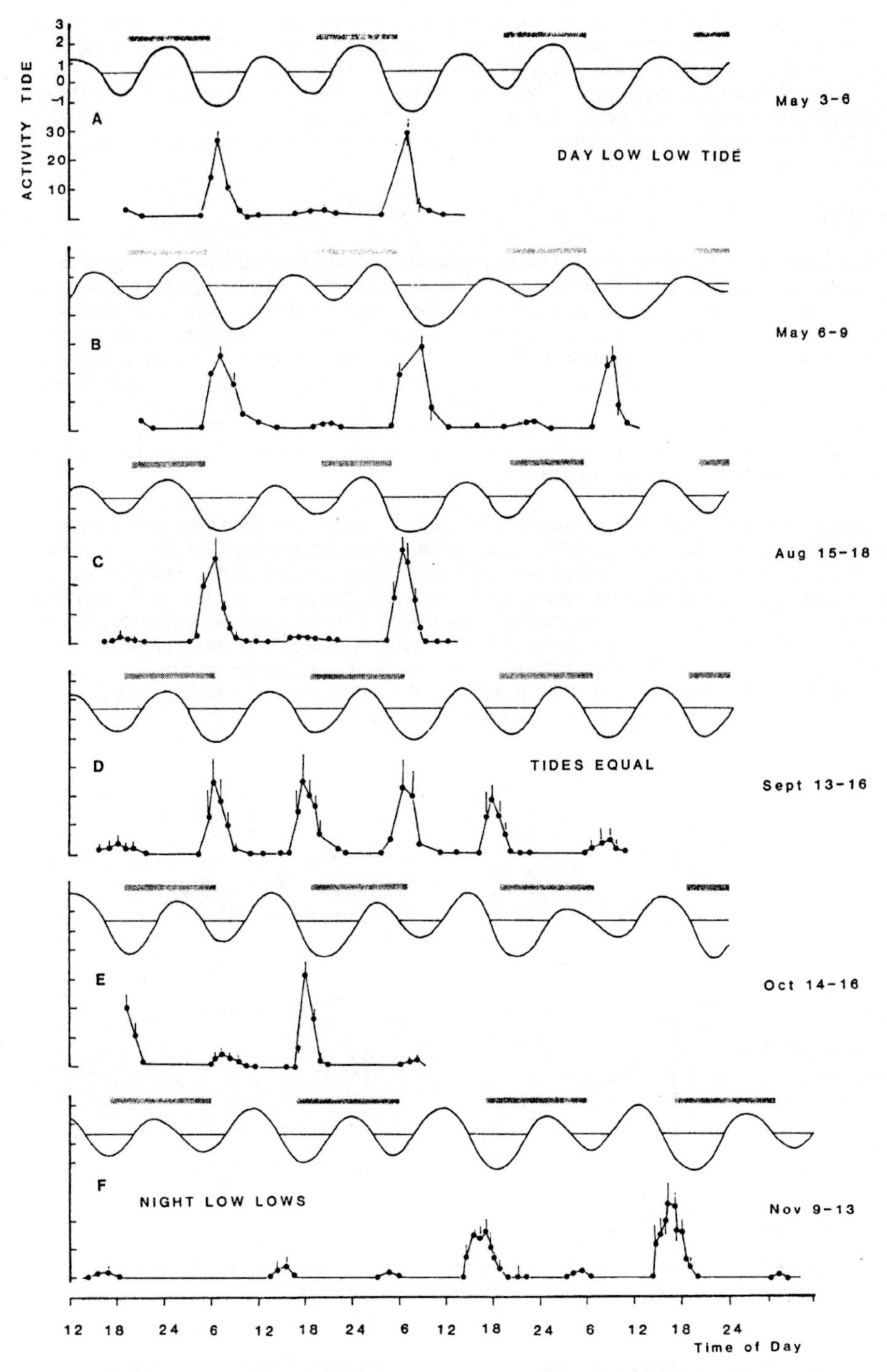

Fig. 2. Activity of S. thersites (mean numbers active·325cm^{-2}±S.D.) relative to tidal cycles (m above chart datum). Horizontal shaded bars indicate night. In A-C the low low tides fell during mid-morning; in D the two daily tides were approximately equal; and in E & F low low tides occurred by night.

The timing of activity rhythms has been analysed for a number of Siphonaria species (see Branch, 1981 for a review; Bertness, Garrity and Levings, 1981; Branch and Cherry, 1985; Hulings, 1985). In general siphonarians are either active during high tide, or during the ebb/flood period when they are awash. Only one species, S. capensis, has been recorded as being active exclusively during low tide (Branch and Cherry, 1985). The pattern demonstrated by S. thersites is thus unusual, particularly in view of its high-shore zonation and the fact that it is active during the day-time low tides in summer, when desiccation-stress will be greatest.

Desiccation

Given the high-shore distribution and the day-time activity during summer low tides, the rapidity with which S. thersites loses water may be critical. Measurements show that S. thersites is extremely prone to water loss: indeed, its rate of desiccation is ten times higher than that recorded for any other limpet (Fig. 3). Its tolerance to water loss is, however, equivalent to that of other species, lethal losses being between 70% and 75% (cf. Branch, 1975; Branch, 1981). Despite the susceptibility of S. thersites to water loss, its activity rhythms are clearly not related to the need to conserve water, sinse (1) activity occurs during low tide and (2) in summer, when desiccation is maximal, activity occurs by day. It is true that in summer activity lasted only about 4h, and ceased by mid-morning. But during winter, when activity occurred by night, the limpets were still active for only about 4h. Thus neither the timing nor the duration of activity seem related to desiccation.

The rate of mortality of S. thersites was very high when it was removed from algal beds and placed on bare rocks free of crevices and algae (Fig. 4). By contrast, all the limpets that were allowed to home to crevices survived the full low-tide period, and 100% survival was also recorded for those left in beds of Iridaea cornucopiae, even when they were physically prevented from returning to their shelters after activity ceased. The high humidity maintained in the algal beds (98%RH) is clearly critical to the survival of the limpets. Coupling these observations with the high rates of desiccation, even relatively small expanses of bare rock (ca. 3m) may act as barriers to the dispersal of S. thersites from one patch of algae to another. This might, in part, explain their patchy distribution, especially as S. thersites undergoes direct development and has no free-swimming larval stage.

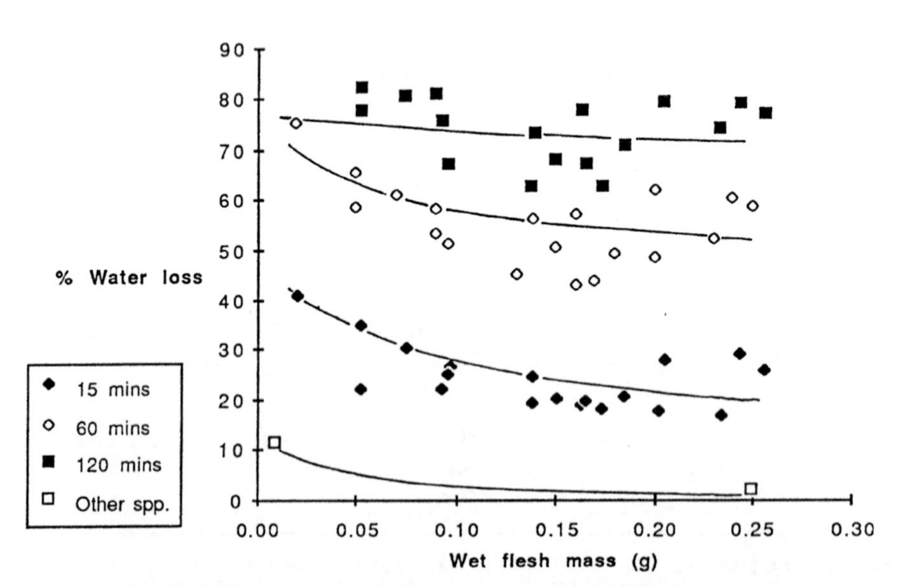

Fig. 3. Water loss from S. thersites after 15, 60 and 120 mins exposure to air. For comparison the lowermost curve (from Branch, 1975) shows water loss for seven Patella spp. after 60 mins. Equations are as follows:
15 mins: $y=15.17x^{-0.225}$; $r^2=0.30$; 60 mins: $y=42.34x^{-0.123}$; $r^2=0.31$
120 mins: $y=72.85x^{-0.012}$; $r^2=0.13$.

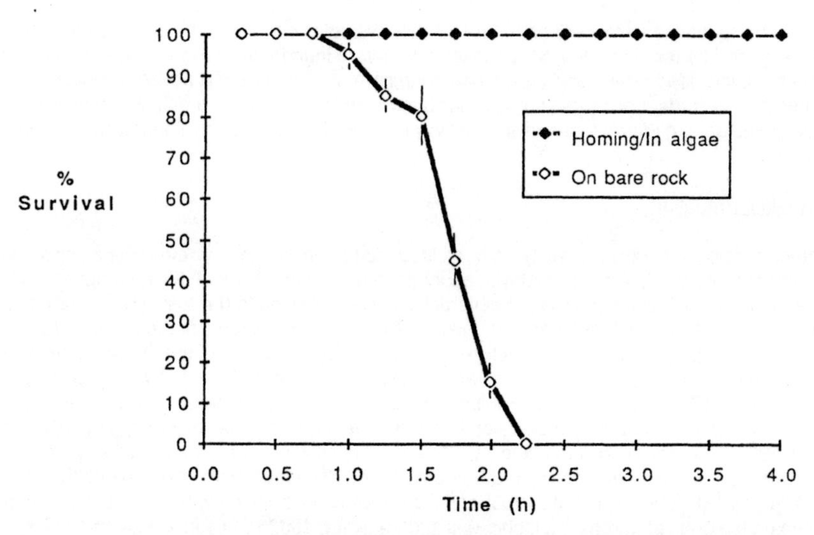

Fig. 4. Survival of S. thersites during low tide when left exposed on bare rock, allowed to home to crevices, or left in beds of Iridaea cornucopiae. Error bars show standard deviations.

Susceptibility to water movement

Siphonaria capensis is vulnerable to wave action because of its low tenacity, and Branch and Cherry (1985) suggest that this is why it is active during low tide. The tenacity of S. thersites is substantially less than that recorded for any other limpet, including S. capensis (Fig. 5). Even when it is stationary and firmly attached to rocks, its powers of attachment are 10 to 40 times lower than those of other species (Branch and Marsh, 1978; Branch, 1981; Branch and Cherry, 1985). Whether moving or stationary, when it is attached to algae its tenacity is so low that it requires <0.02N to dislodge it. Clearly S. thersites has weak capabilities of attachment, and is particularly liable to dislodgement when feeding on macroalgae.

Its survival during high tide was determined when it was a) allowed to home to crevices or gaps between algae; b) allowed to home but then lifted and placed in other crevices (as a control for disturbance); c) lifted from its home crevices and placed between the bases of algae in beds of I. cornucopiae; d) placed on bare rock with no access to crevices or algae; e) prevented from homing by replacing it on its foodplant when homing was attempted. In the first two treatments 100% survival was recorded over the high-tide period. For limpets situated between algae, the loss was only 14%, while limpets left on bare rock experienced losses of 20 to 70% over the high tide. In the last treatment 100% loss was recorded within 15 mins of inundation, and many of the limpets were seen being washed away by the first waves that washed over the algae.

The distribution of S. thersites along a gradient of wave action at Tatoosh supports the contention that it is vulnerable to water movements. At seven stations along a gradient which ran from sites that were moderately sheltered to sites exposed to strong wave action, the following mean numbers of limpets were recorded ($x \pm SD \cdot 400cm^{-2}$): 24.3±9.07; 20.0±8.5; 15.5±6.6; 9.8±3.2; 3.6±2.5; 0.8±0.8 and 0.

These observations on the distribution and zonation of S. thersites, its survival when exposed to wave action and its weak powers of tenacity all suggest that wave action is the primary determinant limiting activity to low tide. It cannot afford to be active, particularly on algae, when the rising tide covers the high-shore. As tidal heights are unpredictable because they depend partly on how rough the sea is, the risks of being washed away will be reduced if activity begins soon after the animals are completely exposed by the receeding tide, ceases well before inundation is likely, and is confined to periods when the low tide is lower and lasts longer than normal. On this basis activity should be concentrated during spring tides and should occur during the lower of the two daily low tides. Re-analysis of the data in Figure 2 shows that activity is, indeed, inversely correlated with the height of low tides (Fig. 6).

Defense Against Predation

Another possible function of activity patterns is to reduce predation. However, the sides of the foot and head of S. thersites are supplied with abundant mucous glands that have a complex multicellular structure: a feature common to several species of Siphonaria (Fretter and Graham, 1954; Marcus and Marcus, 1960; de Villiers and Hodgson, 1984). Acetone extracts of the tissues of S. thersites contain polyproprionates, the structure of which has been analysed and shown to be identical to pectinatone produced by S. pectinata and very similar to the polyproprionates found in S. lessoni and S. diemenensis (D. Manker and D.J. Faulkner, Scripps Institution of Oceanography, pers. comm.). Comparable secondary metabolites have been isolated from eleven species of Siphonaria (Biskupiak and Ireland, 1983; Hochlowski and Faulkner, 1983, 1984; Hochlowski et al., 1983, 1984; Capon and Faulkner, 1984; J. Faulkner, pers. comm.). Some are biologically active, being toxic to goldfish at levels as low as $10 \mu g.ml^{-1}$ (Hochlowski et al., 1983) and suppressing bacterial growth and development of fertilised sea urchin eggs at $1 \mu g.ml^{-1}$ (Hochlowski and Faulkner, 1983). It seems likely that these compounds serve as a defense against predators. Another marine pulmonate, Trimusculus reticulatus, possesses chemicals known to repel predators (Rice, 1985).

Throughout the period of observation, birds were never recorded attacking S.thersites, despite the fact that other species of limpets of comparable size are eaten, e.g. Collisella digitalis, C. pelta, and C. strigatella (personal observations at Tatoosh and San Juan Island, and see Hartwick, 1976; Frank, 1982; Hahn, 1985; Marsh, 1986).

Trials were undertaken with a tide-pool fish (Oligocottus maculatus), and a common scavenging/predatory hermit crab (Pagurus hirsutiusculus) which had been observed feeding on C. strigatella. Comparisons of these predators' responses to damaged or intact S. thersites with their responses to Collisella pelta of comparable size are given in Table 1. Both limpets were mouthed by the fish but, with one exception, all of the S. thersites were then rejected. Of the C. pelta, 85% of the intact limpets and 100% of the damaged animals were consumed. The hermit crabs responded similarly, at first handling the S. thersites and then discarding them, while consuming (and often avidly fighting over) all C. pelta offered to them. However, the hermit crabs returned repeatedly to any damaged S. thersites and, after about 10 mins, eventually shreaded and ate 60% of them. Intact limpets were only eaten on two occasions. Burnupena cattarhacta, a scavenging whelk, responds in a similar way when offered damaged specimens of another siphonarian, S. capensis, initially being repelled by the mucus released by the limpet, but ultimately feeding on it (Branch and Cherry, 1985).

Table 1. Responses of Oligocottus maculatus and Pagurus hirsutiusculus to either intact or damaged Siphonaria thersites or Collisella pelta. N=20

Predator	Prey	Condition	Responses of Predators to Prey (%)			
			Ignored	Handled	Ingested in <10min	Ingested in 10-30min
O. maculatus	S. thersites	Intact	20	80	5	0
		Damaged	15	85	0	0
	C. pelta	Intact	10	90	85	0
		Damaged	5	95	90	10
P. hirsuti-usculus	S. thersites	Intact	5	95	0	10
		Damaged	10	90	20	60
	C. pelta	Intact	5	95	80	20
		Damaged	5	95	100	0

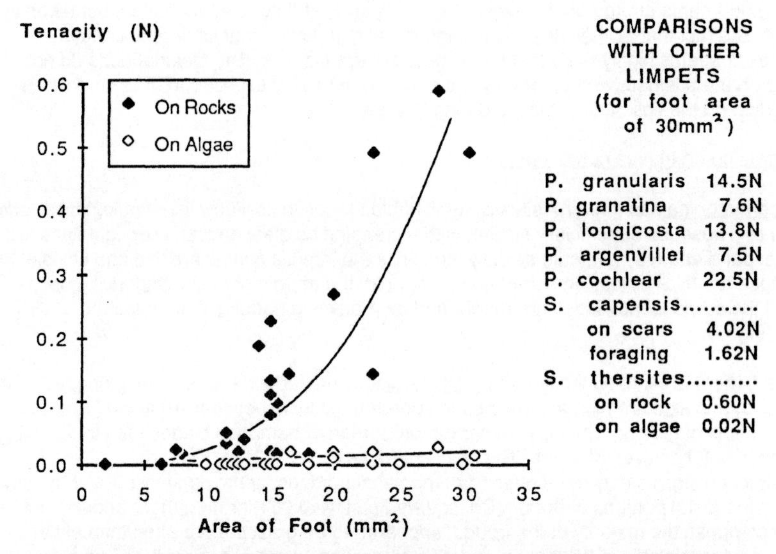

Fig. 5. Tenacity of S. thersites when attached to rocks or on algae. For animals on rocks, $y=8.61E\text{-}6x^{3.28}$; $r^2=0.66$. For comparison, forces sufficient to dislodge other limpets of comparable size are shown (data from Branch and Marsh, 1978; Branch and Cherry, 1985).

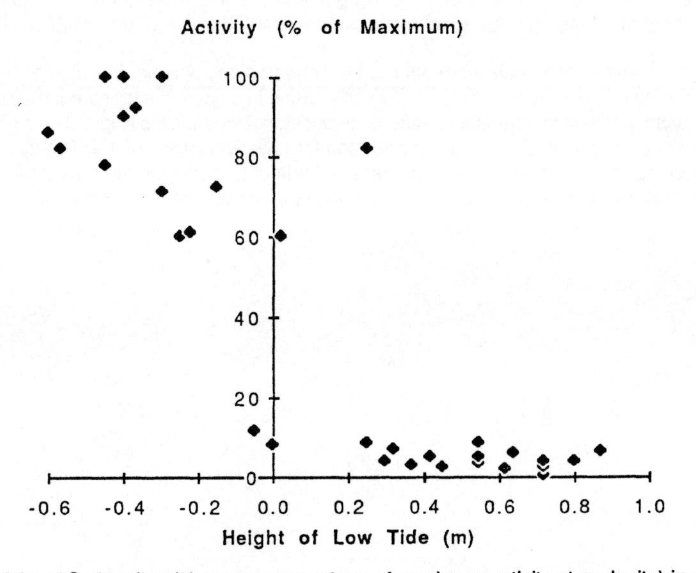

Fig. 6. Activity of S. thersites (shown as percentage of maximum activity at each site) in relation to the height of the low low tide (m above chart datum).

These experiments are too preliminary to be conclusive, and need confirmation; but taken in conjunction with the fact that S. thersites has repugnatorial glands and that its tissues contain polypropionates, it seems highly likely that the limpets are unpalatable. The fact that birds do not appear to feed on them is noteworthy, for they are apparent and very exposed to predators when active and feeding on the tips of algal blades during low tide.

Specificity of grazing on Iridaea cornucopiae

Iridaea cornucopiae has three readily distinguishable stages to its lifecycle: infertile, tetrasporic and carposporic. These are superficially similar, with a creeping holdfast and a few upright flattened blades which expand and may bifurcate apically; but while the infertile blades are thin and flexible, the tetrasporic blades are thicker, less branched and more rigid, the margin of the tetrasporic blades is thickened, and the carposporic blades are roughened by projecting nodules that house the conceptacles (Fig. 7).

In the field, there is clear evidence that I. cornucopiae is grazed extensively. In spring, when the standing stocks of this alga are high and the plants appear most lush, between 10 and 40% of the blades showed signs of grazing. During summer the proportion of damaged blades rose to almost 100%. Grazing is not, however, random. The apical portions of blades were attacked significantly more often than the basal portions, approximately 90% of the grazing taking place on the distal portions of fronds (Chi squared test, $p<0.001$ for $n=250$). In addition, infertile blades, which comprise the majority of the fronds, appeared to be grazed more often than either tetrasporic or carposporic plants. (When their relative proportions were considered, chi squared test $p<0.05$ for $n=100$).

There are, however, several possible grazers that may be responsible for these patterns, and the fact that different parts of the plant are grazed more often that others may simply reflect accessibility. Consequently laboratory experiments were run to test whether S. thersites does graze preferentially on any particular parts of the plant or stages of its life cycle. In the initial trials, limpets were given the choice of whole blades of infertile and tetrasporic plants. These revealed a strong preference for the infertile blades (Fig. 8A). They also showed that the limpets do concentrate on the apical portions of the blades rather than the bases, with 92% of the grazing directed at the distal half of the blades. T and Chi Squared tests confirmed the significance of these results ($P<0.05$ and $P<0.002$ respectively).

In a parallel experiment the limpets were offered separate portions of either apical or basal sections, from either infertile of tetrasporic plants (Fig. 8B), while in a second experiment they were supplied apical or basal portions of carposporic blades as additional alternatives (Fig. 8C). In both cases the infertile fronds were consumed at the highest rate; and for any given stage of the life cycle apical sections were eaten faster than basal sections. In these experiments, however, two-way analyses of variance indicated that only the differences between the rates of feeding on different portions of the

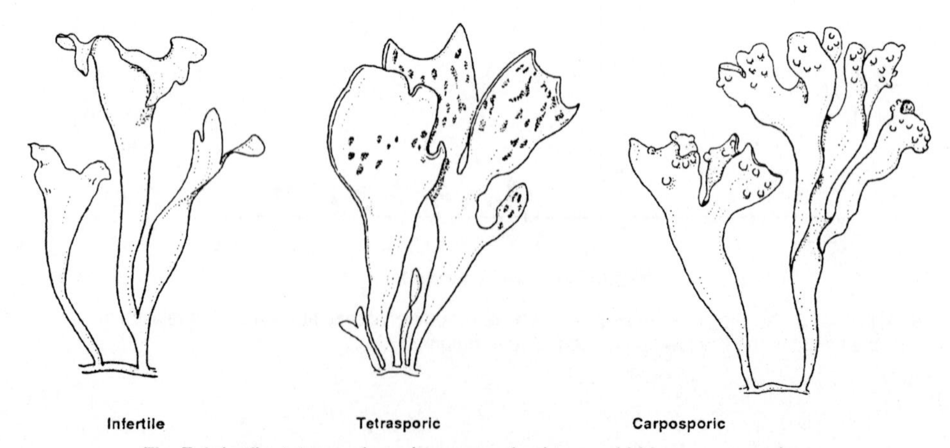

Fig. 7. Infertile, tetrasporic and carposporic phases of Iridaea cornucopiae.

blades were significant (df1,4; p=0.05 and df1,6; p=0.05) while differences between plants in different reproductive conditions were not significant (df1,4; p=0.23 and df1,6; p=0.28). In both cases there was no significant interaction between reproductive condition and portion of blade. In summary, apical portions of infertile blades were eaten faster than any other alternative available to the limpets, maximum rates of consumption being equivalent to 3.2% of the limpets' flesh mass per day; but only the differences between consumption of apical versus basal sections of the blades were statistically significant. While the comparisons between consumption of blades of different reproductive conditions consistently indicated a preference for infertile fronds, this result needs to be interpreted cautiously because the pattern was only significant in two of the three experiments.

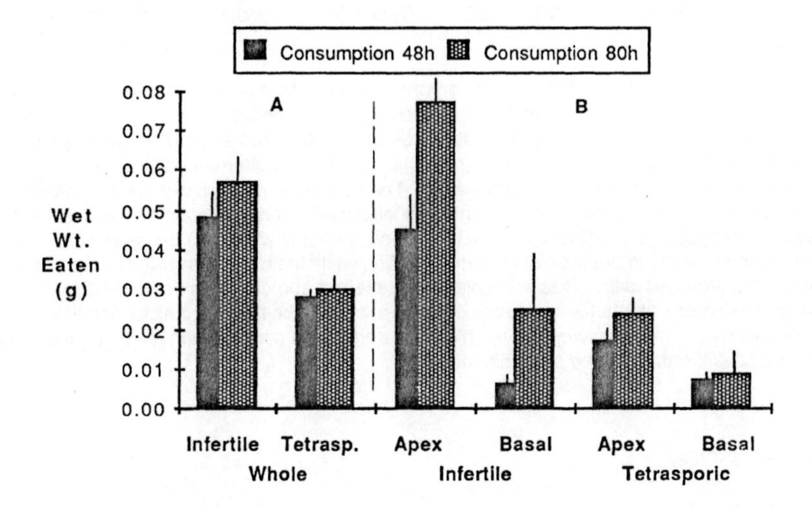

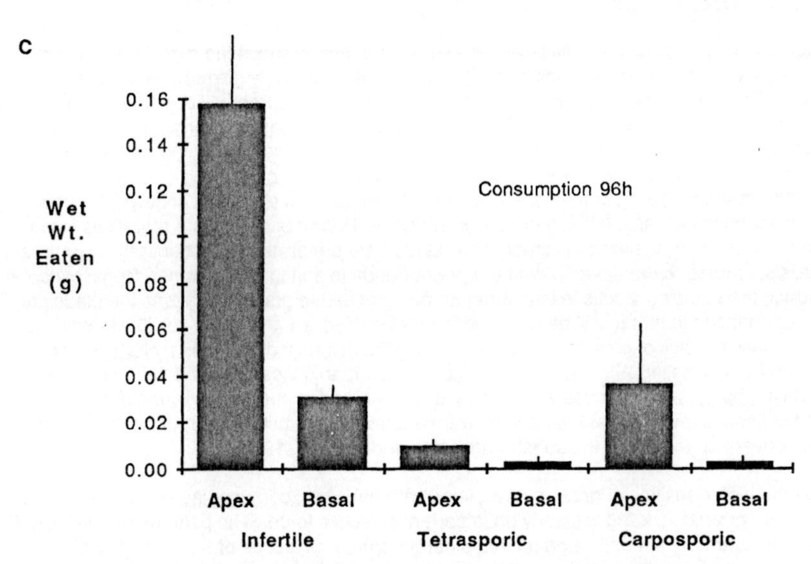

Fig. 8. Rate of consumption of different portions and stages of Iridaea cornucopiae by S. thersites when: (A) the animals are offerred a choice of whole infertile or tetrasporic blades; (B) separate basal or apical portions of infertile or tetrasporic blades are available; and (C) basal or apical portions of carposporic blades are added to the alternatives. Error bars above the means show half the range of the data.

Table 2. Thickness (means±s.d.) of various portions of the blades of Iridaea cornucopiae in relation to the rate of consumption of the blades by Siphonaria thersites (wet g consumed per 6 animals in 96h)

Stage	Portion of blade	Thickness (μm)			Consumption Rate
		Cuticle	Cortex	Total	
Infertile	Apical	1.78 ±0.62	56.2 ± 6.8	312 ±25	0.157
	Basal	6.4 ±0.64	77.0 ±13.0	573 ±38	0.031
Tetrasporic	Apical	5.12 ±2.40	96.0 ± 4.0	727 ±42	0.0090
	Basal	4.21 ±1.10	90.0 ± 9.2	816 ±52	0.0012
Carposporic	Apical	7.1 ±1.06	76.0 ±10.9	670 ±26	0.036
	Basal	6.3 ±0.98	86.0 ±12.6	830 ±43	0.0014

Why S. thersites should prefer the apical portions of infertile fronds is not known. One option is that some portions of the plant are chemically defended against herbivores (e.g. Steinberg, 1985), but given the complexity and diversity of chemical defenses deployed by red algae (Fenical, 1975; Norris and Fenical, 1982), it will not be easy to determine if this causes the pattern of grazing. On the other hand, purely physical properties may be responsible. The thickness of the cuticle and the toughness of the frond may both influence the ease with which a mollusc can feed on a plant (cf. Gaines, 1985). The blades of I. cornucopiae are covered with a thin cuticle, beneath which is a dense cortex of tightly packed cells which give the impression of imparting rigidity, while the bulk of the frond comprises a medulla of loosely arranged cells. Measurements of these layers and of total thickness (Table 2) did reveal that apical sections of infertile fronds are thinnest and have the thinnest cuticle, and there was a significant negative correlation between frond thickness and rate of consumption ($r=-0.95$, $p<0.001$). Whether this is causal remains to be determined.

GENERAL DISCUSSION

Adaptations of Iridaea to grazing

Iridaea cornucopiae supports massive numbers of grazers, of which the most important are Littorina sitkana and amphipods, in addition to S. thersites. It is intensely grazed, and after a seasonal flush of growth in spring, there is an annual regression of the plants, probably due to a combination of grazing and desiccation or thermal stress. The upper limits of the beds of I. cornucopiae at Tatoosh may recede by as much as 30cm over summer (R.T.Paine, pers. comm.), but regrowth occurs by perenation when conditions ameliorate. The major portions of the I. cornucopiae beds are, however, perennial, upright blades being continuously present. Other species of Iridaea undergo marked seasonal variations in biomass, tending to peak in spring and summer. I. cordata regresses in late summer to an overwintering perennial crust. New blades are generated almost entirely by perenation of these crusts, leading to the development of upright fronds in spring and summer. Senescence and grazing reduce the standing stocks in late summer, partly because grazing weakens the blades and increases the chances that they will be removed in storms (Hansen and Doyle, 1976; Hansen, 1977; May, 1986). Similar cycles of growth have been recorded for I. laminarioides and I. ciliata in Chile (Hannach and Santelices, 1985), and for I. cornucopiae in Japan (Hasegawa and Fukuhara, 1955). In southern Chile I. boryana only thrives when grazers are reduced in numbers: experimental removal of limpets and chitons allows the plant to develop from a perenating crust to an erect canopy-forming stage which covers up to 80% of the substratum (Jara and Moreno, 1984).

Given the high densities of grazers associated with beds of I. cornucopiae and the frequency with which fronds are grazed, grazing is clearly an important selective force. The patterns of grazing in the field and in the laboratory strongly suggest that different parts and stages of the plant are differentially defended against grazers. The mechanisms remains unknown, although physical toughness is a possible candidate. The apical portions of infertile blades are probably the least vital to the plant. Grazing on the basal portions could eliminate the entire frond; while attacks on the reproductive fronds will influence the rate of spore release. If the plant does invest in anti-herbivore defenses, it is logical that these defenses should be concentrated in the basal and reproductive portions.

In comparing different species of algae, Littler and Littler (1980) have identified toughness as one means of reducing grazing - achieved by an increase in nonpigmented supportive structures within the plant. Since these structures are nonproductive, the cost of this investment is lower productivity. Investment in defensive structures may change during development. For example, plants may switch from being fast-growing and opportunistic during the juvenile stage, to investing increasing proportions of their tissues to defense as they mature. Heteromorphic life cycles may, in part, be a response to grazing. Various authors have suggested that frondose, fast-growing morphs are productive but vulnerable to grazing, while the alternate crustose phase may be resistant to grazing (Littler and Littler, 1980, 1983; Lubchenco and Cubit, 1980; Slocum, 1980; Dethier, 1981).

While the general theories of Littler and Littler (1980) are attractive, patterns of grazing on other species of Iridaea have yielded results rather different from those described here for I. cornucopiae. Jara (1980) notes that limpets, particularly Siphonaria lessoni, inflict extensive damage to the blades of I. boryana "while preferentially eating reproductive structures". The most heavily grazed stages of I. cordata are the "extremely fertile cystocarpic and tetrasporic plants", and Gerwick and Lang (1977) suggest this is a consequence of the rupture and disintegration of the cuticle over fertile regions of the blade. Gaines (1985) has pursued this question and shown that some grazers, such as the chiton Katherina tunicata, the limpet Collisella pelta and the urchin Strongylocentrotus purpuratus, graze indiscriminately on I. cordata, irrespective of its reproductive condition. He did, however, evince that the gastropod Lacuna marmorata and the isopod Idotea wosnesenski feed selectively on mature plants, avoiding infertile individuals.

As a partial test of the role that the cuticle plays in deterring grazers, Gaines removed the cuticle from one half of a series of fronds of I. cordata, and demonstrated that the rate of attack by grazers was almost fifty times greater on cuticle-free portions of the blades than on intact portions. In the case of I. cornucopiae, S. thersites grazes selectively on the infertile apical portions of plants, where the cuticle is thinnest. While suggestive, it remains to be proven whether this relationship has any causative significance. Gaines (1985) also demonstrated that where the indiscriminate grazers are abundant, I. cordata is scarce or absent ($<1.m^{-2}$), while it is persistent and abundant ($>500.\ m^{-2}$) on vertical surfaces where only the selective grazers (Idotea and Lacuna) are present in significant numbers. Clearly different grazers not only respond differently to anti-herbivore defenses, but have the capacity to profoundly influence the population dynamics of algae by their responses and the specificity of their grazing patterns.

Hannach and Santelices (1985) have examined the relative responses of the isomorphic gametophytic and tetrasporic stages of Iridaea laminarioides and I. ciliata to various environmental factors. Included in their study is an analysis of the rates of grazing of several herbivores on these stages. The urchin Tetrapygus niger and the gastropod Tegula atra grazed equally readily on adults of both stages. Similarly, Siphonaria lessoni fed at the same rate on the juveniles of both reproductive stages of I. laminarioides. On the other hand, Collisella ceciliana consumed juvenile gametophytes ten times faster than juvenile sporophytes. Whether such differential grazing is intense enough to influence the proportions of the two phases in the field is unknown. Strikingly different proportions have been described for I. cordata, the tetrasporic phase being dominant in California and Oregon and the gametangial phase in Washington and British Columbia (Hansen and Doyle, 1976; Dyck, De Wreede and Garbary, 1985; May, 1986). More than 80% of the annual growth of blades is by way of perenation of overwintering crusts, so once one reproductive phase has gained dominance, perenation will tend to perpetuate the dominance. Studies such as those of Hannach and Santelices (1985) could profitably be undertaken on I. cordata.

Adaptations of Siphonaria thersites

Siphonaria thersites combines a number of features that are peculiar for a siphonariid (see Fig. 9). It occurs exclusively in the high-shore, has a small, fragile shell, a large, flexible foot with high mobility, an exceptionally high rate of desiccation and a low tenacity. It is active exclusively during low tide. This behaviour is also unusual, and appears to be a response to the animal's vulnerability to water movement. Because of the limpet's low tenacity, dislodgement by wave action would be a major source of mortality were it to be active during high tide. The limpet's high-shore zonation and the diminution of its numbers along a gradient of increasing wave action at Tatoosh Island are added circumstantial evidence of its vulnerability to waves. The limpets become active as soon as the receeding tide exposes them completely, but ceases well in advance of inundation by the rising tide. Activity is also confined to the lower of the two daily tides and is inversely correlated with the height of the low low tide.

As a consequence, it occurs by night in winter and by day in summer. All these feature point to an avoidance of activity when wave action is a threat.

Coupled with susceptibility to dislodgement, and because of its behavioural and morphological peculiarities, desiccation is also a danger to S. thersites. Between bouts of activity it homes precisely to holes and crevices or shelters under the canopy of its foodplant. Stretches of bare rock are areas of high risk because they provide shelter from neither desiccation nor wave action. Consequently S. thersites is tied closely to perennially available algal beds, notably those of Iridaea cornucopiae, which supply both food and shelter. Its restriction to northern, cooler regions of the coast may be a consequence of the high rate at which it desiccates.

Why I. cornucopiae should be particularly favourable to S. thersites remains unknown. S. thersites does occur on other algae, but never as abundantly as on I. cornucopiae. Despite feeding readily Halosaccion and Ulva, it never sheltered in beds of these algae. The physical structure of algae may be important in determining their suitability as a shelter. Whatever the case, S. thersites gives the impression of being a "prudent predator" by grazing preferentially on the least vital sections of I. cornucopiae, namely the apical portions of infertile blades. In all likelihood this pattern of grazing reflects the palatability of different parts of the plant, the "prudence" being incidental. Harvell (1984) has argued along similar lines to explain the partial predation of a bryozoan by a nudibranch.

Because S. thersites is active during low tides and feeds on the tips of algal fronds, it is very exposed to avian predators. Despite this, birds were never seen preying on it. Furthermore, the preliminary experiments with predators, the possession of mucous glands and the existence of polyproprionates in the limpet's tissues all argue that S. thersites is unpalatable or toxic. Protection of this kind may explain why S. thersites can afford to have a small, fragile shell and low powers of adhesion. These last two features are a liability in terms of desiccation and wave action, but they are associated with a large, flexible foot, which may increase teh efficiency with which the limpets manouevre over the fronds of their foodplant. Yonge (1960) has previously commented that S. thersites is unusually active and mobile for a limpet.

In many respects the behaviour and morphology of S. thersites parallel those of a completely unrelated group of molluscs, namely the opisthobranch genera Onchidium and Onchidella. Lacking shells, these animals are vulnerable to desiccation, and home to crevices and holes between bouts of feeding (Arey and Crozier, 1918; McFarlane, 1980). Their activity is also rhythmic and occurs only during low tide (Hirasaka, 1912; Arey and Crozier, 1921; Fretter, 1943; Marcus and Marcus, 1956; Pepe and Pepe, 1985). Like S. thersites, both these genera possess repugnatorial glands known to repel a number of predators, including starfish, crabs and fish (Arey Crozier, 1921; Ireland and Faulkner, 1978; Ireland et al.,1984; Young, Greenwood and Powell, 1986). The parallels are particularly clear when comparing S. thersites with Onchidella binneyi, for the latter also occurs in the high shore, is weakly attached and easily dislodged by waves, and becomes active once a day shortly after being exposed by the ebbing tide; it too ceases activity well in advance of the time the rising tide covers the shore. Its activity rhythms are endogenous and circalunar (Pepe and Pepe, 1985), and the intensity of its activity is inversely related to the height of the low tide, in a manner comparable to that described for S. thersites. While Onchidella binneyi is active only by night, four other species of the genera Onchidella or Onchidium have been described as being diurnally active. It is possible that these differences are artifacts caused by observations being made over a relatively short period of the year. If these species respond like S. thersites, then they may alter the timing of their activity to coincide with the lower of the two daily low tides; and the timing of these low low tides may shift seasonally between the day and night.

Taken in isolation, many of the behavioural and morphological features exhibited by S. thersites seem inane, considering the biotic and physical threats that can be anticipated in the high-shore. It is only when they are viewed as an entity and in conjunction with the limpet's interaction with its foodplant that their adaptive value becomes apparent. It is striking that a completely unrelated group of opisthobranch molluscs should convergently have developed an almost exactly comparable set of behavioural and morphological adaptations when faced with similar problems.

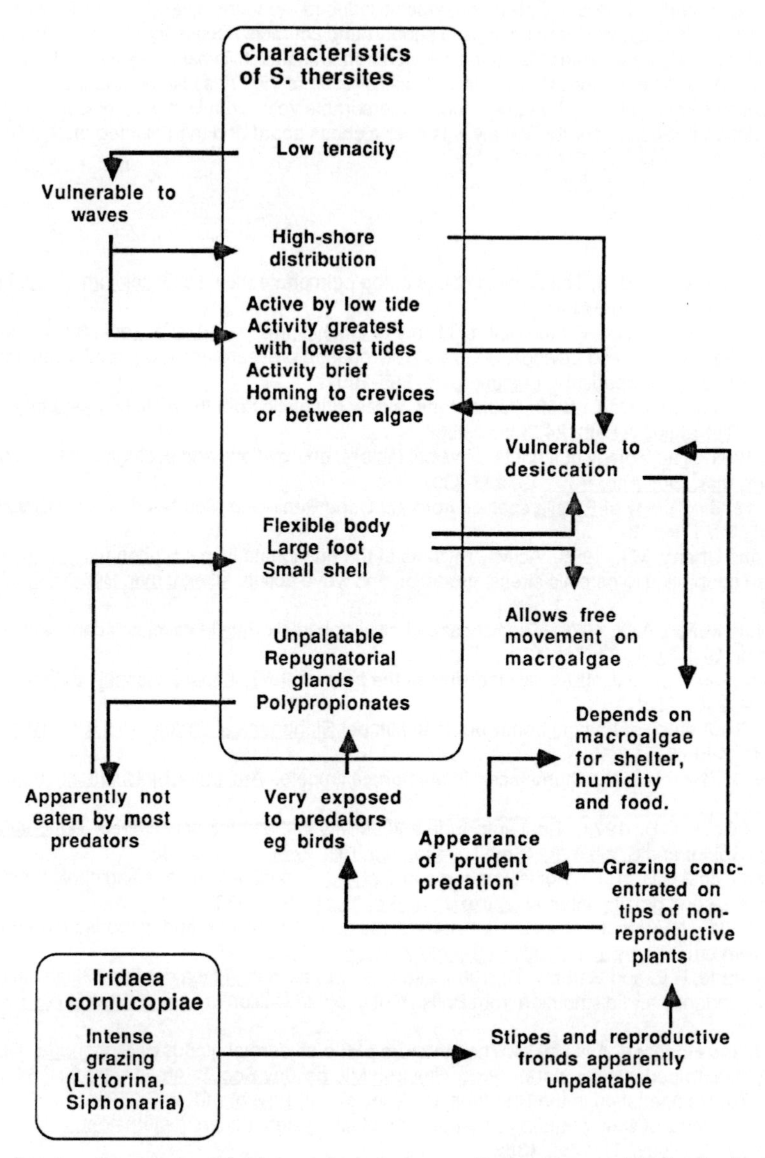

Fig. 9. Summary of the interactions between S. thersites and its food plant I. cornucopiae, and the interdependent adaptations of behaviour and morphology in relation to physical conditions.

ACKNOWLEDGEMENTS

Warmest appreciation is due to Bob Paine, who introduced me to the splendours of Tatoosh, and made everything possible by provided the logistic support and contacts necessary for getting to and fro Tatoosh and Waadah. I also acknowledge permission from the Makah Tribal Council and US Coast Guard to work on Waadah and Tatoosh. Dennis Willows, Director of the Friday Harbor Marine Laboratories accommodated me at the Labs during a memorable year. Tom Daniels provided stimulus and accommodation in Seattle, sharpened my wits over a chess board and then blunted them with B&B.

REFERENCES

Arey, L. and Crozier, W.J., 1918, The homing habits of the pulmonate mollusc Onchidium, Proc. Nat. Acad. Sci. Wash., 4, 319-321.

Arey, L. and Crozier, W.J., 1921, On the natural history of Onchidium, J. exp. Zool., 32, 443-502.

Bertness, M.D., Garrity, S.D. and Levings, S.C., 1981, Predation pressure and gastropod foraging: a tropical-temperate comparison, Evolution, 35, 995-1007.

Biskupiak, J.E. and Ireland, C.M., 1983, Pectinatone, a new antibiotic from the mollusc Siphonaria pectinata, Tetrahedron Lett., 24, 3055-3058.

Branch, G.M., 1981, The biology of limpets: physical factors, energy flow, and ecological interactions, Oceanogr. Mar. Biol. Ann. Rev., 19, 235-380.

Branch, G.M., 1975, Biology of Patella species from the Cape Peninsula, South Africa. V: Desiccation. Mar. Biol., 32, 179-188.

Branch, G.M. and Cherry, M.I., 1985, Activity rhythms of the pulmonate limpet Siphonaria capensis Q. & J. as an adaptation to osmotic stress, predation and wave action, J. exp. mar. Biol. Ecol., 87, 153-168.

Branch, G.M. and Marsh, A.C., 1978, Tenacity and shell shape in six Patella species: adaptive features, J. exp. mar. Biol. Ecol., 34, 111-130.

Capon, R.J. and Faulkner, D.J., 1984, Metabolites of the pulmonate Siphonaria lessoni, J. Org. Chem., 49, 2506-2508.

Cook, S.B., 1971, A study of homing behaviour in the limpet Siphonaria alternata, Biol. Bull. (Woods Hole), 141, 449-457.

Cook, S.B., 1976, The role of the "home scar" in pulmonate limpets, Am. Malacol. Union Inc. Bull., 42, 34-37.

Cook, S.B., and Cook, C.B., 1978, Tidal amplitude and activity in the pulmonate limpets Siphonaria normalis (Gould) and S. alternata (Say), J. exp. mar. Biol. Ecol., 35, 119-136.

Cook, S.B., and Cook, C.B., 1981, Activity patterns in Siphonaria populations: heading choice and the effects of size and grazing interval, J. exp. mar. Biol. Ecol., 49, 69-79.

Dethier, M.N., 1981, Heteromorphic algal life histories: the seasonal pattern and response to herbivory of the brown crust, Ralfsia californica, Oecologia (Berl.), 49, 333-339.

Dyck, L., De Wreede, R.E. and Garbary, D., 1985, Life history phases in Iridaea cordata (Gigartinacaea): relative abundance and distribution from British Columbia to California, Jap. J. Phycol.(Sorui), 33, 225-232.

de Villiers, C.J. and Hodgson, A.N., 1984, The structure of the epidermal glands of Siphonaria capensis (Gastropoda: Pulmonata), Proc. Electron Microscopy Soc. S. Afr., 14, 93-94.

Fenical, W., 1975, Halogenation in the Rhodophyta. A review, J. Phycol., 11, 245-259.

Frank, P., 1982, Effects of winter feeding on limpets by black oystercatchers (Haematopus bachmanni), Ecology, 63, 1352-1362.

Fretter, V., 1943, Studies on the functional morphology and embryology of Onchidella celtica (Forbes & Hanley) and their bearing on its relationships, J. mar. biol. Ass. U.K., 25, 685-720.

Fretter, V. and Graham, A., 1954, Observations on the opisthobranch mollusc Acteon tornatilis (L.), J. mar. biol. Ass. U.K., 33, 565-585.

Gaines, S.D., 1985, Hervivory and between-habitat diversity: the differential effectiveness of defenses in a marine plant, Ecology, 66, 473-485.

Garrity, S.D., and Levings, S.C., 1983, Homing to a scar as a defense against predators in the pulmonate limpet Siphonaria gigas, Mar. Biol., 72, 319-324.

Gerwick, W.H. and Lang, N.J., 1977, Structural, chemical and ecological studies on iridescence in Iridaea (Rhodophyta), J. Phycol., 13, 121-127.

Hahn, T.P., 1985, Effects of predation by black oystercatchers (Haematopus bachmanni Audubon) on intertidal limpets (Gastropoda, Patellacea), M.Sc. Thesis, Stanford University.

Hannach, G. and Santelices, B., 1985, Ecological differences between the isomorphic reproductive phases of two species of Iridaea (Rhodophyta: Gigartinales), Mar. Ecol. Prog. Ser., 22, 291-303.
Hansen, J.E., 1977, Ecology and natural history of Iridaea cordata (Gigartinales, Rhodophyta) growth, J. Phycol.,13, 395-402.
Hansen, J.E. and Doyle, W.T., 1976, Ecology and natural history of Iridaea cordata Rhodophyta (Gigartinacaea): population structure, J. Phycol., 12, 273-278.
Hartwick, E.B., 1976, Foraging strategy of the Black Oystercatcher (Haematopus bachmanni Audubon), Can. J. Zool., 54, 147-155.
Harvell, D., 1985, Why nudibranchs are partial predators: intracolonial variation in bryozoan palatability, Ecology, 65, 716-724.
Hasegawa, Y. and Fukuhara, E., 1955, Ecological studies on Iridophycus cornucopiae (P.&R.) S. et G. 5. On the seasonal and local variations in the weight of fronds, Bull. Hokkaido Fish. Res. Lab., 12, 29-39.
Hiraska, K., 1912, On the structure of the dorsal eye of Onchidium, Dobutsu Gaku Zasshi (Zool. Mag.), 24, 20-35.
Hochlowski, J.E., Coll, J.C., Faulkner, D.J., Biskupiak, J.E., Ireland, C.M., Qi-tai, Z., Cun-heng, H. and Clardy, J., 1984, Novel metabolites of four Siphonaria species, J. Am. Chem. Soc., 106, 6748-6750.
Hochlowski, J.E., and Faulkner, D.J., 1983, Antibiotics from the marine pulmonate Siphonaria diemenensis, Tetrahedron Lett., 24, 1917-1920.
Hochlowski, J.E., and Faulkner, D.J., 1984, Metabolites of the marine pulmonate Siphonaria australis, J. Org. Chem., 49, 3838-3840.
Hochlowski, J.E., Faulkner, D.J., Matsumoto, G.K. and Clardy, J., 1983, The denticulatins, two polyproprionate metabolites from the pulmonate Siphonaria denticulata, J. Am. Chem. Soc., 105, 7413-7415.
Hulings, N.C., 1985, Activity patterns and homing in two intertidal limpets, Jordan Gulf of Aqaba, Nautilus, 99, 75-80.
Ireland, C., Biskupiak, J.E., Hite, G.J., Rapposch, M., Scheuer, P.J. and Ruble, J.R., 1984, Ilikonapyrone esters, likely defense allomones of the mollusc Onchidium verrucullatum, J. Org. Chem., 49, 559-551.
Ireland, C. and Faulkner, D.J., 1978, The defensive secretion of the opisthobranch mollusc Onchidella binneyi, Bioorganic Chem., 7, 125-131.
Jara, H.F., 1980, Herbivoría y dominancia competitiva de Iridaea boryana (Setch. et Gardn.) Skottb. en un frente rocoso semiprotegodo del sur de Chile, Ph.D. thesis, Universidad Austral de Chile, Valdivia.
Jara, H.F. and Moreno, C.A., 1984, Herbivory and structure in a midlittoral rocky community: a case in southern Chile, Ecology, 65, 28-38.
Littler, M.M. and Littler, D.S., 1980, The evolution of thallus form and survival strategies in benthic marine algae: field and laboratory tests of a functional form model, Am. Nat., 116, 25-44.
Littler, M.M. and Littler, D.S., 1983, Heteromorphic life strategies in the brown alga Scytosiphon lomentaria, J. Phycol., 19, 425-431.
Lubchenco, J. and Cubit, J., 1980, Heteromorphic life histories of certain marine algae as adaptations to variations in herbivory, Ecology, 61, 676-344.
MacKay, D.A. and Underwood, A.J., 1977, Experimental studies on homing in the intertidal patellid limpet Cellana tramoserica (Sowerby), Oecologia (Berl.), 30, 215-237.
Marcus, E. and Marcus, E., 1956, On Onchidella indolens (Gould, 1852), Bol. Inst. Oceanogr. S. Paulo, 5, 87-94.
Marcus, E. and Marcus, E., 1960, On Siphonaria hispida, Bol. Fac. Filos. Cien. Letr. Univ. S. Paulo, Ser. Zool., No. 23, 107-140.
Marsh, C.P. 1986, Impact of avian predators on high intertidal limpet populations, J. exp. mar. Biol. Ecol., 104, 185-201.
May, G., 1986, Life history variations in a predominantly gametophytic population of Iridaea cordata (Gigartinaceae, Rhodophyta), J. Phycol., 22, 448-455.
McAlister, R.O. and Fisher, F.M., 1968, Response of the false limpet Siphonaria pectinata Linnaeus (Gastropoda, Pulmonata) to osmotic stress, Biol. Bull. (Woods Hole), 134, 96-117.
McFarlane, I.D., 1980, Trail-following and trail-searching behaviour in homing of the intertidal gastropod mollusc, Onchidium verrucullatum, Mar. Behav. Physiol., 7, 95-108.
Naylor, E, 1988, Clock-controlled behaviour in intertidal animals, in: "Behavioural Adaptations to Intertidal Life", M. Vannini and G. Chelazzi, eds, Plenum Press.
Norris, J.N. and Fenical, W., 1982, Chemical defense in tropical marine algae, Smithsonian Contrib. Mar. Sci., No.12, 417-431.

Pepe, P.J. and Pepe, S.M., 1985, The activity patterns of Onchidella binneyi Stearns (Mollusca: Opisthobranchia), Veliger, 27, 375-380.

Rice, S.H., 1985, An anti-predator chemical defense of the marine pulmonate gastropod Trimusculus reticulatus (Sowerby), J. exp. mar. Biol. Ecol., 93, 83-89.

Slocum, C.J., 1980, Differential susceptibility to grazers in two phases of an intertidal alga: advantages of heteromorphic generations, J. exp. mar.Biol. Ecol., 46, 99-110.

Steinberg, P.D., 1985, Feeding preferences of Tegula funebralis and chemical defenses of marine brown algae, Ecol. Monogr., 55, 333-349.

Thomas, R.S., 1973, Homing behaviour and movement rhythms in the pulmonate limpet Siphonaria pectinata Linnaeus, Proc. Malacol. Soc. London, 40, 303-311.

Underwood, A.J., 1988, Design and analysis of field experiments on competitive interactions affecting behaviour of intertidal animals, in: "Behavioural Adaptations to Intertidal Life", M. Vannini and G. Chelazzi, eds, Plenum Press.

Verderber, G.W., Cook, S.B. and Cook, C.B., 1983, The role of the home scar in reducing water loss during aerial exposure of the pulmonate limpet Siphonaria alternata (Say), Veliger, 25, 235-243.

Yonge, C.M., 1960, Further notes on Hipponix antiquatus with notes on North Pacific pulmonate limpets, Proc. Cal. Acad. Sci., 31, 111-119.

Young, C.M., Greenwood, P.G. and Powell, C.J., 1986, The ecological role of defensive secretions in the intertidal pulmonate Onchidella borealis, Biol. Bull. (Woods Hole), 171, 391-404.

The Timing of Reproduction to Distinct Spring Tide Situations in the Intertidal Insect *Clunio*

Dietrich Neumann

University of Köln
West Germany

INTRODUCTION

At coasts with a strong semidiurnal tidal regime and lunar-semimonthly modulation of the tide's amplitude, characteristic tidal situations reoccur periodically every 14-15 days about a specific time of day which is characteristic for a given location on a coast-line. One of the most effective situations is the time of spring low water when an extremely wide range of the midlittoral zone becomes exposed for some hours during the days around the time of full and new moon. However, the temporal characterics are correspondingly similar for any other phase of the semimonthly tidal pattern, e.g. the neap ebb tide at about the time of the quarters of the moon.

In several intertidal species, the time of reproduction occurs in lunar-semimonthly rhythms, or with twice that period in lunar-monthly rhythms. The mass-concentration of their reproducing partners is obviously correlated with a tidal situation which is optimal for the reproductive success of the species. In the insect Clunio, this situation coincides with spring low water when the egg-deposition of the adults is guaranteed on the exposed substrates of the larval habitat in the lower midlittoral (fig. 1).

The reliable programming of the lunar-rhythmic reproduction to an appropriate species-specific situation of the cycle of spring and neaps might well be the most complex and most fascinating timing problem in marine environments. In order to achieve a profound understanding of the underlying physiological timing mechanisms, several themes seem to merit for careful experimentation. Prerequisite is the possibility of an unrestricted breeding of the species concerned, and this under simulated daily, tidal, lunar and seasonal conditions as well as under their modifications. The main topics of the analysis may be (a) the properties of the physiological clocks controlling developmental processes in relation to the days of the lunar month as well as to the time of day, (b) the exogenous time cues and their perception, (c) the hormonal coupling between the physiological clocks and the developmental processes, (d) the local adaptations of geographical races, and (e) the gene-controlled parameters of the timing mechanisms.

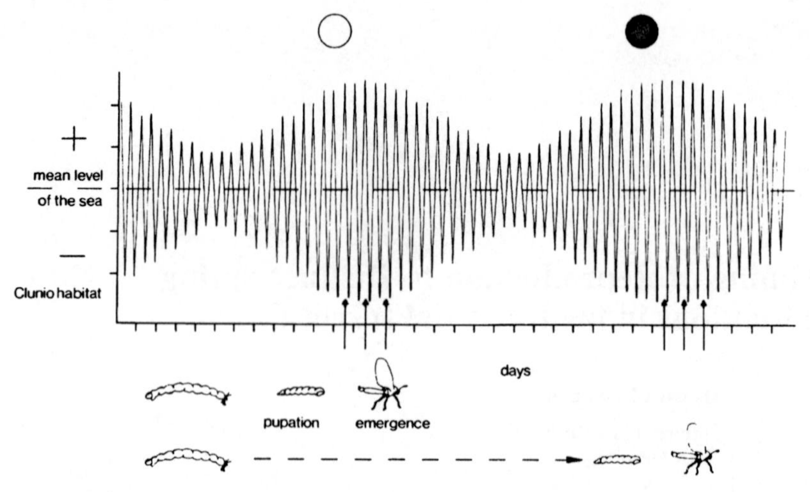

Fig. 1. Schematic representation of the relationship between tides, phases of the moon, the habitat of Clunio marinus, and the lunar-semimonthly emergence of the adults. In this case the emergence occurred just before the afternoon low water on days around spring tides (Neumann, 1981 a).
(O), full moon; (●), new moon.

Three marine species have been analysed up to now both by quantitative field observations and laboratory experiments, and each have demonstrated a complex timing by the combination of two physiological timing mechanisms: a circasemilunar biological clock combined with a circadian one in the midge Clunio inhabiting both the lower midlittoral and the upper sublittoral range of the intertidal zone (Neumann 1966, 1981 b, 1986), a circasemilunar clock and a similar daily direct response in the polychaet Typosyllis (Franke 1985, 1986), and a circasemilunar clock combined with a lunadian timing in 24.8 hr cycles (correlated with the daily shift of the tides) in the semiterrestrial crab Sesarma which releases its larvae at the water's edge of sea-shores or rivers nearby the sea (Saigusa 1978, 1980).

The following review will be concerned with the midge Clunio which has been studied in stocks from populations between subtropic and arctic latitudes for about 25 years. The results have been summarized in schematic graphs demonstrating the main physiological components of the combined circasemilunar and circadian timing mechanisms (figs. 2, 3).

LUNAR AND DAILY TIMING

In stocks of Clunio populations bred in sea-water under a continuous 24 hr light-dark regime, the generation time of the specimens varied considerably between 6 and 14 weeks at temperatures about 20° C. The emergence days were randomly distributed in this interval. In other words, no synchronization of the emergence times of the adults could be detected in relation to the phases of the natural lunar month under these isolated laboratory conditions. However, lunar rhythms of emergence could be simulated in the laboratory when the cultures were additionally confronted with specific time cues such as artificial moonlight or selected conditions of artificial tidal cycles.

For example, in stocks of the subtropical species Clunio tsushimensis a clear-cut semimonthly rhythm was evoked by an artifical moonlight of 0.3 lux, with peaks at about the days of the simulated "full moon" and then two weeks later (during the days of the "new moon"). In the laboratory, this rhythm could be started on all days of the natural lunar month. Therefore, it can be concluded that no other unknown factor of the moon apart from moonlight was controlling the entrainment of the lunar-semimonthly reproduction rhythm.

When synchronized cultures of this kind were released into conditions of continuous LD 12:12 conditions or continuous darkness for up to three months with no further artificial moonlight signal, a freerunning semimonthly rhythm of emergence was observed. In the case of the tsushimensis stock, the mean period of the freerunning rhythm was 14.2 days at 19° C instead of 15 days in the artificial moonlight conditions, or 14.7 days as in the population in nature. Even at three different temperatures (14, 19 and 24 ° C), the period was close to 14-15 days. Only a slight, but significant temperature dependence was realized, with a Q_{10} of only 1,06 (Neumann, in preparation for press). The temperature range tested corresponds to the temperatures of the sea between winter and summer at the place of origin of the Clunio stock (Shimoda, Japan). Thus, these experiments supply evidence for a temperature-compensated circasemilunar oscillator being the decisive component of the physiological timing-mechanism underlying the entrainment to environmental moonlight signals.

The circasemilunar oscillator does not control the emergence of the midges directly as was detected by a careful examination of the metamorphosis of this insect (Neumann 1966, Krüger and Neumann 1983). In total, three physiological switches exist for the temporal control of the midge's development up to the hours of eclosion. The circasemilunar timing mechanism affects the first two: (1) a distinct step in the imaginal disc formation and (2) the pupation two weeks later (compare fig. 2, right side).

After a pupation period of 3-4 days (in the range of 14-24° C) the pharate imagos are prepared for the eclosion at a distinct time of day. The eclosion then represents the third switch. It is under the control of a circadian timing mechanism with the 24 hr light-dark regime as zeitgeber (fig. 2). This kind of circadian timing is well known from the daily eclosion rhythms of many other insects, and from the daily rhythms of behavioural performances of many other animals and plants (Aschoff 1960, Bünning 1977, Pittendrigh 1981).

Consequently, the timing of the short-lived Clunio adults and its reproduction behaviour is based on the combination of two different biological clocks. The lunar timing determines the days at which pharate adults are present every two weeks during spring tides; this is followed by the daily timing which determines the correct time of day correlated in nature with the time of the local spring low tide situation at the place of origin. The phase relationships of this temporal programming, concerning both semimonthly spring tides and the time of day, have to be well adapted both to the occurrence of the reliable time cues as well as to the local tidal pattern at the population's location: genetically controlled physiological properties of both timing mechanisms (components for "distinct phase" in fig. 2., Neumann 1966, 1981 a) achieve this goal.

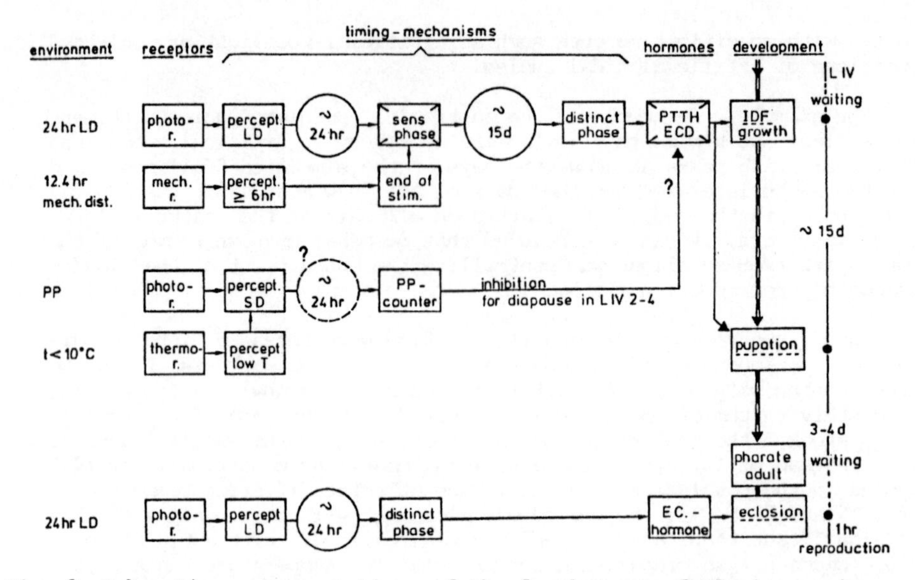

Fig. 2. Schematic representation of the development of Clunio marinus from the last larval instar (LIV) until the eclosion of the imago (right hand side) in combination both with the underlying timing mechanism and its physiological components (middle part), and with the requisite environmental time cues and the receptors (left hand side) indispensible for the synchronisation of the physiological clocks and its developmental processes coupled with the spring tide situations. The lunar-semimonthly timing controlling imaginal disc formation (IDF) and pupation is illustrated in the upper line, the seasonal control of the diapause in the middle part, and the daily control of emergence in the lower line. The hormonal coupling between the timing mechanisms and the development is hypothetical. LD: 24 hr light-dark cycle, mech. dist.: mechanical disturbances of the rising tide, PP: photoperiod, SD: short-day conditions, T: water temperature, r: receptor, end of stim.: end of the mechanical stimulus of 6-10 hr, PTTH-ECD: reaction pathway of prothoracotropic hormone and ecdysteroids, EC: eclosion hormone. The circles represent the physiological clocks involved, both the circadian oscillator (period circa 24 hr) and the circasemilunar oscillator (period circa 15 d).

TIME CUES FOR LUNAR TIMING

It has been established that Clunio populations in subtropical and arctic latitudes differ in their response to environmental time cues of the lunar rhythmic entrainment (Neumann 1968, 1976 a, b, 1978; Neumann and Heimbach 1979, 1984). While stocks of populations from southern latitudes reacted to a monthly moonlight stimulation (Clunio tsushimensis: Shimoda/Japan 34° N; Clunio mediterraneus: Istrien/Jugoslavia 45° N; Clunio marinus: Atlantic coast between Spain and Normandy France, up to 49° N), stocks from the northern fringe of temperate latitudes did not do so (Clunio marinus: Helgoland/North Sea 54° N, stocks from the English Channel coast, from Scotland and Bergen/Norway, all above 50° N).

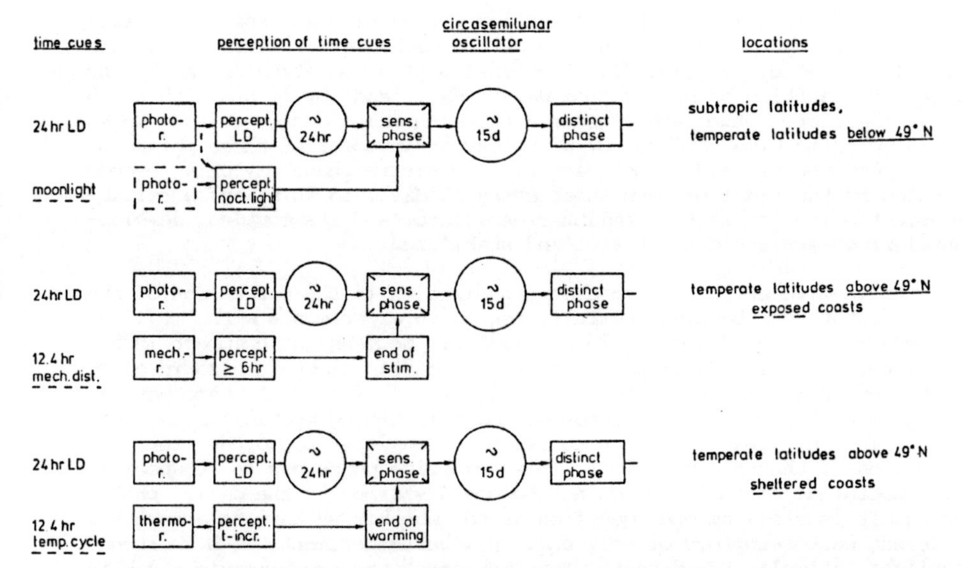

Fig. 3. Comparative view of the different time cues for the entrainment of lunar-semimonthly emergence rhythms at the different locations of the Clunio species and for its geographical races. As suggested by experiments, the main physiological components for the perception of the time cues are added. For further explanation compare the text and fig. 2. sens.phase: sensitive circadian phase for the perception of the lunar or tidal time cue; t.-incr.: temperature increase during the ebbing tide.

In the case of the Helgoland stock it was demonstrated first that lunar-semimonthly rhythms can also be entrained by tidal factors in combination with the 24 hr light-dark cycle. It is interesting to notice that Clunio larvae of this origin did not react to the semimonthly modulations of the tidal regime (e.g. variations of the tidal amplitude), but directly to the 12.4 hr tidal cycle and its variations in mechanical disturbances of the water. In the field, these disturbances are stronger during rising tides than during falling tides, correlating to stronger breaking waves on the intertidal shore. The elevated mechanical stimulation of the larvae might be caused by water turbulence, or vibration of the substrates, or even by underwater sound. However, in the laboratory, the tidal stimulation was effectively simulated in the water basins by the vibrations of synchronous electrical motors (50-200 cps, about 30 dB above background level, 6-10 hr continuously per 12.4 hr cycle). Artificial tidal cycles of changes in hydrostratic pressure were ineffective.

In further experiments, it has been shown (Neumann 1968, 1978) that only a distinct phase relationship between the tidal factor and the 24 hr light-dark regime can be effective for the entrainment of the semilunar rhythms. Distinct phase relationships between both environmental cycles recur periodically every 15 days, in nature this is in strong correlation with the spring-neap-tide cycle and the phases of the moon respectively. Essential components for the perception of a semimonthly phase relationsship between tides and day-night cycle might be (1.) a receptor for the mechanical stimulation and some kind of interval timer for the perception of the effective mechanical stimulus for at least 6 hr duration, and (2.) a circadian clock organization for setting a sensitive daily phase for the perception of the end of the mechanical stimulus (compare upper line in fig. 2, fig. 3). The tidal 6 hr stimulation of a 12.4 hr cycle is shifting by 0.8 hr from day to day, and by 12 hr within 15 days. Therefore, the end of such a stimulus coincides every two weeks with a specific time of day. By a tricky arrangement of these physiological components, a distinct circadian sensitive phase may only become affected by the tidal pattern about every 15 days. In this way, a reliable synchronization of the semilunar oscillator and its coupled developmental processes could be started and stabilized.

The involvement of a circadian organization in the perception of the tidal factor has been successfully tested in experiments with modified light-dark regimes (Neumann 1976 b, 1981 b; Neumann and Heimbach 1985). Therefore, one may additionaly ask if a circadian clock is also involved in the perception of the nocturnal moonlight. In fig. 3 the hypothetical arrangement of components of lunar-semimonthly timing-mechanisms is outlined. In any case, one has to assume a sensitive photophase for the perception of the weak moonlight. This sensitive phase has to be specifically nocturnal so that it does not become disturbed by the daily photoperiod. It is still an open question as to whether we have to assume two different photoreceptors or only one. However,experiments with shortened moonlight stimuli showed that, for instance, the tsushimensis stock is only sensitive for moonlight around the hours of midnight (in a 12 hr dark period from 18 h to 6 h in the range between 21 and 3 h; Neumann 1987). Recent results using modified light-dark cycles between 20 and 30 hr have additionally shown a circadian range of entrainment for a moonlight sensitive Clunio stock (no entrainment in LD 10:10 and LD 15:15, with standard moonlight during 4 successive nights every 30 LD-cycles, clear-cut semilunar entrainment in a corresponding regime close to LD 12:12, e.g. LD 11:11). This result is in convincing conformity with the theory of cicadian clocks and the fact that self-sustained oscillators of the ciradian clock type can be entrained by environmental zeitgeber cycles only within a restricted range of periods near to the 24 hr zeitgeber cycle (Aschoff 1978). Thus, it can be concluded that a circadian physiological organization generally is an essential component in the entrainment of lunar-semimonthly rhythms of the insect Clunio.

Moonlight sensitivity in southern populations and sensitivity for a tidal factor in northern populations represent a successful geographic adaptation of the lunar timing mechanism. While moonlight can be a strong photoperiodic factor during winter and summer at southern latitudes, the influence of moonlight is weakened or even extinguished at the fringe of northern latitudes during summer. Here, the moonlight competes with a relatively short night and with a dusky illumination of the northern horizons during night. In addition, the summer full moon is present only slightly above the horizon so that the weak moonlight intensity may be more often reduced by clouds, as well by light reflections on the water surface of a rough sea.

A further semilunar time cue has been detected in a Northern Clunio population of mid-Norway where the sheltered littoral zone of fjords with low tidal range are found. The populations are exposed in summer to tidal cycles of rising and falling temperatures because during this season the water column of the fjords, even of the upper two metres, can be clearly stratified in temperatures of some degrees Celsius. In the laboratory, the Clunio stock of Bergen could be entrained by the combination of tidal temperature cycles and the 12.4 hr LD (fig. 3, lower line). It is obvious that the lunar-semimonthly timing mechanism has genetically adapted at this place to another reliable tidal zeitgeber factor (Neumann and Heimbach 1984).

HORMONAL COUPLING BETWEEN CLOCKS AND DEVELOPMENT

The reproductive behaviour of the short-lived adults of Clunio is based on a precise temporal control of its development so that time of reproduction can conincide with the spring low water situation every 15 days at a time of day which is specific for any location of the sea coast. So far, we have only considered the environmental times cues and the physiological clocks essential for the timing of the insect's metamorphosis. One may suppose that the physiological coupling between the timing mechanisms and the developmental processes (imaginal disc formation, pupation and eclosion) are controlled by the release of hormones. Particular interest will be focused on finding hormonal factors controlling the imaginal disc formation. Measuring the concentrations of α-ecdyson in small Clunio larvae during the last larval instar and the pupation period (in cooperation with H.D. Spindler, Düsseldorf FRG by means of radioimmunoassay), elevated titres have been recently discussed in the range of both semilunar switches mentioned above, occurring during early imaginal disc formation and just before pupation. Thus, it may be supposed that the reaction pathway of prothoracotropic hormones and ecdysteroids is correlated with the semilunar timing of the metamorphosis (fig. 2).

With regard to the time of day of eclosion it may be suggested that the eclosion hormone is involved (fig. 2), corresponding to the findings of Truman in the butterfly Manduca sexta and other insects (Truman et al. 1981).

LOCAL ADAPTATIONS AND GENE-CONTROLLED PARAMETERS

The regime of semidiurnal tides is charaterized on average by a 12.4 hr cycle of ebb and flow, by a 14.7 d cycle of spring and neaps, and by the recurrence of any tidal situation at the same time of day every 14-15 days in parallel with both the spring-neap-tide cycle and the phases of the moon. On coasts with these tidal influences (e.g. Europe: Atlantic coast, Irish Sea, North Sea), the extreme amplitude of spring tides generally occurs every two weeks about one day after the syzygies (full and new moon). However, time displacements exist in the occurrence of the time of low and high tides, respectively, between different parts of the coast-line. For instance, at the Atlantic coast of France, this time-difference amounts to about 4 hr between the southern Basque coast (43° N) and the northern Normandy coast (49° N). These variations in the phase relationships of tides and day-night cycle may even be close along some parts of the English Channel, or parts of the Irish Sea as well as the North Sea.

Concerning the adaptation of lunar-semimonthly reproduction rhythms to the local tidal regime, one has to expect corresponding time differences in the diurnal emergence times of the local Clunio populations. This has been confirmed by field observations and laboratory experiments (Neumann 1966, 1976 a, Heimbach 1978). In other words, the circadian emergence time is a population-specific property which represents a genetic adaptation of the circadian timing mechanism to the local tidal regime.

In cross-breeding experiments between stocks of the two locations mentioned above (Basque coast and Normandy), it was demonstrated that the different phase relationship of the circadian emergence time and the 24 hr light dark regime as time cue is a gene-controlled property of the circadian timing mechanims of Clunio (Neumann 1966).

Circadian phase-response curves are a physiological indicator for the phase relationship of the circadian oscillator of the timing mechanism and the light-dark cycle. When testing these curves in early and late emerging stocks of Drosophila pseudoobscura, Pittendrigh (1967) did not find any different responses. This means that the circadian oscillator of both stocks have the same phase relationship with the 24 hr time cue and that the stock-specific emergence time can only be controlled by a separate, but coupled, physiological component of the circadian timing mechanism. In the same way, it may be suggested that in Clunio the genetic adaptation of the circadian timing is determined by a coupled "distinct phase" -component (compare fig. 2).

One has to assume similar gene-controlled "distinct phase"-components in the lunar-semimonthly timing mechanisms (figs. 2, 3). These components may represent the control of the correct temporal adaptation of the lunar rhythm to the local tidal regime. In any case, population-specific adaptations have to be expected along the coast lines in correlation with the time-displacements of the tides and the correspondingly changing phase relationships with the time of day. The genetic and population-specific adaptations were demonstrated in several Clunio stocks from different parts of the European coast (Neumann 1978, Neumann and Heimbach 1979).

Finally, seasonal modulations of the lunar-rhythmic reproduction rhythms of Clunio may be mentioned (Neumann 1976 a, 1986). A diapause reaction during the last larval instar had been established in the northern population of Helgoland (54° N); it is controlled by short-day conditions and low temperatures (compare fig. 2; Neumann and Krüger 1985). Obvious diurnal inequalities in the amplitude of two succeeding tides (so-called mixed tides) exist on the Pacific coast of Japan, and this is in combination with a seasonal change. At the location of Shimoda, the lowest spring low water occurr about noon in summer, and 12 hr later about midnight in winter. The diurnal emergence time of the lunar reproduction rhythm of the local Clunio population follows this seasonal phase-jump of the extreme low tides. This can be evaluated as an inevitable adaptation because the egg-deposition on the exposed region succeeds only during the lowest low tide (Oka and Hashimoto 1959). In a Clunio stock from this place, the phase-jump of the circadian emergence rhythm could be simulated in tideless laboratory conditions by changing the photoperiod from long to short days (Neumann 1983). This example impressively demonstrates the adaptability of the Clunio timing mechanisms to the local tidal environment.

CONCLUDING REMARKS

The temporal adaptation of the reproductive behaviour of a marine species to periodically recurring tidal situations at the sea-shore must be based on a well-timed development to mature adults. In the case of a species with short-lived reproducing partners, the timing of the preceding developmental processes has to be correspondingly precise. The midge Clunio perfectly demonstrates all potential possibilities in the temporal control of an insect's development with regard to seasonal, lunar-semimonthly, tidal and daily situations. The oscillatory timing mechanisms controlling the three succeeding events (stage in imaginal disc formation, pupation and eclosion) offer the physiological prerequisites for the selection of distinct phase relationships between these events and reliable environmental time cues. In this way, the final reproduction of Clunio becomes well synchronized with the passing occurrence of favourable spring tide situations at any location on the sea-coast.

Acknowledgments. The research of the biological rhythms was supported over the years by grants of the Deutsche Forschungsgemeinschaft. My thanks are due to Mrs. M. Winter-Bunnenberg for valuable technical assistence and for drawing the graphs, and to Mr. M. Brett for correcting the English text.

REFERENCES

Aschoff, J., 1960, Exogenous and endogenous components in circadian rhythms. Cold Spr. Harb. Symp. quant. Biol. 25, 11-28.

Aschoff, J., 1978, Circadian rhythms within and outside their ranges of entrainment, pp. 172-181 in: "Environmental endocrinology", A. Assenmacher and D. Farmer eds., Springer-Verlag, Berlin-Heidelberg-New York.

Bünning, E., 1977, Die physiologische Uhr. Circadiane Rhythmik und Biochronometrie. 176 p., Springer Verlag, Berlin-Heidelberg-New York.

Franke, H.-D., 1985, On a clocklike mechanism timing lunar-rhythmic reproduction in Typosyllis prolifera (Polychaeta). J. Comp. Physiol. A, 156: 553-561.

Franke, H.-D., 1986, The role of light and endogenous factors in the timing of the reproductive cycle of Typosyllis prolifera and some other polychaetes. Amer. Zool., 26: 433-445.

Heimbach, F., 1978, Emergence times of the intertidal midge Clunio marinus (Chironomidae) at places with abnormal tides, pp. 263-270 in: "Physiology and behaviour of marine organisms", D.S. McLusky and A,.J. Berry ed., Pergamon Press, Oxford and New York.

Krüger, M. u. D. Neumann, 1983, Die Temperaturabhängigkeit semilunarer und diurnaler Schlüpfrhythmen bei der intertidalen Mücke Clunio marinus (Diptera, Chironomidae). Helgoländer Meeresunters. 36, 427-464.

Neumann, D., 1966, Die lunare und tägliche Schlüpfperiodik der Mücke Clunio. Steuerung und Abstimmung auf die Gezeitenperiodik. Z. vergl.Physiol. 53, 1-61 (1966).

Neumann, D., 1968, Die Steuerung einer semilunaren Schlüpfrhythmik mit Hilfe eines künstlichen Gezeitenzyklus. Z. vergl. Physiol. 60, 63-78.

Neumann, D., 1976 a, Adaptations of chironomids to intertidal environments. Ann. Rev. Entomol. 21, 387-414.

Neumann, D., 1976 b, Mechanismen für die zeitliche Anpassung von Entwicklungsleistungen an den Gezeitenzyklus. Verh. Dtsch. Zool. Ges. 1976: 9-28.

Neumann, D., 1978, Entrainment of a semilunar rhythm by simulated tidal cycles of mechanical disturbance. J.exp.mar.Biol.Ecol. 35, 73-85 .

Neumann, D., 1981 a, Synchronization of reproduction in marine insects by tides, pp. 21-35 in: "Advances in Invertebrate Reproduction". Clark Jr., W. H., Adams, T.S. eds., Elsevier/Northholland, New York, Amsterdam, Oxford.

Neumann, D., 1981 b, Tidal and lunar rhythms, pp. 351-380 in: "Biological Rhythms. Handbook of Behavioral Neurobiology", J. Aschoff, Vol. 4, Plenum Press, New York and London.

Neumann, D., 1983, Die zeitliche Programmierung von Tieren auf periodische Umweltbedingungen. Rhein.-Westf. Akademie d. Wissenschaften, Vorträge N 324, 31-68.

Neumann, D., 1986, Life cycle strategy of an intertidal midge between subtropic and arctic latitudes, pp 3-19, in: " The Evolution of Insect Life Cycles. Proceedings in Life Sciences". Taylor, Karban eds., Springer Verlag, Berlin-Heidelberg-New York.

Neumann, D., 1987, Tidal and lunar rhythmic adaptations of reproductive activities in invertebrate species, pp. 152-170 in: "Comparative Physiology of Environmental Adaptations. 8th. ESCPB-Conference in Strasbourg 1986", P. Pevet ed., Karger-Basel, Vol. III.

Neumann, D.; F. Heimbach, 1979, Time cues for semilunar reproduction rhythms in European populations of Clunio marinus. I. The influences of tidal cycles of mechanical disturbance, pp. 423-433 in: "Cylic Phenomena in marine plants and animals", Naylor, E., Hartnoll, R.G. eds., Pergamon Press, Oxford and New York.

Neumann, D.; F.Heimbach, 1984, Time cues for semilunar reproduction rhythms in European populations of Clunio marinus. II. The influence of tidal temperature cycles. Biol. Bull. 166: 509-524.

Neumann, D.; F. Heimbach, 1985, Circadian range of entrainment in the semilunar eclosion rhythm of the marine insect Clunio marinus. J. Insect Physiol. 31: 549-557.

Neumann, D.; Krüger, M., 1985, Combined effects of photoperiod and temperature on the diapause of an intertidal insect. Oecologia (Berl.) 67, 154-156.

Oka, H.; Hashimoto, H., 1959, Lunare Periodizität in der Fortpflanzung einer pazifischen Art von Clunio (Diptera, Chironomidae). Biol. Zbl. 78, 545-559.

Pittendrigh, C.S., 1967, Circadian systems. I. The driving oscillation and its assay in Drosophila pseudoobscura. Prod.Nat.Acad.Sci. 58: 1762-1767.

Pittendrigh, C. S., 1981, Circadian systems: general perspective, pp. 57-80 in: "Biological Rhythms. Handbook of Behavioral Neurobiology", Vol. 4. J. Aschoff, ed., Plenum Press, New York, London.

Saigusa, M.; Hidaka, T., 1978, Semilunar rhythm in the zoea-release activity of the land crab Sesarma. Oecologia (Berl.) 37: 163-176 .

Saigusa, M., 1980, Entrainment of a semilunar rhythm by a simulated moonlight cycle in the terrestrical crab, Sesarma haematocheir. Oecologia (Berl.) 46: 38-44.

Patterns of Movement in Intertidal Fishes

Robin N. Gibson

Dunstaffnage Marine
Research Laboratory
Argyll, Scotland, U.K.

INTRODUCTION

The intertidal zone frequently represents a rich source of food and shelter and as such is inhabited by a wide variety of animals. Many of them are able to withstand the changing environmental conditions and live there continuously. Others, lacking this ability, avoid the zone at low tide and occupy the area only when it is immersed. This basic difference allows intertidal animals to be broadly categorized as either residents or visitors. The greater motility of fishes compared with most other intertidal animals enables them to employ a wide range of behaviour patterns for exploiting the intertidal zone, although the majority tend to be visitors. There are, however, large numbers of fishes whose movements are more limited in extent and which can be regarded as resident intertidal animals for most of their lives. This paper briefly reviews the activity patterns of both categories of fishes in relation to their function and discusses whether such patterns can be regarded as special adaptations to intertidal life. The examples used to illustrate particular behaviour patterns are not exhaustive because full reviews of the biology and behaviour of intertidal fishes are given elsewhere (Gibson, 1982a, 1986a).

A CLASSIFICATION OF INTERTIDAL FISHES

Although differing in detail the basic distinction between fishes as either resident or temporary inhabitants of the intertidal zone is generally agreed upon by most workers (Gibson, 1982a). The system used here is that outlined by Potts (1980) based on the earlier version proposed by Gibson (1969). In this system species are classified either as residents or transients and the latter category is further subdivided into visitors which occur on an accidental, tidal or seasonal basis. The seasonal visitors consist of both juveniles and adult migrants. Although not allowing for the description of some categories (e.g. tidal visitors which are only present seasonally) such a scheme provides a convenient framework for discussing patterns of movement and will be followed here.

Although few quantitative comparisons have been made (e.g. Thomson & Lehner, 1976) most observations suggest that visitors outnumber residents in both numbers and biomass.

FISHES AS VISITORS TO THE INTERTIDAL ZONE

Movements in Space

Visitors enter and leave the intertidal zone on a regular basis and although they may move around within it and even return to the same place on successive high tides (Carlisle, 1961; Potts, 1985) their movements are predominantly in and out of the zone. Such movements may cover considerable distances (up to several hundred metres, Tyler, 1971; Kuipers, 1973; Wirjoatmodo & Pitcher, 1984) depending on the tidal range and the slope of the shore.

Movements in Time

The movement in and out of the intertidal zone can be purely by chance (accidentals), on a regular tidal basis (tidal visitors) or only at certain seasons of the year (seasonal visitors). The seasonal visitors may be either juveniles which recruit into the zone as a nursery ground and then move into deeper water as they grow or adults depositing their eggs during the reproductive season. The juveniles may migrate with each tide as on sandy (Tyler, 1971; Gibson, 1973, 1986b; Kuipers, 1973) or muddy shores (Summers, 1980; Kneib, 1984) and tidal creeks (Shenker & Dean, 1979; Kleypas & Dean, 1983) or remain at low tide as in salt marshes (Kneib, 1987) or on most rocky shores. Even on sand or mudflats, however, some fishes can survive in tidal pools until the next high tide (Anderson et al., 1977) and recently Berghahn (1983) and van der Veer & Bergman (1986) have described how young plaice (*Pleuronectes platessa*) stay in such pools until conditions become unsuitable and then begin migrating with each tide.

FISHES AS RESIDENTS IN THE INTERTIDAL ZONE

Movements in Space

Compared with visitors, the movements of residents all take place *within* the intertidal zone, are much more restricted in extent and basically two-dimensional. Some species are known to limit their movements to less than a metre from a refuge (Abel, 1973) whereas others wander over a home range of several square metres. The movements of these more mobile species have been measured by marking them in rock pools at low tide and then following their movements by low tide collections at subsequent intervals. Such a technique, although useful in determining the general area of movement over a given time, provides little information as to the extent and direction of movement over the important high tide period. The presence of a fish in a pool on subsequent low tides, for example, could be due either to the absence of movement from the pool or the ability to return to it as the tide ebbs. Such information can only be gained by direct observation (Williams, 1957; Green, 1971b; Graham et al., 1985; Cancino & Castilla, 1987) or telemetry (Ralston & Horn, 1986) and these techniques have shown a variety of movement patterns, from a complete shift in distribution at high tide as in *Girella* and *Clinocottus* (Williams, 1957) to a very limited degree of movement over an area of about 2 m^2 as in *Cebidichthys* (Ralston & Horn, 1986).

Movements in Time

Even more than visitors, the timing of the movements of residents is strongly influenced by the tidal cycle. Resident species on rocky shores are mostly, but not always (Green 1971c), active over the high tide period, remaining quiescent at low tide but few quantitative measurements have been made of the change in activity level over the tidal cycle in the field.

The study by Ralston & Horn (1986) is the most detailed so far undertaken and showed that all the six individuals of *Cebidichthys violaceus* investigated were primarily active during the flooding tide, whether this occurred during the day or night. In contrast, the mudskippers are most active at low tide and some retire to their burrows at or above the high tide mark at high tide (Brillet, 1975). Most residents seem to be diurnal, although nocturnal movements take place in some species (e.g. *Blennius pavo*, Fishelson, 1963) and others are active by day or night (*Girella*, *Clinocottus*, Williams, 1957; *Oligocottus*, Green, 1971c).

FACTORS AFFECTING PATTERNS OF MOVEMENT IN SPACE

The predominant factor controlling the extent of movement of an individual must be its powers of locomotion; large pelagic species can obviously travel farther and faster than small benthic ones and hence can range over wider distances. Given comparable locomotor capabilities, however, many other factors act together to influence the actual movement patterns observed. The nature of the habitat, and particularly the amount of cover available and the degree to which the shore is subjected to wave action, is of prime importance. Where cover is absent, on sandy beaches for example, the fishes are mostly tidal visitors and move up and down the shore with the tide. Where cover is present as on rocky shores, the residents are strongly thigmotactic and keep close to cover. Such thigmotactic tendencies, which are absent in all tidal visitors, are advantageous in resisting displacement by turbulence. Turbulence and wave action themselves can have marked effects on movement either by determining the location and extent of activity (e.g. Cancino & Castilla, 1987) or sometimes by preventing it altogether (Taborsky & Limberger, 1980).

Two further characteristics of resident species which restrict their movement are their frequent limitation to particular levels on the shore, at least at low tide (Green, 1971a; Gibson, 1972; Horn & Riegle, 1981; Yoshiyama, 1981), by various habitat selection mechanisms (e.g. Nakamura, 1976) and their ability to return ("home") to places formerly occupied. Again, such characteristics are rarely present in visitors particularly on featureless shores, although Riley (1973) described the return of young plaice to their original location on a sandy beach when experimentally displaced. The homing capabilities of several rocky shore fishes have been investigated (Williams, 1957; Gibson, 1967; Green, 1973) and in the most intensively studied species, *Oligocottus maculosus*, homing ability apparently depends on olfactory and visual senses (Green, 1971b; Khoo, 1974; Craik, 1981). One explanation for the ability of fishes to home is that they must at some time in their life, acquire and retain a topographic knowledge of their immediate environment. Experiments by Aronson (1951, 1971) with *Bathygobius soporator* showed that such knowledge can be acquired over one high tide period and retained for several weeks. *O. maculosus* can retain its ability to home for several months, but in this species spatial familiarity with an area may be acquired in the wide ranging juvenile stages because the extent of movement of adults is insufficient to account for learning of the area over which they home when experimentally displaced (Green, 1971b, Craik, 1981). The juveniles of many other species are responsible for colonising new habitats and several studies (Bussing, 1972; Marsh et al., 1978) but not all (Beckley, 1985) have indicated that depopulated areas are usually first recolonised by juveniles, although the speed of such recolonisation is strongly dependent on the time of year relative to the breeding season (Beckley, 1985). Recruitment and colonisation takes place in the later juvenile stages, but recent work has suggested that behavioural factors may also operate at an earlier stage in the life history to retain the planktonic larvae of many species close to the shore rather than allowing them to be dispersed widely offshore (Marliave, 1986; Potts & McGuigan, 1986).

A final factor affecting the movements of fish in space is the presence or absence of territoriality. Reproductive territoriality is common during the breeding season in the resident species which lay their eggs on the substratum and guard them until they hatch, but is variously developed at other times of year. It seems to be completely absent in sculpins (Cottidae) for example but highly developed in mudskippers (Brillet, 1975; Clayton & Vaughan, 1982). Visitors, except for rare cases, are completely non-territorial.

FACTORS AFFECTING PATTERNS OF MOVEMENT IN TIME

The factors controlling the patterns of movement in time are all related to the two predominant cycles in the environment: those of the day-night cycle and the tidal cycle. The varying responses to the stimuli associated with these two interacting cycles are responsible for the differences in movement patterns observed. The change in light intensity over the 24 hour light-dark cycle controls the state of activity of diurnal/ noctural species but superimposed on this is the response to those factors associated with the tides, and if anything represents the behavioural adaptation of intertidal animals to their environment it is this ability to synchronise their activity/inactivity with the tidal cycle.

Tidal visitors are in continual movement but their level of activity may vary with the state of the tide (Gibson, 1980, 1986b), particularly if they are migrating long distances each tide. The actual stimuli governing the movements of visitors are not known although for benthic species any of the factors associated with the changing tide (pressure, light intensity, temperature, currents etc) could be responsible (see Gibson, 1973 for discussion). For resident species the signals for starting or finishing activity may be much more distinct, as for example when the tide enters or leaves a rock pool or submerges a previously emersed fish. Here again, however, the relative importance of individual stimuli is virtually unknown although detailed observations by Green (1971c), have shown that for *Oligocottus maculosus* at least, changes in temperature, light and turbulence are all important whereas salinity is not. Such external factors may evoke a direct response but it is now evident that many species are capable of controlling their level of activity internally because endogenous rhythmicity has been demonstrated in all the main families of resident intertidal fishes and some tidal visitors (Gibson, 1978, 1982a). In controlled laboratory conditions fishes possessing such a "biological clock" mechanism exhibit rhythms of activity whose persistence and phase relative to the local tide cycle varies considerably. The limited evidence available suggests that in rocky shore residents the endogenous activity peaks are centred on the predicted time of high water, whereas in tidal visitors activity maxima occur on the ebb (Gibson, 1983). It might be expected that mudskippers which are most active at low tide would have an endogenous rhythm with peaks at low tide but the results for the only species that has been tested (*Boleophthalmus chinensis*) indicates that its endogenous activity peaks are also centred on the ebb tide period (Ishibashi, 1973). The use of activity patterns recorded in controlled conditions in the laboratory to predict activity patterns in the wild where direct responses to environmental factors are also present is not necessarily, however, a valid procedure. Unfortunately, few species have been observed both in the wild and the laboratory to enable the contribution of endogenous rhythmicity to be elucidated (Green, 1971c; Gibson, 1975, 1978) and critical experiments to test the survival value of endogenous rhythmicity have yet to be performed. Nevertheless, some further clues as to the importance of environmental factors controlling the timing of activity may be gained from the results of experiments designed to isolate the factors responsible for the entrainment of endogenous rhythms. So far, only pressure change

(Gibson, 1971, 1982b; Gibson et al., 1978), fluctuations in water level (Ishibashi, 1973) and periodic feeding (Nishikawa & Ishibashi, 1975) have been implicated in this respect. In general, though, the level and type of activity exhibited at any one time must be the resultant of the relative strengths of the endogenous and exogenous stimuli.

THE FUNCTIONS OF THE MOVEMENT PATTERNS

The patterns of movement so far described can be interpreted as solutions to three main problems. The first is that the food supply in the intertidal zone is only periodically accessible and so must be exploited when available. Fishes not adapted for conditions at low water must therefore migrate to and from their intertidal food source with each tide. Such tidally synchronised movements lead to a tidal rhythm of food intake (e.g. Healey, 1971; Summers, 1980). Those which are adapted to low water conditions could, in theory, feed over the whole tidal cycle provided that their low tide location (in tide pools for example) or powers of locomotion out of water (mudskippers, amphibious blennioids) allowed them to do so. For most resident species however, feeding when the tide is in allows them access to a wider range of food organisms and the medium for normal locomotion. In addition, the prey organisms themselves may be more active and hence more accessible, over the high tide period.

High tide movement away from the location occupied at low tide provides both visitors and residents alike with the second problem - how to avoid being stranded in an unfavourable position on the shore when the tide ebbs. For visitors, avoidance takes the form of simple downshore migration with the tide and the ebb-phased endogenous rhythm recorded in several species may be an environment-independent mechanism for ensuring that such movements are timed to take place at the correct phase of the tidal cycle. For rocky shore residents the problem is just as acute because failure to find a suitable low tide refuge before they are stranded by the outgoing tide could also be fatal. The observed restricted extent of the movements of such residents, their ability to learn the topographic details of their immediate environment and the use of this knowledge to return to suitable low tide refuges can all be seen as adaptations to avoid being stranded in unsuitable locations at low water. How such locations are chosen initially remains to be discovered, but learning from past experience may be important here also. In this context, Williams (1957) has observed on the ebbing tide that pools that do not drain at low tide contained many fishes whereas those that dry out had none.

Furthermore, all observations on homing and the fidelity to specific sites have shown that such behaviour is sufficiently flexible to allow for change in the fishes' currently occupied low tide refuge. Consequently, rather than remain in or continue to home to, a site which has been rendered unsuitable by storms for example, an individual can relocate its centre of activity to another site within its usual range of movement. Such flexibility of behaviour seems essential for species living in an environment where changes in habitat topography may be frequent.

The third problem is that the small size of many intertidal fishes, whether visitors or residents, means they are particularly vulnerable to predators. For many of the young fishes that undertake tidal feeding migrations, the movement up the shore close to the advancing waterline may have the additional advantage of keeping them in shallow water and reducing the risk of predation from those larger predators which also move in with the tide. Where cover is available as a refuge some species have reduced this risk further by becoming a resident during the vulnerable juvenile stages. Although the assumption that the intertidal zone acts as a refuge from predation seems reasonable, it has only been demonstrated experimentally for juveniles of the saltmarsh killifish *Fundulus heteroclitus* (Kneib, 1987). As an adult this species is a tidal visitor to marshes (Kneib, 1984) but rather than spawning below low water mark provides protection for its offspring by laying its eggs intertidally (Taylor et al., 1977). Such reproductive behaviour is the rule among residents, of course, but the intertidal zone is also used as a refuge for the eggs of several other basically subtidal species, of which the Californian grunions with their habit of spawning at high water mark of spring tides on sand beaches (Walker, 1952; Thomson & Muench, 1976) and the Pacific herring (Blaxter & Hunter, 1982) are the best known.

For residents, their restricted movement, intimate knowledge of their environment and the location of suitable refuges can both be seen as aids in avoiding predators and there is good evidence for small fishes that the presence of, and access to, suitable shelter is important in reducing predation (Phillips & Swears, 1979; Behrents, 1987). Extreme cases of the avoidance of predators can be seen in those rocky shore residents (e.g. *Alticus*, Magnus, 1963; *Mnierpes*, Graham, 1970; *Entomacrodus*, Graham *et al*., 1985) which have become amphibious in their lifestyle and inhabit the shallow turbulent region at the water's edge which is difficult to exploit by terrestrial and aquatic predators alike.

EVOLUTION OF INTERTIDAL FISHES

Exploitation of the intertidal zone as a feeding ground and refuge from predation for small fishes can be regarded as two of the factors which may have been responsible for the gradual evolution of a resident intertidal fish fauna. A speculative series of stages in such an evolutionary scheme could be as follows.

1. Use of shallow water as a feeding ground with resultant decrease in predation pressure from larger predators unable to function efficiently in such depths.

2. Increasing tendency to remain in suitable locations at low tide further reducing both the time exposed to aquatic predators and the energetic costs associated with tidal migration. Development of morphological and physiological adaptations to the increased fluctuation in environmental conditions encountered.

3. Increasing population density of resident species leads to greater competition for food and space so that increasingly higher zones on the shore are occupied by those species capable of doing so, culminating in the adoption of an amphibious life style. Such speculation may provide a partial explanation for the fact that most amphibious species are tropical in distribution, reflecting the greater time for the evolutionary scheme proposed to have been in operation. In the younger faunas (in evolutionary terms) of temperate regions, amphibious species are rare although early stages in the scheme may be represented by the voluntary nocturnal excursions onto land of the Mediterranean blennies described by Zander (1983) and the ability of others such as *Blennius pholis* to leave the water if necessary (Davenport & Woolmington, 1981).

REFERENCES

Abel, E.F., 1973. Zur Oko-Ethologie des amphibisch lebenden Fisches Alticus saliens (Forster) und von Entomacrodus vermiculatus (Val.) (Blennioidea, Salariidae) unter besondere Berucksichtigung des Fortpflanzungsverhaltens. Sbr. ost. Akad. Wiss., Abt. I, 181, 137-153.

Anderson, W.D. Jr., J.K. Dias, D.M. Kupka & N.A. Chamberlain, 1977. The macrofauna of the surf zone off Folly Beach, South Carolina. NOAA Tech. Rep. NMFS SSRF-704, 1-23.

Aronson, L.R., 1951. Orientation and jumping behaviour in the gobioid fish Bathygobius soporator. Am. Mus Novitat., No. 1486, 1-22.

Aronson, L.R., 1971. Further studies on orientation and jumping behaviour in the gobioid fish Bathygobius soporator. Ann. N.Y. Acad. Sci., 188, 378-392.

Beckley, L.E., 1985. Tide-pool fishes: recolonization after experimental elimination. J. exp. mar. Biol. Ecol., 85, 287-295.

Behrents, K.C., 1987. The influence of shelter availability on recruitment and early juvenile survivorship of Lythrypnus dalli Gilbert (Pisces: Gobiidae). J. exp. mar. Biol. Ecol., 107, 45-59.

Berghahn, R., 1983. Untersuchungen an Plattfischen und Nordseegarnelen (Crangon crangon) im Eulittoral des Wattenmeeres nach dem Ubergang zum Bodenleben. Helgolander Meeresunters., 36, 136-181.

Blaxter, J.H.S. & J.R. Hunter, 1982. The biology of the clupeoid fishes. Adv. mar. Biol., 20, 3-223.

Brillet, C., 1975. Relations entre territoire et comportement aggressif chez Periophthalmus sobrinus Eggert (Pisces, Periophthalmidae) au laboratoire et en milieu natural. Z. Tierpsychol., 39, 283-331.

Bussing, W.A., 1972. Recolonisation of a population of supratidal fishes at Eniwetok atoll, Marshall Islands. Atoll Res. Bull., No. 154, 1-4.

Cancino, J.M. & J.C. Castilla, 1987. Emersion behaviour and foraging ecology of the common Chilean clingfish Sicyases sanguineus (Pisces: Gobiesocidae). J. nat. Hist., in press.

Carlisle, D.B., 1961. Intertidal territory in fish. Anim. Behav., 9, 106-107.

Clayton, D.A. & T.C. Vaughan, 1982. Pentagonal territories of the mud-skipper Boleophthalmus boddarti (Pisces, Gobiidae). Copeia 1982, 232-234.

Craik, G.J.S., 1981. The effects of age and length on homing performance in the intertidal cottid, Oligocottus maculosus Girard. Can. J. Zool., 59, 589-604.

Davenport, J. & A.D. Woolmington, 1981. Behavioural responses of some rocky shore fishes exposed to adverse environmental conditions. Mar. Behav. Physiol., 8, 112.

Fishelson, L., 1963. Observations on littoral fishes of Israel. I. Behaviour of Blennius pavo Risso (Teleostei, Blenniidae) Israel J. Zool., 12, 67-80.

Gibson, R.N., 1967. Studies on the movements of littoral fish. J. Anim. Ecol., 36, 215-234.

Gibson, R.N., 1969. The biology and behaviour of littoral fish. Oceanogr. Mar. Biol. Annu. Rev., 7, 367-410.

Gibson, R.N., 1971. Factors affecting the rhythmic activity of Blennius pholis L. (Teleostei). Anim. Behav., 19, 336-343.

Gibson. R.N., 1972. The vertical distribution and feeding relationships of intertidal fish on the Atlantic coast of France. J. Anim. Ecol., 41, 189-207.

Gibson, R.N., 1973. The intertidal movements and distribution of young fish on a sandy beach with special reference to the plaice (Pleuronectes platessa L.). J. exp. mar. Biol. Ecol., 12, 79-102.

Gibson, R.N., 1975. A comparison of field and laboratory activity patterns of juvenile plaice. In H. Barnes (ed.) Proc. 9th Europ. mar. Biol. Symp., Aberdeen University Press, Aberdeen, 13-28.

Gibson, R.N., 1978. Lunar and tidal rhythms in fish. In J.E. Thorpe (ed) Rhythmic activity of fishes. Academic Press, London, 201-213.

Gibson, R.N., 1982a. Recent studies on the biology of intertidal fishes. Oceanogr. Mar. Biol. Annu. Rev., 20, 363-414.

Gibson, R.N., 1982b. The effect of hydrostatic pressure cycles on the activity of young plaice Pleuronectes platessa. J. mar. biol. Ass. U.K., 62, 621-635.

Gibson, R.N., 1983. Hydrostatic pressure and the rhythmic behaviour of intertidal marine fishes. Trans. Am. Fish. Soc., 113, 479-483.

Gibson, R.N., 1986a. Intertidal teleosts: Life in a fluctuating environment. In T.J. Pitcher (ed.) The behaviour of teleost fishes. Croom Helm Ltd., Beckenham, 388-408.

Gibson, R.N., 1986b. Observations on the behaviour of young plaice on sandy beaches. Prog. Underwat. Sci., 11, 27-32.

Gibson, R.N., J.H.S. Blaxter & S.J. De Groot, 1978. Developmental changes in the activity rhythms of the plaice (Pleuronectes platessa L.). In J.E. Thorpe (ed.) Rhythmic activity of fishes. Academic Press, London, 169-186.

Graham, J.B., 1970. Preliminary studies on the biology of the amphibious clinid Mnierpes macrocephalus. Mar. Biol., 6, 136-140.

Graham, J.B., C.B. Jones & I. Rubinoff, 1985. Behavioural, physiological and ecological aspects of the amphibious life of the pearl blenny Entomacrodus nigricans Gill. J. exp. mar. Biol. Ecol., 89, 255-268.

Green, J.M., 1971a. Local distribution of Oligocottus maculosus Girard and other tidepool cottids on the west coast of Vancouver Island, British Columbia. Can. J. Zool., 49, 1111-1128.

Green, J.M., 1971b. High tide movements and homing behaviour of the tidepool sculpin Oligocottus maculosus. J. Fish. Res. Bd Can., 28, 383-389.

Green, J.M., 1971c. Field and laboratory activity patterns of the tidepool cottid Oligocottus maculosus Girard. Can. J. Zool., 49, 255-264.

Green, J.M., 1973. Evidence for homing in the mosshead sculpin (Clinocottus globiceps). J. Fish. Res. Bd Can., 30, 129-130.

Healey, M.C., 1971. The distribution and abundance of sand gobies, Gobius minutus, in the Ythan estuary. J. Zool., Lond., 163, 177-229.

Horn, M.H. & K.C. Riegle, 1981. Evaporative water loss and intertidal vertical distribution in relation to body size and morphology of stichaeoid fishes from California. J. exp. mar. Biol. Ecol., 50, 273-288.

Ishibashi, T., 1973. The behavioural rhythms of the gobioid fish Boleophthalmus chinensis (Osbeck). Fukuoka Univ. Sci. Rep., 2, 69-74.

Khoo, H.W., 1974. Sensory basis of homing in the intertidal fish Oligocottus maculosus Girard. Can. J. Zool., 52, 1023-1029.

Kleypas, J. & J.M. Dean, 1983. Migration and feeding of the predatory fish, Bairdiella chrysura Lacepede, in an intertidal creek. J. exp. mar. Biol. Ecol., 72, 199-209.

Kneib, R.T., 1984. Patterns of invertebrate distribution and abundance in an intertidal saltmarsh: causes and questions. Estuaries, 7, 392-412.

Kneib, R.T., 1987. Predation risk and use of intertidal habitats by young fishes and shrimp. Ecology, 68, 379-386.

Kuipers, B., 1973. On the tidal migration of young plaice (Pleuronectes platessa L.) in the Wadden Sea. Neth. J. Sea Res., 6, 376-388.

Magnus, D.B.E., 1963. Alticus saliens, ein amphibisch lebender Fisch. Natur Mus., 93, 128-132.

Marliave, J.B., 1986. Lack of planktonic dispersal of rocky intertidal fish larvae. Trans. Am. Fish. Soc., 115, 149-154.

Marsh, B., T.M. Crowe & W.R. Siegfried, 1978. Species richness and abundance of clinid fish (Teleostei; Clinidae) in intertidal rock pools. Zoologica Afr., 13, 283-291.

Nakamura, R., 1976. Experimental assessment of factors influencing microhabitat selection by the two tidepool fishes Oligocottus maculosus and O. snyderi. Mar. Biol., 37, 87-104.

Nishikawa, M. & T. Ishibashi, 1975. Entrainment of the activity rhythm by the cycle of feeding in the mudskipper, Periophthalmus cantonensis (Osbeck). Zool. Mag. Tokyo, 84, 184-189.

Phillips, R.R. & S.B. Swears, 1979. Social hierarchy, shelter use, and the avoidance of predatory toadfish (Opsanus tau) by the striped blenny (Chasmodes bosquianus). Anim. Behav., 27, 1113-1121.

Potts, G.W., 1980. The littoral fishes of Little Cayman (West Indies). Atoll Res. Bull., No. 241, 43-52.

Potts, G.W., 1985. The nest structure of the corkwing wrasse, Crenilabrus melops (Labridae: Teleostei). J. mar. biol. Ass. U.K., 65, 531-546.

Potts, G.W. & K.M. McGuigan, 1986. A preliminary survey of the distribution of postlarval fish associated with inshore reefs with special reference to Gobiusculus flavescens (Fabricius). Prog. Underwat. Sci., 11, 15-25.

Ralston, S.L. & M.H. Horn, 1986. High tide movements of the temperate zone herbivorous fish Cebidichthys violaceus (Girard) as determined by ultrasonic telemetry. J. exp. mar. Biol. Ecol., 98, 35-50.

Richkus, W.A., 1981. A quantitative study of intertidepool movement of the wooly sculpin Clinocottus analis. Mar. Biol., 49, 277-284.

Riley, J.D., 1973. Movements of O-group plaice Pleuronectes platessa as shown by latex tagging. J. Fish Biol., 5, 323-343.

Shenker J.M. & J.M. Dean, 1979. The utilization of an intertidal salt marsh by larval and juvenile fishes: abundance, diversity and temporal variation. Estuaries, 2, 154-163.

Summers, R.W., 1980. The diet and feeding behaviour of the flounder Platichthys flesus (L.) in the Ythan estuary, Aberdeenshire, Scotland. Estuar. cstl. mar. Sci., 11, 217-232.

Taborsky, M. & D. Limberger, 1980. The activity rhythm of Blennius sanguinolentus Pallas, an adaptation to its food source. Mar. Ecol., P.S.Z.N., 1, 143-153.

Taylor, M.H., L. DiMichele & G.J. Leach, 1977. Egg stranding in the life cycle of the mummichog Fundulus heteroclitus. Copeia 1977, 397-399.

Thomson, D.A. & C.E. Lehner, 1976. Resilience of a rocky intertidal fish community in a physically unstable environment. J. exp. mar. Biol. Ecol., 22, 1-29.

Thomson, D.A. & K.A. Muench, 1976. Influence of tides and waves on the spawning of the Gulf of California grunion Leuresthes sardina (Jordan & Evermann). Bull. Sth Calif. Acad. Sci., 75, 198-203.

Tyler, A.V., 1971. Surges of winter flounder, Pseudopleuronectes americanus, into the intertidal zone. J. Fish. Res. Bd Can., 28, 1727-1732.

van der Veer, H.W. & M.J.N. Bergman, 1986. Development of tidally related behaviour of a newly settled O-group plaice (Pleuronectes platessa) population in the western Wadden Sea. Mar. Ecol. Progr. Ser., 31, 121-129.

Walker, B.W., 1952. A guide to the grunion. Calif. Fish Game, 38, 409-420.

Williams, G.C., 1957. Homing behaviour of California rocky shore fishes. Univ. Calif. Publ. Zool., 59, 249-284.

Wirjoatmodo, S. & T.J. Pitcher, 1984. Flounders follow the tide to feed: evidence from ultrasonic tracking in an estuary. Estuar. cstl. Shelf Sci., 19, 231-242.

Yoshiyama, R.M., 1981. Distribution and abundance patterns of rocky intertidal fishes in central California. Env. Biol. Fish., 6, 315-332.

Zander, C.D., 1983. Terrestrial sojourns of two Mediterranean blennioid fish (Pisces, Blennioidea, Blenniidae). Senckenberg. Marit., 15, 19-26.

Predation of Intertidal Fauna by Shorebirds in Relation to Time of the Day, Tide and Year

Peter R. Evans
University of Durham, U.K.

INTRODUCTION

Shorebirds or waders, members of the Sub-Orders Scolopaci and Charadrii, are the most important bird predators of intertidal areas during their periods of exposure between successive high tides (Baird et al 1985). Some gulls (Laridae) and shelduck (Tadornidae) also forage intertidally, both when such areas are exposed and when they are covered by the tide. Certain diving ducks, particularly eiders (Somateridae), also feed on benthic intertidal animals, but only at high water when the rocks or flats are covered; and several heron spp. (Ardeidae) forage there.

Shorebirds forage by two distinct methods. Plovers Charadrii have short beaks but large eyes in relation to head size. They search for food visually, looking from a vantage position for signs of invertebrates active at or near the surface of the substratum. They then run forward and attempt to capture the prey before it moves out of reach e.g. by retreating into a soft sediment (Pienkowski 1983). Plovers take chiefly annelids, small crustacea and small gastropods, rather than bivalves, as foods. In contrast, other waders, both sandpipers Scolopacidae and oystercatchers Haematopodidae, have long beaks in relation to head size. Although they too use visual cues by day to locate prey at the surface of the substratum, they also feed by touch within soft sediments. When using this method, Oystercatchers *Haematopus ostralegus* and Knot *Calidris canutus* rely on making direct contact between the bill and the prey (usually a bivalve) to achieve prey detection, but other small Calidris sandpipers can detect prey without probing directly on to them, perhaps by substrate dislocation and chemoreception (Gerritsen et al 1983). These long-billed species take a variety of polychaetes, some oligochaetes, and many smaller crustacea and molluscs. Prey animals possessing hard exoskeletons or protected by shells are usually swallowed whole and ground up in the birds' gizzards, but oystercatchers hammer open larger bivalves and Curlews *Numenius arquata* penetrate the defences of large clams *Mya arenaria* through the siphon hole (Zwarts and Wanink 1984). The long-billed waders, unlike the plovers, are able to detect and capture prey at considerable depths in soft substrates (Curlew bills may reach 17 cm in length).

Most shorebirds that have been studied so far are capable of feeding at night. Although plovers continue to forage visually during darkness, sandpipers seem normally to switch to tactile foraging, at least in the absence of moonlight (Hulscher 1976). The relative importance of prey detection by direct touch, as opposed to chemoreception, at night has not yet been studied, nor have field studies been undertaken specifically to test the possible importance of audible cues at night. Circumstantial evidence suggests that the proven ability of some plovers to detect buried prey by sound cues in laboratory conditions (Lange 1968) is probably not relevant to the field situation (Pienkowski 1983).

This paper reviews present knowledge of the extent to which the behaviour of foraging shorebirds either influences, or is influenced by, the behaviour of their prey. It is concerned particularly with the timing of behavioural events on a seasonal, tidal and daily basis and examines first the possibilities that seasonal movements of birds might be timed ultimately, i.e. in an evolutionary sense, by seasonal changes in abundance of their food resources or, conversely, that seasonal changes in abundance and availability of prey might be timed to minimize predation by shorebirds.

SEASONAL CHANGES IN PREDATION BY SHOREBIRDS ON INTERTIDAL ANIMALS

In the temperate zone of the northern hemisphere, the intensity of predation by waders on the intertidal fauna of soft and rocky shores is highly seasonal. This arises for three reasons:

(i) the total numbers of birds present vary from month to month.

(ii) their daily food requirements vary according to the physiological and behavioural activities in which they are engaged at different seasons

(iii) the species composition of the birds present, and hence their average size and food requirements, also varies seasonally.

Most shorebirds are long-distance migrants and many nest in the arctic. Of the latter, most adults leave as soon as their breeding attempts have been completed or have failed. They move to the northern temperate zone in late summer to moult their plumage, though a few species moult at more northerly sites and a few delay moult until reaching the end of their migrations. Juveniles leave the breeding grounds after their parents, but many catch up with them on the moulting grounds, which, for many species, are restricted to a few very large intertidal sites where they are relatively safe from mammalian predators. Total numbers of birds present on the northern temperate intertidal areas are thus usually greatest at this time of year, i.e. in late summer. After this, many birds continue migration southwards, e.g. from north to central and south America (Morrison 1984), and from Europe to Africa. However many also move westwards from continental Europe to the British Isles for the winter and yet others stay on those European coasts that are climatically least severe (Pienkowski and Evans 1984). In late winter and early spring many return eastwards from Britain, or north from Africa, to the international Wadden Sea areas to moult into breeding plumage (Boere 1976), before departing northwestwards to Greenland or northeastwards to their Siberian nesting grounds. Fig. 1 emphasizes that the seasonal patterns of variation in total numbers of waders in three almost adjacent but large geographical areas of Europe are very different, largely because of differences in numbers of birds present in mid-winter, along a climatic gradient from Britain (relatively mild) to the Danish Wadden Sea, where intertidal areas are often covered by ice for periods of many days so that waders would die because they could not feed. Allied to these geographical differences in numbers, seasonal variation in the intensity of predation by birds on intertidal fauna also differs from area to area, except that in all areas, predation is least in mid-summer. (Conversely, predation of intertidal fauna by fish and the larger invertebrate predators, while areas are covered by the tide, will be highest in the summer months).

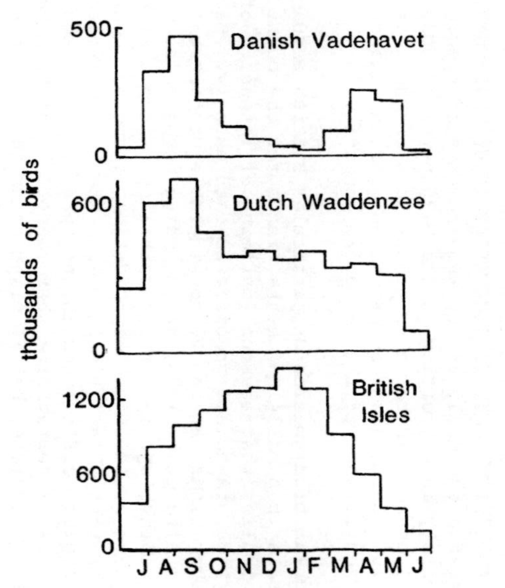

Fig. 1. Average monthly counts of total numbers of shorebirds in three regions of western Europe (compiled from data in Prater (1981) and Smit and Wolff (1981))

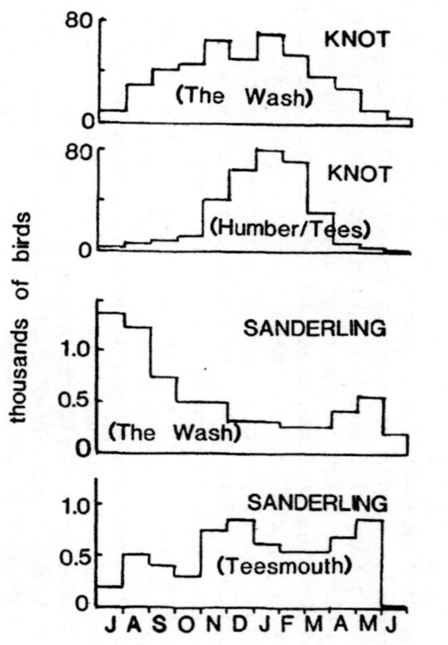

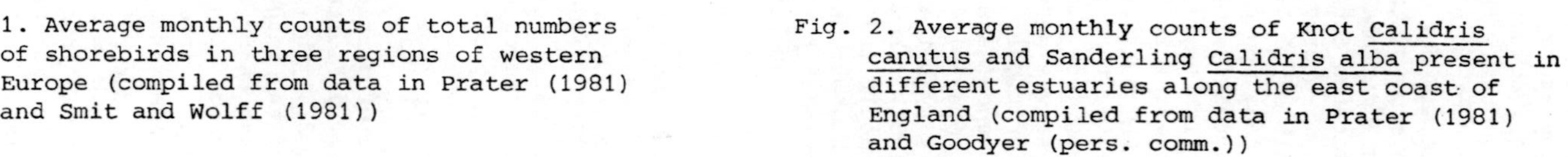

Fig. 2. Average monthly counts of Knot *Calidris canutus* and Sanderling *Calidris alba* present in different estuaries along the east coast of England (compiled from data in Prater (1981) and Goodyer (pers. comm.))

Seasonal changes in daily food requirements accentuate the patterns resulting from changes in bird numbers. Birds require food for maintenance of body temperature, synthesis and storage of new materials (such as fat and feathers), covering the costs of foraging methods, and for other activities, notably those involving flight e.g. to reach safe roost sites at high water or to escape predators. To some (unknown) extent, these requirements are not strictly additive, since energy used in locomotion can provide some of the heat needed to maintain body temperature. Also, some of the additional energy needed to maintain the slightly elevated body temperatures found in many bird species during moult may be offset by a reduction in energy used for flight activity at this time. In most waders, moult of all their plumage takes place in late summer and early autumn, whereas moult of only the body feathers occurs in early spring. Migration normally follows moult at both seasons. In preparation for prolonged flights, fat is stored as fuel. The quantities of fat deposited are often greater in spring than in autumn, as the distances flown, non-stop, from the north temperate zone to the arctic breeding grounds are usually greater than the stages flown on migration southwards, after moult in the autumn (see e.g. Davidson and Evans, in press). Waders also lay down fat reserves during the late autumn, once they have reached the end of their migrations. (These reserves are maintained through the winter as an 'insurance' against short periods of difficulty in obtaining adequate daily food intakes). The amounts of fat stored by different species before the winter vary but are always less than those accumulated before migration (Davidson 1981); they are also stored over much longer periods. Thus the periods of moult and preparation for migration in spring and autumn are the seasons of highest daily food intake and therefore also of greatest daily predation by each bird on the intertidal fauna. It will be noted that these are also the seasons of greatest shorebird abundance on the intertidal areas of the Danish Wadden Sea, but not around the British Isles (Fig. 1). Severe weather in some winters hardly alters shorebird distributions.

Although the total numbersof waders tend to be highest in Britain in mid-winter, seasonal changes in numbers of different individual species can differ markedly from the overall pattern. Table 1 lists some of the commoner species, together with their periods of greatest abundance (taken from Prater 1981). Because the daily energy requirements of species probably vary approximately in proportion to the 0.75 power of body mass (Lavigne 1982), seasonal changes in relative proportions of the different wader species present in intertidal areas can cause profound changes in the total daily food requirements of the shorebird community.

Numbers of particular bird species on particular British estuaries do not necessarily fluctuate in parallel with the trends for the country as a whole (Fig. 2). This implies that seasonal patterns of intensity of predation on the intertidal fauna are specific to each estuary. Because overall predation by shorebirds can remove considerable proportions of the annual production (Baird et al 1985) and standing crops (Goss-Custard 1980) of several invertebrate species, the possibility exists that any behavioural adaptations to avoid shorebird predation, shown by a particular prey species, may differ in seasonal timing or in extent in different estuaries. Such geographical differences in anti-predator behaviour are much less likely in coastal invertebrates where genetic isolation between populations is probably incomplete, but it might occur in the first five species in Table 3, i.e. those normally found only in estuaries. Another feature that could lead to differences in anti-predator adaptations amongst different estuarine populations of a prey species is the flexibility of diet of many shorebird species, each of which tends to feed on whatever prey is most abundant and available to its particular method of feeding in a particular place. Two examples are given in Table 2 and others by Pienkowski (1981).

Table 1. Seasonal changes in abundance of the commonest shorebirds on British coasts[a].

Species	Months of greatest abundance	Range of body mass (g)
[b]Curlew *Numenius arquata*	8-10; 2	700-1000
Oystercatcher *Haematopus ostralegus**	9-2	500-800
Black-tailed Godwit *Limosa limosa*	9-4	250-500
Bar-tailed Godwit *Limosa lapponica*	11-2	250-450
Grey Plover *Pluvialis squatarola*	9-2	200-300
Knot *Calidris canutus*	11-2	100-200
[b]Redshank *Tringa totanus*	9-12	120-180
Turnstone *Arenaria interpres**	8-3	90-180
Purple Sandpiper *Calidris maritima**	11-4	50-100
Ringed Plover *Charadrius hiaticula*	8-9; 5	50-90
Sanderling *Calidris alba*	7-8; 5	50-100
Dunlin *Calidris alpina*	11-2	40-80

[a]Based on counts on British estuaries (Prater 1981), but representative also of main periods of use of rocky intertidal areas by those species that also use them (indicated by an asterisk)

[b]Data from intertidal areas only. Both species make extensive use of coastal and inland fields for feeding in mild winters (Townshend 1981).

Table 2. Main prey species of Knot *Calidris canutus* and Bar-tailed Godwit *Limosa lapponica* in different geographical areas

Estuary:	Wash	Tees	Morecambe	Wattenmeer
Knot prey:	Macoma	Mytilus Hydrobia	Mytilus Macoma	Macoma Littorina Hydrobia
Estuary:	**Wash**	**Tees**	**Lindisfarne**	**Mauritania**
Bar-tailed Godwit prey:	Arenicola Lanice	Nereis	Arenicola Scoloplos Notomastus	Polychaetes Abra

In Table 3 are listed some of the main shorebird predators of the more abundant intertidal invertebrates. Although several shorebirds take the same prey species, each species of predator tends to take a different size range of the prey, in accordance with profitability theory (see e.g. Zwarts and Wanink 1984 for detailed analysis of predation on *Mya arenaria* by Oystercatcher and Curlew). The outcome of such behaviour is that small shorebird species, i.e. those towards the bottom of Table 1, tend to take younger age-classes of a particular prey than do larger shorebird species. This is reflected in the summary in Table 3, but was treated in more detail by Evans et al (1979) and Zwarts and Wanink (1984).

Table 3. Known main shorebird predators of the more abundant intertidal invertebrates in western Europe

Species	(Age)	Shorebird predators
Hydrobia ulvae		Dunlin, Knot, Redshank, Grey Plover
Macoma balthica	0-1	Dunlin, Knot
	1+	Knot, Bar-tailed Godwit, Oystercatcher
Scrobicularia plana	1+	Oystercatcher, Curlew
Nereis diversicolor	0-1	Dunlin, Knot, Redshank
	1+	Grey Plover, Redshank, Avocet[a] Bar-tailed Godwit, Curlew
Corophium volutator		Dunlin, Redshank
Littorinids		Turnstone, Purple Sandpiper, Knot, Oystercatcher, Curlew
Patella vulgaris		Oystercatcher
Mytilus edulis	0-1	Sanderling, Turnstone, Knot
	1+	Knot, Oystercatcher, Curlew
Mya arenaria	0-1	Knot
	1+	Oystercatcher, Curlew
Cerastoderma edule	0-1	Knot
	1+	Oystercatcher, Curlew
Arenicola marina	0-1	Ringed Plover
	1+	Grey Plover, Bar-tailed Godwit, Oystercatcher, Curlew
Nereis virens		Grey Plover, Curlew
Nephtys spp.	1+	Redshank, Curlew
Nerine cirratulus		Sanderling, Ringed Plover
Small polychaetes: Heteromastus, Notomastus, Manayunkia, Scoloplos		Dunlin, Ringed Plover, Grey Plover, Bar-tailed Godwit
Small oligochaetes: Tubifex		Dunlin
Carcinus maenas		Redshank, Greenshank[b], Oystercatcher, Bar-tailed Godwit, Curlew
Crangon crangon		Redshank, Greenshank
Talitrus saltator		Dunlin, Ringed Plover
Bathyporeia spp		Sanderling, Ringed Plover
Haustorius arenarius		Sanderling, Ringed Plover

[a]Avocet Recurvirostra avosetta
[b]Greenshank Tringa nebularia

Although shorebirds eat large quantities of many of the invertebrates listed in Table 3, some prey are made available by adverse weather conditions which would have caused their deaths in any case, e.g. wrecks of Hydrobia and of spat of Mytilus and Macoma, following heavy seas; or the washing away by spring tides of banks of wrack deposited on the previous series, thereby making available to predators such species as Talitrus, and Coelopid spp. larvae, that normally lie buried safely within the decaying seaweed banks. Furthermore, shorebirds do not always kill their prey, e.g. Bar-tailed Godwits always extract Arenicola tail-first from their burrows and frequently obtain only the tail portion of the worm (Smith 1975); presumably each head portion regenerates another tail.

As mentioned earlier, shorebird numbers and daily demand are highest during the periods of autumn and spring migration in many estuaries. Dugan (1981) examined whether any correlations exist between the times of use of estuaries by different shorebirds and the periods of maximum densities, reproduction and growth of the five common invertebrate species listed at the head of Table 3. He was searching for evidence that shorebirds might time their spring migrations to follow periods of renewed growth of their invertebrate prey after the winter, but he found no consistent relationships. In North America, Schneider (1981) has argued that the timing of northward movement of birds follows a spring bloom of littoral benthos production, but this does not apply in Europe. Many of the larger shorebird species leave the British Isles in March to go to the Wadden Sea areas, several months before densities of invertebrates rise in either of these geographical regions (Dugan 1981, Prokosch 1984). The timings of spring migration of the different breeding populations and species of arctic nesting shorebirds are highly predictable from year to year, and almost certainly related to the earliest dates that the high latitude breeding grounds can be occupied (Pienkowski and Evans 1984). Thus most of the population of Knots breeding in Greenland and arctic Canada utilize the German Wattenmeer in the month of April,whilst preparing for migration, whereas most of those breeding in Siberia (birds which have spent the winter in W. Africa) utilize the same general area as a migration staging post later in April and in May. Clearly the later migration of the Siberian population through Germany could not be adapted to the same temporal features of the prey populations as the earlier migration of the Greenland birds, and it seems probable that neither timing is actually influenced by characteristics of the prey.

Further evidence that the timing of shorebird migration in spring is not controlled by the timing of any increase in prey size, abundance or biomass density is provided by the observations that Turnstone achieve the increased rates of prey intake, necessary for storage of fat reserves before migration, only by decreasing the proportion of time spent in vigilance during foraging (Metcalfe and Furness 1984) and that Sanderling increase the total duration of foraging during each day to achieve the same result (Gudmundsson 1986).

The timing of autumn migration of arctic-breeding shorebirds is much less consistent from year to year and depends primarily on the success or otherwise of breeding attempts. Whatever the outcome, adults of one sex normally leave before those of the other. Several suggestions have been advanced regarding the adaptive nature of this behaviour. These have been reviewed by Myers (1981) and include the possibility that the chance of survival of those adults that reach migration staging posts early is enhanced, because substantial reductions in densities of invertebrate prey occur eventually at these sites, apparently as a result of shorebird predation (Schneider and Harrington 1981). However, it is not known whether these reductions in prey density depress food intake rate of the later arrivals, as would be necessary if there were to be a selective advantage to early migration from the arctic.

Even if the timings of shorebird migrations are not adapted to particular temporal features of invertebrate life-cycles, the routes used may well be. The most spectacular example known so far is the concentration of migrant birds in Delaware Bay each May, coinciding with the reproduction there of the horseshoe crab _Limulus_. More subtle explanations of choice of migration staging posts may eventually be discovered, perhaps relating to changes in the relative availability of prey at different sites. Many polychaetes are known to become more active and to bury less deeply in the substratum as temperatures rise in spring and this renders them more vulnerable to predation (Pienkowski 1983a). Thus the densities of available prey rise, even though absolute densities continue to fall.

Some intertidal invertebrates may become more vulnerable to predation during reproduction, e.g. by coming to the surface of soft sediments. It is not known to what extent shorebirds capitalize on this behaviour and whether predation by birds might have exerted any selective effect on the timing of reproduction, or its synchrony, in each invertebrate prey species. Indeed, it is not clear whether predation by shorebirds on pre-reproductive animals is ever sufficiently intense to depress the reproductive output, recruitment and subsequent population density of the prey, or whether (as Reise 1986 has argued) omnivorous invertebrate predators of larval and newly settled life-stages exert over-riding effects on subsequent prey population levels.

Seasonal changes in zonation on the shore and in depth distribution within the substratum have usually been interpreted as adaptations to avoid freezing conditions. The possibility that either or both of these phenomena might also be adaptations to avoidance of predation has not received much attention. In an earlier paper (Evans 1979), I suggested that the seasonal changes in burying depth of *Macoma balthica* described from the Wash in eastern England by Reading and McGrorty (1978) might be an adaptation to minimize predation risk. This hypothesis was tested by Dugan (1981) who examined seasonal changes in burying depth on other estuaries, notably the Humber and Morecambe Bay (Fig. 3). He found that, although there were considerable differences in behaviour of *Macoma* between estuaries, these might be related to avoidance of predation only in autumn and not in spring. Fig. 3 shows that *Macoma* buried deeper later in the autumn on the Humber than on the Wash, in parallel with the later arrival of Knot on the Humber (Fig. 2). In spring, numbers of Knot decrease slightly later than the months in which *Macoma* begin to move towards the sediment surface on the Humber and the Wash. However, on Morecambe Bay, numbers of Knot remain very high until May, so the upward movements of *Macoma* there (Fig. 3) cannot be attributed to cessation of predation. It is thus uncertain whether seasonal changes in invertebrate behaviour are influenced by seasonal changes in risk of predation by shorebirds, or whether they represent a compromise merely between responses to seasonal changes in abiotic factors and feeding opportunities.

TIDAL RHYTHMS OF PREDATION BY SHOREBIRDS

Because shorebirds, unlike ducks, never dive to obtain their invertebrate prey, it is not surprising to find that most of their foraging is restricted to those times of tide when at least some of the intertidal is exposed. Rhythms of shorebird foraging have been reviewed at some length by Burger (1984) and only certain aspects will be highlighted here.

On soft shores, at least by day, many wader species feed while following the tide-edge as it retreats and subsequently returns, in order to remain in the zones where prey activity is highest, prey are closest to the surface, or both (Evans 1979). Because the tide-edge moves most rapidly at around the mid-tide period, Reise (1986) assumed that invertebrates were less at risk of capture in the mid-tidal zones, and at greater risk towards the high and low water marks where the tide advances and retreats more slowly. This is a considerable over-simplification; different shorebird species differ in the extents to which they use particular intertidal zones, both spatially and temporally, as will

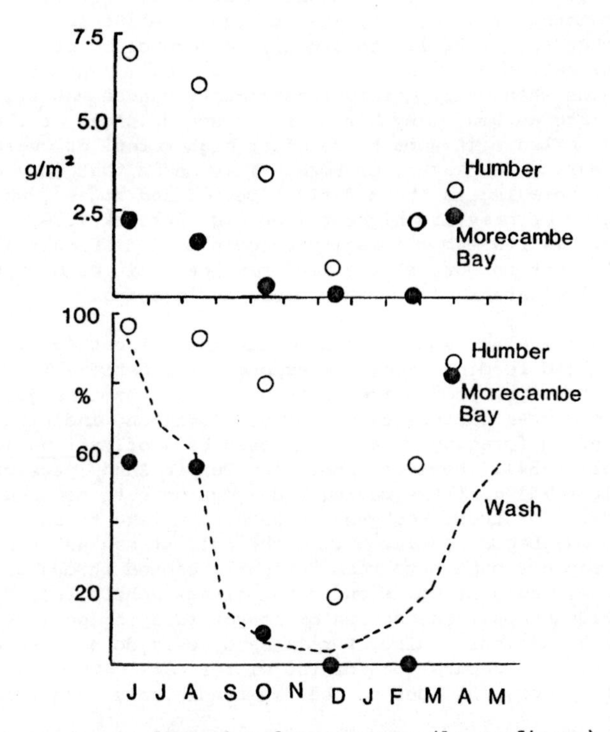

Fig. 3. Biomass (upper figure) and percentage (lower figure) of *Macoma balthica* of sizes 9-13mm, preferred by Knot *Calidris canutus*, in mud at depths of less than 33mm in different months in three British estuaries (redrawn from Dugan (1981) and Reading and McGrorty (1978)).

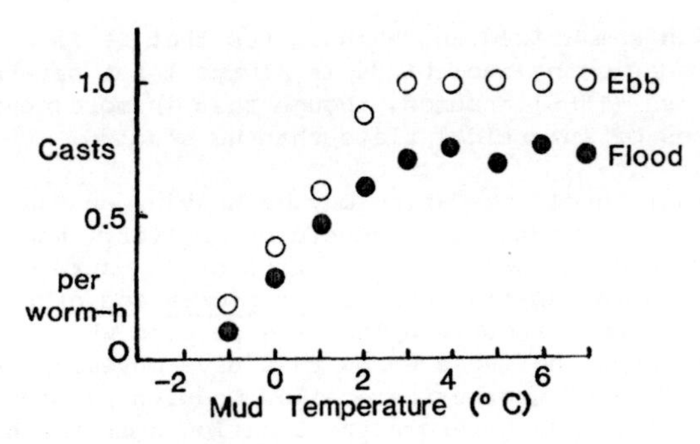

Fig. 4. The frequency of cast formation by the lugworm *Arenicola marina* on the ebb and flood tides, at different substrate temperatures (redrawn from Smith (1975)).

be shown. Below mid-tide level, certain birds that defend feeding territories, e.g. some Curlew and Grey Plover, are left behind by the retreating tide (Townshend 1982, Townshend et al 1984). Indeed, many plovers (whether territorial or not) stay to feed, well spaced-out, at mid- or even higher shore levels, to avoid interference by other birds with their visual method of foraging (Pienkowski 1981). Furthermore, during those months when daily energy requirements can be satisfied most easily, e.g. in late autumn, many birds either spend longer at the high water roost (and so leave it some hours after high water) or cease feeding for a few hours during low water, or both. This means that they often concentrate their foraging in the mid-tidal period and zones , which is often where the densities of prey are highest (see e.g. Puttick 1979). In winter, when daily demands increase but foraging becomes more difficult as temperatures fall, shorebirds on soft shores feed for longer in each daytime tidal cycle and eliminate any low-water pause in foraging.

On rocky shores also, some waders follow the retreating tide only until their preferred feeding areas are exposed at mid-tidal level or slightly below. Most Turnstones are faithful to such home ranges for many months during the winter (Metcalfe and Furness 1985) and individual Oystercatchers defend foraging areas on exposed beds of *Mytilus* (Goss-Custard and Durell 1984). However, Knots and Purple Sandpipers often concentrate their feeding at the moving tide-edge on both the ebb and flood of the tide. For these rocky-shore waders, unlike those on soft shores, foraging may become easier during the coldest months, since bivalves such as *Mytilus* may not manage to remain tightly closed throughout their period of exposure, and gastropods on rocks perhaps adhere less firmly. There is less evidence on rocky shores of longer foraging periods by waders in colder conditions. Also, Purple Sandpipers do not increase their body mass in autumn in preparation for the winter (Atkinson et al 1981) which suggests that foraging does not become predictably more difficult at that season.

In general, larger species of shorebirds forage for lower proportions of each period of exposure of their intertidal feeding grounds (Evans 1981). This arises in part from their need to take larger prey items in order to feed profitably, these larger prey tending to occur only at lower tidal levels. Thus large shorebirds often leave their high water roosts slightly later than do small species, and regularly fly to roost earlier than the smaller waders.

The information summarized above indicates that it is not possible to conclude that predation by shorebirds is always least severe at mid-tidal levels, as Reise (1986) assumed, though this is more nearly true on the large expanses of intertidal flats that he studied.

At a given tidal level, predation may be heavier on the ebb than the flood tide. This may be due to lower levels of activity, and thus availability, of the prey after a longer period of exposure (Fig. 4); but it may also result, in the special case of *Arenicola* and other cast-forming polychaetes, from a reduction in the ease with which shorebirds can detect the cues revealing those worms that have moved close to the substrate surface. To test whether Bar-tailed Godwits have more difficulty in detecting a new cast, and therefore the location near the surface of an *Arenicola*, on a substratum already covered by casts, Smith (1975) removed casts from stretches of beach shortly before they were reflooded by the tide. He then compared the foraging successes of the Godwits there with those on the same stretches of beach one day later, at the same stage of the tide, but with the casts formed during the ebb tide left intact. Both male and female godwits made significantly more successful attempts to capture lugworms on the simulated ebb (♂ 73%, ♀ 75%) than on the natural flood tide (♂ 70%, ♀ 71%). This does not fully account for the difference Smith found in capture rates per minute of Arenicola at equivalent tidal levels on the ebb (1.8-1.9) and flood (1.2), but these measurements were made in a site where Arenicola densities (and therefore the potentially confusing effect of previously formed casts) were considerably higher than on the experimental sites. Information on differences in predation rate, at the same tidal level, between the ebb and flood tides is scarce for other shorebird/invertebrate interactions. Puttick (1979) noted that Curlew Sandpipers Calidris ferruginea fed faster on the flood than ebb tide, but did not determine why. It is therefore impossible to assess whether the frequency of activity bouts by different prey species during exposure of the intertidal might be adapted in any way to minimize risks of being taken by shorebirds.

Predation by waders on the shore fauna during the high water period occurs only to a limited extent. Sanderling is the only species that feeds at all regularly in Europe on strand lines of dried, buried, seaweeds, taking Talitrus and wrack-flies. However, during migration periods, when birds are replenishing fat reserves, many smaller wader species feed on the invertebrate fauna of decaying damp wrack-beds at high water. Turnstone also continue to feed during migration periods over the high waters of neap tides, when some prey are still accessible on rocky shores. Outside the migration seasons, Turnstone, Sanderling, Oystercatchers and Knot often feed along the tide-line at high water after storms, when large Mytilus or their spat have been torn from subtidal rocks and washed ashore.

Whether shorebird predation has in any way shaped the present-day zonation patterns of sedentary intertidal invertebrates is not certain. But clearly the upshore/downshore movements (see e.g. McLachlan et al 1979) of some of the fauna of sandy beaches during each tidal cycle, under the cover of water, provides an effective method of outwitting predatory waders.

NOCTURNAL PREDATION BY SHOREBIRDS IN THE INTERTIDAL

Most shorebird species have been recorded at night on feeding grounds that they are known to use by day. This is true even of plovers that are known to forage visually (Pienkowski 1983). The question arises as to the relative impact of predation by night and by day on the intertidal fauna. Pienkowski (1982) estimated that in mid-winter, Ringed and Grey Plovers obtain more than half of their daily food requirements by night at Lindisfarne, northern England, and subsequent studies by Wood (1984) also support a similar conclusion for Grey Plovers at Teesmouth, also in northern England. Wood was able to measure the time-budget of the same individual bird on its territory by day and by night, and found no significant differences in the total time spent foraging and in the bird's peck rate in these two periods. Other workers have observed or assumed that the rate of prey intake tends to be less at night, but it is very difficult to ascertain what proportion of attempts to capture prey are successful at night (see e.g. Hulscher 1974, 1976). What is clear is that shorebirds prefer to obtain their food requirements by day if possible, but that, particularly during the short and cold days of mid-winter, this is often impossible and night-feeding is undertaken regularly, even in the absence of moonlight (Pienkowski et al 1984). It is not known to what extent night feeding occurs at more southerly latitudes.

It is well-known that several Crustacea of sandy beaches become more active at night (e.g. Eurydice pulchra, Jones and Naylor 1970; Talitrus saltator, Geppetti and Tongiorgi 1967). Changes from day to night in the activity patterns at the substrate surface of Corophium volutator,

Hydrobia ulvae, Nereis virens and N. diversicolor were reported by Dugan (1981a) and of other polychaetes by Pienkowski (1983a). Thus it would be wrong to assume that the invertebrate prey taken by shorebirds at night are exactly the same as those known to be taken by the same species by day, even in the same site. Differences might occur in the prey spectrum, the relative proportions of each prey taken, the sizes taken, or all three. These could arise not only through changes in the relative availabilities of prey, but also through changes in the foraging techniques used by the birds, particularly the change by sandpipers from use of a mixture of cues to locate prey by day to the use of tactile foraging by night. Additionally, some individual predators (radio-tagged Grey Plovers) have been found to feed in different sites on an estuary by day and by night (Dugan 1981a, Wood 1986) which could lead to a change in prey taken.

The concentration of activity into the hours of darkness by many invertebrates of the intertidal zone has often been interpreted as an anti-predator adaptation, and this seems a reasonable conclusion in general terms. However, a great deal more needs to be known about the sensory capabilities of different shorebirds; their quantitative success in nocturnal foraging; and seasonal and weather-induced changes in the nocturnal behaviour of the invertebrates. Only when these have been established will it be possible to evaluate the importance of predation in the maintenance of the daily (and nightly) rhythms of behaviour in intertidal invertebrates that we see today.

CONCLUSION

Reise (1986) points out that shorebirds and aquatic predators of tidal flat invertebrates are unlikely to adapt to each other to minimize prey overlap because they are separated by the tidal cycle and because most of the bird populations are migratory. He also suggests that 'no close adaptive reciprocal coherence' is to be expected between predators and prey, that coastal birds will not adjust their population size to the local strength of prey populations to avoid over-exploitation, and that tidal flats differ from rocky shores in that predation is relatively unimportant in the latter. It is clear, however, that the timing of predation by shorebirds is closely matched to exploit prey behaviour during the tidal cycle and by day and night, and that at least some aspects of prey behaviour minimize predation risk. It is also clear that the mobility of birds gives them an opportunity to adjust the numbers that utilize a given coast or estuary to the density of available prey in each site, on a seasonal basis, even though this process may be over-ridden during spring by the need for rather precise timings of migrations to arctic breeding grounds.

REFERENCES

Atkinson, N.K., Summers, R.W., Nicoll, M. and Greenwood, J.J.D., 1981, Populations, movements and biometrics of the Purple Sandpiper Calidris maritima in eastern Scotland, Ornis Scand., 12:18.

Baird, D., Evans, P.R., Milne, H. and Pienkowski, M.W., 1985, Utilization by shorebirds of benthic invertebrate production in intertidal areas, Oceanogr. Mar. Biol. Ann. Rev., 23:573.

Boere, G.C., 1976, The significance of the Dutch Waddenzee in the annual life cycle of arctic, subarctic and boreal waders, Ardea, 64:210.

Burger, J. 1984, Abiotic factors affecting migrant shorebirds, in "Shorebirds: migratory and foraging behavior", J. Burger and B.L. Olla, eds, Plenum Press, New York.

Davidson, N.C., 1981, Seasonal changes in the nutritional condition of

shorebirds (Charadrii) during the non-breeding season, Unpublished Ph.D. thesis, University of Durham, U.K.
Davidson, N.C. and Evans, P.R., in press, Pre-breeding accumulation of fat and muscle protein by arctic-breeding shorebirds, Proc. 19th Int. Ornithol. Congress.
Dugan, P.J., 1981, Seasonal movements of shorebirds in relation to spacing behaviour and prey availability, Unpublished Ph.D. thesis, University of Durham, U.K.
Dugan, P.J., 1981a, The importance of nocturnal feeding in shorebirds: a consequence of increased invertebrate prey activity, in "Feeding and survival strategies of estuarine organisms", N.V. Jones and W.J. Wolff, eds., Plenum Press, New York.
Evans, P.R., 1979, Adaptations shown by foraging shorebirds to cyclical variations in the activity and availability of their intertidal invertebrate prey, in "Cyclic Phenomena in Marine Plants and Animals", E. Naylor and R.G. Hartnoll, eds., Pergamon Press, Oxford.
Evans, P.R., 1981, Reclamation of intertidal land: some effects on shelduck and wader populations in the Tees estuary, Verhandlung Ornithol. Ges. Bayern, 23:147.
Evans, P.R., Herdson, D.M., Knights, P.J. and Pienkowski, M.W., 1979, Short-term effects of reclamation of part of Seal Sands, Teesmouth, on wintering waders and shelduck, Oecologia, Berlin, 41:183.
Geppetti, L. and Tongiorgi, P., 1967, Nocturnal migrations of Talitrus saltator (Crustacea Amphipoda), Monitore Zool. Ital. (N.S.), 1:37.
Gerritsen, A.F.C., van Heezik, Y.M. and Swennen, C., 1983, Chemoreception in two further Calidris species, with a comparison of the relative importance of chemoreception during foraging in Calidris species, Netherlands J. Zool., 33:485.
Goss-Custard, J.D., 1980, Competition for food and interference amongst waders, Ardea, 68:31.
Goss-Custard, J.D. and Durrell, S.E.A. le V, 1984, Feeding ecology, winter mortality and the population dynamics of Oystercatchers on the Exe estuary, in "Coastal Waders and Wildfowl in Winter", P.R. Evans, J.D. Goss-Custard and W.G. Hale, eds., Cambridge University Press, Cambridge.
Gudmundsson, G.A., 1986, Aspects of pre-migratory feeding ecology of Sanderling Calidris alba at Teesmouth, NE England, Unpublished M.Sc. dissertation, University of Durham, U.K.
Hulscher, J.B., 1974, An experimental study of the food intake of the Oystercatcher Haematopus ostralegus L. in captivity during the summer, Ardea, 62:155.
Hulscher, J.B., 1976, Localisation of cockles (Cardium edule L.) by Oystercatchers (Haematopus ostralegus L.) in darkness and daylight, Ardea 64:292.
Jones, D.A. and Naylor, E., 1970, The swimming rhythm of the sandbeach isopod Eurydice pulchra, J. exp. mar. Biol. Ecol., 4:188.
Lange, G., 1968, Uber Nahrung, Nahrungsaufnahme und Verdaungstrakt mitteleuropäischer Limikolen, Beitr. Vogelkunde, 13:225.
Lavigne, D.M., 1982, Similarity in energy budgets of animal populations, J. Anim. Ecol., 51:195.
McLachlan, A., Wooldridge, T., and van der Horst, G., 1979, Tidal movements of the macrofauna on an exposed sandy beach in South Africa. J. Zool., Lond., 187:433.
Metcalfe, N.B. and Furness, R.W., 1984, Changing priorities: the effect of pre-migratory fattening on the trade-off between foraging and vigilance. Behav. Ecol. Sociobiol., 15:203.
Metcalfe, N.B. and Furness, R.W.,1985, Survival, winter population stability and site fidelity in the Turnstone Arenaria interpres, Bird Study, 32:207.
Morrison, R.I.G., 1984, Migration systems of some New World shorebirds, in "Shorebirds: migration and foraging behavior", J. Burger and

B.L. Olla, eds., Plenum Press, New York.
Myers, J.P., 1981, Cross-seasonal interactions in the evolution of sandpiper social systems, Behav. Ecol. Sociobiol., 8:195.
Pienkowski, M.W., 1981, Differences in habitat requirements and distribution patterns of plovers and sandpipers as investigated by studies of feeding behaviour, Verhandlung Ornithol. Ges. Bayern, 23:105.
Pienkowski, M.W., 1982, Diet and energy intake of Grey and Ringed Plovers, Pluvialis squatarola and Charadrius hiaticula, in the non-breeding season, J. Zool., Lond., 197:511.
Pienkowski, M.W., 1983, Changes in the foraging pattern of plovers in relation to environmental factors, Anim. Behav., 31:244.
Pienkowski, M.W., 1983a, Surface activity of some intertidal invertebrates in relation to temperature and the foraging behaviour of their shorebird predators. Mar. Ecol. Prog. Ser. 11:141.
Pienkowski, M.W. and Evans, P.R., 1984, Migratory behavior of shorebirds in the western Palaearctic, in "Shorebirds: migratory and foraging behavior", J. Burger and B.L. Olla, eds., Plenum Press, New York.
Pienkowski, M.W., Ferns, P.N., Davidson, N.C. and Worrall, D.H., 1984, Balancing the budget: problems in measuring the energy intake and requirements of shorebirds in the field, in "Coastal Waders and Wildfowl in Winter", P.R. Evans, J.D. Goss-Custard and W.G. Hale, eds, Cambridge University Press, Cambridge.
Prater, A.J., 1981, "Estuary Birds of Britain and Ireland", Poyser, Calton, England.
Prokosch, P., 1984, The German Wadden Sea, in "Coastal Waders and Wildfowl in Winter", P.R. Evans, J.D. Goss-Custard and W.G. Hale, eds., Cambridge University Press, Cambridge.
Puttick, G., 1979, Foraging behaviour and activity budgets of Curlew Sandpipers, Ardea 67: 111.
Reading, C.J. and McGrorty, S., 1978, Seasonal variations in the burying depth of Macoma balthica (L.) and its accessibility to wading birds, Estuarine Coastal Mar. Sci., 6: 135.
Reise, K., 1986, "Tidal Flat Ecology", Springer Verlag, Berlin and Heidelberg.
Schneider, D.C., 1981, Food supplies and the phenology of migratory shorebirds: an hypothesis, Wader Study Group Bull., 33:43.
Schneider, D.C. and Harrington, B.A., 1981, Timing of shorebird migration in relation to prey depletion, Auk 98:801.
Smith, P.C., 1975, A study of the winter feeding ecology and behaviour of the Bar-tailed Godwit (Limosa lapponica), Unpublished Ph.D. thesis, University of Durham, U.K.
Smit, C.J. and Wolff, W.J. 1981, "Birds of the Wadden Sea", Wadden Sea Working Group Report 6, Balkema, Rotterdam.
Townshend, D.J., 1981, The importance of field feeding to the survival of wintering curlews Numenius arquata on the Tees estuary, in "Feeding and Survival Strategies of Estuarine Organisms", N.V. Jones and W.J. Wolff, eds, Plenum Press, New York.
Townshend, D.J., 1982, The use of intertidal habitats by shorebird populations, with special reference to Grey Plover Pluvialis squatarola and Curlew Numenius arquata, Unpublished Ph.D. thesis, University of Durham, U.K.
Townshend, D.J., Dugan, P.J. and Pienkowski, M.W., 1984, The unsociable plover: use of intertidal areas by Grey Plovers, in "Coastal Waders and Wildfowl in Winter", P.R. Evans, J.D. Goss-Custard and W.G. Hale, eds, Cambridge University Press, Cambridge.
Wood, A.G., 1984, Time and energy budgets of the Grey Plover Pluvialis squatarola at Teesmouth, Unpublished Ph.D. thesis, University of Durham, U.K.
Zwarts, L. and Wanink, J., 1984, How Oystercatchers and Curlews successively deplete clams, in "Coastal Waders and Wildfowl in Winter", P.R. Evans, J.D. Goss-Custard and W.G. Hale, eds, Cambridge University Press, Cambridge.

Zonal Recovering in Equatorial Sandhoppers: Interaction Between Magnetic and Solar Orientation

Leo Pardi, Ali Said Faqi (*), Alberto Ugolini
Felicita Scapini and Antonio Ercolini

University of Florence, Italy
(*) National University, Somalia

INTRODUCTION

An orienting mechanism based exclusively on a sun compass runs into considerable difficulties in the Tropics (see Braemer, 1960; Lindauer 1960; Ercolini, 1964) on account of the annual change in the sun's declination.

Circumequatorial populations of the sandhopper Talorchestia martensii Weber show a sufficiently correct zonal orientation along the sea-land axis, (Y axis), during both the clockwise and anti-clockwise path of the sun (Ercolini, 1964; Pardi and Ercolini, 1966). On a dry substrate many populations of circumequatorial sandhoppers, unlike mediterranean species, will usually orient seawards, but they may also orient landwards albeit less frequently (Ercolini, 1964; Pardi and Ercolini, 1966). This two-way escape direction is nevertheless perfectly compatible with a damp zone destination, since strands of wet detritus are left on the eulittoral due to the wide tidal range. Even young inexperienced sandhoppers, born in the laboratory and reared under artificial light, show a fairly correct orientation which coincides with that of their parents, the first time they are exposed to the sun (Ercolini, 1964; Pardi, 1967).

Pardi (1967) attempted to explain these results with the hypothesis that sandhoppers only used parametres they could gather from the sun, but this supposition was not backed up by experiments. Van der Bercken et al. (1967) observed that Talitrus saltator from the North Sea showed a non-visual and ecologically correct orientation, which Arendse (1980) and Arendse and Kruysvijk (1981) later identified as magnetic.

Thus the question arose whether magnetic orientation plays an equally important role in the zonal orientation of equatorial amphipods. Scapini and Ercolini (1973) were unable to demonstrate the existence of non-visual orientation in Talorchestia martensii from the analysis of results from group releases. On the other hand, many experiments proved that the sun and the moon had to play some part in the zonal orientation of the species, i.e.: 1) the Santschi mirror test gave at least partially positive results; 2) zenithal reaction: the sandhoppers are disorientated and heap together in groups when the sun (Ercolini, 1964) or moon (Pardi and Ercolini, 1965) pass close to the zenith; 3) at least partially positive results from LD cycle shifting experiments (Ercolini, 1964; Pardi et al., 1985); 4) a temporary increase in disorientation if thick cloud hides the sun or moon.

Recently, group trials in a dark room were performed on a Somalian population of sandhoppers, with the magnetic field as normal or deflected by 90°. Results revealed that equatorial adult sandhoppers also possess a magnetic orientation (Pardi et al., 1985), which enables them to assume a good orientation along the sea-land axis. Some ambiguous results (points 1 and 3) could be explained by admitting that these crustaceans are endowed with two compass mechanisms, which are normally synergetic. If these two mechanisms are forced to compete with each other, by changing the solar azimuth (point 1), temporal information (point 3) or (still in sunlight) magnetic information, some individuals will use one compass, some the other, and some alternate between the two.

It has been proved that the ability to assume an ecologically correct orientation is inherited in Mediterranean species of T. saltator (Pardi et al., 1958; Pardi, 1960; Pardi and Ercolini, 1986) and that it is based on a chronometrical sun compass (Marchionni, 1963; Scapini, 1986).

This study aims to 1) extend research to magnetic orientation in equatorial sandhoppers by testing individuals from different populations; 2) establish whether inexperienced young sandhoppers (born in the laboratory) possess the capacity for magnetic orientation; and 3) establish, by means of LD phase shifting experiments, whether the correct orientation shown by inexperienced young sandhoppers (already proved) is based exclusively on a sun compass, a magnetic compass or both.

1. MATERIALS AND METHODS

Three populations of T. martensii were used: A) Mogadishu: theoretical escape direction towards the sea=TED=145°, TED for land=325°; B) Sar Uanle: TED (sea)=132°, TED (land)=312°; C) Ras Mntoni: TED (sea)=15°, TED (land)=195°.

The Mogadishu population was kept in the laboratory on a 12:12 LD cycle in phase with the natural cycle, and tested the day after collection. Ras Mntoni's individuals were kept in the laboratory under an artificial 12:12 LD cycle and tested 20 days after collection. The Sar Uanle population was sifted to select males and females over a certain size (adults) which were kept for breeding. Part of this population was kept under an artifical LD cycle in phase with natural conditions (sunrise at 6.00 hours), and the rest (whose internal clock was set forward) were given a 12:12 LD cycle three hours ahead of normal time (sunrise at 3.00 hours). The inexperienced offspring born from these adults were kept under the same conditions as their parents. The adults were tested 26 days after collection, and the young 7-25 days after hatching.

Recordings were made by testing single specimens in a glass bowl (see Pardi and Papi, 1953), and the directions of the first 60 positions each individual assumed were measured by a goniometer from below. Each trial lasted about 7 minutes. Error in readings can be estimated to be in the order of ±2.5°. The radial and non-radial position of the longitudinal body axis was taken into account during readings. "Radiality" is given by the frequency percentage with which radial positions are assumed in the 60 readings.

All experiments were carried out at Mogadishu (Somali Republic) at the "Centro di Faunistica ed Ecologia Tropicali del CNR", in January and February, 1987. The adults and the young born in the laboratory were released in the dark room in a bowl with a light of about 11000 lux centred directly over the middle (zenithal light). Two minutes were allowed to elapse after the adults had been introduced into the bowl before the first readings were taken, and one minute for the young sandhoppers. Readings were taken every 5 seconds.

A first series of tests (1 day) was carried out with the normal magnetic field (i.e. leaving the Helmoltz spool switched off), and a second with the magnetic field deflected (Magnetic North to the West) but kept at the same intensity as the magnetic field at that spot. In the experiments under the sun the sandhoppers were tested with the natural geomagnetic field and the bowl was screened to hide the surrounding landscape from view. The period between introducing each sandhopper into the bowl and taking the first reading was reduced to one minute, and for young sandhoppers the interval between one reading and the next to 3 seconds. Controls and experimentals (clock shifted) were tested simultaneously in two identical apparatus set next to each other. In the experiment on the Mogadishu population, results for the controls tested in sunlight were taken at random from an extensive series of trials for studies on bidirectionality.

Statistical analysis was by the circular statistics methods proposed by Batschelet (1965; 1981). Every release gives the distribution of the positions for each individual in the order they were assumed. We call this the "individual distribution" (IND). Rao's test was applied to test if the INDs differ statistically from randomness. We used an automatic calculus to determine uni or plurimodality in the distributions, as used for calculating the first to fourth trigonometric moment (duplication-quadruplication of angles), thus leaving the decision, based on mean vector length for all four possibilities, to the computer. The mean angles for INDs in the same series of trials give another distribution which we shall call "interindividual distribution" (INT). Given the remarkable clarity of our results and the difficulties involved in a statistical analysis of the second order (see Batschelet, 1978; 1981) without loss of information on individual behaviour so essential in this type of study, we preferred to give only the frequencies for IND mean angles in the quadrants corresponding to the X and Y axes.

2. RESULTS

2.1. Experiments in the dark room: zenithal artificial light

Natural geomagnetic field: adults from the Mogadishu population (Fig. 1A-E). INT show a marked tendency for the Y axis (sea-land) but no preference for either of the two opposite (sea-land) directions. A few resultants lie along the X axis. 9 out of 35 individual distributions are statistically non-significant (open symbols). Clustering in the sea and land quadrants is mostly due to some individuals changing from one direction to the other (Fig. 1C). 9 individuals (Fig. 1B) seem to prefer one direction only, others behave in more than one way (Fig. 1D,E). One individual shows a significant tetramodal distribution (Fig. 1E). Radiality varies from 11% to 100%.

Artificial deflected geomagnetic field (magnetic North to the West): adults from the Mogadishu population (Fig. 2A-E). The effect of the deflected magnetic field is exactly as expected: the mean angles of the INDs are deflected by about 90° with respect to those with the natural magnetic field in the dark room (Fig. 1A) and the majority fall into quadrants I and III. There is no preference for either of the two directions along the deflected Y axis, and just the slightest sign of transversal orientation, as occurred with the natural magnetic

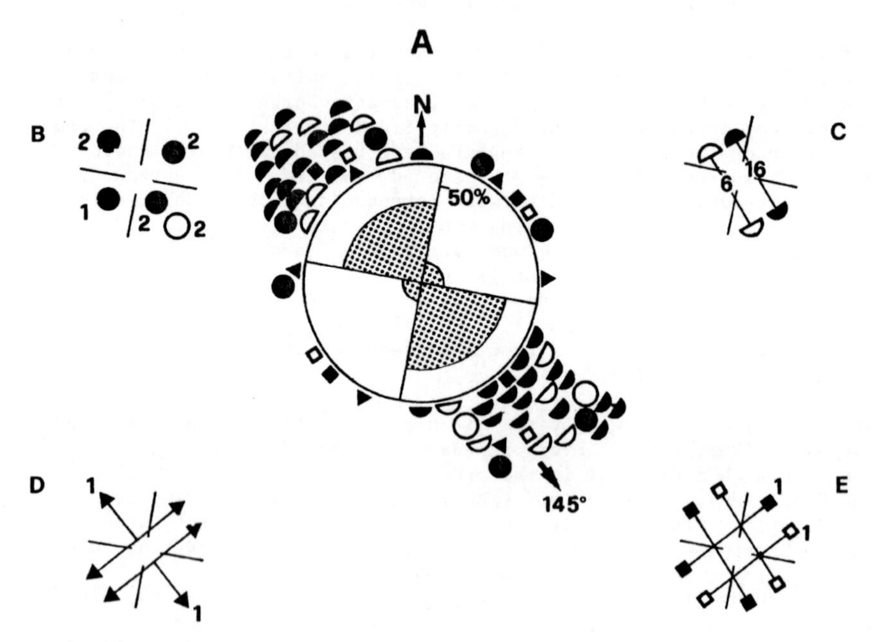

Fig. 1. Adults from Mogadishu population. Artificial zenithal light in the dark room. A) interindividual distributions. Each symbol represents the resultant direction for each individual distribution. Dots, unimodal; half dots, bimodal; triangles trimodal; squares, tetramodal individual distributions. Open and filled symbols, non-significant and significant individual distributions respectively. The percentage of mean directions of individual distributions falling in each quadrant is given for graphical convenience. B-E, uni-tetramodal individual distributions. The number for each type of individual distribution is also given. The arrows outside the distribution represent the North and the TED.

field. INDs were non-significant for 4 sandhoppers out of 35. In 11 sandhoppers the resultant for INDs fell within quadrants I or III (one fell into the second quadrant) (Fig. 2B). In all cases of bimodal distribution, the resultant fell within quadrant I and III, as expected (Fig. 2C). Only the choices of 1 individual resulted in trimodal distribution (Fig. 2D) and 4 in tetramodal distributions (Fig. 2E). Radiality varies from 36% to 100%.

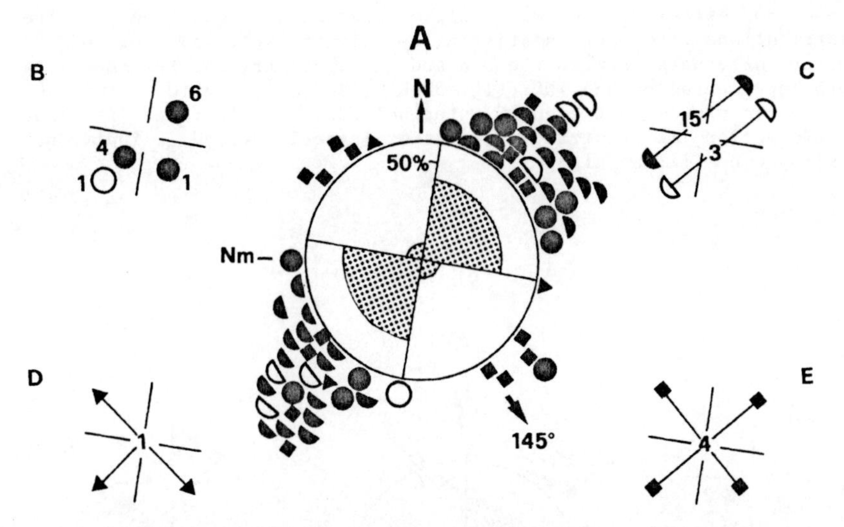

Fig. 2. Adults from Mogadishu population. Artifical zenithal light in dark room with artificially deflected geomagnetic field (Nm). For further explanations see Fig. 1.

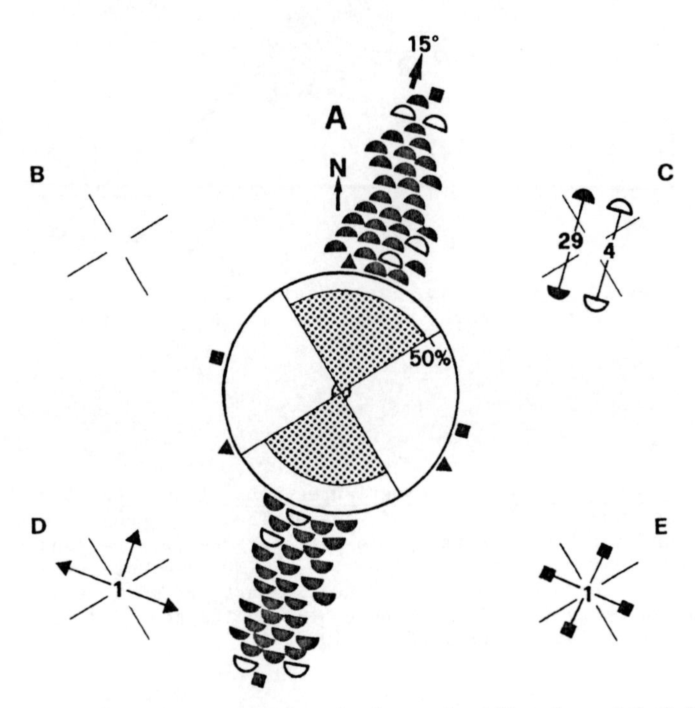

Fig. 3. Adults from Ras Mntoni population. Artificial zenithal light in dark room. For further explanations see Fig. 1.

Natural geomagnetic field: adults from the Ras Mntoni population (Fig. 3A-E). The INT is very clustered round the sea-land axis, with no preference for either of the two possible directions. Only 4 out of the 35 distributions are not statistically significant. 33 out of 35 sandhoppers alternate between the sea and the land: tri and tetramodality are both represented by one IND (Fig. 3D,E). Radiality varies from 10% to 100%, which proves that even if interindividual variability is low, when information is restricted to the magnetic field, individual variability can still be high.

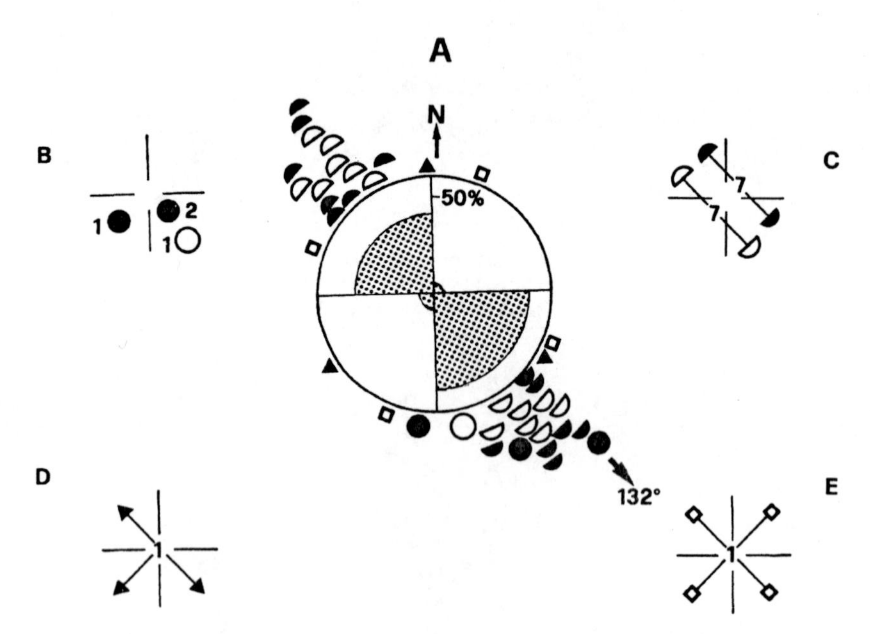

Fig. 4. Inexperienced young from Sar Uanle population. Artificial zenithal light in dark room. For further explanations see Fig. 1.

Natural geomagnetic field: inexperienced young from the Sar Uanle population (Fig. 4A-E). The INT of inexperienced young shows a clustering around the Y axis with no clear preference for the land or the sea, and a good clustering in both directions. INDs are not significant in about half of the sandhoppers (9/20). Apart from 4 individuals (Fig. 4B), the majority (14/20) alternate between the sea and the land (Fig. 4C). Radiality varies from 55% to 100%.

2.2. Experiments under the sun

Adults from the Mogadishu population (Fig. 5A-E). The INT for adults tested under the sun (=controls of Fig. 1A-E) shows a marked preference for the sea. The groups of mean angles in the sea and land

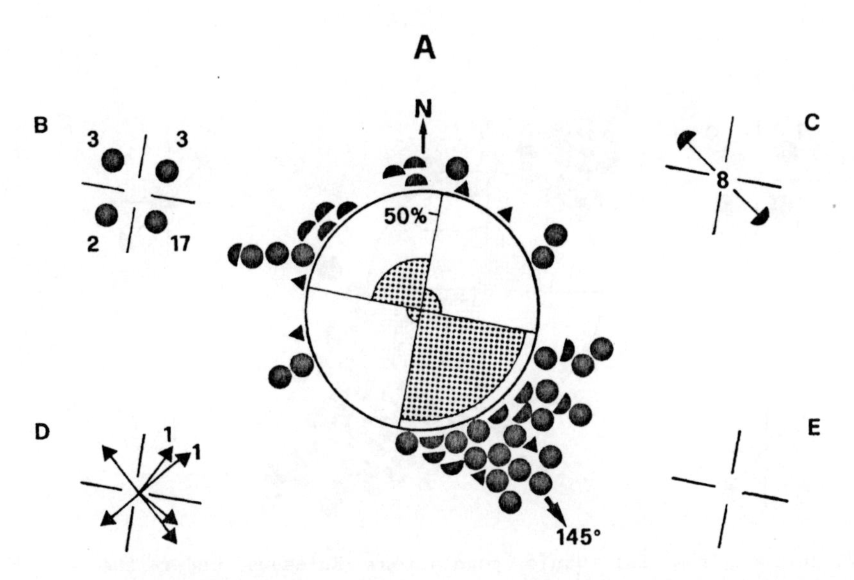

Fig. 5. Adults from Mogadishu population. Releases under the sun. For further explanations see Fig. 1.

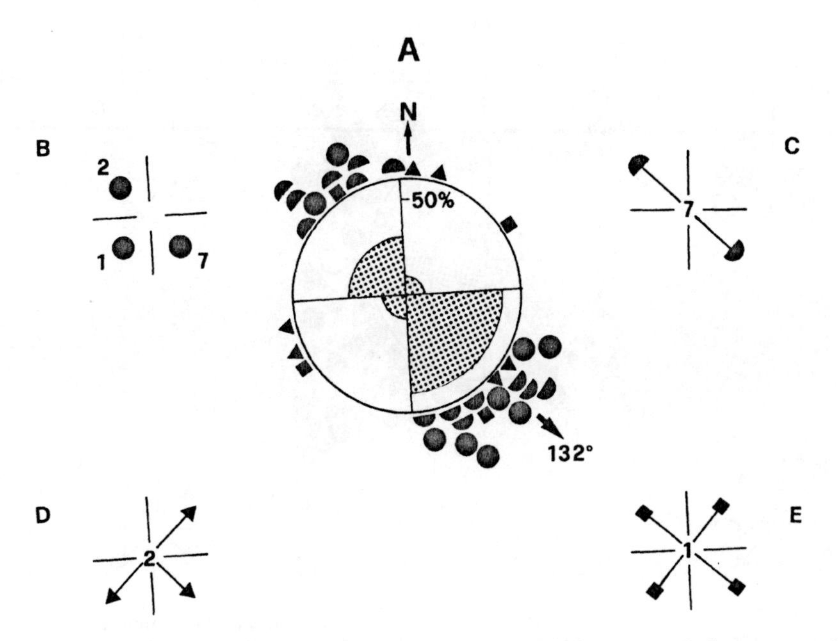

Fig. 6. Inexperienced young from Sar Uanle population. Releases under the sun. For further explanations see Fig. 1.

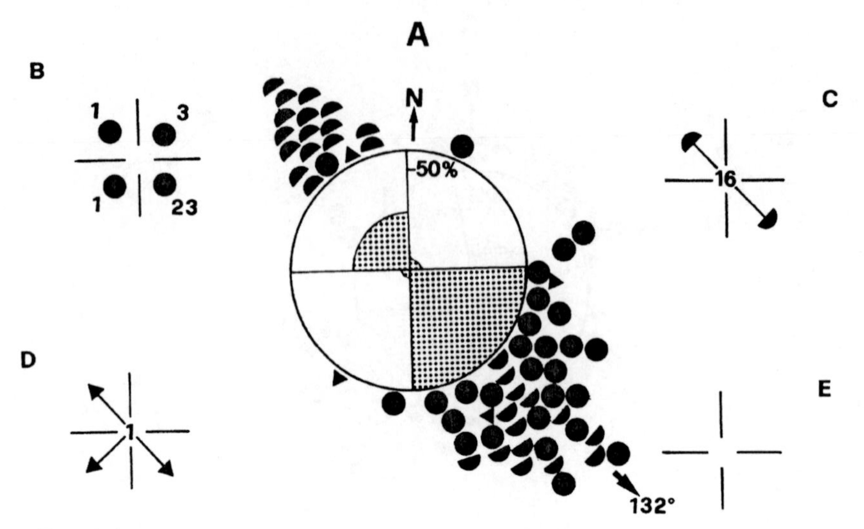

Fig. 7. Adults from Sar Uanle population. Releases under the sun. Controls of Fig. 8.

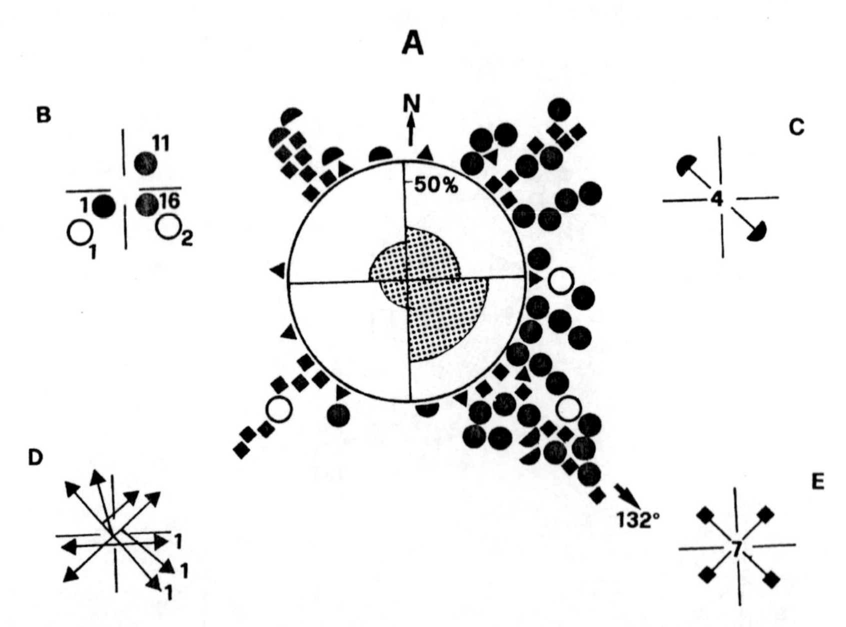

Fig. 8. Adults from Sar Uanle population. Releases under the sun. Experimentals (clock-shifted). See text and Fig. 1 for further explanations.

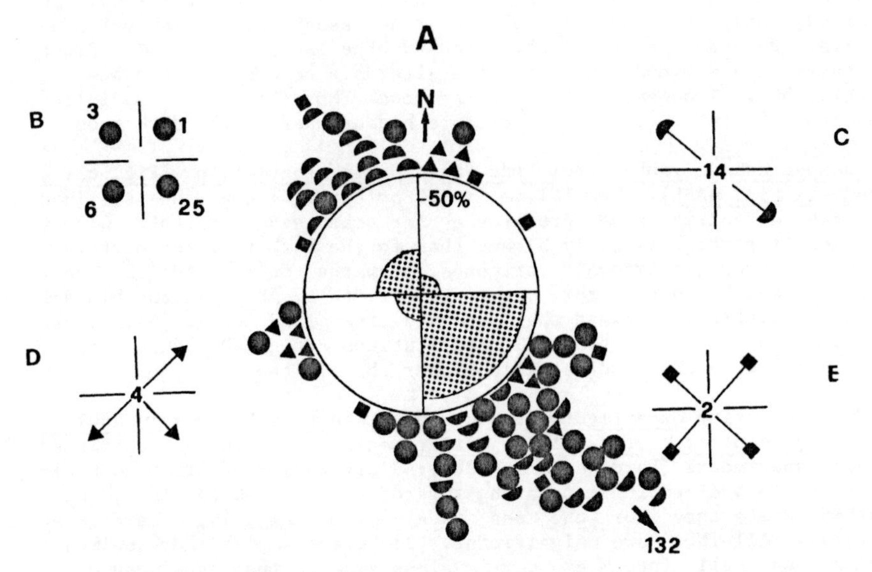

Fig. 9. Inexperienced young from Sar Uanle population. Releases under the sun. Controls of Fig. 10.

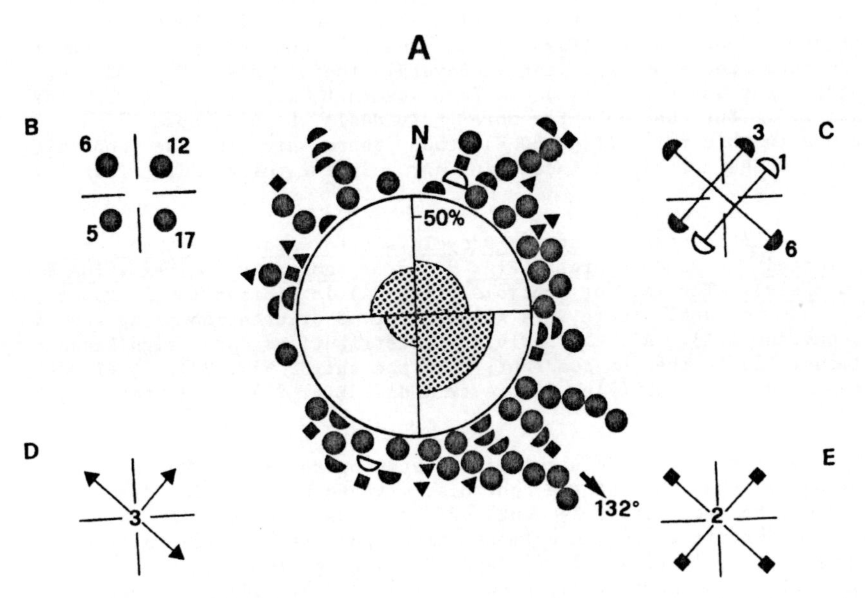

Fig. 10. Inexperienced young from Sar Uanle population. Releases under the sun. Experimentals (clock-shifted). See text and Fig. 1 for further explanations.

quadrants are clearly more dispersed than in the dark room, but the 35 INDs are all significant. If orientation for each individual is considered, half (17 out of 35) of the sandhoppers are oriented exclusively towards the sea, (only 3 towards the land) (Fig. 5B). Eight individuals with a bimodal distribution alternate between the sea and the land (Fig. 5C). Two sandhoppers show trimodal INDs (Fig. 5D). Radiality varies from 75% to 100%, and is therefore higher than in the dark room.

Inexperienced young from the Sar Uanle population (Fig. 6A-E, controls of Fig. 4A-E). The INT is well oriented along the sea-land axis, with a fairly good preference for the sea quadrant. Global dispersion is perhaps slightly higher than in the dark room but certainly not less. Seven individuals oriented seawards only, and 2 only landwards, and 1 opted for quadrant III (Fig. 6B). Seven bimodal individuals alternate between the sea and the land (Fig. 6C). Three individuals show tri or tetramodal distributions (Fig. 6D,E). Radiality varies from 77% to 100% and is again higher than in the dark room.

Clock shifting experiments (L-D cycle put forward by 3hrs): adults. (Fig. 7A-E, controls; Fig. 8A-E, experimentals). Results confirmed previous experiments (Pardi et al., 1985): distribution of the controls (Fig. 7A-E) is bidirectional pointing towards the sea and the land, with a marked preference for the sea (see also Fig. 1C, Pardi et al., 1985). All INDs are significant. 23 mean angles for unimodal distributions fall into, or very close to, the sea quadrant (Fig. 7B). The 16 bimodal distributions fall exclusively in a sea-land direction (Fig. 7C).

Distribution in the clock-shifted experimental adults (Fig. 8A-E) is tetradirectional with one group of mean angles lying along the Y axis with a clear preference for the sea, and another along the X axis with a preference for quadrant I. Distribution along the X axis seems far more evident here than above (Figs. 1-4). Only 3 out of 45 individual distributions are non-significant. Several individuals (Fig. 8B) opt exclusively for quadrant I (11) or for quadrant II (18), and 2 for quadrant III. But there are far more individuals which choose 3 or 4 directions in this trial (Fig. 8D-E) than there are in the controls (Fig. 7) (10 against 1), or indeed in any of the other distributions examined so far.

Clock-shifting experiments (L-D cycle set forward by 3hrs). Young inexperienced sandhoppers. (Fig. 9A-E, controls; Fig. 10A-E, experimentals). The INT for controls (Fig. 9A) is bidirectional towards the sea and the land. There is a slight hint of orientation along the X axis (quadrant III). All the individual distributions are significant: 35 unimodal (25 in the sea quadrant, 6 in the third, Fig. 9B), 14 bimodal in a sea-land direction (Fig. 9C), 4 trimodal (Fig. 9D) and 2 tetramodal (Fig. 9E).

The INT for inexpert clock-shifted sandhoppers (Fig. 10A) is clearly more dispersed than that of the controls, with no preferential direction, in spite of the fact that 54 out of 55 INDs are significant. The majority of distributions are unimodal with the resultants falling in quadrants I (12), II (17), III (5) and IV (6) (Fig. 10B). Ten resultants are bimodal (quadrants I-III and II-IV); Fig. 10C), 3 trimodal (Fig. 10D), and 2 tetramodal (Fig. 10E).

3. DISCUSSION AND CONCLUSIONS

Under artificial zenithal light in the dark room (and consequently with no information on the horizontal plane) circumequatorial populations of T. martensii show a remarkable ability to orient themselves precisely along the Y axis (sea-land), with no obvious preference for either of the two possible quadrants (bidirectionality) (Fig. 1 for the Mogadishu population). This type of orientation is magnetic, as proved in the experiments with an artificial magnetic field (Fig. 2 for the same population). These data confirm previous results obtained for the same population using a different recording technique (Pardi et al., 1985). However, the methods adopted in this experiment demonstrated the phenomenon even better, and proved that bidirectionality is also intraindividual.

The bidirectional magnetic orientation for each population agrees with the direction of the Y axis of the home shore (Arendse, 1978). The Ras Mntoni population (Fig. 3), whose sea-land axis almost coincides with N-S axis, exhibit a far more precise bidirectional orientation than the other two (Mogadishu and Sar Uanle), whose two axes lie almost 90° apart. Presumably it is easier to assume an angle of only a few or zero degrees from the magnetic field than it is a much wider one; this could be of interest for future research into the receptor or receptors involved. The two populations of adults with a "wide" angle (Mogadishu and Sar Uanle), showed a more frequent transverse orientation, but still bidirectional (along the X axis parallel to the shore; this is difficult to explain ecologically, but perhaps it has to do with zonal maintenance, Arendse, 1980).

Bidirectional magnetic orientation must be innate, since even young sandhoppers born in the laboratory and with no experience of the lay of their beach with respect to the earth's magnetic field, show this ability.

The most striking result in the INTs for controls under the sun, in the adults (Fig. 5) as well as inexpert young (Fig. 6), is the greater dispersion with respect to distributions in the dark room. Under this aspect it appears that the sun is a disturbing factor. Moreover, distributions in sunlight show two other equally obvious characteristics (see Figs. 7 and 9 as well as Figs. 5 and 8): A) prevalence of the sea quadrant with respect to the land, i.e. a reduction in the marked bidirectionality seen in the darkroom; B) the INDs are constantly significant.

As far as point A is concerned, it should be noted that sea-land equivalence in the darkroom is as much inter as intraindividual. If bimodal INDs are considered (all of which are always along the sea-land axis), the number of recordings of positions in the two quadrants is more or less the same: in 36 individual choice was more for the sea, and in 29 for the land, in 4 they were exactly the same (P=NS, χ^2 test). On the other hand, under the sun, the sea component prevailed in as many as 31 INDs and the land in only 6 ($P<0.01$, χ^2 test). Under the sun, the sandhoppers tend to opt more for the sea quadrant, since the sun provides them with additional information regarding the direction to take along the Y axis, the position of which is already given by the magnetic compass. This can be an advantage to the sandhoppers: by heading towards the sea, they will always end up finding the right conditions of humidity if they are in danger of dehydration, but this is not always the case if they choose the opposite direction.

The fact that the INDs are constantly significant under the sun also shows that the solar reference, as we have already seen, increases interindividual route variability, but lowers the intraindividual one. In short, routes are more dispersed under the sun but more straighter. Since the sandhoppers' goal (damp sand, stranded detritus) stretches along the beach, a straight route, but not strictly perpendicular to these shore, would be more advantageous to them than a twisted one even though it may still be in line with the Y axis. Obviously this effect of route stabilization is not connected with the sun's role as a chronometric compass, but the predominance of seaward headings implies that the latter must be involved to some extent, since the sun's azimuth changes during the day. This use of the sun as a compass (alongside the magnetic one) is proved in the clock-shifting experiments, which confirm previous trials in full (Pardi et al., 1985; Pardi, 1987). The INT of clock-shifted adults (3 hours ahead, Fig. 8) is clearly tetramodal (remembering that 7 INDs were tetramodel, Fig. 8E) compared to the controls (Fig. 7). The only explanation for this lies in treatment the experimentals received. It is true that transverse orientation (along the X axis) also occurred in the dark room (Figs. 1,2,3) and in the controls under the sun (Figs. 5,6,7) but it is plainly more evident in the clock-shifted trials (with almost half along the Y axis compared to less than 17% in the other cases). The most plausible explanation (see also Pardi et al., 1985; Pardi, 1987) is that as a result of clock-shifting, the information on the sea-land axis from the two compasses are thrown into conflict: the sun compass gives a Y axis lying 90° to the same axis determined magnetically (and thus coinciding with the X axis). On the other hand, a similar result was obtained for sandhoppers tested under the sun, by turning the magnetic field in an anti-clockwise direction for about 90°, i.e. manipulating the magnetic and not the sun compass (Pardi et al., 1985). Why, under these contradictory conditions, some individuals are exclusively "solar", and others exclusively "magnetic" while yet others alternate from one system to the other, we do not know. Pardi (1987) noticed a size difference: "magnetics" usually seemed larger, but this hypothesis needs to be confirmed. Leaving this problem to the future, for now we may conclude that adults (or least part of them) glean information the whole day through about changes in the sun's azimuth for the time of the year in which they are tested (see also Fig. 2 in Pardi et al., 1985). In adults, therefore, the two compasses can work together the entire day.

This interpretation can give a satisfactory answer, once and for all, to the problems arising from the annual change in the sun's declination in intertropical zones. Adult sandhoppers, having had previous experience on the shore, are able to calibrate and adjust day by day the chronometrical compass information they receive from the sun, with the help of their magnetic compass (which we have seen is innate); they can learn to do this during life (for the ability of calibration of solar information by other references, for amphipods in temperate regions, see Ugolini and Scapini, 1988; Ugolini et al., 1984 and this Vol.).

Total calibration of sun, moon, and star compasses from the magnetic one is known to exist in other groups, according to Wiltschko and Wiltschko (1975; 1976) and Wiltschko et al. (1987) (orientation by the stars in the genus Syvia), Keeton (1971) and Wiltschko (1983) (solar orientation in homing pigeons), Baker (1987) (moon orientation in the moth Agriotis exclamationis).

If magnetic information for the sea-land axis is innate and solar information only secondary, we would not expect clock-shifting to have any effect on the sandhoppers, and we would expect them to orient by their magnetic compass alone. However, our experiments prove that the young do feel the effects of clock-shifting (Figs. 9,10). Bearing in mind that inexpert young are often more dispersed than experienced adults, we could give Fig. 10 the same interpretation as Fig. 8, i.e.

that inexperienced sandhoppers also use solar-compass information to know how the sun-arc changes during the day. Obviously this proposition brings up once more the difficulties regarding the annual change in the sun's declination: how can inexperienced young have innate information on the sun-arc for the whole year? We think we can give an "economical" solution to this problem: the solar azimuth, for some periods of the year, does not change its position much during the early hours of the morning (from ENE to ESE) and in the late hours of the afternoon (from WNW to WSW). Huge differences, however, occur around noon, depending on whether the sun's azimuth passes over the Northern or Southern hemicycle. The inexpert young from circumequatorial zones only need innate information on the sun's approximate position in relation to their home beach sea-land axis for the early morning and late afternoon. Only experienced individuals, as well as using the above information, can compensate the apparent movement of the sun along the complete sun-arc (clockwise or anti-clockwise) by calibrating it on the magnetic compass. This hypothesis explains the disorientation in inexpert sandhoppers and why they were affected by clock-shifting: if they were shifted forward 3 hours, they were tested at practically just the same time as "their" circum-meridian period (subjective).

A final problem concerns the so-called "zenithal-reaction", i.e. the total disorientation of sandhoppers when the sun is near the zenith (±5°) (see Pardi, 1986): why do not the sandhoppers use their magnetic compass? So far this reaction has only been observed in adults tested in groups, so one possible explanation could be the following: if the sandhoppers are under conditions of great stress, it is probably more advantageous to them to seek immediate cover (under the sand in nature, and under their companions in the experiments) rather than calculate the right direction with their magnetic compass.

REFERENCES

Arendse, M.C., 1980, Non-visual orientation in the sandhopper Talitrus saltator (Mont.), Netherl. J. Zool., 30:535.

Arendse, M.C. and Kruyswijk, C.J., 1981, Orientation of Talitrus saltator to magnetic fields, Netherl. J. Sea Res., 15:23.

Batschelet, E., "Statistical methods for the analysis of problems in animal orientation and certain biological rhythms," Am. Inst. Biol. Sci., Washington D.C. (1965).

Batschelet, E., 1978, Second-order statistical analysis of directions, in "Animal migration, navigation, and homing," K. Schmidt-Koenig and W.T. Keeton eds., Springer Verlag, Berlin Heidelberg New York.

Batschelet, E., "Circular Statistics in Biology," Academic Press, London (1981).

Baker, R.R., 1987, Integrated use of moon and magnetic compass by the heart-and-dart moth, Agriotis exclamationis, Anim. Behav., 35:94.

Bercken, J. van den, Broekhuizen, S., Ringenbelrg, J. and Velthuis, H.H.W., 1967, Non-visual Orientation in Talitrus saltator, Experientia, 23:44.

Braemer, W., 1960, A critical review of the sun-azimuth hypothesis, Cold Spring Harb. Symp. quant. Biol., 25:413.

Ercolini, A., 1964, Ricerche sull'orientamento astronomico di anfipodi litorali della zona equatoriale. I. L'orientamento solare in una popolazione somala di Talorchestia martensii Weber, Z. vergl. Physiol., 49:138.

Keeton, W.T., 1971, Magnets interfere with pigeon homing, Proc. nat. Acad. Sci. U.S.A., 68:102.

Lindauer, M., 1960, Time-compensated sun orientation in bees, Cold Spring

Harb. Symp. quant. Biol., 25:371.
Marchionni, V., 1963, Modificazione sperimentale nella direzione innata di fuga in Talorchestia deshayesei Aud. (Crustacea Amphipoda), Boll. Ist. Mus. Zool. Univ. Torino, 6:29.
Pardi, L., 1960, Innate components in the solar orientation of littoral amphipods. Cold Spring Harb. Symp. quant. Biol., 25:394.
Pardi, L., 1967, Studi sull'orientamento astronomico dei Crostacei Anfipodi in regioni equatoriali. Atti Accad. naz. ital. Entomol. Rc., 14:44.
Pardi, L., 1987, Wenn die Strandflohkrebse den Himmel anschauen: Ein Beitrag zur tierischen Orientierung an der Meeres-Landesgrenze, in: "Information processing in animals, Vol. 4," M. Lindauer ed., Gustav Fischer Verlag, Stuttgart New York.
Pardi, L. and Ercolini, A., 1965, Ricerche sull'orientamento astronomico di Anfipodi della zona equatoriale. II. L'orientamento lunare in una popolazione somala di Talorchestia martensii Weber, Z. vergl. Physiol., 50:225.
Pardi, L. and Ercolini, A., 1966, Ricerche sull'orientamento astronomico di Anfipodi della zona equatoriale. III. L'orientamento lunare in una popolazione somala di Talorchesta martensii Weber a Sud dell'Equatore (4° Lat. S.), Monitore zool. ital. (Suppl.), 74:80.
Pardi, L. and Ercolini, A., 1986, Zonal recovery mechanisms in talitrid crustaceans, Boll. Zool., 53:139.
Pardi, L., Ercolini, A., Ferrara, F., Scapini, F. and Ugolini, A., 1985, Orientamento zonale solare e magnetico in Crostacei Anfipodi di regioni equatoriali. Atti Accad. naz. Lincei Rc. (Cl. Sci. fis. mat. nat.), 76:312.
Pardi, L., Ercolini, A., Marchionni, V. and Nicola, C., 1958, Ricerche sull'orientamento degli Anfipodi del litorale: il comportamento degli individui allevati in laboratorio sino dall'abbandono del marsupio, Atti Accad. Sci. Torino, 92:1.
Pardi, L. and Papi, F., 1953, Ricerche sull'orientamento di Talitrus saltator (Montagu) (Crustacea, Amphipoda). I. L'orientamento durante il giorno in una popolazione del litorale tirrenico, Z. vergl. Physiol., 35:459.
Scapini, F., 1986, Inheritance of solar direction finding in sandhoppers. 4. Variation in the accuracy of orientation with age. Monitore zool. ital. (N.S.), 20:53.
Scapini, F. and Ercolini, A., 1973, Research on the non-visual orientation of littoral amphipods: experiments with young born in captivity and adults from a Somalian population of Talorchestia martensii Weber (Crustacea Amphipoda), Monitore zool. ital. (N.S.) Suppl., 5:23.
Ugolini, A. and Scapini, F., 1988, Orientation of the sandhopper Talitrus saltator (Amphipoda Talitridae) living on dynamic sandy shores, J. Comp. Physiol. A, 162:in press.
Ugolini, A., Scapini, F. and Pardi, L., 1984, Solar orientation in Talitrus saltator Montagu from retro-dunal lagoons, Monitore zool. ital. (N.S.), 18:181.
Ugolini, A., Scapini, F., Beugnon, G. and Pardi, L., 1988, Learning in zonal orientation of sandhoppers, This Volume.
Wiltschko, R., 1983, The ontogeny of orientation in young pigeons, Comp. Biochem. Physiol., 76A:701.
Wiltschko, W. and Wiltschko, R., 1975, The interaction of stars and magnetic fields in the orientation system of night migrating birds. Part I. Autumn experiments with European warblers (Gen. Sylvia), Z. Tierpsychol., 37:337.
Wiltschko, W. and Wiltschko, R., 1976, Die Bedeutung des Magnetkompasses fur die Orientierung der Vogel, J. Ornithol., 117:362.
Wiltschko, W. and Wiltschko, R., 1987, The development of the star compass in the garden warblers, Sylvia borin, Ethology, 74:285.

Aspects of Direction Finding Inheritance in Natural Populations of Littoral Sandhoppers (*Talitrus saltator*)

Felicita Scapini, Alberto Ugolini and Leo Pardi
University of Florence, Italy

Littoral sandhoppers such as Talitrus saltator Montagu maintain the damp sand strip they live in by means of zonal orientation: they crawl or hop seaward if dehydrated and landward if doused with sea water, along the y-axis perpendicular to the shoreline. For Mediterranean populations the sun is the most important cue while polarization of the sky can orient the sandhoppers when the sun is hidden (Pardi and Papi, 1953). Substrate slope, landscape vision, and local sky factor act as coadiuvant factors or, in the absence of the sun, as vicariants (Ercolini and Scapini, 1974; Hartwick, 1976; Pardi and Scapini, 1979; Ercolini et al., 1983; Ugolini et al., 1986).

Both solar orientation and the direction in which each population heads have been shown to be innate (Pardi et al., 1958; Pardi, 1960; Scapini and Pardi 1979). Expert individuals collected on differently oriented shores (Tyrrhenian and Adriatic, Fig. 1) and their inexpert laboratory-born offspring were tested in dry conditions under the sun far inland (Florence). Tests were performed at 9.00, 12.00 and 15.00 hours in a flat circular arena (Ø 40 cm) which permitted only sight of the sun and sky. Groups of ten individuals each were released from below center in the arena and collected in pit-fall traps at the rim. Both experts and inexperts of each population headed seaward (Fig. 1). The inexpert specimens were more dispersed than the experts (compare the mean vector lengths in Fig. 1).

It is hardly conceivable that natural selection can maintain this inherited directional tendency adapted to the shoreline position considering how greatly sandy beaches vary in space and time. In fact, these can alter direction within a few chilometers depending on changes in dominant winds and sea currents or the presence of artificial barriers. Near river mouths shorelines are even more fluid due to the effect of the seasons, rains, and cultivations along the upper river. Moreover, sandhoppers migrating from differently oriented points may interchange their genes.

To ascertain how correctly natural populations are adapted to their shorelines, both adults collected a few chilometers apart on arcuated sandy beaches and their laboratory-born inexpert offspring were tested far inland (Scapini et al., 1985). As shown (Fig. 2) by the collecting points and theoretical escape directions, both experienced and inexperienced individuals headed towards the theoretical escape direction (TED) of their collecting point (V tests). The experienced individuals were always more concentrated and better oriented towards their TED than the inexperienced

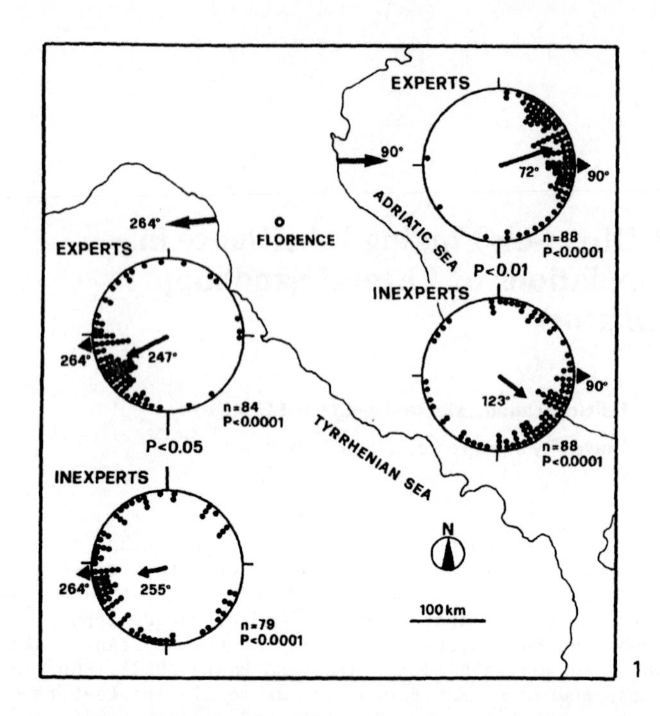
EXPERTS
90°
72°
90°
ADRIATIC SEA
n=88
P<0.0001
P<0.01
INEXPERTS
123°
90°
n=88
P<0.0001
264°
FLORENCE
EXPERTS
264°
247°
n=84
P<0.0001
P<0.05
TYRRHENIAN SEA
INEXPERTS
264°
255°
n=79
P<0.0001
N
100 km
1

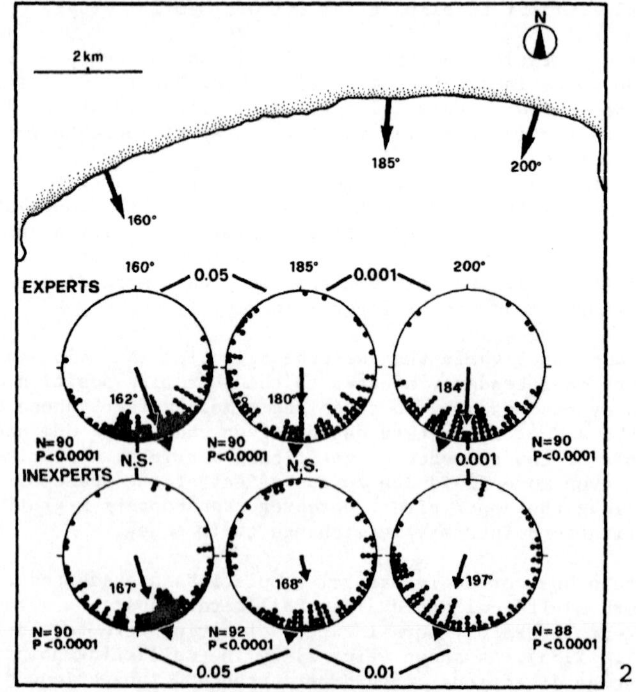
N
2 km
185°
200°
160°
160°
0.05
185°
0.001
200°
EXPERTS
162°
180°
184°
N=90
P<0.0001
N=90
P<0.0001
N=90
P<0.0001
N.S.
N.S.
0.001
INEXPERTS
167°
168°
197°
N=90
P<0.0001
N=92
P<0.0001
N=88
P<0.0001
0.05
0.01
2

Fig. 1. Talitrus saltator from the Vecchiano beach (Pisa, Theoretical Escape Direction=264°) and Casal Borsetti (Ravenna, TED=90°). Collecting points with TEDs, sun orientation of expert adults collected on the beach and their laboratory-born inexpert offspring, both tested inland (Florence). Dots, individual bearings; arrows in the circles, mean vectors; triangles, population's TED; n, number of individuals tested. Clustering around the TED was tested with the V test; distributions were compared with Watson's U^2 test (Batschelet, 1981)

Fig. 2. Talitrus saltator from the Feniglia beach (Grosseto, Tyrrhenian shore). Collecting points with TEDs, sun orientation of expert adults collected on the beach and their laboratory-born inexpert offspring, both tested inland (Florence). Explanations in Fig. 1. (Scapini et al., 1985, redrawn).

ones (compare the lengths and directions of the mean vectors). There were differences in orientation between collecting points in both experienced and inexperienced individuals. These differences, statistically significant for the Watson's U^2 test, are in scatter and/or rotated mean vectors which point towards the respective TEDs. The results for a different beach (Fig. 3) were comparable except for the collecting point near the mouth of the river (TED 280°) where expert adults were more scattered than the inexpert young. We may conclude that natural populations are genetically heterogeneous.

An exchange of genes is possible because these beaches are continuous, but talitrids generally migrate along the sea-land y axis (see Geppetti and Tongiorgi, 1976). Furthermore, genetic changes are theoretically more limited in these one-dimensional habitats than in two- or three-dimensional ones (Cavalli-Sforza, 1977). A sort of genetic cline could occur.

In order to reveal the transmission mechanism, mass-crosses between differently oriented populations were performed (Pardi and Scapini, 1983;Fig. 4A). Adult males and females were collected on the shores and mass-crossed in the labortory, after careful control of the females to prevent intra-population breeding. The cross-results (Fig. 4B) showed 1) no differences between the reciprocal crosses, and 2) headings in a direction intermediate between the parental ones. Apparently the mechanism is a simple one with no maternal heredity or dominance. Totally separated populations (with no genetic flow) from the Tyrrhenian and Adriatic shores (Fig. 5A) were also crossed. Only one cross produced any brood (Tyrrhenian females and Adriatic males). As in the previous crosses, the F1 offspring (Fig. 5B) tended to concentrate in a direction intermediate between the parental ones, and were more dispersed than the inexpert brood of the intrapopulation crosses (compare with Fig. 1). Apparently the Tyrrhenian and Adriatic populations share the same transmission mechanism.

Pardi and Scapini (1983) hypothesized that the transmission mechanism of direction finding in sandhoppers was a simple oligogenic one (Fig. 6). The mechanism proposed was two genes with two alleles each (Fig. 6A), or one gene with four alleles (Fig. 6B). Different genotypes were assumed for 8 directions 45° apart. A direction need not be very precise as young talitrids do not cover long distances in their migrations. Thus an error of about 45° would not have dramatic consequences. Furthermore, it has been shown that - in nature - local cues can improve the orientation of the talitrids (see Ugolini et al., present volume). One genotype in the first formulation (A) of the hypothesis or two in the second (B) would not determine any direction. A cross between opposite genotypes, such as east- and west-oriented ones, should produce "scattered" individuals and, in fact,

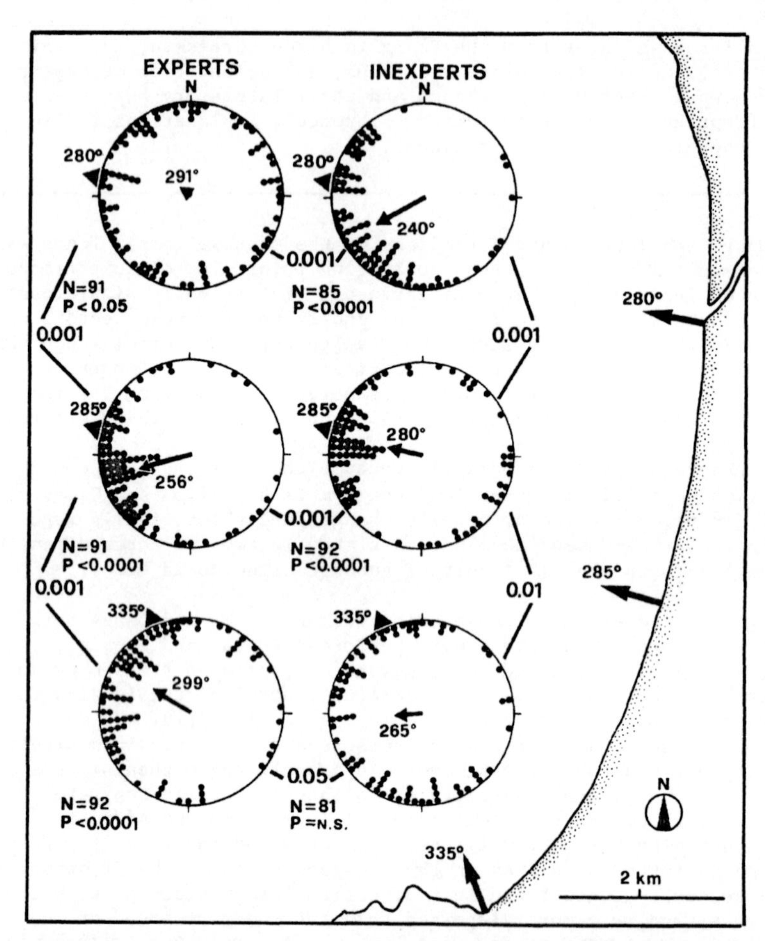

Fig. 3. Talitrus saltator from the Giannella beach (Grosseto, Tyrrhenian shore). Collecting points with TEDs, sun orientation of expert adults collected on the beach and of their laboratory-born inexpert offspring, both tested inland (Florence). Explanations in Fig. 1. (Scapini et al., 1985, redrawn)

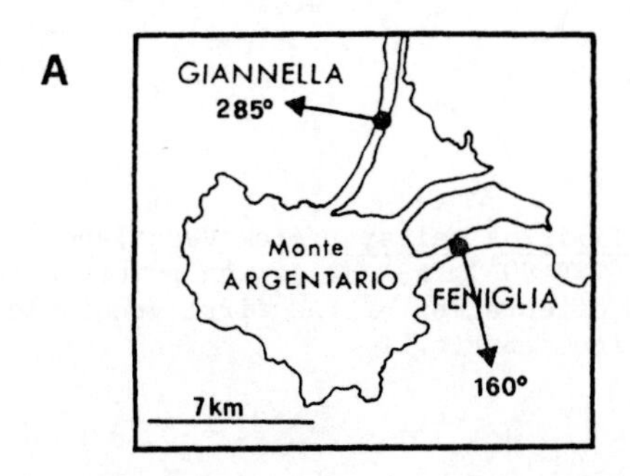

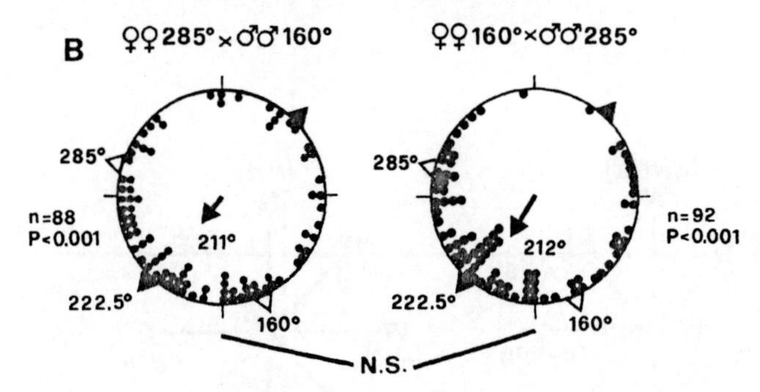

Fig. 4. Mass-crosses between Talitrus saltator from Feniglia (TED=160°) and Giannella (TED=285°). A, collecting points and TEDS of the parents; B, sun orientation of the first generation offspring of the two reciprocal crosses, tested inland (Florence). Explanations in Fig. 1. (Pardi and Scapini, 1983, redrawn)

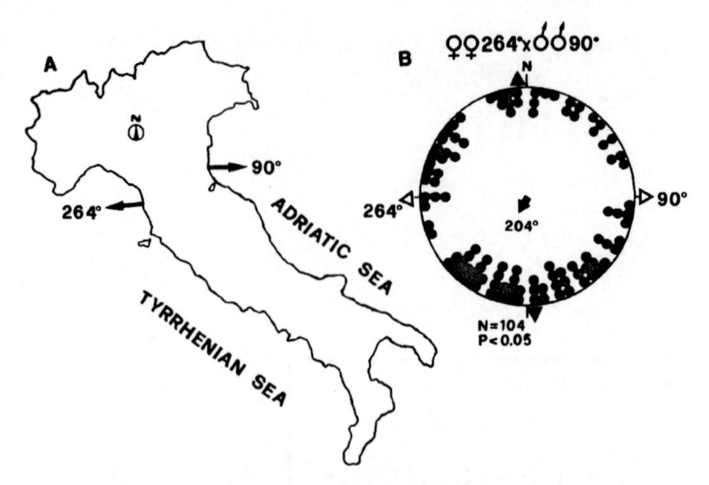

Fig. 5. Mass-cross between <u>Talitrus saltator</u> from Vecchiano (TED=264°) and Casal Borsetti (TED=90°). A, collecting points and TEDs of the parents; B, sun orientation of the first generation cross-offspring. Explanations in Fig. 1

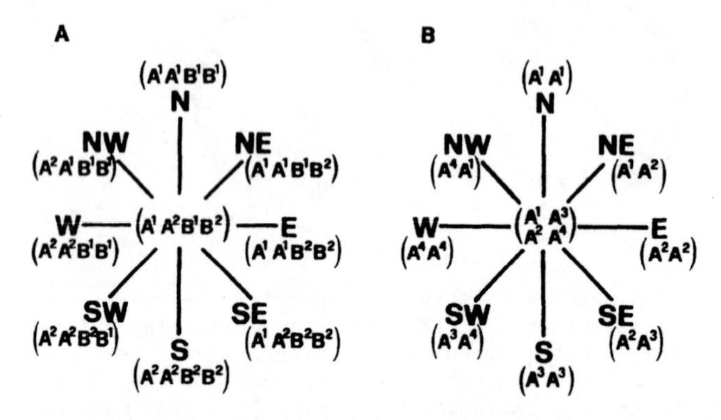

Fig. 6. Oligogenic transmission hypothesis of direction finding. The genotypes and determined directions are shown. The central genotypes would not determine any directional preference. (Pardi and Scapini, 1983, redrawn)

the offspring of opposite-oriented populations (Fig. 5) were scattered in comparison to the first cross shown (Fig. 4). On the other hand, a cross between - for example - a southeast and northwest oriented pair would

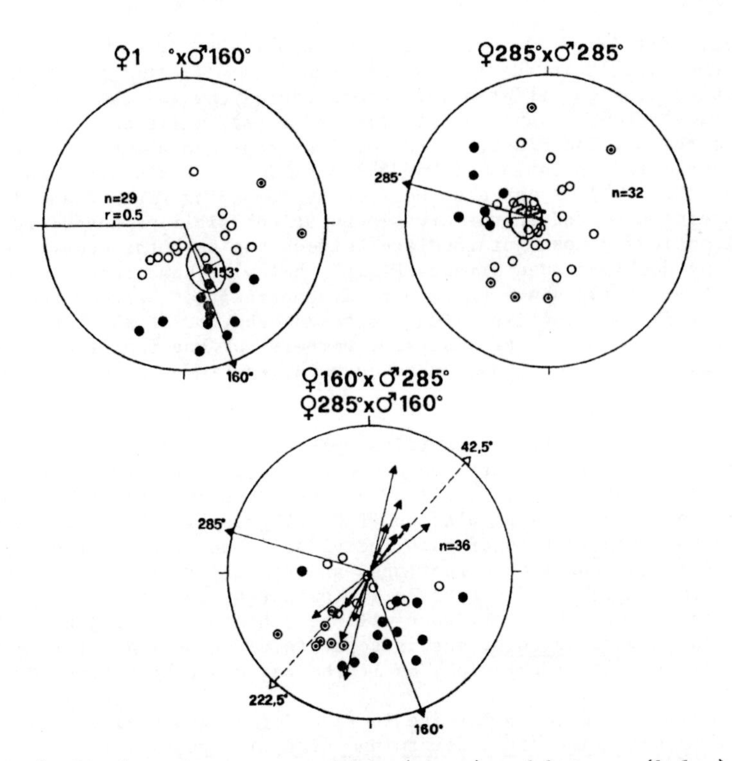

Fig. 7. Single pair crosses within (above) and between (below) the populations of Talitrus saltator from Feniglia (TED=160°) and Giannella (TED=285°). Dots, family mean vectors; confidence ellipses, population means (P=0.05: the second order analysis was not performed for the cross-population because this presented bimodal behavior); n, number of families tested. Black dots, oriented families (V tests $P<0.05$); circles, scattered families (V and z tests not significant); circled dots, families oriented in a direction different from the TED (the mean direction confidence limits do not contain the TED); double-headed arrows, bimodal families (z test significant in the second trigonometric moment) (Batschelet, 1981). Triangles outside the circles, expected escape directions (Scapini and Buiatti, 1985, redrawn)

produce northeast and southwest oriented offspring, in other words a bimodal progeny.

Such progeny have arisen from single pair crosses. (Scapini and Buiatti, 1985; Fig. 7). Pairs chosen at random from the natural populations each produced 10 to 20 offspring which were tested in the same arena device as above. Each dot (Fig. 7) represents the mean value for siblings born from parents collected in the field. Siblings from different pairs in the same population behaved differently: some oriented towards the TED (black dots), others about 90° away from it (circled dots), while some were scattered (circles). The center of the ellipse represents the population mean. The Giannella population (TED=285°) showed a bigger scatter than that of Feniglia (TED=160°) as can also be seen by comparing Figs. 2 and 3. Bimodal behavior appeared in the cross-population: siblings headed in opposite directions almost intermediate between the parental ones. They are represented by double-headed arrows (Fig. 7, below). These results indicate an oligogenic mechanism since there were discontinuities between the behavior of different families. The scattered behavior of the "scattered" families (circles) could be because some members of one family oriented in different individual directions, or because the specimens were individually scattered.

Therefore siblings of some families were tested singly in a glass bowl experimental device for 9 minutes each and recorded at intervals of 15 seconds. Figure 8 shows the orientation of the offspring born from two different pairs in the same population (TED=264°: compare Fig. 1, left). The dots are the mean vectors of each individual, the arcs are the confidence limits of the mean directions, and the ellipses are the means of each family. The length of the vectors indicate that even in the "scattered" family (B, the confidence ellipse contains the origin of the axex) individuals were oriented but in separate directons. Scatter in one family may result from segregation, as if the parents were heterozygous.

The experimental data so far are in favor of heterogeneity in natural populations. Is selection for a determined direction possible? A mass-selection experient was performed using the same population as above. Individuals collected in the field and tested in Florence in the arena device were selected for a TED of 264°±22.5 and of 144°±22.5 (120° apart: Fig. 9, above). Tests on the selected lines showed no differences between lines in the first-selected generation (Fig. 9, center). This result

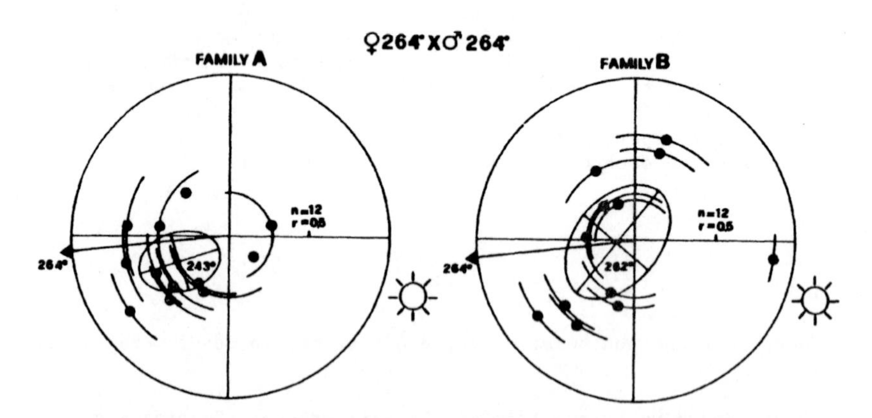

Fig. 8. Single pair crosses of Talitrus saltator from Vecchiano (TED=264°). Orientation of the offspring of two pairs (A, B) tested individually and recorded repeatedly. Dots, individual mean vectors; arcs, confidence limits of the individual mean directions (P=0.05: Batschelet, 1981); ellipses, family means (P=0.05, Hotelling's test); n, number of specimes tested in each family

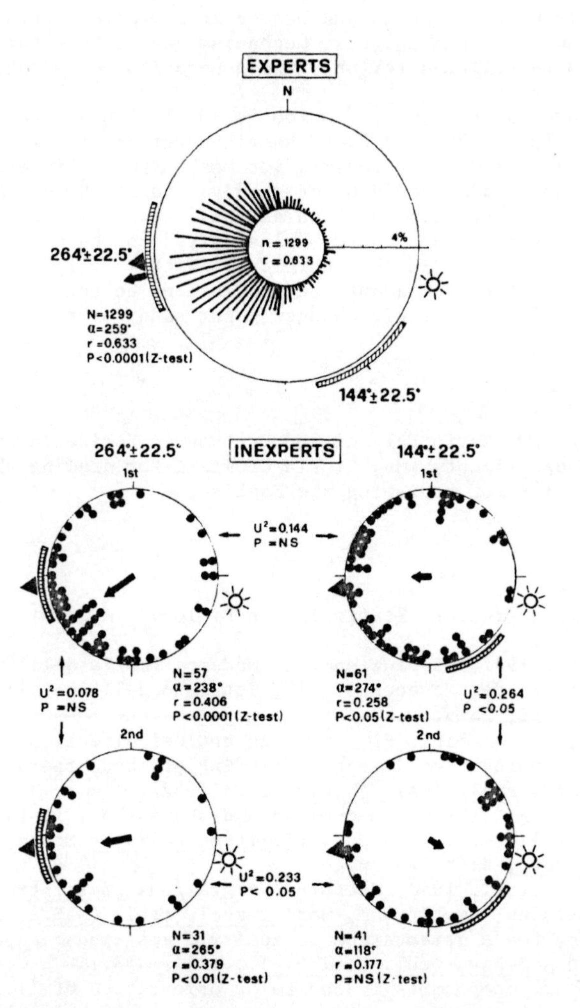

Fig. 9. Mass selection experiment on Talitrus saltator from Vecchiano (TED=264°). Orientation of expert adults (histogram, above) collected on the beach and selected for a TED of 264° and of 144°, and of the two selected lines in the first and second generations (inexperts). Dots, individual bearings. Explanations in Fig. 1

can be explained by the following facts: 1) experienced individuals were selected which could have learned a different direction from their innate one (see Ugolini et al., present volume), and 2) heterozygous individuals are present in nature. Instead, the second selected generations (Fig. 9, below) showed statistically different orientations. The individuals selected for the TED headed in that direction, while those selected for the other direction (144°) were more scattered and headed in a mean direction very close to the expected one. An oligogenic mechanism can account for a positive response to selection after only two generations of selection.

In summary, there is experimental evidence for 1) inheritance of the directional tendency in sandhoppers 2) a genetic heterogeneity in natural populations, and 3) an oligogenic transmission mechanism. An inborn mechanism of direction finding could be of survival value for sandhoppers, which are in danger of dehydrating as soon as they leave the mother's brood pouch. Genetic heterogeneity can confer "plasticity" to the natural populations in case of changes in the shoreline. With an oligogenic mechanism, natural selection can adapt the populations to the shoreline even when this undergoes relatively rapid changes - for sandhopper generations.

ACKNOWLEDGEMENTS

We wish to thank Dr. Mario Fallaci and Dr. Donato Fasinella for their contribution to the experiments, Ms. Cinzia Giuliani for drawing the tables and Ms. Sarah Mascherini for reviewing the English.

REFERENCES

Batschelet, E., 1981, "Circular Statistics in Biology", Academic Press, London.

Cavalli-Sforza, L. L., 1977, Evoluzione. La Moderna Teoria dell'Evoluzione, in: "Enciclopedia del Novecento", II, Istituto dell'Enciclopedia Italiana Treccani, Roma.

Ercolini, A., Pardi, L., Scapini, F., 1983, An optical directional factor in the sky might improve the direction finding of sandhoppers on the seashore, Monit. zool. ital. (N.S.), 17:317-327.

Ercolini, A., Scapini, F., 1974, Sun compass and shore slope in the orientation of littoral amphipods (*Talitrus saltator* Montagu), Monit. zool. ital.(N.S.), 8:85-115.

Geppetti, L., Tongiorgi, P., 1967, Nocturnal migrations of *Talitrus saltator* Montagu (Crustacea, Amphipoda), Monit. zool. ital. (N.S.), 1:37-40.

Hartwick, R.F., 1976, Beach orientation in talitrid amphipods: capacities and strategies, Behav. ecol. sociobiol., 1:447-458.

Pardi, L., 1960, Innate components in the solar orientation of littoral amphipods, Cold Spring Harbor Symp. Quan. Biol., 25:394-401.

Pardi, L., Ercolini, A., Marchionni, V., Nicola, C., 1958, Ricerche sull'orientamento degli anfipodi del litorale: il comportamento degli individui allevati in laboratorio sino dall'abbandono del marsupio, Atti Accad. Sci. Torino (Cl. Sci. Fis. Mat. Nat.), 92:1-8.

Pardi, L., Papi, F., 1953, Ricerche sull'orientamento di *Talitrus saltator* Montagu (Crustacea, Amphipoda). I. L'orientamento durante il giorno in una popolazione del litorale tirrenico, Z. vergl. Physiol., 35:458-489.

Pardi, L., Scapini, F., 1979, Solar orientation and landscape visibility in *Talitrus saltator* Montagu (Crustacea, Amphipoda), Monit. zool. ital. (N.S.), 13:210-211.

Pardi, L., Scapini, F., 1983, Inheritance of solar direction finding in sandhoppers: mass-crossing experiments, J. Comp. Physiol., 151:435-440.

Scapini, F., Buiatti, M., 1985, Inheritance of solar direction finding in sandhoppers, III, Progeny tests, J. Comp. Physiol., A, 157:433-440.

Scapini, F., Pardi, L., 1979, Nuovi dati sulla tendenza direzionale innata nell'orientamento solare degli anfipodi litorali, Atti Accad. Lincei. Rend. Cl. Sci. Fis. Mat. Nat., (VIII) 66:592-597.

Scapini, F., Ugolini A., Pardi, L., 1985, Inheritance of solar direction finding in sndhoppers. II. Differences in arcuated coast-lines, J. Comp. Physiol., 156(6):729-735.

Ugolini, A., Scapini, F., Pardi, L., 1986, Interaction between solar orientation and landscape visibility in Talitrus saltator (Crustacea, Amphipoda), Mar. Biol., 90:449-460.

Learning in Zonal Orientation of Sandhoppers

Alberto Ugolini, Felicita Scapini
Guy Beugnon (*) and Leo Pardi

University of Florence, Italy
(*) Paul Sabatier University
Toulouse, France

INTRODUCTION

It has long been known that amphipods living on sandy shores use a sky compass (the sun or moon) to find their way back to the damp belt of the splash zone (Pardi and Papi, 1952, 1953; Papi and Pardi, 1953). It has also been proved that the mechanism governing solar orientation in mediterranean sandhoppers is innate, and that these sandhoppers inherit the ability to assume a specific escape direction perpendicular to their own particular shore (Pardi et al., 1958; Pardi, 1960; see also Pardi and Ercolini, 1986 for a review).

More recently, different populations of sandhoppers living along a continuous stretch of a curved sandy coastline several kilometres long were studied to analyse the precision of their astronomical orientation (Scapini et al., 1985). Results confirm that the sandhoppers inherit the escape direction relative to their own population. Individuals born in the laboratory from adults collected from different points along the same coastal arc showed significantly different orientation, and their mean direction satisfactorily agreed with the expected theoretical escape direction (TED) of the sea facing the point where their parents were collected.

However, very few studies have been carried out to investigate whether sandhoppers can adjust their inherited orientation if the sea-land axis changes abruptly in space or time (Ugolini et al., 1984a; Ugolini and Scapini, 1987), and whether they can learn a new, ecologically efficient escape direction to suit the new conditions on their home beach or on another they might reach through active or passive displacement.

1. MODIFICATION IN ORIENTATION ON THE HOME BEACH

It has often been noticed (Pardi, 1960; Scapini et al., 1985; Scapini, 1986a; Ugolini and Scapini, 1987), that on their home beach sandhoppers can improve their seaward orientation, probably with the help of previous experience they have gained on their daily excursions up and down the Y axis at dusk and dawn, (Geppetti and Tongiorgi,1967a;b).

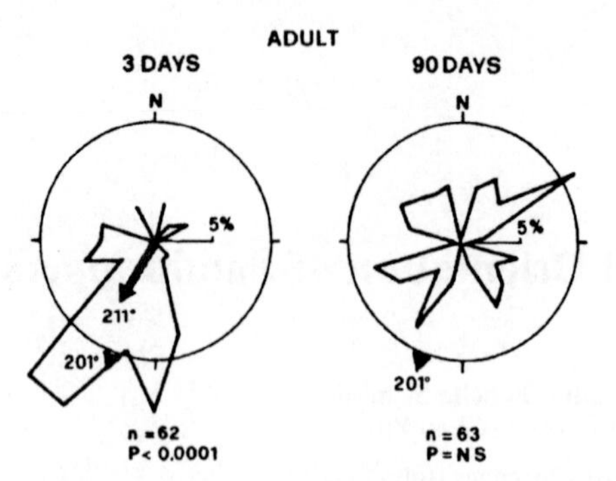

Fig. 1. Effect of captivity (3-90 days) on solar orientation. Black triangles outside the distributions represent TED; arrows, mean vectors (circle radius corresponds to mean vector length=1); n, sample size; P, probability level for V test (see Batschelet, 1981). (From Scapini, 1986 redrawn).

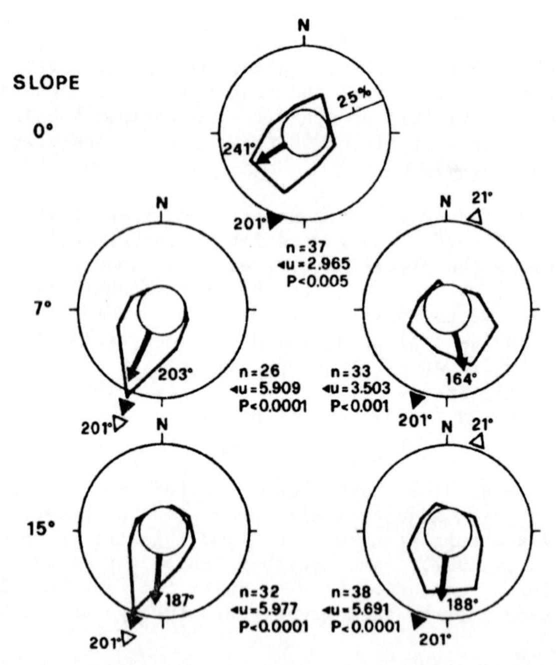

Fig. 2. Influence of ground slope on solar orientation. Dry substratum. Top distribution, controls (slope=0°); left, concordant stimuli; right discordant stimuli. Black triangles, TED for solar orientation; white triangles, TED for slope orientation. The V test (u) has been calculated assuming as expected direction the TED for solar orientation. For further explanations see Fig. 1. (From Ercolini and Scapini, 1974 redrawn).

Table 1. Mean vector length and difference (in absolute value) between TED and mean angle for expert adults and their laboratory born offsprings (inexpert). S, Scapini; U, Ugolini. See text for further explanations.

TED	MEAN VECTOR LENGTH Expert	MEAN VECTOR LENGTH Inexpert	$\lvert\Delta\alpha\rvert$ Expert	$\lvert\Delta\alpha\rvert$ Inexpert	REFERENCE
90	0.725	0.475	18	33	S et al. 1987
160	0.844	0.586	6	7	S 1986b
160	0.769	0.639	2	7	S et al. 1985
160	0.692	0.617	4	10	U unpubbl.
185	0.502	0.371	5	17	S et al. 1985
200	0.750	0.478	16	3	S et al. 1985
201	0.747	0.677	9	13	S et al. 1985
201	0.640	0.490	10	18	S 1986b
211	0.513	0.406	1	16	S et al. 1985
245	0.533	0.407	14	23	S et al. 1985
245	0.499	0.355	17	21	U & S 1987
260	0.557	0.210	5	34	U & S 1987
264	0.625	0.400	17	9	S et al. 1987
264	0.657	0.389	9	15	U & S 1987
265	0.448	0.153	4	16	U unpubbl.
268	0.763	0.407	12	27	U & S 1987
268	0.581	0.413	32	64	U unpubbl.
280	0.162	0.607	11	40	S et al. 1985
280	0.661	0.361	10	16	U & S 1987
285	0.622	0.144	2	8	S 1986b
285	0.650	0.376	29	5	S et al. 1985
335	0.494	0.311	36	70	S et al. 1985

Results from a number of experiments carried in Florence (and therefore considerably far from the sea) from 1979 to 1986 were compared. The experiments consisted in releasing the sandhoppers in a circular arena (20cm in radius) fitted with a screen to hide the landscape from view and shield them from the wind. The resultant mean vector length and angles of distribution were considered for experienced adult sandhoppers and for their offspring born in the laboratory (Tab. 1). 14 different populations from beaches relatively constant in direction were tested in a total of 22 experiments. The experienced adults proved to be far better clustered round their TED than their inexperienced young in 21 experiments out of 22 ($P<0.001$, χ^2 test) and their mean angles were nearer their TED than those of their offspring in 19 experiments ($P<0.001$, χ^2 test). Since the mechanism behind solar orientation in sandhoppers does not seem to develop with age (Scapini, 1986b), the difference in orientation can only be explained by experience. Previous experiments demonstrated how the length of time spent in captivity can also affect orientation in sandhoppers (Ercolini, 1964; Scapini, 1986a). After 3 months captivity in tanks containing uniformly damp sand and under a LD cycle in phase with and corresponding to the natural cycle, sandhoppers proved to be far more dispersed on release than freshly collected specimens (Fig. 1).

2. MODIFICATION IN ORIENTATION BASED ON LOCAL, NON-CELESTIAL FACTORS

The sandhoppers may be able to adjust their orientation by referring to local cues typical of their own collecting point, as well as by aid of previous experience from their daily excursions.

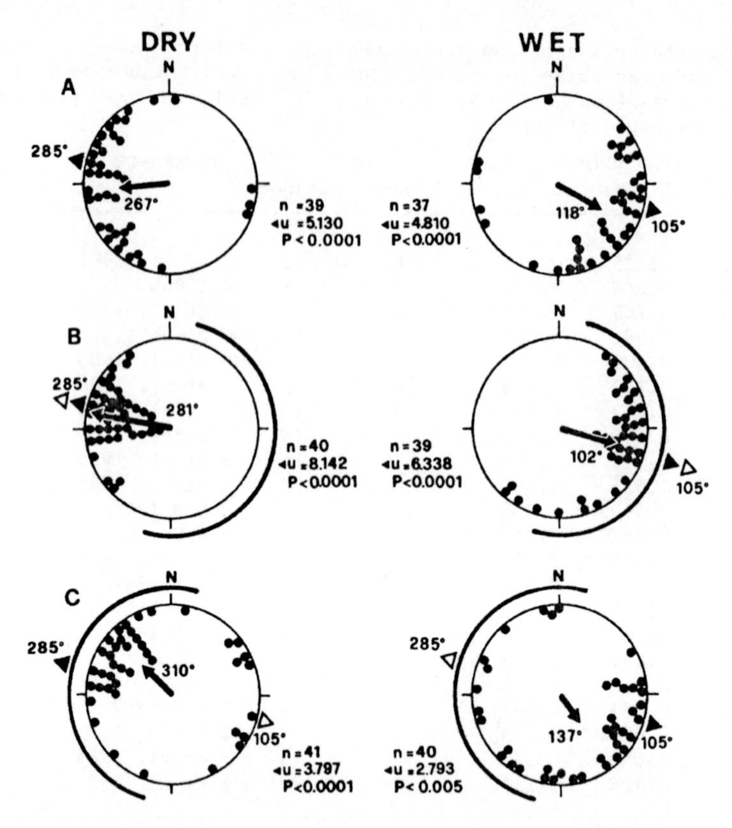

Fig. 3. Influence of sight of artificial landscape (semicircles outside the distributions) on solar orientation. Left, experiments on dry substratum; right, experiments on wet substratum. A, controls; B, concordant stimuli; C, discordant stimuli; white triangles, TED for landscape orientation. For further explanations see Fig. 1.

2.1. Slope (Fig. 2)

In the absence of any other orientating stimuli, sandhoppers will use the slope of the substratum to find their way back to their desired spot, as Craig (1973) proved in *Orchestoidea corniculata*, and Ercolini and Scapini (1974) in *Talitrus saltator*.

Dehydrated sandhoppers tested in the dark will always head for the lower end of a tilted box, but will opt for the higher end if the substratum is wet. In the circular arena under the sun, they show a marked improvement in orientation if the arena slopes towards the TED (Fig. 2 left). If the arena slopes in the opposite direction to the TED (Fig. 2 right) their orientation is worse, although they are still directed towards the sea.

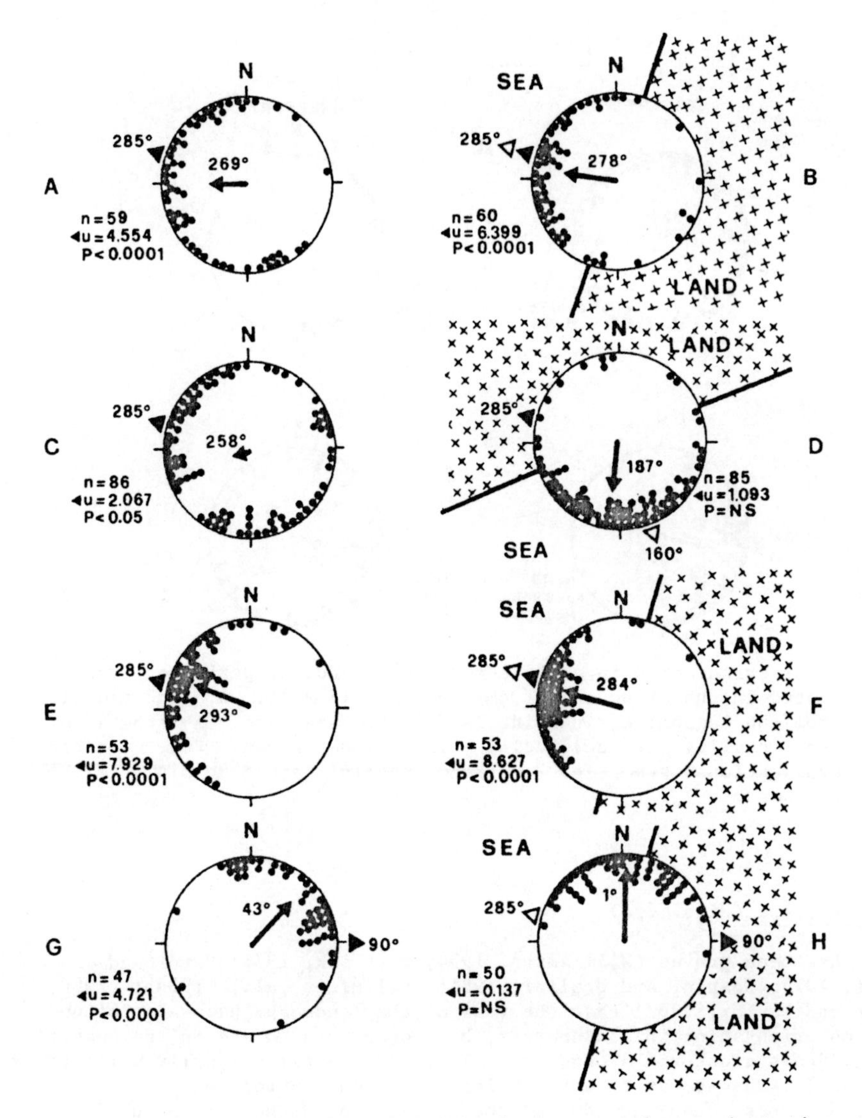

Fig. 4. Influence of sight of natural landscape on solar orientation. A and B, experiments on home beach in arena with and without screen, respectively. C and D, experiments on differently orientated beach, with and without screen, respectively. E-H, experiments on home beach using controls (E,F) and clock-shifted individuals (G,H), with screen (E,G) and without screen (F,H). For further explanations see Fig. 1. (From Ugolini et al., 1986 redrawn).

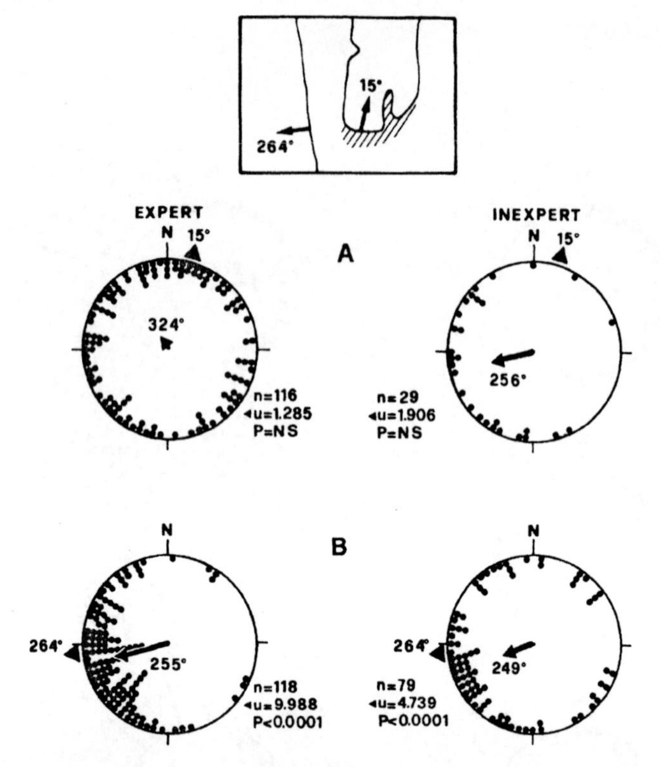

Fig. 5. Solar orientation in experienced adults and inexperienced young, tested in the arena away from the sea. TED=15°. A, individuals from the lagoon; B, individuals from the seashore. A schema of the locality of collecting is also given. For further explanations see Fig. 1. (From Ugolini and Scapini, 1987 redrawn).

2.2. Landscape (Figs. 3,4)

It has been proved (Williamson, 1954; Hartwick, 1976; Pardi and Scapini, 1979; Scapini and Ugolini, 1981; Ugolini et al., 1984b; 1986; Edwards and Naylor, 1987) that the view of the landscape has a profound effect on orientation in sandhoppers, both on precision and on the number of individuals clustered around their TED. This applies equally well to the natural landscape as to an artificial one constructed out of black cardboard shapes occupying 180 of the horizon round the arena used in the experiments (Ugolini et al., 1986). If the sun and the landscape convey the same directional information, the sandhoppers show a marked improvement in orientation - towards the sea on a dry substrate (Fig. 3B left), towards the land on a wet one (Fig. 3B right). If the two stimuli conflict each other (with the artificial landscape placed in the hemicycle of the sea), orientation in the sandhoppers deteriorates (Fig. 3C).

Two experiments were carried out to test how sandhoppers react when the natural landscape seems to contradict solar information: 1) sandhoppers were released on their home beach, in the screened arena with the landscape hidden from view (Fig. 4A) and in the unscreened arena

(Fig. 4B): view of the landscape improves orientation. A second batch from the same population was released at exactly the same time under the same conditions on a strange shore with a different geographical orientation (Figs. 4C,D): whilst the controls are still orientated in the direction of their home beach TED (Fig 4C), the experimentals in the unscreened arena turn towards the TED of the alien shore (Fig. 4D).

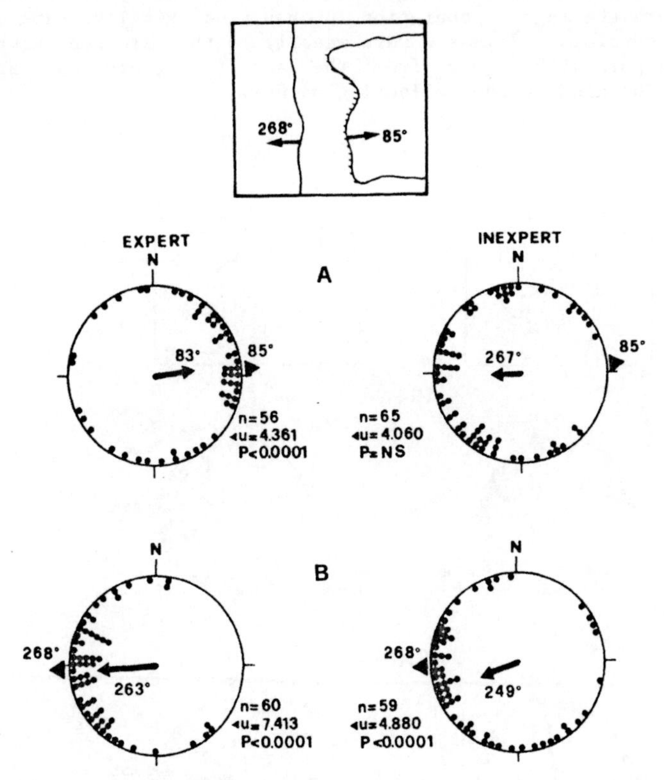

Fig. 6. Solar orientation in experienced adults and inexperienced young, tested in the arena away from the sea. TED=85°. For further explanations see Fig. 5.

2) Controls and experimentals (clock-shifted back 9 hours) were released on their home beach in the screened and unscreened arena. The controls in the screened arena are well clustered around their TED (Fig. 4E) but orientation improved in the arena without a screen (Fig. 4F). The mean vector of the experimental screened sandhoppers points towards the TED (Fig.4G), whilst without the screen points halfway between the TED and the direction indicated by the landscape (Fig. 4H).

Even when T. saltator is heading seawards it can still to see the dune and the retrodunal vegetation on account of its wide visual field (146 for each eye, Beugnon et al. 1987).

3. MODIFICATION IN ORIENTATION AFTER NATURAL DISPLACEMENT TO A DIFFERENT BEACH

After active displacement (daily migrations in search for food,

Geppetti and Tongiorgi, 1967a;b), or passive displacement (e.g. caused by sea storms or currents), some individuals may find themselves on a differently orientated beach from the first. In these circumstances, it is of vital importance for the individual to adapt its own orientation to the new sea-land axis. To investigate this possibility, a series of experiments was carried out in the summer of 1985 and 1986 on two populations of sandhoppers living on the banks of lagoons behind the sand dunes. The populations living on the seashore immediately in front of the dunes, where the lagoon population probably originally came from, were taken as controls. Releases were made 1) in the screened arena at the collecting points, 2) away from the sea (in Florence), and 3) directly onto the sand at the collecting points.

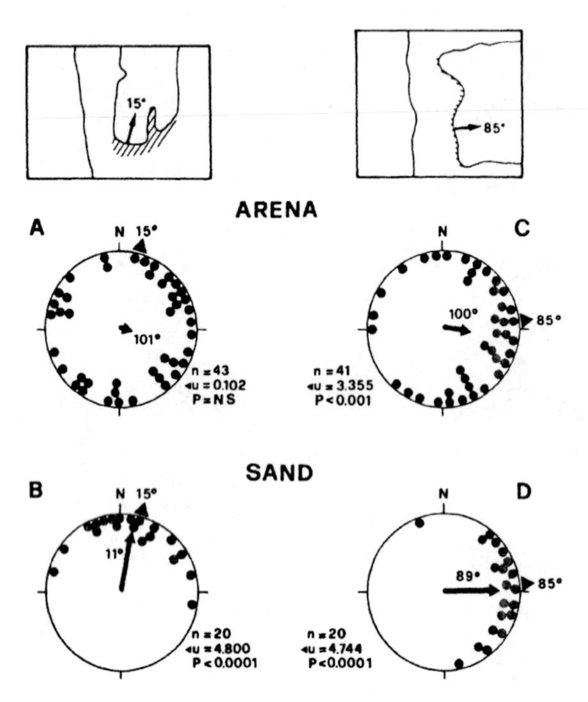

Fig. 7. Solar orientation in adults from the lagoons. A and C, tests in screened arena; B and D, releases on sand at collecting points. For further explanations see Fig. 1.

The experienced adults from the banks of the lagoon (TED=15°) near the River Serchio Estuary (Province of Pisa) (Fig. 5A left) are dispersed, but the sandhoppers from the sea shore immediately in front of the lagoon are well clustered round their TED (=264°) (Fig. 5B). The inexpert young born in the laboratory from both populations (Fig. 5 right) are clustered round the TED corresponding to the seashore. However, it must be stressed that this lagoon has a wide damp belt (approximately 5m wide along the Y axis) (Fig. 5 top) and that its banks are not stable throughout the year. A survey in November, 1986

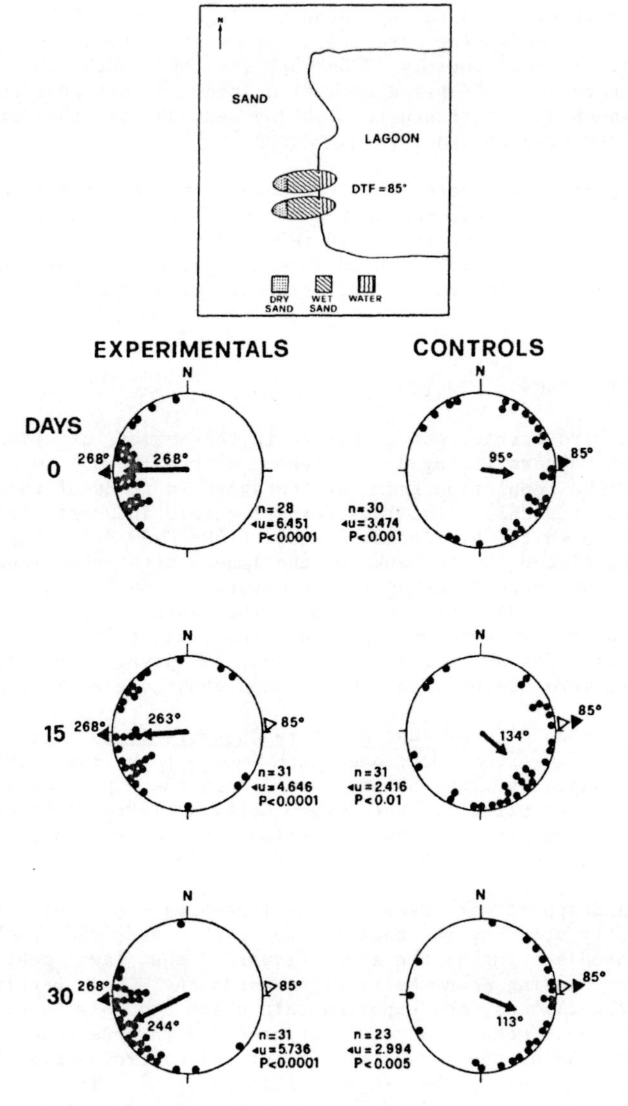

Fig. 8. Learning of new TED in adults. Experimentals, individuals collected on the seashore; controls, individuals collected on the lagoon banks. Left, length of time spent in the tanks on the lagoon shore. White triangles, new TED. The V test is calculated referring to the TED of home beach. Top, schematic representation of the tanks on the bank of the lagoon with TED=85°. For further explanations see text and Fig. 1.

found that the River Serchio had completely destroyed the banks of the lagoon at that point.

Sandhoppers from another small lagoon behind the dunes near Grosseto, South Tuscany, tested in Florence (Fig. 6A, left) showed a statistically significant clustering round the lagoon TED (=85°), whilst their laboratory born offspring were still orientated towards the sea (Fig. 6A, right). In the controls (TED=268°, Fig. 6B), both the adults and their laboratory born offspring showed a seaward orientation. The banks of this lagoon do not fluctuate with the seasons, and the damp belt is narrow (approximately 1m wide) (Fig. 6 top).

When the lagoon sandhoppers were released in the arena at the collecting points (Fig. 7A,C), results were similar to those of the previous experiments, but when they were released directly onto the sand (Fig. 7B,D), those with TED=15° pointed in a mean direction corresponding to the TED of the collecting point, whilst those with TED=85° improved their orientation.

4. THE ABILITY TO LEARN A NEW TED

Experiments were carried out in nature in the summer of 1986. The population of sandhoppers living by the lagoon with a TED=85° were used as controls, and the population from the sea shore in front of the dunes (TED=268°) as experimentals. Two lots (experimentals and controls) were put separately into white plexiglass tanks (140x40x30 cm). The tanks (Fig. 8 top) were placed on the banks of the lagoon with the long axis perpendicular to the shore, keeping to the natural slope of the ground; one end of the tank always contained water. The adults were tested in the screened arena before they were put into the tank, then again after they had been inside for approximately 15 and 30 days. The offspring born in the tanks were tested when they reached about one week of age.

The results (Fig. 8) show that adult sandhoppers do not alter their original orientation. After about one month in the tank, the difference between the mean angle and the TED was only 24° in the experimentals, and clustering remained practically the same (Fig. 8 left). The control adults showed a constant escape direction and clustering, in the direction of the water of the lagoon (Fig. 8 right).

When the sandhoppers were made to flee landwards in their typical escape reaction, by stirring the sand in the tank, and the number of individuals counted heading up the slope (towards the land behind the lagoon) and those heading down the slope (towards the land backing the sea = water in the lagoon), the experimentals which had only been in the tank for one hour were seen to move significantly downwards (i.e. in the direction of the land on the seashore), whilst the controls opted for the landward direction behind the lagoon (Fig. 9 top). The difference between experimentals and controls proved to be statistically significant. After 10 days in the tank, the experimentals behaved in more or less the same way as the controls and headed towards the land behind the lagoon (Fig. 9 centre). After about one month they still behaved in the same way (Fig. 9 bottom).

The young sandhoppers born in the tanks from the experimentals and tested in the arena, turned in a different mean direction from their parents, pointing towards the lagoon TED and not that of the seashore (Fig. 10 left). The offspring born in the tanks from lagoon-dwelling controls were also significantly clustered round the lagoon TED (Fig. 10 right), and not round the seashore TED as those born in the laboratory were (see Fig. 6A, right).

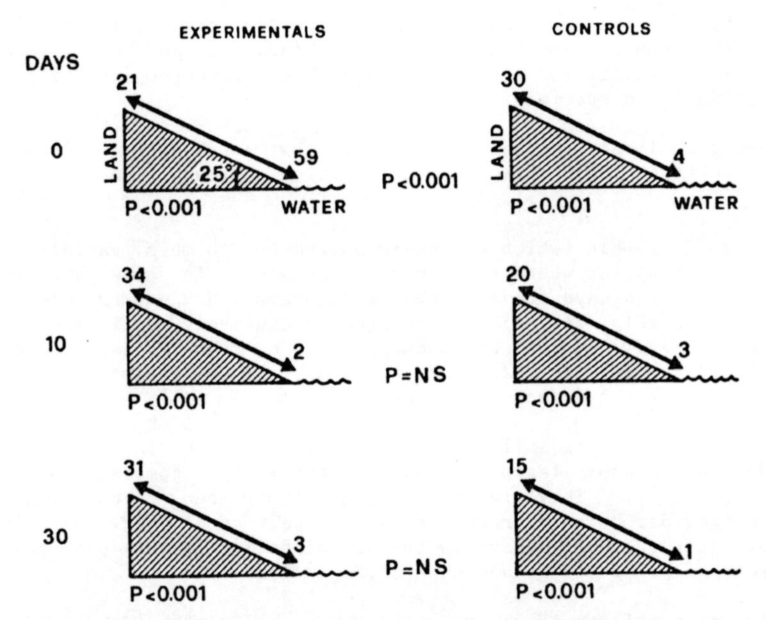

Fig. 9. Escape reaction in adults provoked by stirring the sand in the tanks. The numbers over the arrows represent the number of sandhoppers moving up (land) or down (water) in the tanks. Left, time spent in the tanks. Comparisons between frequencies were made with X^2 test (probability level, P, is also given). See text for further explanations.

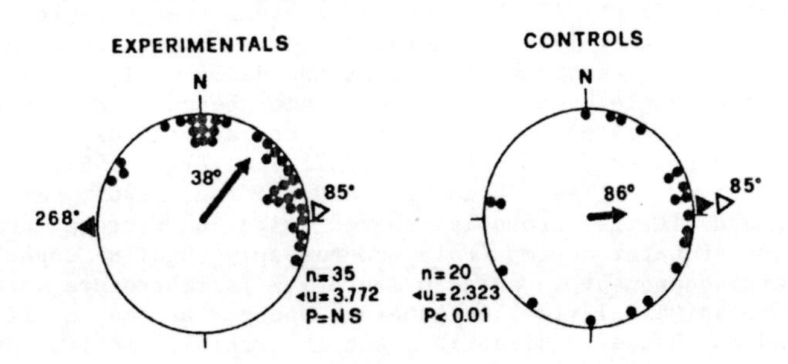

Fig. 10. Learning of new TED in young born in the tanks (about one week old). For further explanations see Figs. 1,9.

CONCLUSIONS

Results lead to the conclusion that the sandhoppers can use local, non-celestial cues (landscape and slope of the substratum) to improve their solar orientation along the sea-land axis, regardless of any learning processes (Figs. 2, 3, 4 and 7). They also learn how to perfect their sun compass mechanism when they refer to these cues. In fact young sandhoppers born in the laboratory are significantly worse in their solar orientation than their experienced parents (Tab. 1), and experienced adults kept in captivity for a long time are more dispersed on release than freshly gathered specimens (Fig. 1).

Depending on local physical conditions, the sandhoppers can also drastically change their original orientation through the learning process.

On a wide damp belt (which decreases motivation to move rapidly up and down the y axis), or where the shore fluctuates in time and the sandhoppers are not always present, the sandhoppers learn not to rely on their sun compass (Fig. 5A, 7A), but to steer exclusively by local, non celestial cues (Fig. 7C). If the sandhoppers come from a narrow damp belt on a stable shore, they can learn to find their new flight direction with the aid of the sun compass (Fig. 6A).

As for their learning ability, first results seem to indicate that adult sandhoppers cannot learn to adjust their sun compass to new situations (Fig. 9). They learn to find their new, ecologically efficient flight direction by referring to local cues (Fig. 10). The ability to adjust the sun compass to the new direction of the sea seems to be a prerogative of young sandhoppers (Fig. 11).

The learning ability in young Talitrus may at first appear redundant with the fact that escape direction is genetically controlled, which should be sufficient in itself to guarantee a mean orientation in line with their own TED, even for populations living several kilometres apart on the same stretch of curved sandy coast line (Scapini et al., 1985). Local, non-celestial cues and compass references would be coadjutant factors in solar orientation, or could substitute the sun-compass if the need arises. However, it must be remembered that 1) the sun still remains the most important factor in orientation (see also Pardi and Ercolini, 1986), and since it is totally unaffected by local changes it can always be relied upon; and 2) the boundary line between water and the land is the least stable of all the ecotonal systems (see Hernkind, 1972; 1983; Vannini and Chelazzi, 1985). Therefore it is advantageous for the sandhoppers to be able to learn how to adapt the daily cycle of variation in their orientation angle even on their own home beach. Moreover, it has already been demonstrated that other riparian arthropods also have this capacity (wolf spiders, Papi et al., 1957; Tongiorgi, 1961; Papi and Tongiorgi, 1963; rove beetles, Ercolini and Badino, 1961; crickets, Beugnon, 1986), and it is probably shared with Dermaptera (Labidura riparia) and a fresh-water shrimp (Palaemonetes sp.) (Ugolini unpublished data). The learnt component in solar orientation is therefore probably more developed in animals living on shores where the coast line is subject to a sudden change in direction, and in rapidly moving animals which could often find themselves on differently orientated shores.

If future experiments prove that only young sandhoppers can adjust their sun compass to a new escape direction, then this ability would seem to be restricted to a limited period in the life of the sandhopper when it is least resistent to dehydration. Adult sandhoppers which actively or passively find themselves on a alien shore could still survive by referring to the landscape, ground slope, or some other orientating factor such as the earth's magnetic field (Arendse, 1980; Arendse and Kruyswijk, 1981) to maintain their zone. Their offsping, on the other hand, could learn to use their sun compass regardless of what their genetically inherited information tells them about the direction of their home shore.

REFERENCES

Arendse, M.C., 1980, Non-visual orientation in the sandhopper Talitrus saltator (Mont.). Netherl. J. Zool. 30:535-554.

Arendse, M.C. and Kruyswijk, C.J., 1981, Orientation of Talitrus saltator to magnetic fields. Netherl. J. Res. 15:23-32.

Batschelet, E., 1981, Circular statistic in biology. Academic Press, London.

Beugnon, G., 1986, Development of cross-shoreline orientation in crickets, in: "Orientation in space", G. Beugnon, ed., Privat, Toulouse.

Beugnon, G., Lambin, M. and Ugolini, A., 1987, Visual and binocular field size in Talitrus saltator Montagu (Crustacea, Amphipoda). Monitore zool. Ital. (NS) 21: 151-155.

Craig, P.C., 1973, Behaviour and distribution of the sand-beach amphipod Orchestoidea corniculata. Mar. Biol. 23: 101-109.

Edwards, J.M. and Naylor, E., 1987, Endogenous circadian changes in orientational behaviour of Talitrus saltator. J. mar. biol. Ass. U.K. 67:17-26.

Ercolini, A., 1963 Ricerche sull'orientamento solare degli anfipodi. La variazione dell'orientamento in cattivita'. Arch. zool. ital.48:147-179.

Ercolini, A. and Badino, G., 1961, L'orientamento astronomico di Paederus rubrothoracicus Goeze (Coleoptera, Staphylinidae). Boll. Zool. 28:421-432.

Ercolini, A. and Scapini, F., 1974 Sun compass and shore slope in the orientation of littoral amphipods (Talitrus saltator Montagu). Monitore zool. ital. (NS) 8:85-115.

Geppetti, L. and Tongiorgi, P., 1967a, Nocturnal migration of Talitrus saltator (Montagu) (Crustacea, Amphipoda). Monitore zool. ital. (NS) 1:37-40.

Geppetti, L. and Tongiorgi, P., 1976b, Ricerche ecologiche sugli artropodi di una spiaggia sabbiosa del litorale tirrenico. II. Le migrazioni di Talitrus saltator (Montagu) (Crustacea, Amphipoda). Redia 50:309-336.

Hartwick, R.F., 1967, Beach orientation in talitrid amphipods: capacities and strategies. Behav. Ecol. Sociobiol. 1:447-458.

Herrnkind, W.F., 1972, Orientation in shore-living arthropods, especially the sand fiddler crab, in: "Behavior of marine animals. Vol. 1. Invertebrates", H.E. Winn and B.L. Olla, eds., Plenum Press, New York London.

Herrnkind, W.F., 1983, Movement patterns and orientation, in: "The biology of Crustacea. Vol 7. Behavior and Ecology", D.E. Bliss, F.J. Vernberg and W.B. Vernberg, eds., Academic Press, New York.

Papi, F. and Pardi, L., 1953, Ricerche sull'orientamento di Talitrus saltator (Montagu) (Crustacea Amphipoda). II. Sui fattori che regolano la variazione dell'angolo di orientamento nel corso del giorno. L'orientamento di notte. L'orientamento diurno di altre popolazioni. Z. vergl. Physiol. 35:490-518.

Papi, F., Serretti,L. and Parrini, S., 1957, Nuove ricerche sull'orientamento ed il senso del tempo in Arctosa perita (Latr.) (Araneae, Lycosidae). Z. vergl. Physiol. 39:531-561.

Papi, F. and Tongiorgi, P., 1963, Innate and learned components in the astronomical orientation of wolf spiders. Ergebn. Biol. 26:259-280.

Pardi, L., 1960, Innate components in solar orientation of littoral amphipods. Cold Spring Harb. Symp. quant. Biol. 25:394-401.

Pardi, L. and Ercolini, A., 1986, Zonal recovery mechanisms in talitrid crustaceans. Boll. Zool. 53:139-160.

Pardi, L., Ercolini, A., Marchionni, V. and Nicola, C., 1958, Ricerche sull'orientamento degli anfipodi del litorale: il comportamento degli individui allevati in laboratorio sino dall'abbandono del marsupio. Atti Accad. Sci. Torino (Cl. Sci. fis. mat. nat.) 92:1-8.

Pardi, L. and Papi, F., 1952, Die Sonne als Kompass bei Talitrus saltator (Montagu) (amphipoda Talitridae). Naturwissenschaften 39:262-263.

Pardi, L. and Papi, F., 1953, Ricerche sull'orientamento di Talitrus saltator (Montagu) (Crustacea Amphipoda). I. L'orientamento durante il giorno in una popolazione del litorale tirrenico. Z. vergl. Physiol. 35:459-489.

Pardi, L. and Scapini, F., 1979, Solar orientation and landscape visibility in Talitrus saltator. Monitore zool. ital. (NS) 13:210-211.

Scapini, F., 1986a, Inheritance of solar direction finding in sandhoppers. 4. Variation in the accuracy of orientation with age. Monitore zool. ital. (NS) 20:53-61.

Scapini, F., 1986b, Inheritance of solar direction finding in sandhoppers, in: "Orientation in space", G. Beugnon, ed., Privat, Toulouse.

Scapini, F. and Ugolini, A., 1981, Influence of landscape on the orientation of Talitrus saltator. Monitore zool. ital. (NS) 15:324-325.

Scapini, F., Ugolini, A. and Pardi, L., 1985, Inheritance of solar direction finding in sandhoppers. II. Differences in arcuated coastlines. J. Comp. Physiol. 156:729-735.

Scapini, F., Ugolini, A. and Pardi, L., 1987, Aspects of directional findigs inheritance in natural populations of littoral sandhoppers (Talitrus saltator). This Vol., pp.

Tongiorgi, P., 1961, Sulle relazioni tra habitat ed orientamento astronomico in alcuna specie del gen. Arctosa (Araneae, Lycosidae). Boll. Zool. 28:683-689.

Ugolini, A. and Scapini, F., 1987, Orientation of the sandhopper Talitrus saltator Montagu (Amphipoda, Talitridae) living on dynamic sandy shores. J. Comp. Physiol. A, IN PRESS.

Ugolini, A., Scapini, F. and Pardi, L., 1984a, Solar orientation in Talitrus saltator Montagu (Crustacea Amphipoda) from retro-dunal lagoons. Monitore zool. ital. (NS) 18:181-182.

Ugolini, A., Scapini, F. and Pardi, L., 1984b, Importanza della visione del paesaggio nell'orientamento zonale di Talitrus saltator Montagu (Crustacea Amphipoda). Boll. Zool. (Suppl.) 51:109.

Ugolini, A., Scapini, F. and Pardi, L., 1986, Interaction between solar orientation and vision of landscape in Talitrus saltator Montagu (Crustacea Amphipoda). Mar. Biol. 90:449-460.

Vannini, M. and Chelazzi, G., 1985, Adattamenti comportamentali alla vita intertidale tropicale. Oebalia (NS) 11:23-37.

Williamson, D.I., 1954, Landward and seaward movements of the sand-hopper Talitrus saltator. Adv. Sci. 41:71-73.

Foraging Excursion and Homing in the Tropical Crab *Eriphia smithi*

Marco Vannini and Francesca Gherardi
University of Florence, Italy

INTRODUCTION

Crabs of the genus Eriphia are large and common Xanthoids, inhabiting most temperate and tropical rocky shores (Reynolds and Reynolds, 1977, Zipser and Vermeij, 1978).

Eriphia smithi McLeay, a common Indo-Pacific species, is an active intertidal predator (Vannini et al., 1987), which lives well hidden in the cliff holes during HW and migrates along the intertidal rocky platform in search for food as far as 50 m from the cliff at LW. Migrations are longer and the number of migrating crabs is higher at ST than at NT, at night than in the day, and among adults than among young (Vannini, 1987).

The hole is a very important resource for avoiding predation during HW, for moulting and for copulation; some crabs were seen to occupy the same burrow for as long as 20 days (Vannini, 1987). The hypothesis was made that the crabs must be endowed with a refined homing system for getting back to the cliff, after foraging excursions (especially on the nocturnal moonless LWST). The study of this mechanism was the aim of this research.

MATERIAL AND METHODS

Experiments were made in October 1986, in a small gulf (Awadaxan, 19 Km South of Mogadishu, Somalia) where previous observations on feeding habits and migration periodicity were made in July 1984 (Vannini et al., 1987, Vannini, 1987).

The locality is bordered by a rocky cliff about 4 m high (Fig. 1). Most of the crabs inhabit the many holes and crevices on the lower portion of the cliff surface while only few of them were found dwelling in the scattered holes along the intertidal platform.

Two kind of experiments were set up to i) describe the paths made by a small number of marked crabs, spontaneously wandering on the platform or after passive displacement, ii) measure the orientation (i.e. the final azimuth) of a larger number of animals, released after displacement.

i) Path description. By applying different coloured LEDs, encapsulated in an epoxide resin cast together with a small battery, to the carapace cardiac region, it was possible to observe the crabs from as far as 100 m. In many cases yellow-green fishing lights were also used. The crabs were captured and, after the light was attached, immediately released in the same capture site. The whole procedure did not take more than one minute. The light lasted much longer than a single LW.

Two observers, using special binoculars with a compass incorporated, could simultaneously observe the same crab and record the respective angle from which the crab was seen.

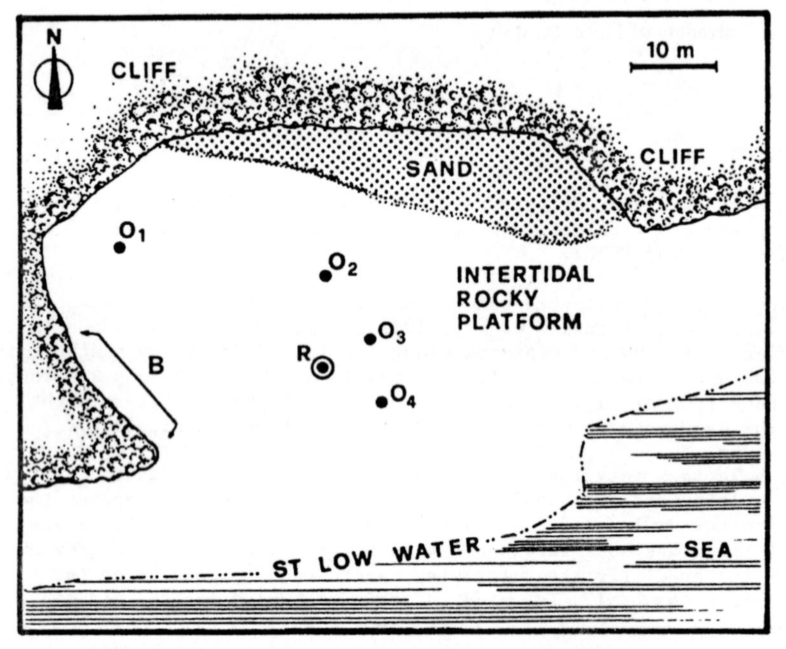

Fig. 1. Awadaxan bay: R, release point; 01-04, observation points; B, the cliff section with the highest hole density and where all the Awadaxan experimental crabs were collected.

Thanks to the different coloured lights, 4-5 crabs could be followed during the same LW. The LED colour, both angles and the exact observation time were continually registered on a tape recorder, for all the crabs in turn, for the entire duration of LW (about 3-4 hours). The two observers were sitting in the geometrically most suitable position, i.e. at roughly 90° from each other. Four observation points had been previously defined on the platform and one or both of the observers could easily switch from one point to the other, in order to continually maintain optimal observing distance and angle.

The binocular reading error was about +/- 1° ; since the average observation distance was about 25 m, the average error was about 0.5 m (25 x sin(1) = 0.44). Combining the errors from both observers gives an average incertitude area of less than 1 sqm.

Knowing the relative position of the cliff and of the various observation points, the sequences of angle measurements could easily be transformed into cartesian coordinates (using a program with a few elementary trigonometric rules), which finally allowed the actual path to be plotted.

ii) Experimental releases. Crabs were taken from the experimental area or other areas and, once the light attached on their carapace, released in the centre of the platform (Fig. 1).

The crabs were released 2-3 hours before the tide rose. Their position was recorded just before they reached the cliff or, if they were still wandering around, as soon as the first waves reached the platform; crabs which did not walk more than 5 m from the release point were not considered.

Some crabs were blinded using dental resin; the eyestalks were gently lain on the bottom of the orbit and then covered with the resin.

RESULTS

i) Path description. Our first intent was to have an accurate description of the path the crabs spontaneously left from the outset of their journey on the platform to their re-entry at the cliff.

Three out of the 17 recorded paths are described at ST (Fig. 2) and three at NT (Fig. 3). ST excursions are longer (time spent in foraging at ST is about twice as long as at NT, Vannini, 1987) but basically very similar. It is impossible to say if the crabs, on their return to the cliff, go back exactly the same spot from where they left on their migration; we could only put lights on crabs which had already started the migratory phase and we never had any definite information on the holes from which they emerged.

In some cases the backward excursion overlaps the outward journey but in most cases this is not true. The estimated error in the path reconstruction (+/- .5 m) is often largely lower than the distance between the two paths (es. Fig. 2, a). The average possible path overlapping ratio, among the 12 longest paths, was estimated at 27.6% (+/- 10.6) which would suggest that the crabs' homing performances do not rely on chemical cues left on their outward journies.

Experiments were also made by removing eight crabs from the cliff and releasing them from the centre of the platform (Fig. 4). These crabs were well oriented towards the cliff and moved straight back to it. The direct experience of walking during the outwards phase thus seems superflous for correct orientation during the return phase.

The above experiments suggest that visual cues play a fundamental role in orienting the crabs towards the cliff.

To test the existence of skototactic orienting mechanisms, five crabs from a shore oriented 170° , about 100 m away from the study area, were taken and tested together with normal local crabs and bilaterally and monolaterally blinded local crabs (Fig 5).

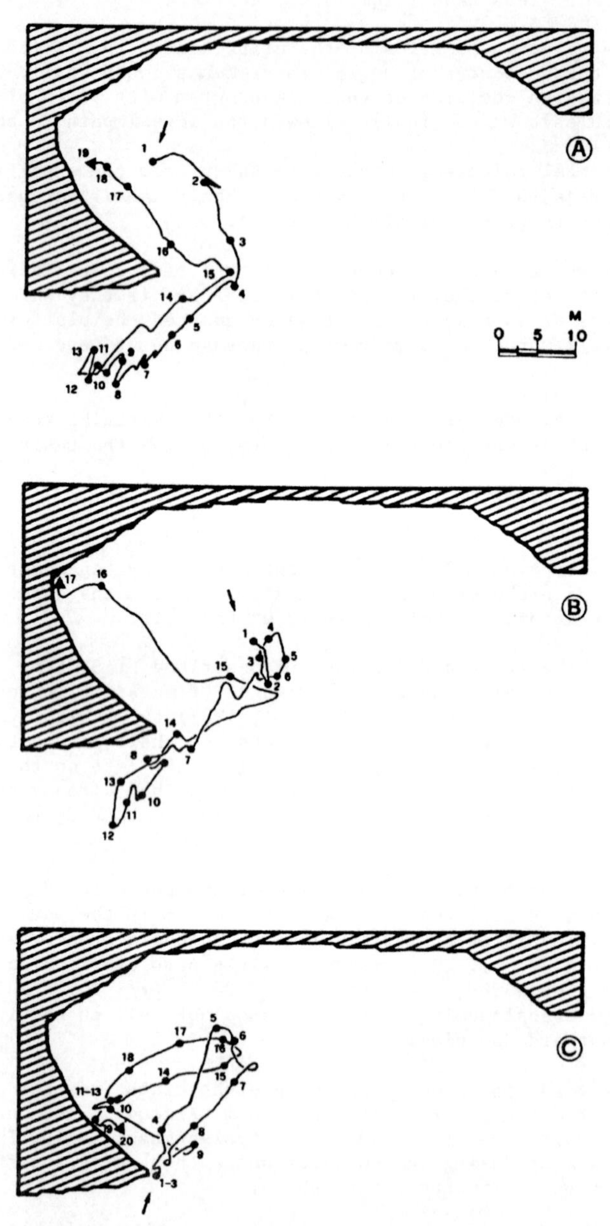

Fig. 2. Awadaxan crabs. Spontaneous path left on a moonless ST. The arrow indicates the light attachment; a dot every ten minuts. A female, 35 mm carapace length; B, female 37.2; C, female 40.1.

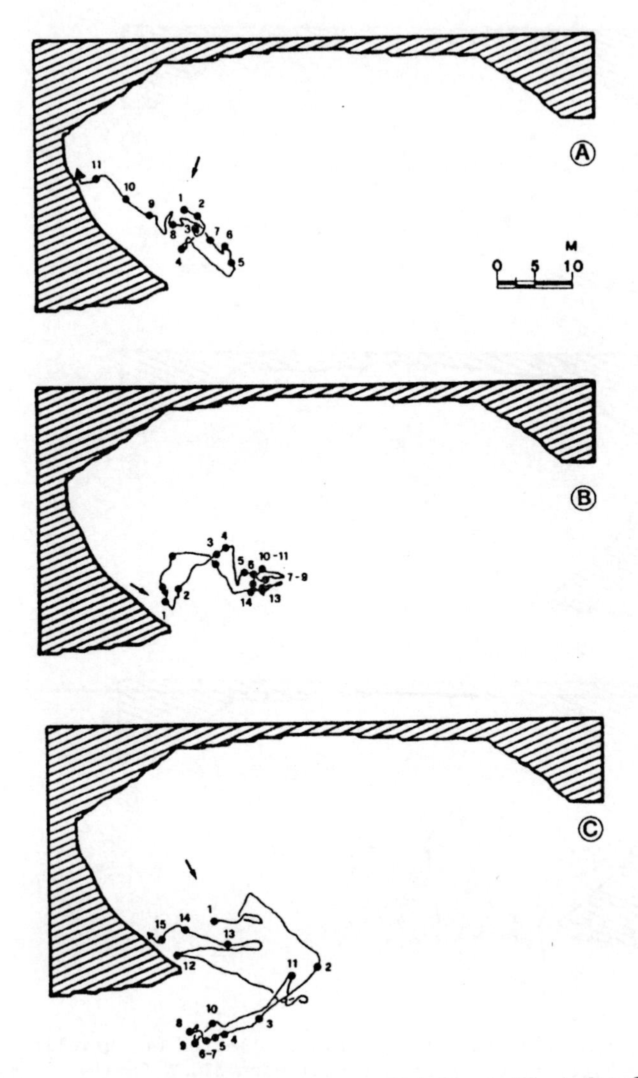

Fig. 3. Awadaxan crabs. Spontaneous path left at NT. A female, 35 mm carapace length; B, female 37.2; C, female 40.1.

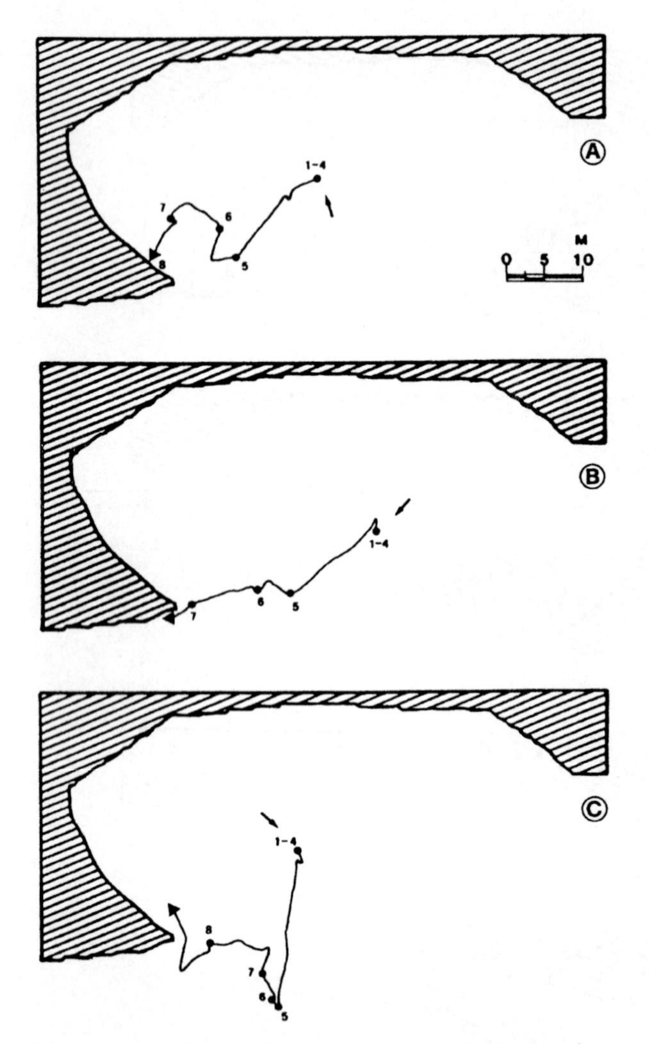

Fig. 4. Awadaxan crabs. Paths left by crabs collected on the cliff and experimentally traslocated, at full moon ST. A female, 35 mm carapace length; B, female 37.2; C, female 40.1.

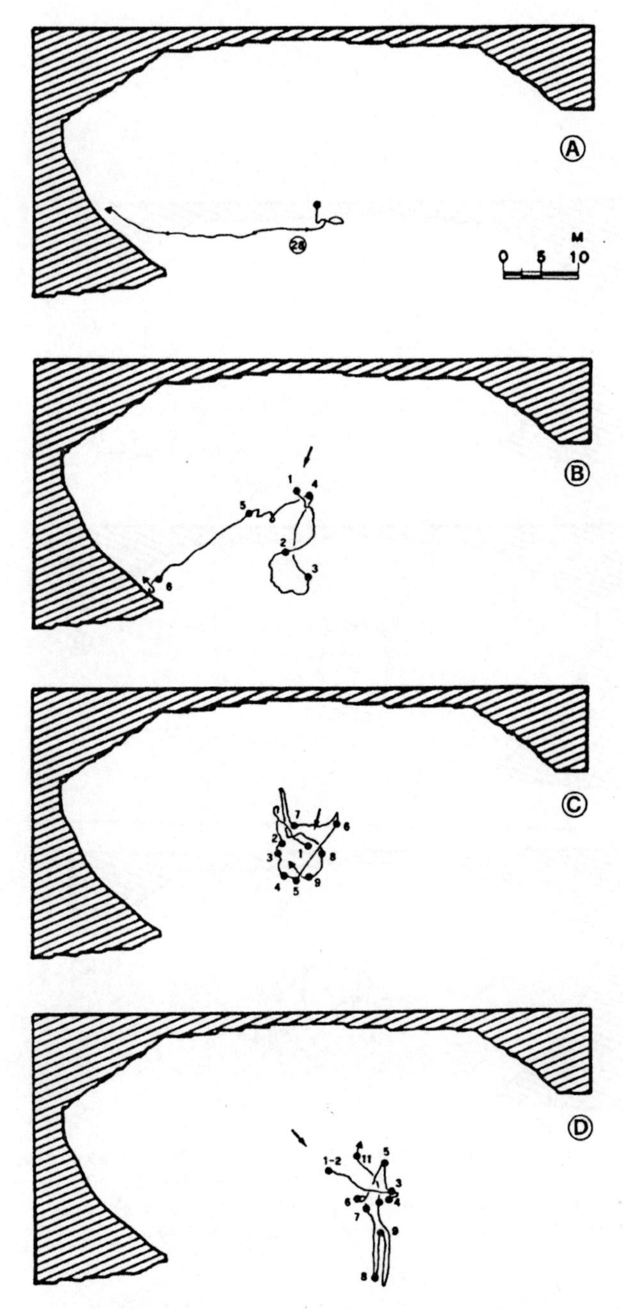

Fig. 5. Awadaxan crabs. Paths left by crabs collected on the cliff and experimentally traslocated, at full moon ST. A, intact female, 35 mm carapace length; B, monolaterally blinded female 37.2; C and D, bilaterally blinded females 40.1 and 39.7, respectively.

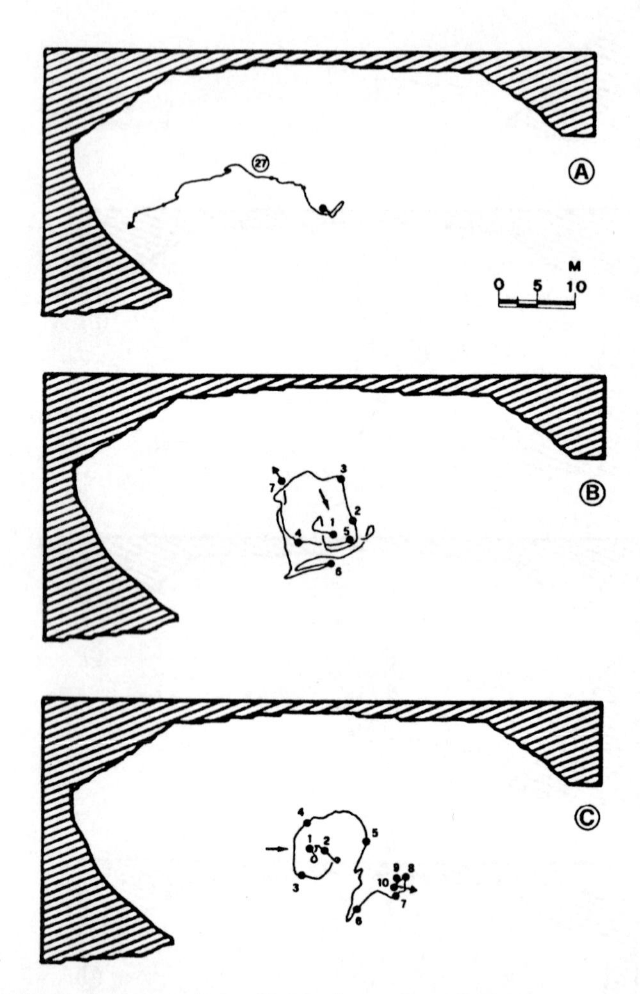

Fig. 6. Paths left by crabs collected on the cliff and experimentally displaced, at full moon ST. A, Awadaxan female, 35 mm carapace length; B and C, females 40.1 and 39.7, respectively, from a straight cliff, 100 m far from Awadaxan bay.

Five crabs from each category were in fact tested but crabs other than local ones, did not usually walk very far from the release point and immediately tried to bury themselves in the seaweed.

Blind crabs and intact foreign crabs appear totally disoriented, while intact local crabs are well oriented as usual; the single monolaterally blinded animal which left the release point, after a straight outward excursion, assumed a correct orientation.

The fact that foreign crabs and blind local animals both appear disoriented, obviously can not be explained by assuming that skototaxis plays a relevant role in the homing behaviour of E. smithi.

ii) Orientation. Leaving aside the actual tracks left by the crabs and recording, after a certain time, only the final azimuth and distance from release point of several crabs at the same time, it is possible to collect information (albeit less precise) on a much larger number of animals. Experiments were made between ST and NT.

Local crabs were released together with a sample from a population coming from an area with a different orientation and cliff shape (Fig. 7).

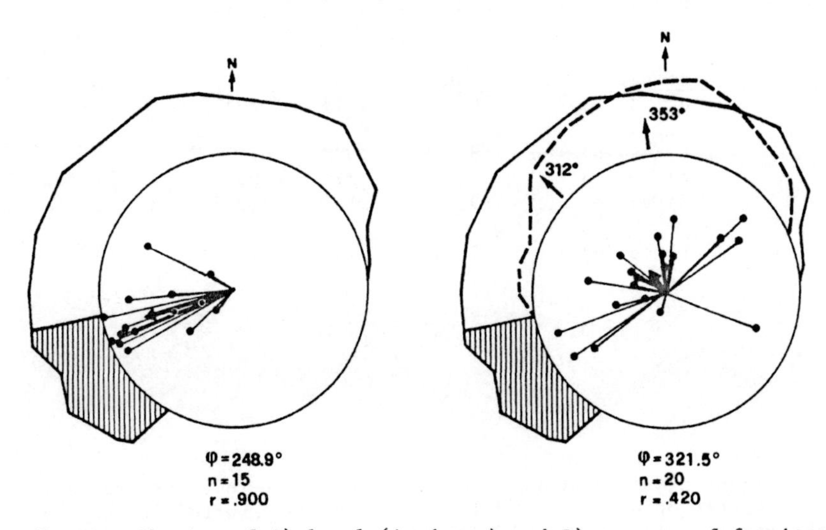

Fig. 7. Distribution of A) local (Awadaxan) and B) a group of foreigner crabs from a straight cliff, after release in the centre of Awadaxan bay. Lines outside the circles represent the height of the cliff of Awadaxan bay (continuous line) and of the straight cliff (dashed line). Shaded area, the cliff sector from which the Awadaxan crabs were collected. The dark arrows inside the circles represent the mean vectors while the external arrows represent the skototactic 'baricenter' of the two cliffs. Circle diameter, 30 m; cliff height in arbitrary angular units. The lines connecting the dots with the center of the circle do not represent the actual crab paths but the distance from the center at recapture.

If the crabs were guided by simple skototaxis they should have moved towards the cliff, no matter the population they belonged to.

The local crabs did not simply walk towards the cliff but were very concentrated (Table 1, a) around the direction of 249° corresponding to the cliff sector from which they were removed (Fig. 7, shaded area), as if they could identify that particular area from the rest of the cliff, and were not just relying on a simple skototactic mechanism.

Table 1. Statistics of orientation experiments: Rayleigh test (r), Moore test (D) and Watson test (U) (Batschelet, 1981).

tests :	n	angle	r	P	D	P
a) Awadaxan	15	248.8	.900	<.001	1.939	<.001
b) foreigners	20	321.4	.418	<.05	.910	n.s.
c) intact	9	236.7	.942	<.001	1.607	<.001
d) monolat. blinded	9	248.3	.653	<.02	1.230	<.01
e) bilater. blinded	9	263.2	.028	n.s.	.928	n.s.
f) Awadaxan	7	258.4	.897	<.001	1.426	<.001
g) Southern bay	13	258.7	.693	<.001	1.531	<.001

comparisons:	n	U	P
a) vs b)	20, 15	.343	<.005
c) vs d)	9, 9	.088	n.s.
c) vs e)	9, 9	.340	<.002
d) vs e)	9, 9	.162	>=.05
f) vs g)	7, 13	.102	n.s.

The foreigner crabs, although they are very scattered, seem to walk towards the center of the cliff (321°) as would be expected if they were simply skototactic. Their distribution is anyway partly an artifact. Crabs moving seawards sometimes reached the water and thus were "rebounding" back in another direction. How much this effect is responsible for the slight crab concentration (Table 1, b, r=.418) on the land side of the platform and what may be due to a possible feeble skototaxis is not possible to ascertain. If the Moore test is applied, i.e. taking into account not only the azimuths of the crabs but also their distances from the release points, the significance disappears (D=.910). The foreigner group distribution was in any case very different from that of the Awadaxan crabs (Table 1, a vs b).

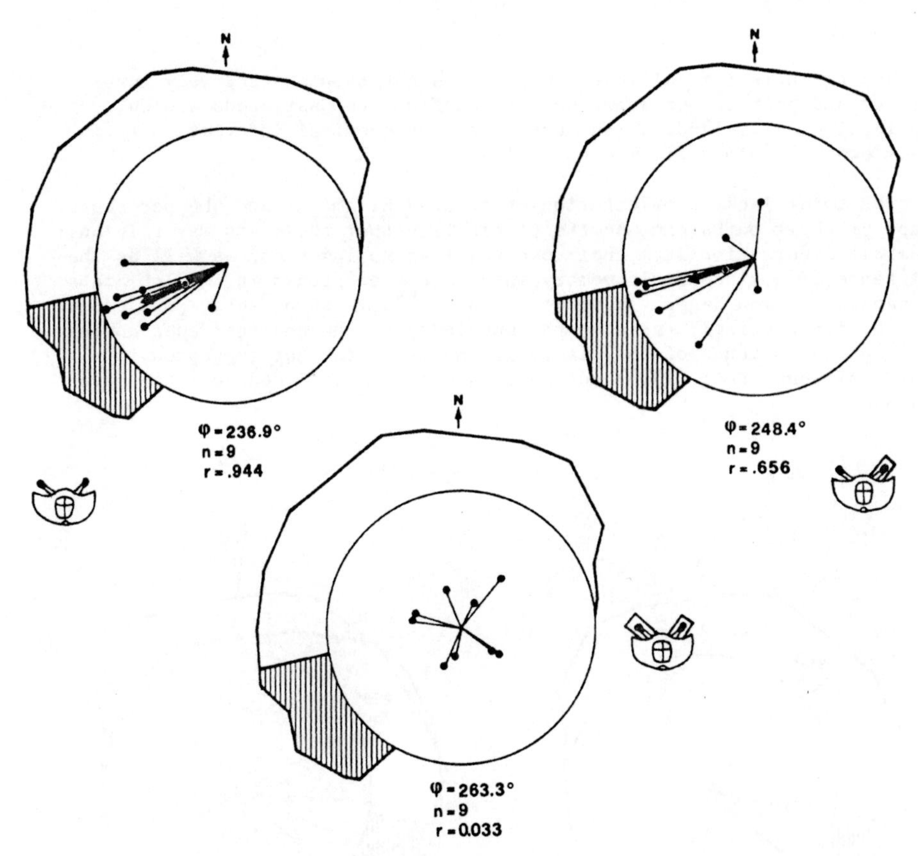

Fig. 8. Distribution of Awadaxan crabs, A, intact, B, monolaterally blinded and C, bilaterally blinded.

A second test was made by using local crabs which were blinded, monolaterally blinded or intact (Fig. 8).

Blinded crabs were dispersed while most of the crabs belonging to both control groups moved correctly towards the cliff thus, once again, excluding any chemical mapping of the substratum or any sort of chemical trail following (Table 1, c, d, e).

A further experiment was set up with local crabs and two populations of foreigner crabs coming from a small bay where the shape of the cliff and its orientation were similar to that of Awadaxan; the cliff baricenter was 295 and the highest crabs concentration was in the southernmost cliff section.

Local crabs (Fig. 9) still behave as in previous experiments, while foreigners now seem less dispersed than those in Fig. 7, which inhabited a differently shaped area. In this case, whilst less concentrated than locals, foreigner crabs orient towards the same part of the cliff as the former (Table 1, f, g).

DISCUSSION

Homing in intertidal invertebrates has not been exstensively investigated and most of our knowledges is confined to Gastropods and Chitons (Chelazzi et al., 1988) which mostly rely on chemical information from their own or conspecific mucous trails.

The total lack of orientation of blinded E. smithi and the particular shape of their paths exclude the possibility that the crabs may rely on chemical information from their own trail as molluscs do, as well as the existence of a kinesthetic memory which has been proved in fiddler crabs (Altevogt and von Hagen, 1964; Herrnkind, 1972). An olfactive map could also be hypothesized, based on the knowledge of the chemical cues coming from various patches of the platform, analogous to that suggested in homing pigeons (Papi, 1976) but this also can be excluded for the above reasons.

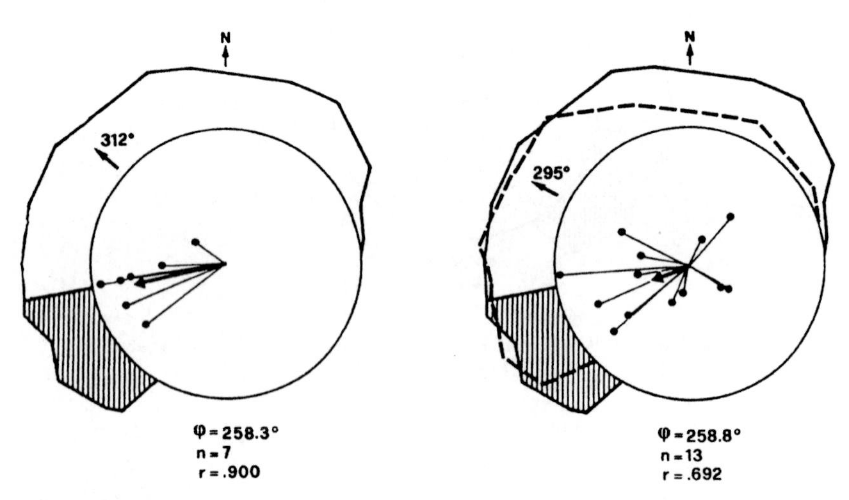

Fig. 9. Distribution of A, Awadaxan crabs and B, crabs from a bay South of Awadaxan, with similarly shaped cliff; these crabs also inhabited the extreme SW sector of their cliff.

Another common orienting mechanism in intertidal animals is skototaxis. Skototaxis can always lead a shore animal, if dislodged by waves or through some other accidents, back to the coast by the shortest route. The darkest part of the landscape (even on a moonless night) is always represented by the shore line, and moving towards the darkest side will invariably guide the animal ashore. In exactly the same conditions of the present experiments, skototaxis alone proved to be sufficient to orient intertidal gastropods towards the cliff (Chelazzi and Vannini, 1976). In similar tests crustaceans as sandhoppers (Williamson, 1951; Ugolini et al., 1986), fiddler crabs (Herrnkind, 1972) and terrestrial hermit crabs (Vannini, 1975) could also orient by skototaxis.

The present results cannot exclude the existence of a skototactic influence in E. smithi behaviour, but they show that, if present at all, it represents nothing but an accessory orienting mechanism.

Terrestrial hermit crabs, on a large flat beach, also orient easily towards rocks and trees where they can find shade and refuge or, in their absence, they will try to climb the beach and cross the dunes; in any case a difference was observed between hermit crabs from the beach where the tests were made, and foreign animals (Vannini, 1975; Vannini and Chelazzi, 1981).

Expert E. smithi do not move towards the darkest sector of their landscape but towards a defined part of the cliff, the only part rich in holes and the one where they actually live.

Foreigner crabs, coming from a differently shaped (straight) cliff are lost when tested in the Awadaxan semicircular bay; instead of assuming a skototactic orientation and walking straight towards the land (and then searching for any suitable hole) they keep wandering around on the platform (sometimes immediately hiding among the seaweed roots and remaining immobile for 2-3 hours) without assuming any special orientation unless, maybe, a slight skototactic one.

Crabs collected from other semicircular bays seem to 'recognize' the Awadaxan bay as their own, and are not as lost like the crabs from the straight cliff.

Tests in the latter case are exiguous, but generally they do confirm that the whole alimentary excursion of these crabs is controlled by visual cues which refer to the special shape of the place they permanently inhabit.

Many intertidal animals living on coherent substrata go on periodical foraging excursions, even though they are basically isospatial, according to recently suggested terminology (Vannini and Chelazzi, 1985). No survival strategy is possible, in these cases, without a refuge where to avoid the negative phase of the tidal wave. One should expect complex homing mechanisms to be the rule in most of these animals.

Suitable, empty holes are rare on the cliff where E. smithi was studied; crabs prefer to shorten their feeding excursions so as not to risk loosing their refuges. Males can only copulate if they own a burrow where they can attract a female; fights for hole possessions were frequently seen and the owner is usually the winner. We can conclude that suitable holes are a powerful limiting factor for these crab populations (Vannini, 1987).

It has been shown that the Awadaxan population is very stable and, in 1984, none of 120 marked animals, checked twice a day for about month, was ever seen to abandon the bay and join populations from adjacent bays (Vaninni, 1987).

In these conditions, it is not surprising that crabs had to develop such a refined orienting mechanism as the memorization and the visual identification of their 'home-cliff'.

How refined this mechanism is, how the crabs can switch and re-learn other landscape (if dislodged) and if they ever do, how other guiding cues (astronomical, chemical, anemotactic, reotactic, etc.) could be integrated in the crabs' homing has yet to be investigated.

ACKNOWLEDGEMENTS

Part of the responsability for this work lies with the local shark population which obliged the first author to interrupt his previous studies on subtidal crustaceans and to dedicate his attention on a safer intertidal species such as Eriphia smithi is. Many thanks are due to A. Ugolini for helpful suggestions in analyzing the data. The field work was supported by the Centro di Faunistica ed Ecologia Tropicali of C.N.R. and by the Ministero della Pubblica Istruzione ("40%" grant).

ABSTRACT

Eriphia smithi is an intertidal Xanthoid (Menippidae) crab very common along the Indo-west Pacific rocky shores. It lives in cliff holes from which it emerges at LW, foraging amomg the intertidal platform. During the nocturnal feeding excursions the crabs wander as far as 50 m from their burrows and their paths were studied by means of coloured LED, stuck on their carapaces. The kind of path they took showed that they find their way back to the cliff without using chemical cues nor information from the outward journey; blinded crabs cannot find their way back and crabs non living in the experimental area (Awadaxan bay) are not oriented and hardly reach the coast if released in the center of the bay. Releasing several crabs in the center of the bay and simply measuring their azimuth after a certain time confirms that blinded crab are not oriented while local crabs are very concentrated in the cliff direction. Crabs collected from a bay with a semicircular cliff, very similar to that of Awadaxan, seem less scattered than crabs coming from an area where the cliff was straight. The hypothesis is made that the crabs homing performance, during spontaneous migration or after passive displacement, may be based on specific recognition of familiar landscape.

LITTERATURE

Altevogt, R. and von Hagen H.-O., 1964, Ueber die Orientierung von Uca tangeri im Freiland, Z. Morphol. Oekol. Tiere, 53: 636-656.

Batschelet, E., 1981, Circular Statistics in Biology, Academic Press, London, New York.

Chelazzi, G., Focardi, S. and Deneubourg, J.-L., 1988, Analysis of movement patterns and orientation mechanisms in intertidal chitons and gastropods, in: "Behavioral adaptations to intertidal life", G. Chelazzi and M. Vannini, Eds., Plenum Press, New York.

Chelazzi, G. and Vannini, M., 1976, Researches on the coast of Somalia. The shore and the dune of Sar Uanle. 9. Coastwards orientation after displacement in Nerita textilis Dillwyn (Gastropoda Prosobranchia). Monitore zool. ital. (N.S. Suppl.), 8: 161-178.

Herrnkind, W. F., 1972, Orientation in shore-living arthropods, especially the sand fiddler crab, in: "Behavior of marine animals", Vol. 1, "Invertebrates",: 1-59, H. E. Winn and B. L. Olla, Eds., Plenum Press, New York.

Papi F., 1976, The olfactory navigation system of the homing pigeon, Verh. dt. zool. Ges., 69: 184-205.

Reynolds, W. W. and Reynolds, L. J., 1977, Zoogeography and the predator-prey 'arms race': a comparison of Eriphia and Nerita species from three faunal regions, Hydrobiologia, 56: 63-67.

Ugolini, A., Scapini, F. and Pardi, L., 1986, Interaction between solar orientation and vision of landscape in Talitrus saltator Montagu (Crustacea Amphipoda), Mar. Biol., 90: 449-460.

Vannini, M., 1975, Researches on the coast of Somalia. The shore and the dune of Sar Uanle. 4. Orientation and anemotaxis in the land hermit crab Coenobita rugosus (Milne Edwards), Monitore zool. ital. (N.S. Suppl.), 6: 57-90.

Vannini, M., 1987, Notes on the ecology and behaviour of Eriphia smithi, Monitore Zool. Ital. (N.S.), Suppl., 22: (in press).

Vannini, M. and Chelazzi, G., 1981, Orientation of Coenobita rugosus (Crustacea: Decapoda): a field study in Aldabra, Mar. Biol., 64: 135-140.

Vannini, M. and Chelazzi, G., 1985, Adattamenti comportamentali alla vita intertidale tropicale, Oebalia, 11: 23-37.

Vannini. M., Chelazzi, G. and Gherardi, F., 1987, Eriphia smithi feeding habits and gastropod predation, Mar. Biol.,(in press).

Williamson, D. I., 1951, Studies in the biology of Talitridae (Crustacea, Amphipoda): visual orientation in Talitrus saltator, J. Mar. Biol. Assoc. U. K., 30: 91-99.

Zipser, E. and Vermeij, G. J., 1978, Crushing behavior of tropical and temperate crabs, J. exp. Biol. Ecol., 31: 155-172.

When Limiting Factors aren't: Lessons from Land Crabs

Thomas G. Wolcott and Donna L. Wolcott
North Carolina State University
U.S.A.

INTRODUCTION

Marine animals invading the intertidal zone face a number of potentially stressful or lethal physical factors (e.g. extreme temperatures, desiccation). As physiological ecologists, we have set out to investigate the various sorts of mechanisms by which animals cope with intertidal and supratidal existence, and our research has a larger physiological emphasis than most other programs represented in this symposium. However, many of the adaptations for coping with the challenges of the physical environment are behavioral, instead of (or in addition to) physiological. In this paper we discuss how the concept of physical "limiting factors" has been used to elucidate the coping mechanisms of a series of intertidal and terrestrial crabs. Two types of behavioral adaptation are included in the discussion: 1) behaviors which contribute to homeostasis during stress and might be considered part of physiological mechanisms (e.g., behaviors for taking up water); and 2) behaviors which remove the animal from stressful microenvironments (e.g., selection of specific microhabitats, locomotion, behavioral range limitation).

"Limiting factors" are physical conditions which exceed the physiological tolerances of the species in question. In this discussion we are dealing with "proximate", not "ultimate" limiting factors. The distinction may be illustrated by an extreme example from a non-crustacean group. Intertidal limpets require rock surface for attachment. The vegetation and loose soil in meadows above sea cliffs represent an "ultimate" limiting factor for these animals; they could not live there if they tried. This is not particularly interesting, even though true, because the limpets do not "try" to live in meadows. The effect of the "ultimate" limiting factor is only theoretical. In contrast, "proximate" limiting factors have measurable, not just potential, effects on ecology. They stress or kill individuals attempting to colonize marginal habitat or (more typically, under fluctuating conditions) occasionally kill the portion of the population that has expanded into marginal habitat. The range of a species is thus in dynamic balance with a "proximal" limiting factor. An example of such a factor, for the limpets, is desiccation during tidal exposure. When the sea is unusually calm, no spray is thrown onto high intertidal rocks. Limpets that had expanded the species' range by moving up into the splash zone while it was wetter are killed, and the "range limit" of the population retreats.

A number of recent reviews contain descriptions of crustacean behavior used in coping with physical factors. Behavior in general is reviewed by Dunham, and ecology by T. G. Wolcott, in a forthcoming volume on the biology of land crabs (Burggren and McMahon, in press). Migratory behavior of terrestrial crustaceans has been reviewed by T. G. and D. L. Wolcott (1986). "Limiting factors" are often invoked, implicitly if not explicitly, in reviews of the behavioral and physiological adaptations used by crustaceans in colonizing land (Little, 1983; Powers and Bliss, 1983). Powers and Bliss (1983) state "For most species in most habitats...the major factors limiting density and distribution appear to be mainly physical...".

These reviews show that the isopods, amphipods and reptant decapods are the crustacean groups that have been particularly successful at invanding intertidal zones, and thence terrestrial environments. Anomuran and brachyuran crabs, by virtue of their large size, are the best known, and the semi-terrestrial and terrestrial brachyurans will be the principal focus of this paper. Several physical factors obviously represent potential stressors for crustaceans, and a number of typical adaptations have been studied; a brief overview follows.

Heat balance and water balance are intimately related in air, where evaporation can occur, and several adaptations affect both at once. Temperatures in subaerial habitats are generally more extreme, and fluctuate more rapidly, than those under water. Mechanisms of coping include evaporative cooling in species that can afford the water loss (e.g., mesic, but not xeric isopods). Behavioral thermoregulation is particularly evident among fiddler crabs, which modify absorbance of heat with chromatophoral color changes, and shuttle between cool burrows and the hot sediment surface.

Balance of both heat and water budgets is favored by behavioral selection of benign spatial and temporal microhabitats. Small species tend to be cryptic, and large ones to burrow. As illustrated in other papers of this symposium, behavioral rhythms are common (tidal near shore, diel inland), with animals becoming active during times when physiological risks are relatively low and resources most available.

Water balance is complicated by evaporation during exposure to air. Intertidal and terrestrial crustaceans compensate for evaporative losses by drinking or otherwise taking up water. Transpiration is minimized by various physiological and morphological adjustments which reduce overall permeability. Behavioral adaptations include kinetic responses to humidity, selection of humid microhabitats, conglobation (rolling into a ball), or aggregation. On the other hand, excess water is problematic for crustaceans adapted to living in air. Water contains relatively little oxygen, and in terrestrial environments is likely to be strongly hypo-osmotic to the animals. The more terrestrial crustaceans drown if immersed, and not surprisingly avoid excess water.

Crustaceans meet in various ways the challenge of maintaining gas exchange and acid-base regulation while exposed to air. Many retain a small volume of water in the gill chamber, and several species of crabs recirculate water from the branchial chambers over tracts of setae on the body surface. This apparently facilitates exchange of gases, and may permit excretion of nitrogenous waste into the branchial water and its subsequent loss to the atmosphere as gaseous ammonia. In air, ion exchange with the medium is not available for maintaining acid-base balance, and subaerially active crustaceans rely on respiratory mechanisms which involve the branchial water (Burnett and McMahon, 1987).

Reproduction while exposed to air must succeed without the buoyant support and osmotic buffering of the sea. A common behavioral adaptation is that copulation and ovulation become independent of the molt cycle. There is little in the way of premolt associations like the attendance and protection given aquatic premolt female crabs by the males who will mate with them immediately after they molt, while they are soft. The more terrestrial crustaceans typically copulate while both individuals are in the hard intermolt state.

Larvae and juveniles are less tolerant of environmental extremes than are adults. Brooding behavior is a "preadaptation" maintained in terrestrial amphipods and isopods. Virtually all semiterrestrial and terrestrial crabs are dependent on water for reproduction. The species which invaded land through the intertidal release planktonic larvae into the sea. Those of freshwater origin often have abbreviated development, carry larvae to advanced stages (Rabalais and Gore, 1985), and may have a short aquatic juvenile existence. Release of larvae or juveniles requires migrations from the adult habitat to the appropriate aquatic habitat (reviewed in T. G. and D. L. Wolcott, 1986).

The spawning migrations tend to be synchronous within a local population, typically peaking near the time of spring tides (full or new moons). Synchrony presumably places larvae in the plankton at optimal times for tidal transport, and may contribute to saturation of predators that attack ovigerous females. The incubation and egg-release behaviors associated with breeding migrations also tend to minimize losses of females and egg masses; females typically remain in burrows during incubation, avoid water of inappropriate salinity (which might cause premature hatching and/or death of larvae), and approach the sea in ways that minimize risks of damage by waves or marine predators (T. G. and D. L. Wolcott, 1982a).

Crustaceans that are active subaerially encounter physical conditions under which locomotion and sensory modalities can be quite different from those in water. The low density of the fluid medium requires more skeletal and muscular support, but its low viscosity and inertia permit very rapid movements like jumping from rock to rock ("Sally Lightfoot" crab, *Grapsus grapsus*) or running up to 2 m/s (ghost crabs, *Ocypode ceratopthalmus*; Hafeman and Hubbard, 1969). Sound can be detected directionally, either as airborne pressure waves or as substrate vibrations, and is used extensively for acoustic signaling and detection of prey (e.g., Klaassen, 1973; Horch, 1975). Objects are visible for long distances through air, and terrestrial crustaceans often have better vision than their aquatic relatives. In flat, open habitats crabs have developed long eyestalks with 360° vision. Flicker fusion rates and acuity (which is concentrated along the plane of the horizon) are comparable to those of insects (Zeil et. al., 1986). Visual information (polarized light, sun compasses, other celestial cues) is commonly used in orienting behavior relative to the shoreline and tidal gradient. Examples of oriented behavior include the daily and tidal migrations by talitrid amphipods and *Eriphia* presented in this symposium.

Very little of the existing literature deals with larval or juvenile crustaceans. The problems of studying vast numbers of tiny, dispersing, "disposable" organisms are obvious. Nevertheless, arriving at any complete understanding of behavioral adaptations of crustaceans to intertidal life will require study of all life history stages (see "future directions", below).

BEHAVIORAL ADAPTATIONS OF CRABS TO "LIMITING FACTORS"

How do physical factors affect distributions of inter-and supratidal decapod crustaceans? According to Little (1983), "...one would expect the major problems faced by decapods colonising the terrestrial environment to be concerned with salt and water balance and with reproduction...". The reproductive issues have been mentioned above, and the majority of this section will deal with salt and water balance.

Examination of intertidal zonation in limpets on rocky shores led T. G. Wolcott (1973) to hypothesize that behaviors resulting in observed distributions were predicated on the ratio between resources and risks accruing to individuals that moved outside the existing distribution.

If by extending the range animals gain access to abundant additional resources, while modestly increasing the risk of exceeding their physiological tolerances, then their fitness should be increased. The high reproductive output made possible by the additional resources will increase expected lifetime reproductive success despite the greater risk of death (shorter life expectancy). Under these conditions, natural selection should favor behaviors which lead to range extension to the limits of physiological tolerance, and we should expect to find physical "limiting factors".

On the other hand, if risks increase faster than availability of additional resources as the range is extended, then individuals that move into marginal habitat will have lower expected lifetime reprodutive success (fitness). Under these circumstances taking risks does not pay, and natural selection should favor conservative bahavior. Animals should select "safe", physiologically benign microhabitats, the range should be behaviorally limited, and physical "limiting factors" should not be evident.

To test if these principles are general to other intertidal systems, and have predictive value, we have examined another series of species showing vertical zonation: three species of "land crabs". The ghost crab *Ocypode quadrata* is principally an inhabitant of ocean beaches. It excavates its burrows in damp sand and has a high evaporation rate and low desiccation tolerance (Bliss, 1968; Herreid, 1969). The large (to 500g) Caribbean land crab *Cardisoma guanhumi* burrows in heavy lowland soils, typically near mangroves. The burrows extend below the water table and contain a reservoir of water (Herreid and Gifford, 1963). The evaporation rate of this species is lower, and the desiccation tolerance higher, than that of ghost crabs (Bliss, 1968; Herreid, 1969). The third species, the red land crab *Gecarcinus lateralis*, inhabits burrows in dry upland soils of Caribbean islands and Bermuda. It has the lowest evaporation rate and highest desiccation tolerance of the three (Bliss, 1968; Herreid, 1969).

LIMITATION BY WATER

The differences in evaporation rates and desiccation tolerances of these three species are correlated with apparent differences in availability of water in their respective habitats, suggesting that water is the "limiting factor". We tested the hypothesis that differences in distributions are determined by decreasing availability of water farther from shore, and differing abilities of crabs to obtain and conserve water.

O. quadrata was the first species examined. According to the literature, this crab is restricted to beaches and requires nightly immersion in the sea to wet its gills. The principal risk associated with extending its range thus appeared to be desiccation. The likely resources appeared to be water and food suitable for this "scavenger". The hypothetical gradients of resource availability and risk may be visualized as in Fig. 1. Risk is assumed to be low on the beach and rises sharply inland of the foredune as distance from the water increases. Availability of resources for scavenging is presumed to be highest at the drift line, and somewhat lower inland where various plant and animal foods are present but not concentrated.

These initial assessments of risks and resources for ghost crabs were substantially modified by field and laboratory observations. The presumed risk of desiccation, and requirement for nightly immersion, became doubtful when we found ghost crabs in the interior of North Carolina barrier islands, up to 400m from the nearest shore. Tracking of marked individuals by radio or "ball-and-chain" (lead-tipped wire pendants that left a track in sand) showed that crabs remained away from the beach and any other surface water for a week or more. Despite the reportedly high evaporation rate and absence of evident sources of water, they showed much diurnal activity. A different risk became apparent on the beach: cannibalism (see below).

If desiccation is the most serious risk for ghost crabs, then the most critical resource is water. The only possible source of water appeared to be interstitial moisture from the damp sand in the burrows. In the laboratory crabs were found to extract water from sand containing as little as 3% water by weight. Similar water content is available up to a meter above the water table in barrier island sands. The water collected from the sand by hydrophilic setae surrounding the opening between the 2nd and 3rd walking legs, and drawn into the branchial chamber by partial vacuum (as much as 70mm Hg below ambient) (T.G. Wolcott, 1976, 1984). It is either absorbed via the gills or swallowed. By this mechanism ghost crabs are able to maintain water balance wherever they can burrow down to damp sand, even far from the shore.

Food is a second important resource. During the day, when most observations had been made, ghost crabs are scavengers on plant and animal matter, and it seemed reasonable to assume that suitable food was available in the vegetated island interior as well as on the bare shore. However, nocturnal observation and tracking with an image intensifier scope

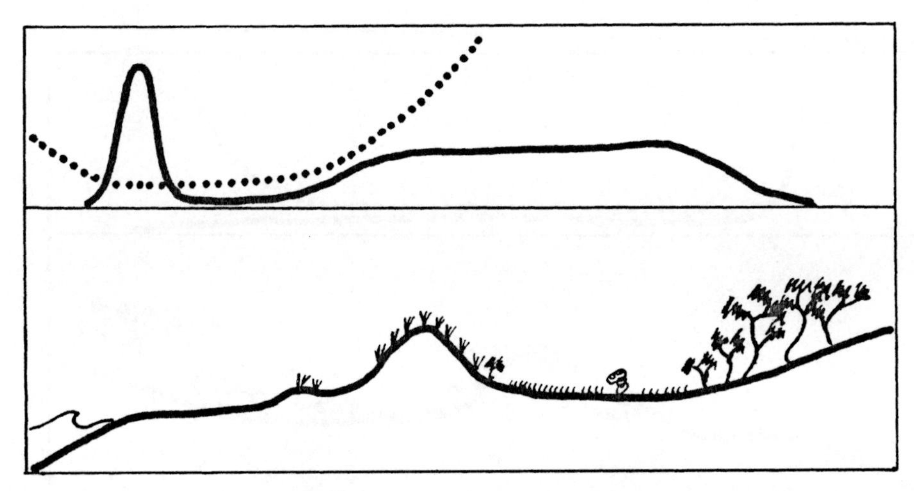

Fig. 1. Hypothetical gradients of risk and resources for ghost crabs on Norh Carolina barrier beaches, based on information in the literature. Risk was assumed to increase sharply inland on the beach because of the animals' high desiccation rates and reported requirement for nigthly immersion. Food resources for these "scavengers" were assumed to be most abundant at the drift line, but also present in the vegetated areas inland of the bare backshore

showed that ghost crabs living near the beach do most of their foraging on the foreshore at night. Over 90% of the items in the diet are live prey, mostly Emerita talpoida and Donax variabilis (T.G.Wolcott, 1978). Stomachs of beach crabs are typically fuller, and contain higher-quality food (i.e., more animal matter) than do stomachs of inland crabs.

These data led us to revise our assessments of resource and risk gradients. Inland habitats are not particularly risky. Although evaporation rates are high, the crabs rapidly replace evaporative losses from soil moisture. Sufficient water is available for ghost crabs anywhere in low-lying sandy soil, including areas beyond the crabs' distribution. Water is not the proximal "limiting factor" for that distribution. On the other hand, inland habitats do not offer additional resources. The beach affords the richest food supply, but presents some risk of intraspecific aggression and cannibalism. Thus, the revised picture of resource and risk gradients (Fig. 2) suggests that there is no overriding reason for ghost crabs to move farther inland. Adaptive behavior should keep most of the population on the beach, and that is where it is. Small numbers of crabs (usually subadults) are found inland, possibly as a response to aggression by dense populations of adults on the beach. As winter approaches a larger proportion of the population seems to burrow in the dunes rather than on the beach. This behavior may have selective value because in cold winters the 6-8°C isotherm (the lower lethal limit; T.G. Wolcott unpubl.) can go below the depth of the water table, hence of the deepest burrow, on the beach. In the dunes the water table lies deeper, and burrows can extend below the critical isotherm (T.G. Wolcott, unpubl.).

C. guanhumi was not examined in detail because it requires habitat in which it can burrow to below the water table. In this sense, therefore, availability of water does serve as an "ultimate" limiting factor for the crabs' range. However, there is no evidence that all suitable habitat is occupied and water is therefore the "proximate" limiting factor. This would be extremely difficult to quantify because both the availability of

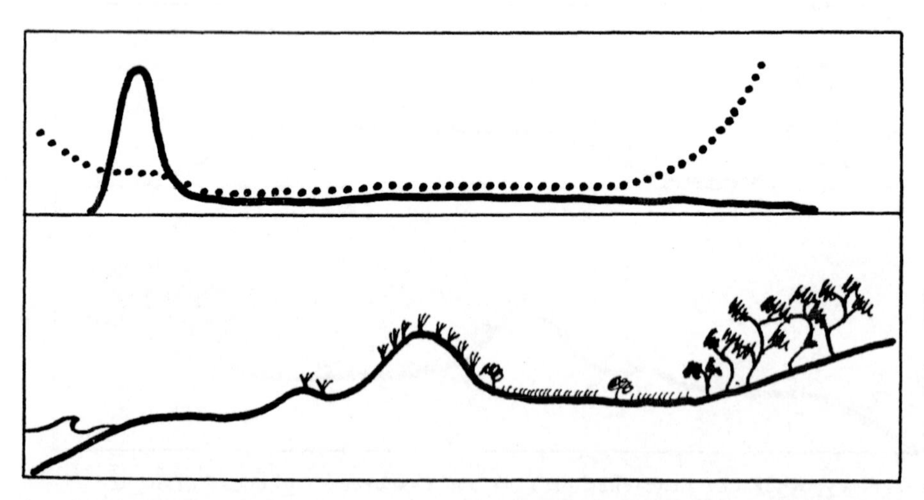

Fig. 2. Gradients of risk and resources, modified by new information. Risk is elevated on the beach due to cannibalism in dense populations of ghost crabs, but the threat of desiccation inland of the beach is minimal, because crabs can obtain ample water from damp sand. Resources for the predatory ghost crabs are high only on the beach

near-surface groundwater and the distribution of C. guanhumi are localized around marshes and watercourses, and both are spatially complex on small scales.

G. lateralis, in contrast, occupies a broad band from near the shore to upland habitats 200-400m inland. It makes burrows in dry soil and is described as largely herbivorous (Bliss et al., 1978, but see below). The principal risk for these crabs appears to be desiccation. Two principal resources are water (form damp sand, rain or dew), and food (forage plants). We tested the hypothesis that G. lateralis is more terrestrial than ghost crabs, and lives in drier habitats, because it is better at extracting water from soil.

As with the ghost crabs, laboratory and field observations caused us to revise our assessments of resources and risks. The risk of desiccation does not appear to vary greatly with increasing distance from shore, because activity is restricted to periods when humidity is high, through circadian rhythms and direct responses to humidity and rainfall. During inactive periods the burrow provides a high-humidity microenvironment that minimizes evaporation.

The availability of resources is also different than we originally supposed. Water is tightly held by the highly organic soil in which G. lateralis burrows. Under normal field conditions it is completely inaccessible to these "highly terrestrial" crabs because they are actually less able than ghost crabs to extract soil moisture (T.G. Wolcott, 1984). G. lateralis must rely on rain, dew and forage plants for its water. Field observations do not indicate any gradient in precipitation or water content of plants between the shore and several hundred meters inland. The second postulated resource, food, is superabundant both within the crabs' range and beyond its limit.

We conclude that water does not limit G. lateralis, since there is no apparent gradient in either risk of desiccation or availability of water. Abundant additional resources (food) are available inland of the existing range. Based on these gradients of resources and risks, the crabs would be expected to extend their range inland to all areas where suitable food, and soil for burrowing, are available. In fact, the natural population is limited to within about 250m of shore. A new hypothesis is needed.

LIMITATION BY SALTS

Our new hypothesis was that differences in availability of salts, and in the crabs' abilities to conserve salts, determined the differences in distributions. The water resources available to the crabs (groundwater, precipitation, forage plants) all appeared likely to become markedly hypo-osmotic to the crabs' hemolymph with increasing distance from the sea. Crabs cannot produce urine less concentrated than their hemolymph, and could be expected to experience high rates of ion loss. The principal risk for these animals in inland habitats thus appeared to be hemodilution, and the critical resource to be salts (ions). The hypothesis was tested by comparing distributions of three species with gradients in availability of salts in the field, and with differences in their respective osmoregulatory abilities.

Field measurements show that ghost crabs do not experience a marked gradient in salt resources near their range limit. The salinity of interstitial water in sand declines steeply within the beach, becoming essentially fresh water even before the foot of the first dune is reached (Fig. 3). Over most of their range, ghost crabs seem to live in very hypo-

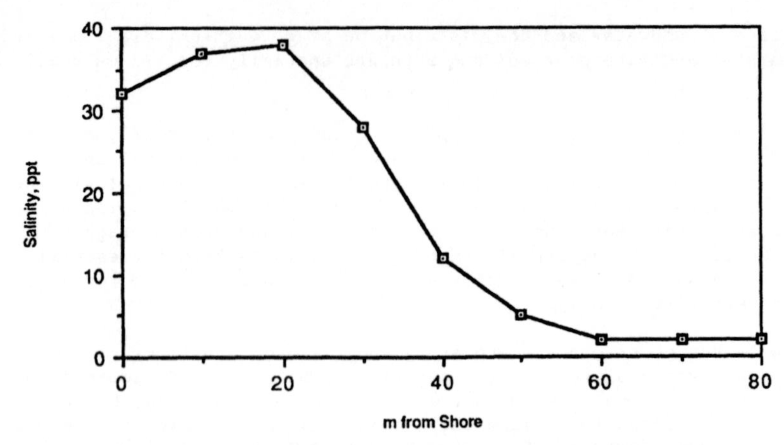

Fig. 3. Salinity of groundwater on a North Carolina barrier beach. Water available to ghost crabs (O. quadrata) becomes fresh within 60m of the swash zone

osmotic conditions. The risks involved, however, are lower than anticipated. In the laboratory, ghost crabs held under simulated "inland conditions" (sand dampened with deionized water or 1% seawater, at temperatures of 10-20°C) survived for up to 4 weeks (Hall, 1982). Rates of ion loss were low for most animals, although under the most extreme conditions hemolymph concentrations declined by 20-50% over 4 weeks.

Although C. guanhumi was not examined in detail, similar conditions seem to pertain. The steep portion of the salinity gradient in ground- and burrow-water appears immediately adjacent to the sea; it declines to fresh water within a few meters of the shore and remains low from there on past the inland range limit of the crabs (Herreid and Gifford, 1963). Despite producing only isosmotic urine, the crabs clearly are able to tolerate freshwater conditions. Herreid and Gifford's (1963) "fresh" water samples were obtained from occupied burrows, and we have found individuals living above 150m elevation along streams on St. John, U.S. Virgin Islands. Osmoregulation in "fresh water" (1% seawater) has also been demonstrated in the Pacific species C. carnifex (Wood and Boutilier, 1985).

G. lateralis also seems to experience no gradient in availability of salt resources near its range limit. Concentrations of ions in rain and dew are variable but generally low within a few meters of the surf, where salt spray is deposited on surfaces, but in the rest of the crabs'range (and beyond) ion content of available water is consistently less than 1% seawater (12 mEq/l). Forage plants show a similar pattern: ion content (including adherent salts) drops in the first few meters from the sea, to low values which remain essentially constant to well beyond the inland limit of the crabs' range (T. G. and D. L. Wolcott, in review).

The estimated risk of hemodilution depends on the technique used. The traditional regime of immersing crabs in test salinities (e.g. Flemister, 1958) is inappropriate for this species, which normally does not enter water. Accordingly, we tested osmoregulatory abilities of G. lateralis under three more ecologically reasonable treatments.

When the source of salts was sand dampened (7.5% by weight) with various test salinities, the crabs showed slow declines in hemolymph osmotic concentration at salinities below 50% sea water. Even in sand dampened with deionized water or 1% seawater, however, there was only about a 20% decrease in concentrations of major inorganic ions in hemolymph after 31 days (T. G. and D. L. Wolcott, in review).

The ability of crabs to osmoregulate with "drinking water" as the salt source was tested after it was shown that soil water was inaccessible in nature. Again, the crabs showed minimal declines in hemolymph concentration when provided strongly hypo-osmotic water. Even after six weeks with access only to deionized water, hemolymph osmotic concentration, and concentrations of major ions, had decreased by only about 13% (Fig. 4, redrawn from T. G. and D. L. Wolcott, in review).

The role of vegetation as the sole ion source for restoring and maintaining salt balance was tested in ion-depleted crabs that were offered deionized water and natural forage plants collected from beyond the range limit. Hemolymph concentrations recovered toward normal values in vegetation-fed crabs, but not in those fed ion-free "flavored" filter paper (Fig. 5).

We must conclude that none of the three species is limited by salts, despite the low concentrations of ions available in the environment and the high rates of salt loss expected through urine. There is no gradient in availability of salts near the range limit, and all the crabs could, based on their osmoregulatory abilities, live beyond their observed ranges. It appears that crabs must conserve ions very tightly. How can this occur in animals that produce only isosmotic urine?

BEHAVIORAL MECHANISMS OF ION CONSERVATION

Land crabs have behavioral options unavailable to aquatic crabs, beacause they are immersed in air rather than water. In particular, they appear to have the possibility of further manipulating urine after it exits the nephropores, because it is not immediately carried away by the medium.

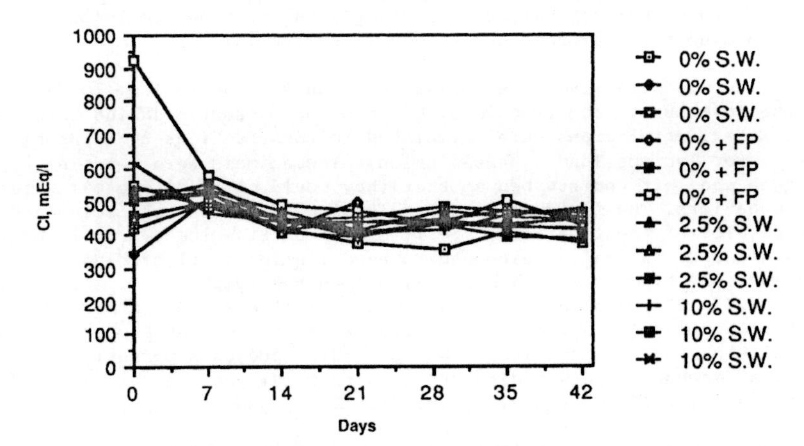

Fig. 4. Chloride concentration in hemolymph of G. lateralis given only drinking water of the concentrations indicated. "0% + FP" denotes animals given de-ionized water and fed low-salt filter paper "food" to assess effects of ion loss through feces

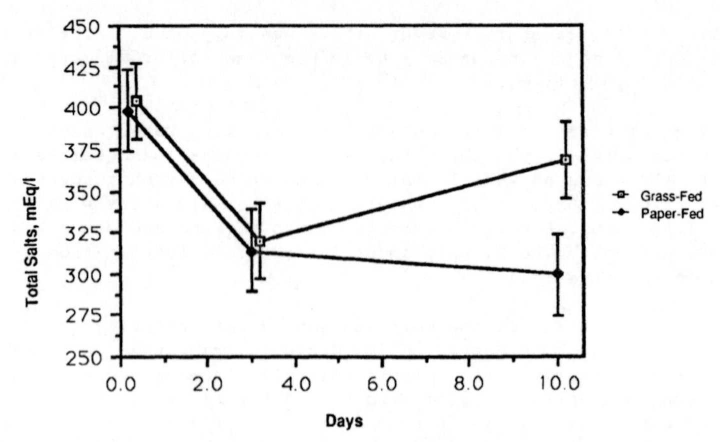

Fig. 5. Total salts (determined by conductivity) in hemolymph of G. lateralis from the field (left), after ion depletion (center), and after 7 days feeding on inland grass or salt-free filter paper. Crabs fed inland grass showed significant recovery (bars=95% confidence intervals)

We hypothesized that the crabs pass the originally isosmotic urine to other organ systems (gills or gut) that are capable of producing strong ionic and osmotic gradients. In other words, they extrarenally reprocess the urine to produce a dilute "final excretory product". (This "product" is denoted by the abbreviation "P" to distinguish it from urine in the discussion that follows.)

The hypothesis was first tested with the dry-burrow species, O. quadrata and G. lateralis, because their nephropores are located within the mouth field and transfer of urine to the gut or branchial chambers would be easily accomplished. Nephropores of C. guanhumi are located outside the mouth field. Furthermore, we reasoned that because this species has a pool within its burrow, it might osmoregulate by the mechanism usually used by hyper-regulating crabs: actively transporting ions into the hemolymph from the dilute medium via the osmoregulatory gill epithelia.

Control crabs were given a stimulus to produce urine (volume load), without the stimulus to conserve ions, by infusing isosmotic saline into the hemocoel. Experimental crabs were stimulated to conserve ions by infusing deionized water (volume load + hemodilution). The animals were held in funnel cages and left undisturbed so that they could carry out normal water-handling behaviors. The fluid ultimately discarded ("P") was collected under oil (for details of the technique, see T. G. and D. L. Wolcott, 1985). Both ghost crabs and G. lateralis responded to hemodilution by discarding "P" that was markedly hypo-osmotic and hypo-ionic to hemolymph (T. G. and D. L. Wolcott, 1982b, 1985, and in prep.), whereas control crabs discarded "P" isosmotic to hemolymph (Fig. 6). These results led us to test C. guanhumi as well. Despite having the pool in its burrow, this species apparently uses the identical mechanism, discarding "P" less than 10% the concentration of hemolymph and urine (T. G. and D. L. Wolcott, 1984, and in prep.).

The most likely mechanism for production of dilute "P" seemed to be reclamation of ions from the urine by active transport, but extrarenal secretion of water has been postulated for other crabs (Shaw, 1959; Greenway, 1980, 1981). The alternatives were explored by monitoring fluxes of water as well as ions. "Backpacks" volumetrically substituted "pseudo-

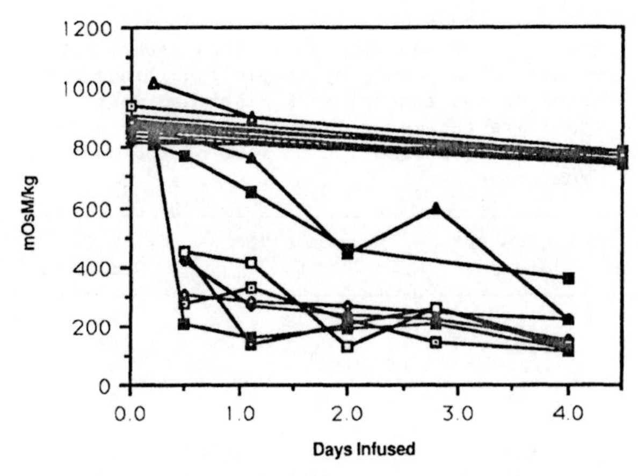

Fig. 6. Osmotic concentration in hemolymph (upper straight lines) and final excretory product ("P", lower lines) of ghost crabs infused with de-ionized water, showing that ion-depleted crabs conserve salts by discarding dilute fluid. Control crabs infused with isosmotic saline produced "P" isosmotic with hemolymph

urine" ("PU") of known, constant composition into the urine-reprocessing pathway. When crabs urinated, the urine passed through tubes cemented over the nephropores into one chamber of the backpack, displacing a rubber septum an delivering an equivalent volume of "PU" that was deposited adjacent to the nephropores. The "PU" was ionically similar to urine, and was labelled with inulin. Antibiotics prevented microbial degradation of inulin. Analyzing changes in both ions and inulin allowed us to distinguish between extrarenal reclamation of ions and secretion of water; addition or subtraction of water from the urine would result in decreases or increases (respectively) in the concentration of inulin.

Extrarenal reclamation of ions is the principal mechanism for conservation. While ion concentrations decreased dramatically, inulin concentrations actually rose, indicating that there is no net secretion of water. In fact, some of the water appears to be reabsorbed along with the salts. The net result is that over 90% of the chloride is reclaimed from the urine before it is discarded as "P" (Table 1).

What behaviors and organs are involved in producing this remarkable hypo-osmotic excretory product? Addition of a vital stain to the "PU" permitted tracing the pathway taken by urine during reprocessing by C. guanhumi. Crabs were frozen as soon as the first stained "P" appeared, and dissected while frozen to prevent migration of dye. Superficial inspection showed that the location of the nephropores does not hinder transfer of urine to the branchial chamber; the "bristle patches" lateral to the mouth field act as capillary conduits from the nephropores to the Milne-Edwards openings at the base of the chelipeds. Not surprisingly, the gills were intensely stained. Fluid could presumably be pumped from the branchial chambers up to the mouth, swallowed, and reprocessed in the gut; however, no dye was found in the gut. We conclude that the gills are the principal site of ion reclamation in the land crabs, and that water-handling behaviors are adapted to moving urine from the nephropores to the branchial chambers.

Table 1. Reclamation of Chloride from urine of _C. guanhumi_ infused with de-ionized water. Both volume (deter mined by inulin labelling) and chloride concentration of discarded waste ("P") are reduced relative to those of urine, reflecting reclamation of most of the salts from urine

Crab #	Relative "P" volume	Relative "P" ·Cl	% Cl Reclaimed
11	.455	.138	91.9
	.641	.122	91.3
12	.565	.064	96.1
14	.538	.150	87.6
	.690	.155	88.1
15	.343	.073	94.9
16	.310	.032	97.6
	.550	.093	94.1
Average	.476	.103	92.6

The consequence of this behavioral adaptation is that land crabs are able to produce, in effect, a dilute urine. This results in very tight conservation of ions, which in turn permits existence in areas with very limited salt supply. Because of this adaptation, availability of salt is adequate for crabs even inland of the habitats they currently occupy.

FUTURE DIRECTIONS FOR RESEARCH

In our studies of "zonation" in land crabs, we have failed to identify a physical "limiting factor". The "limiting factor" approach to physiological ecology seems to have some limiting factors itself. Such a good _a priori_ understanding of the animal's ecology is apparently required to recognize relevant "resources" and "risks" that the "resource/risk gradient" paradigm has little predictive value. Furthermore, it appears that the number of situations in which physical limitation would be predicted is even smaller than anticipated because of unexpected adaptations to "risks". Is this approach worth pursuing?

We submit that despite its apparent limitations, it remains an efficient and direct way to identify and attack critical questions in physiological ecology. The first step is to formulate hypotheses about interactions between the organism and its environment. Based on existing information, what would be expected to proximally limit the species' distribution? Field observations and laboratory experiments are used to compare the animals' behavioral and physiological coping mechanisms with conditions in the microhabitats occupied by, and available to, the animals. The data are used to test the hypothesis; either the distribution is dictated by the "limiting factor", or it isn't. Either answer is valid and provides significant insights.

If a predicted "limiting factor" is indeed limiting, then the investigator has an ecological answer about the distribution of the species. He/she also has another clue about the conditions under which one might expect to find physical limitation in other systems. If the hypothetical factor turns out not to be limiting (as is likely to be the usual case), the investigator is directed to the crucial behavioral and physiological questions that will define the adaptations permitting the animal to circumvent the limitation. With more complete information about microhabitat, behavior and physiology, one can reexamine what environmental factors are likely to be most important, and proceed to the next hypothesis. The information we have acquired about our three species of land crabs has brought us to a different stage of this process for each of the species.

Ghost crabs are not physically limited. They do not appear to encounter severe risks by extending their range farther inland. Their ability to extract water from damp sand and to conserve salts would allow them to exploit additional habitat. However, the richest food resources are on the beach. The absence of any "payoff" for moving inland is probably adequate to account for the ghost crabs being found predominantly on beaches.

In the case of C. guanhumi the situation is less clear. The crabs are physiologically and behaviorally well equipped to deal with the problems of water and salt balance in terrestrial environments near marshes or watercourses. Access to a shallow water table is possibly a "limiting factor", but we have insufficient data on what microhabitats are suitable, and on whether C. guanhumi actually exploits all such areas or is behaviorally limited to a narrower range. In some areas it seems likely that restricted recruitment, or reproductive failure in most years, limits populations. The northernmost population of these crabs (Hungry Bay, Bermuda) appeared to be composed of only about 2 size classes in 1982 (T. G. and D. L. Wolcott, unpubl. obs.). Both were large; sketchy data on growth rates (Feliciano, 1962; Henning, 1975) suggest that successful recruitment had occured only twice in the last 10-15 years. If population densities are kept low by low recruitment rates ("supply-side ecology" of Roughgarden and others), competition for resources among adults should be low. Consequently, the "payoff" for moving into risky marginal habitat in search of more available resources should be low, and so should be the expectation of finding physical limitation.

The most terrestrial species, G. lateralis , presents the most intriguing situation. Its behavioral and physiological adaptations seem to reduce the risks of inland existence to the extent that the Bermudian population could live almost anywhere on the islands. Superabundant resources, in the form of forage plants, exist inland. Nevertheless, the species is behaviorally limited to near the shore. Why hasn't it expanded its range to the limits of its physiological tolerances? An obvious explanation is the necessity of breeding migrations to the shore. However, migrating crabs move rapidly during the night and take shelter by day. They are typically exposed to (presumably) increased predation for only a few nights, and energetic costs of locomotion are modest (Herreid, 1982). These costs and risks could account for the restriction of land crabs to the nearshore zone if the "payoff" for moving inland is slight.

Other information suggests that restriction of recruitment, or even feedback control, mantains population density below a level that would "push" crabs farther inland. In the Nonsuch Island Nature Preserve, Bermuda, castor bean pomace was spread as a fertilizer for new plantings. The pomace was avidly consumed by G. lateralis as a high protein food (D. Wingate, Bermuda Ministry of Agriculture and Fisheries, pers. comm.), and a population outbreak followed. Because there is a planktonic larval phase, the outbreak cannot be attributed to increased reproductive output by the adult crabs on this small island in response to the additional food.

Instead, it appeared that increased recruitment occurred. This could be because more food was available for recruits. It could also be attributable to a decrease in cannibalism. Although forage plants often appear to be superabundant, they contain little protein, and N is a limiting nutrient for both the herbivorous species we have examined (D. L. and T. G. Wolcott, 1984, 1987). Adult crabs seek high-nitrogen foods (e.g., carrion). They readily cannibalize juveniles, and the intensity of cannibalism is modulated by quality of the vegetable diet (D. L. and T. G. Wolcott, 1984).

This led us to hypothesize that when food quality is low, recruitment is controlled by cannibalism, population density remains low, and little pressure exists to extend the range inland. Under these conditions, factors which favor remaining near shore (shorter breeding migrations, and the input of protein from marine sources via cannibalism of recently-settled recruits) become dominant, and the range is behaviorally limited.

Cannibalism is a common behavior among herbivores (Fox, 1975), which have access to limited nitrogen. As a behavioral mechanism of population regulation, it is particularly suited to the intertidal (and other environments) where sedentary adults have highly dispersive larvae. Regulating reproductive output in response to conditions in the local adult habitat would be inappropiate, since the larvae are broadcast and the conditions where they ultimately settle may be quite different. Cannibalism, in contrast, provides an appropriate response to changing local conditions, by modulating the addition of individuals to the local population. During locally low availability of N, recruitment could be restricted and population density controlled by cannibalism, while breeding adults continued injecting high reproductive output into the planktonic "lottery". The fitness of the adults would not be compromised, because the recruits they cannibalized would be unlikely to be their own offspring. Indeed, cannibalistic adults would increase their own fitness during times of low N availability both by indirectly harvesting additional N (derived from the plankton) and by decreasing potential competition.

During locally high availability of N, intensity of cannibalism would presumably decrease, more recruits survive, and the population grow. An obvious question is what selective forces would ever cause adults to become less cannibalistic, if they would not be eating their own offspring. Possible explanations include energetics (high-quality plant food, when available, is easier to catch than little crabs or other animal food); increased availability of mates; and potential for saturating predators during breeding migrations. Insufficient data are available to test any of these hypotheses, and the need for field research is obvious.

Another major gap in our understanding of behavioral adaptation to intertidal life by crustaceans is the paucity of information about the behavior, ecology, and habitats of life history stages other than adults.

What behaviors are used by larvae, which may have been developing while drifting in the plankton for 3-6 weeks, to return inshore and find suitable habitats for settling? How do they succeed in replenishing adult stocks on oceanic islands where unidirectional current regimes appear likely to sweep all larvae away? Vertical migration is a common behavior of zooplankton, and appropriate patterns of vertical movement would allow the selection of current regimes for transport of larvae inshore. Recent studies indicate that larvae, by choosing the right depth, can even "surf" inshore on internal waves (Shanks, 1987). The difficulty of marking and tracking vast numbers of tiny larvae makes testing of hypotheses difficult. To attack some of these questions, we are developing a computerized buoy that will be programmed to move vertically according to models of larval behavior (either theoretical, or empirically derived from laboratory studies). Releasing the buoys into the sea at appropriate sites and times and tracking them will allow us to examine the consequences of the postulated behavior patterns. Comparison of buoy movements with fine-scale oceanographic data may eventually allow predictions of larval transport and success of recruitment.

Behavioral adaptations of juvenile terrestrial and semiterrestrial crabs are even less well known than those of larvae. For most species, the juveniles are seldom seen and may not even share the habitat of adults. Where are they? For instance, in Puerto Rico, the juvenile *C. guanhumi* are found near mangrove leaf litter remote from adult habitats (H. Rojas, Univ. of Puerto Rico, Mayaguez, pers. comm.). What are their ecological roles? What relationships exist between density of adult stocks and recruitment? What other factors affect survivorship of recruits? Current work by Nancy O'Connor in our laboratory suggests that differences in distribution of two fiddler crabs (*Uca pugilator* and *U. pugnax*) are already apparent in the distribution of recently settled juveniles. To what degree are distributions of mobile species determined by passive deposition of larvae, or by settling behavior of larvae, or differential mortality of recruits, or post-settling movement to preffered habitat for adults?

Intensive field observations are needed to determine which of these questions are most critical, and most tractable. Observation needs to be followed by formulation of testable hypotheses. Rigorous, replicated field experiments should manipulate such variables as density of adults, and supply of high-N food to adults and/or recruits. Such experiments are notoriously difficult to perform properly, but would provide significant contributions to our understanding of intertidal ecology, as well as new insights into the ecology of organisms subsisting on low-N diets. We hope that this workshop will encourage some researchers to accept these formidable challenges, and will stimulate some of the needed research.

ACKNOWLEDGMENTS

Most of the research reported here has been supported by National Science Foundation grants DEB 77-16631 and PCM 82-10465 to the authors. Dan Bosso and Laura Hall assisted in the laboratory.

REFERENCES

Bliss, D. E., 1968, Transition from water to land in decapod crustaceans, *Amer. Zool.*, 8:355-392.

Bliss, D. E., van Montfrans, J., van Montfrans, M., and Boyer, J. R., 1978, Behavior and growth of the land crab *Gecarcinus lateralis* (Freminville) in South Florida, *Bull. Amer. Mus. Nat. Hist.*, 160(2):111-152.

Burggren, W., and McMahon, B. R. (eds.), "The Biology of Land Crabs", Cambridge University Press, N.Y. (In Press).

Burnett, L. E., and McMahon, B. R., 1987, Gas exchange, hemolymph acid-base status, and therole of branchial water stores during air exposure in three littoral crab species, *Physiol. Zool.*, 60:27-36.

Feliciano, C., 1962, Notes on the biology and economic importance of the land crab *Cardisoma guanhumi* Latreille of Puerto Rico, *Spec. Contr., Inst. of Marine Biol., Univ. of Puerto Rico*, 29 pp.

Flemister, L.J., 1958, Salt and water anatomy, constancy and regulation in related crabs from marine and terrestrial habitats, *Biol. Bull.*, 115:180-222.

Fox, L. R., 1975, Cannibalism in natural populations, *Annu. Rev. Ecol. Syst.*, 6:87-106.

Greenaway, P., 1980, Water-balance and urine production in the Australian

arid-zone crab Holthuisana transversa, J. exp. Biol., 87:237-246.
Greenaway, P., 1981, The fate of glomerular filtration markers injected into the hemolymph of the amphibious crab Holthuisana transversa, J. exp. Biol., 91:339-347.
Hafeman, D. H., and Hubbard, J. L., 1969, On the rapid running of ghost crabs (Ocypode ceratopthalma), J. exp. Zool., 170:25-31.
Hall, L. A., 1982, Osmotic and ionic regulation in overwintering ghost crabs, Ocypode quadrata, MS thesis, North Carolina State University, 135 pp.
Henning, H. G., 1975, Kampf-, Fortpflanzungs- und Hautungsverhalten -- Wachstum und Geschlechtsreife von Cardisoma guanhumi Latreille (Crustacea, Brachyura), Forma et Functio, 8:463-510.
Herreid, C. F., 1969, Integument permeability of crabs and adaptation to land, Comp. Biochem. Physiol., 29:423-429.
Herreid, C. F., 1982, Energetics of pedestrian arthropods, in: "Locomotion and Energetics in Arthropods", C. F. Herreid and C. R. Fourtner, eds., Plenum, N.Y.
Herreid, C. F., and Gifford, C. A., 1963, The burrow habitat of the land crab, Cardisoma guanhumi (Latreille) in South Florida, Ecology, 44:773-775.
Horch, K., 1975, The acoustic behavior of the ghost crab, Ocypode cordimana Latreille, 1818 (Decapoda, Brachyura), Crustaceana, 29:193-205.
Klaassen, F., 1973, Stridulation und Kommunication durch Substratshall bei Gecarcinus lateralis (Crustacea Decapoda), J. comp. Physiol., 83:73-79.
Little, C., 1983, "The Colonisation of Land: Origins and Adaptations of Terrestrial Animals", Cambridge University Press, N.Y.
Powers, L. W., and Bliss, D. E., 1983, Terrestrial Adaptations, in: "Biology of Crustacea", D. E. Bliss, ed. in chief, Vol. 8, "Environmental Adaptations", F. J. Vernberg and W. B. Vernberg, eds., Academic Press, N.Y.
Rabalais, N. N., and Gore, R. H., 1985, Abbreviated development in decapods, in: "Crustacean Issues 2: Larval Growth", A. M. Wenner, ed., A. A. Balkema, Rotterdam.
Shanks, A. L., 1987, The onshore transport of an oil spill by internal waves, Science, 235:1198-1200.
Shaw, J., 1959, Salt and water balance in the East African fresh water crab, Potamon niloticus (Milne Edwards), J. exp. Biol., 36:157-176.
Wolcott, D. L., and Wolcott, T. G., 1984, Food quality and cannibalism in the red land crab, Gecarcinus lateralis, Physiol. Zool., 57:318-324.
Wolcott, D. L., and Wolcott, T. G., 1987, Nitrogen limitation in the herbivorous land crab Cardisoma guanhumi, Physiol. Zool., 60:262-268.
Wolcott, T. G., 1973, Physiological ecology and intertidal zonation in limpets (Acmaea): a critical look at "limiting factors", Biol. Bull., 145:389-422.
Wolcott, T. G., 1976, Uptake of soil capillary water by ghost crabs, Nature (London), 264(5588):756-757.
Wolcott, T. G., 1978, Ecological role of ghost crabs, Ocypode quadrata (Fabricius), on an ocean beach: scavengers or predators?, J. exp. mar. Biol. Ecol., 31:67-82.
Wolcott, T. G., 1984, Uptake of soil interstitial water: mechanisms and ecological significance in the ghost crab Ocypode quadrata and two gecarcinid land crabs, Physiol. Zool., 57:161-184.
Wolcott, T. G., and Wolcott, D. L., 1982a, Larval loss and spawning behavior in land crabs, Gecarcinus lateralis Freminville, J. Crustacean Biol., 2:477-485.
Wolcott, T. G., and Wolcott, D. L., 1982b, Urine reprocessing for salt conservation in terrestrial crabs (Abstract), Amer. Zool., 22:897.
Wolcott, T. G., and Wolcott, D. L., 1984, Salt conservation by extrarenal resorption in Cardisoma guanhumi (Abstract), Amer. Zool., 25:78A.
Wolcott, T. G., and Wolcott, D. L., 1985, Extrarenal modification of urine

for ion conservation in ghost crabs, *Ocypode quadrata* (Fabricius), *J. exp. mar. Biol. Ecol.*, 91:93-107.

Wolcott, T. G., and Wolcott, D. L., 1986, Factors influencing the limits of migratory movements in terrestrial crustaceans, *in*: "Migration: Mechanisms and Adaptive Significance", M. A. Rankin, ed., (Proceedings of the University of Texas Centennial Symposium). *Contr. mar. Sci. Univ. Texas mar. Sci. Inst. Suppl.*, 27:257-273.

Wolcott, T. G., and Wolcott, D. L., Availability of salts as a limiting factor for the land crab *Gecarcinus lateralis*, (In Review).

Wood, C. M., and Boutilier, R. G., 1985, Osmoregulation, ionic exchange, blood chemistry and nitrogenous waste excretion in the land crab *Cardisoma carnifex*: a field and laboratory study, *Biol. Bull.*, 169:267-290.

Zeil, J., Nalbach, G., and Nalbach, H. O., 1986, Eyes, eyestalks and the visual world of semiterrestrial crabs, *J. comp. Physiol. A*, 159:801-811.

Barnacle Larval Settlement: the Perception of Cues at Different Spatial Scales

Edwin Bourget

Laval University, Quebec, Canada

ABSTRACT

I review here some recent work carried out (1) to characterize differences in habitat selection by the larvae of the barnacle, Semibalanus balanoides, in two regions, the Gulf of St. Lawrence and the Bay of Passamaqquoddy on the Atlantic coast of Canada, and (2) to examine the mechanism used by the larvae to select a microhabitat to settle. The focus is placed on the influence of scales of substratum heterogeneity, scales in the exploration of substratum characteristics by the larvae, and its perception of cues. The spatial scales considered range from 1 m down to a few um. A model summarizing the sequence of events taking place during the settlement process is presented.

INTRODUCTION

The most important event in the life cycle of a sessile invertebrate is probably that of selecting where to settle. Since this choice will determine the conditions for its juvenile and adult life, the mechanisms leading to the selection of a settling site must be precise. Learning is not involved since the activity is not repeated. The precision involved is demonstrated by the fact that the larvae of many sessile invertebrates exhibit consistent microhabitat selection both in space and time despite the multiplicity of factors potentially influencing the settlement process.

Barnacles, and in particular Semibalanus balanoides, have been intensively studied with respect to larval settlement behaviour and responses to settling cues (see reviews by Crisp, 1974; 1976; Newell, 1979; and recent papers by Strathmann and Branscomb, 1979; Strathmann et al., 1981; Moyse and Hui, 1981; Hudon et al., 1983; Yule and Walker, 1985). Barnacle larvae and those of many other marine invertebrates (Gray, 1974; Hannan, 1984) actively examine the characteristics of the substratum before selecting the exact microsite to settle (see Crisp 1974, 1976). According to the current theory this exploration of the substratum takes place sequentially. First, the competent cyprid larvae (those capable of settling) drift to the bottom (Bousfield, 1955; Knight-Jones and Quasim, 1955). The path is presumably mostly determined by water turbulence and current patterns near the bottom (Hannan, 1984). Such events as the frequency of contacts with the bottom and drift distances between contacts with the bottom are unknown, but the larvae must, if they are to select favourable microsites, touch the substratum often enough to identify the conditions (e.g. shore level, exposure to waves and sunshine). If conditions are favourable they remain to explore the surface in greater detail (see Newell, 1979). Three epibenthic phases are involved in examining the substratum (Crisp and Meadows, 1963; Crisp, 1974): (1) a "broad exploration phase" during which the larva takes a more or less straight path on the substratum, turning infrequently, then (2) after a long enough stay on the substratum, another behaviour pattern is adopted, that of "close exploration", characterized by short steps with frequent changes in direction. Finally, (3) as attachment nears, the larva enters the "inspection phase" during which there are at most to and fro movements within its own length" (Crisp, 1974). In doing so it leaves traces of adhesive on the substratum surface (Walker and Yule, 1984; Yule and Walker, 1985), presumably to maintain the cypris larvae on the substratum. Each of these phases take place at a different spatial scale and presumably also at different time scales, though no information documents the latter.

Numerous studies examine local variations in the distribution and abundance of barnacles (Lewis, 1964; Hawkins and Hartnoll, 1983; Underwood and Denley, 1984; Connell, 1985; Gaines and Roughgarden, 1985; Bergeron and Bourget, 1986). Both differential mortality (Connell, 1961 a,b; 1970; and differential settlement (Gaines and Roughgarden, 1985; see Connell, 1985) have been used to account for local abundance, the latter mechanism probably having received the most attention. Many factors influence the settlement of intertidal barnacles. These include light quality, orientation and intensity (Weiss, 1947; Barnes et al., 1951), surface reflectance, colour and background illumination (Gregg, 1945; Smith, 1948), surface angle (Pomerat and Reiner, 1942), current (Crisp, 1955), water agitation and exposure to waves (Smith, 1946), surface contour (Crisp and Barnes, 1954), surface texture (Barnes, 1956; Yule and Walker, 1985), water depth (Bousfield, 1954), shore level (Strathmann and Branscomb, 1979), stage of tide (Weiss, 1947), the presence and abundance of conspecifics, fragments or extracts of conspecifics or related species (see Crisp, 1974; Larman and Gabbott, 1975; Larman et al., 1982; Yule and Crisp, 1983; Rittschof et al., 1984), algae, primary films (Miller et al., 1948; Daniel, 1955; Strathmann et al., 1981; Hudon et al., 1983), and the nature of molecular films (Rittschof et al., 1984). Further, physiological factors (eg. age of cyprid larvae (Yule and Crisp, 1983; Rittschof et al., 1984), can affect the ability of the larvae to select surfaces.

On a geographic scale, numerous reports document variations in microhabitat occupancy. For instance, *Semibalanus balanoides* is found only in cracks and crevices in

Spitsbergen (Feyling-Hanssen, 1953), and mainly on vertical walls, crevices and the undersurfaces in Greenland (Petersen, 1962). No information is available on the selectivity of the larvae for different types of microhabitat. In the northwestern Gulf of St. Lawrence, barnacles settle mostly in crevices and sides of boulders; horizontal surfaces are avoided. By contrast, further south on the shores of New Brunswick and northern New England, all rocky surfaces are colonized, a situation comparable to that on the European coast. Thus, the question of what causes differences in microhabitat selection in different geographical regions is raised. Considering the many factors influencing microhabitat selection by barnacle larvae, it could be argued that such differences in microhabitat selection are environmentally mediated. However, differences in microhabitat occupancy or selection are often maintained over widely varying local environmental conditions (e.g. sheltered and exposed shores, shaded and exposure to sunshine, granite and shale, etc.), and despite large variations in abundance of competent larvae (Gaines et al., 1985). Thus, the hypothesis of genetically mediated differences is suggested. This idea is supported by the studies by Knight-Jones et al. (1971) and Doyle (1976) on genetically mediated habitat selection differences by the larvae of another sessile marine invertebrate, Spirorbis.

Barnacle larval settlement is most frequently studied by examining individual components of the complex combination of physical and biotic conditions observed in the field (usually those presumed to be important), using univariate controlled experiments in the laboratory. While providing evidence of the influence of individual factors on settlement, this approach cannot elucidate the respective influences of factors acting simultaneously. The data on the effects of individual factors on larval settlement should aid in developing techniques for determining the component variables which are important in nature.

I will here summarize recent experiments on habitat selection by barnacle larvae in two regions, the northwestern Gulf of St. Lawrence and the Passamaquoddy Bay along the Atlantic coast of Canada (see Chabot and Bourget, in press), and to examine the mechanism of habitat selection used by the larvae in these two regions (see Letourneux and Bourget, in press). I will focus on the influence of scales of substratum heterogeneity, scales in the exploration by the larvae and the larvae's perception of cues.

MATERIALS AND METHODS

Study Locations

Barnacle settlement patterns in two ecological regions, the Gulf of St. Lawrence and the Atlantic coast of Canada, specifically in Passamaquoddy Bay, were examined (Fig. 1). The Gulf is a semi-enclosed subarctic sea (Dunbar, 1979) measuring 214,000 km² (El-Sabh, 1976) while Passamaquoddy Bay in the Bay of Fundy measures 225 km². Both regions exhibit large local variations of topography, exposure to sunlight, wave action and geology.

Animal Distribution and Scale of Substratum Heterogeneity

Four stations in the Gulf and three stations in the St. Andrews region were selected for the analysis of the effect of scales of substratum heterogeneity on barnacle distribution. All stations were selected either because other ecological studies were being carried out at these stations or because of the accessibility to the shore. The two stations used for experimental studies (Capucins and St. Andrews) were chosen for their accessibility.

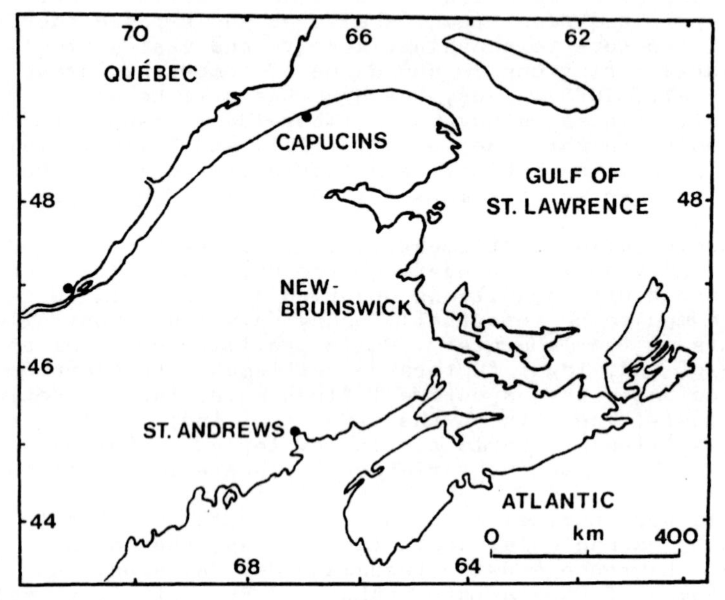

Fig. 1. Location of study sites in the Gulf of St. Lawrence (Capucins) and the Bay of Passamaquoddy (St. Andrews).

The relationship between the scales of substratum heterogeneity and the abundance and distribution of animal populations was quantified using a custom made electronic curvimeter. This apparatus, using wheels of different diameters (1, 2, 8, 16, 32, 64, 96 cm), determined the distance along the rock surface over a linear distance of 4 m on a transect line, and the corresponding distances actually occupied by various species (Fig. 2). Hence, the change in the relationship of distance accessible to distance occupied by each species in relation to a gradient of scales (as indicated by different wheel diameters) provided a means of measuring the influence of scales of substratum heterogeneity on the distribution of different species.

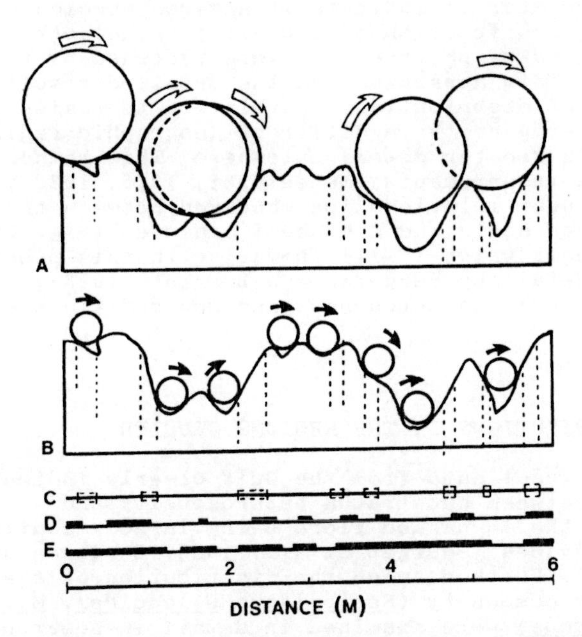

Fig. 2. A, B, Diagrams illustrating the measurement of substratum heterogeneity and distance accessible with wheels of different diameters. C. Distance occupied by Semibalanus balanoides. D. Distance accessible with a wheel of 64 cm diameter. E. Distance accessible using a wheel of 8 cm diameter.

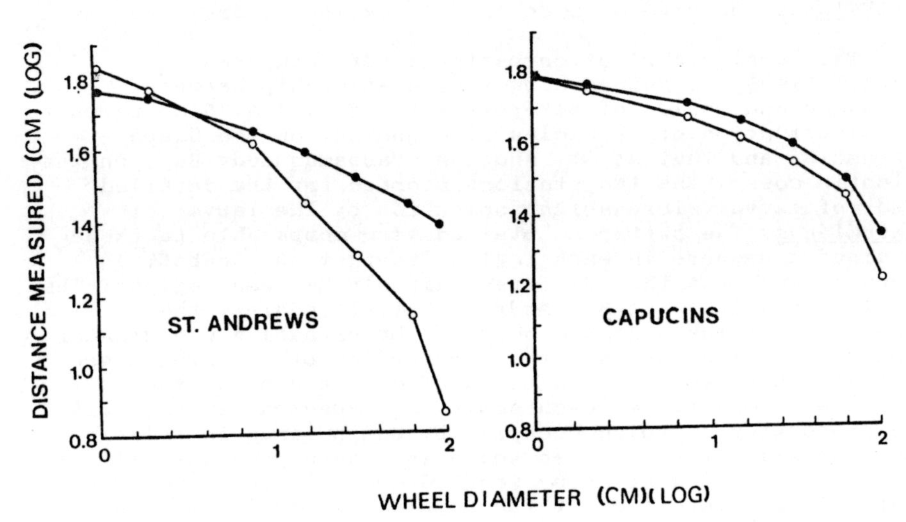

Fig. 3. Relationship between the distance measured on the substratum and wheel diameter.

The wheels (calibrated to the nearest 0.5 cm) were connected electronically to a portable Epson HX-20 micro-computer which allowed the operator to indicate if a given species was present. With each fraction of a wheel turn, electronic signals triggered by photosensors were recorded by the microcomputer. This apparatus and the detailed results on animal and plant distributions in relation to scales of substratum heterogeneity in different geographic regions will be described in greater detail eslewhere. This method is akin to the fractal measurements (Mendelbrot, 1975; 1982), except for the non-linear relationships observed between the distance measured at each scale and the wheel diameter (Fig. 3). The other experiments which I will summarize in this paper are described in detail by Bergeron and Bourget (1986), Chabot and Bourget (in press) and Letourneux and Bourget (in press).

BARNACLE DISTRIBUTIONS IN THE REGIONS STUDIED

Distributional data from the Gulf clearly indicated a relationship between substratum heterogeneity and distribution, the fauna and flora being largely confined to cracks and crevices (Bourget et al., 1985; Bergeron and Bourget, 1986). Further, when the distribution of the dominant species of the community (*Semibalanus balanoides*, *Mytilus edulis*, and *Fucus*) were examined in detail in numerous crevices, one pattern of space partitioning was frequently observed. The upper portion of crevices was devoid of organisms, the middle portion occupied by the barnacle *Semibalanus balanoides* associated with fucoids, and the basal portion of the crevices was occupied by the mussel *Mytilus edulis*. While there are no comparable data for the intertidal community on the Atlantic Coast, it was obvious that there *Semibalanus* occurred outside as well as inside crevices.

The local distribution patterns were compared quantitatively by determining the relationship between abundance and scales of heterogeneity. Fig. 4 A, B compares the distribution of barnacles at Capucins, on the Gaspe Peninsula, and that at St. Andrews, Passamaquoddy Bay, on the Atlantic coast, the two stations adopted for the detailed study of larval microhabitat selection by the larvae of *Semibalanus*. The patterns obtained were comparable to those obtained elsewhere in each region (Bourget and Chabot, in preparation), but they differed markedly between regions. The results show that, at St. Andrews, at all scales, the proportion of the distance occupied by barnacles was greater than at Capucins, whether this proportion of occupied space was calculated in relation to total distance or to the distance accessible at each scale. As expected, in the Gulf, the proportion of space occupied by *Semibalanus* in relation to space accessible at a given scale increased progressively at smaller scales (as given by wheel diameter). This indicates that a large proportion of the population was located in crevices rather than on open surfaces, while in the St. Andrews region the opposite trend (a slight reduction in the proportion of surface accessible was occupied at small scale) was observed, indicating greater abundance on exposed surfaces than on crevice surfaces.

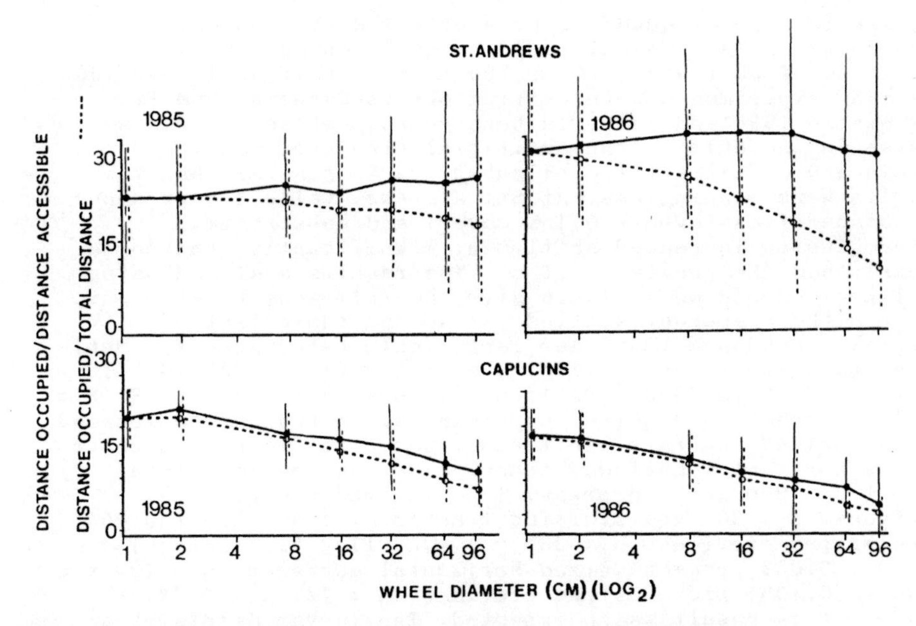

Fig. 4. Proportion of space occupied by adult Semibalanus balanoides at different scales of heterogeneity, as measured using wheels of different diameter. (-) Proportion of space accessible to a wheel of a given diameter occupied by barnacles. (--) Proportion of total space (as measured with a wheel of 1 cm diameter) occupied by barnacles.

LARVAL SELECTION FOR CRYPIC AND NON CRYPTIC HABITATS

Site Selection at Large Scales of Heterogeneity
(10-30 cm deep crevices)

Based on (1) the observed differences in habitat occupancy by adult Semibalanus, (2) the apparent greater settlement of larvae in crevices than outside crevices (Bergeron and Bourget, 1986), (3) the absence of predators (e.g. Thais lapillus) in the mid-intertidal zone in the Gulf, the prediction was made that in the absence of adult conspecifics and other macroscopic organisms, the larvae of Semibalanus would preferentially select crevices (10-30 cm deep) in the Gulf of St. Lawrence, but that this preference would be less pronounced on the Atlantic coast. Two sets of field experiments were carried out to test this prediction. A first set of experiments was carried out using artificial substrata and a second set using natural substrata. In the first set of experiments two scales of heterogeneity (30 cm deep crevices and cracks smaller than 15 mm) were used. In experiments using large scale heterogeneity 8 large artificial V-shaped crevices each with two adjacent horizontal surfaces abutting the upper part, were placed in the field prior to barnacle settlement. On each horizontal and inclined surface 10 x 10 cm quadrats were either kept uncolonized (controls) or were precolonized by adult conspecifics to densities of 50 (1984), 50 or 100 (1985) individuals per 100 cm^2. The experiments were carried out at Capucins and St. Andrews in 1984 and 1985. In both years, settlement was so sparse in the Gulf in the artificial crevices, adjoining panels and on all artificial substrata in general that the results were meaningless. At St. Andrews, while the presence of conspecifics (even < 0.1 % cover) and substratum heterogeneity increased settlement significantly, the former clearly had the greatest effect. The results analysed using multiple regression analysis give the following relationship between the number of settled larvae and the effect of barnacle abundance (Bar) and large scale heterogeneity (Het) (log transformed): $Y = 1.196$ (s.e. = 0.056) + 0.036 (0.04) (Bar) - 0.002 (0.0003) (Het); n = 96 quadrats, $R^2 = 63.8$, $F = 82.1$). Further, and quite surprisingly, in both pre-colonized and colonized quadrats, settlement was significantly greater on the horizontal surfaces than inside the crevices (see Chabot and Bourget, in press) (uncolonized horizontal surfaces, n = 36: x + s.e.(log transformed) = 1.229 + 0.054; uncolonized crevices, n = 36: x + s.e. (log transformed) = 0.619 + 0.059; pre-colonized horizontal surfaces, n = 12: x = 2.015 + 0.109; precolonized crevices, n = 12, x = 1.421 + 0.123). This result was unexpected. The larvae displayed a definite preference for exposed horizontal surfaces.

Owing to the difficulty of obtaining settlement for the Gulf, in 1986 the relative influence of heterogeneity and adult conspecifics on larval settlement was examined using natural crevices. Twenty crevices of approximately the same size (15-20 cm deep), shape (angle at the base 45-60°) and orientation (parallel to the shoreline), were selected at St. Andrews and Capucins. The initial density of adults on crevice sidewalls and adjacent surfaces were determined in ten quadrats (10 x 10 cm) per surface. One and a half month prior to settlement, at each site, ten crevices and adjacent horizontal surfaces were totally denuded and cleaned using blow torches and/or quicklime to remove barnacles, other fouling organisms, and potential cues. Each denuded crevice and its adjoining horizontal surfaces was paired with a control undenuded crevice similar in size, shape and initial density of adults, and adjoining exposed horizontal surfaces. The intensity of settlement was measured 10 days after the beginning of the settlement in ten predetermined quadrats (10 x 10 cm) per surface. The results, from Chabot and Bourget (in press), which are summarized in Fig. 5, clearly indicate significantly greater settlement in crevices than outside crevices in the Gulf and the opposite situation on the Atlantic coast. The results support the hypothesis of greater cryptic behaviour in the Gulf, and indicate a net preference to settle outside crevices on the Atlantic coast.

A major question pertaining to the behaviour of the barnacle larvae is raised by the results of this experiment. The larvae settling preferentially inside crevices in the Gulf is consistent with the hypothesis of larval selection for cryptic habitats resulting from strong selective pressures occurring on the population settling outside crevices (Bourget et al., 1985; Bergeron and Bourget, 1986). The most astonishing result, however, is preferred settlement outside

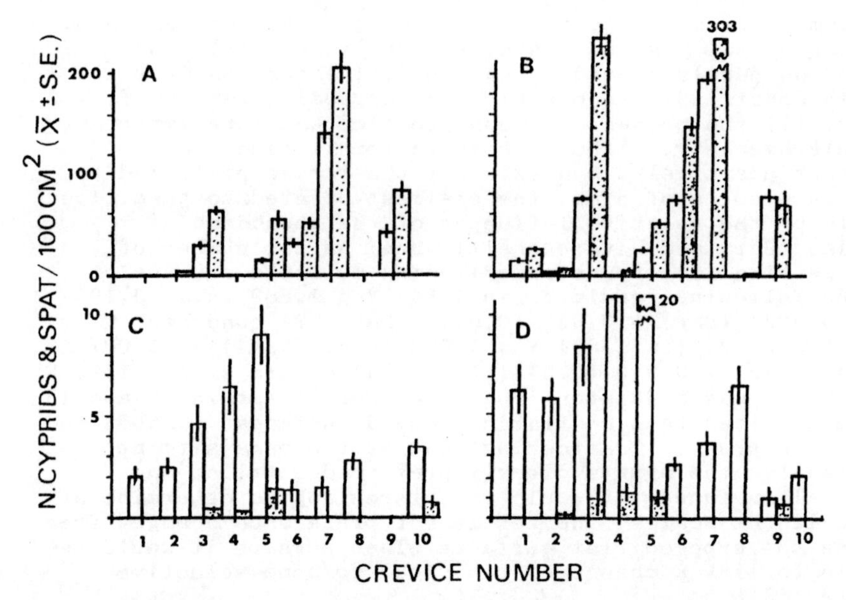

Fig. 5. Settlement in crevices (open bars) and adjacent horizontal surfaces (dotted) at St. Andrews (A, B) and Capucins (C, D). (A) and (C) denuded surfaces (10 crevices, 10 adjacent surfaces per site), (B) and (D) undenuded surfaces (10 crevices and 10 adjacent surfaces per site).

crevices, on the Atlantic coast. The main problem for interpreting these results is that all known reports of barnacle larval preferences in relation to substratum heterogeneity are related to small scale heterogeneity (contour and texture) (see above). Thus, preference in relation to large scale heterogeneity was unknown. Thais lapillus, the main predator of Semibalanus, is often associated with crevices, and forages in and around this microhabitat (Menge, 1978). Thus, at St. Andrews the avoidance of large crevices by the barnacle larvae may be an adaptive response to limit mortality from predation. In the Gulf, Thais is limited to the very low intertidal zone (dominated by mussels) and to the undersides of boulders from which it can launch predatory excursions on boulder sides. It is never observed in crevices of the mid-intertidal zone (Bourget et al., 1985, Bergeron and Bourget, 1986). Thus, Thais showed no predatory pressures in the same niches in the two regions studied.

Site Selection and Small Scales of Heterogeneity (5-15 mm grooves)

To examine the relative importance of small scales of heterogeneity and the abundance of conspecifics (0, 50, 100 individuals / 100 cm^2) other experiments were carried out using slate panels grooved at depths of 15, 10, 5 mm and ungrooved control panels. In 1984, the experiments included 5 and 10 mm grooves, while in 1985 15 mm grooves were added to the experimental design. Again, virtually no settlement occurred on panels placed in the Gulf. The results from the Atlantic coast indicate clearly that at scales ranging from 5 to 15 mm (1) the presence of conspecifics was more important than heterogeneity, although both factors influenced settlement positively, and (2) that the larvae preferred to settle in grooves of 5 mm, the smallest offered to them. The analysis of the relative influence of the abundance of barnacles (Bar) and heterogeneity (Het) on the number of larvae settled (log transformed) using multiple regression gave the following results: (in 1984) Y = 1.809 (s.e.:0.109) + 0.018 (0.003) (Bar) + 0.056 (0.014) (Het), 24 quadrats, R^2 = 0.714, F = 26.2; (in 1985) Y = 1.678 (s.e.: 0.141) + 0.007 (0.0004) (Bar) + 0.040 (0.014) (Het), 51 quadrats, R^2=.159, F = 4.5. There was consistently greater settlement in 10 and 15 mm size crevices than on flat ungrooved surfaces (Chabot and Bourget, in press). Considering that at the same site at larger scales the cypris clearly preferred settling on horizontal surfaces, it would be interesting to determine at what scale (or scales) the settlement preference changes from crevices to exposed flat surfaces. Then perhaps it would be possible to link a change of behaviour to some selective pressure.

SETTLING CUES IN RELATION TO SPATIAL AND TEMPORAL SCALES

There is still confusion concerning (1) the processes taking place during settlement, and (2) how they are influenced by various factors. Crisp and Meadows (1963) state that "each releaser opens the way to the next phase of behaviour". The evidence supporting this "sequential or chronological" hypothesis comes from experiments in which the competent cyprid larvae were given two very strong settling stimuli (grooves and settling factor) either singly or together and the settling response was compared with those obtained in other situations where no signal or negative stimuli were presented. However, this model may not adequately describe field situations. For instance, Crisp and Meadows (1963) report that when confronted with single factors individually (grooves and settlement factor), larvae are not strongly stimulated to settle. They state that "in our experiments neither the settling factor nor the contour stimulus alone was enough to cause more than a very few cyprids to pass through this stage and settle... However, when both stimuli were encountered simultaneously they re-inforced each other, though frequently the cyprid explored several pits before finally settling". Here, Crisp and Meadows clearly indicate the synergistic effects of two predominant settling factors. Further, Strathmann et al. (1981) indicate, from experiments carried out in the field, that the larvae may be using other cues, presumably algae, for settlement, at least at certain scales. The natural situation being immensely more complex than the laboratory situation, the larval settling response could be different from that anticipated from laboratory experiments. A surface immmersed in the sea is rapidly colonized by bacteria, various species of diatoms and detritus of all shapes, sizes, and habits (prostrate to erect, individual cells to colonial clones) (Hudon et al., 1983) offering a variety of surface characteristics ranging from

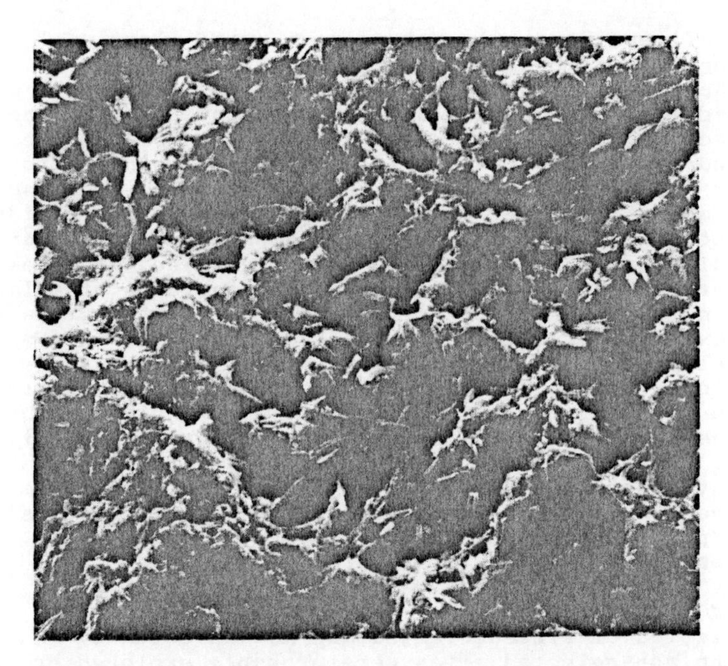

Fig. 6. Intertidal substratum heavily colonized by benthic diatoms and detrital matter. Scale bar = 10 um.

relatively clean surfaces to surfaces heavily fouled by a microscopic community (Fig. 6). Thus, presumably the larvae when exploring such surfaces receive positive as well as negative stimuli, but whether they make binary decisions in a sequential manner or integrate all stimuli as vectorial elements over time or space remains purely conjectural. Rather than determine the relative importance of certain factors over others by examining their individual influence on settlement, we set out to rank the importance of surface characteristics (physical and biotic) on larval settlement in the field, at sites selected and unselected by the larvae.

The various epibenthic (time) phases during settlement (broad exploration, close exploration and inspection phases) can also be coupled to different spatial scales. For instance, for study purposes, we assumed that the passive suprabenthic phase of contacts with the substratum takes place at distances of about 1 m from one another. This arbitrary distance is of little significance in absolute terms in the present context. The description of the close exploration phase allows us to assume that the events occurring during this phase take place at scales of the order of 1 mm, while the final phase immediately prior to settlement, the inspection phase, most likely takes place at scales much smaller than 1 mm, and perhaps smaller than 0.1 mm. Given this general hypothetical behavioural scheme, and the fact that the larvae appear to change their preference with spatial scales (see above), we set out to rank by order of relative importance, the factors strongly influencing settlement at each spatial scale. The larvae cannot explore all possible settling sites, so to examine the factors influencing settlement at the larger scale we compared the physical and biotic characteristics of the sites selected by the larvae (as near as possible to the frontal end of the settled cypris) to those of other sites avoided by the larvae. For this purpose, surfaces from the zone just above the upper limit of Semibalanus, (and avoided by the settling larvae in 1984) were compared with colonized sites within the zone occupied by barnacles. The two sites were 1 m from each other. The microscopic features of a large number of surfaces (selected sites = 84; unselected sites = 60) were examined quantitatively and semiquantitatively using scanning electron microscopy. The data were analysed in relation to settlement using a stepwise logistic regression (see Letourneux and Bourget, in press). The results indicated that at the large scale the cover of Urospora wormskjoldii (Mertens), detritus and of Achnantes parvula, as well as general patches of detritus, were the most likely settlement cues. Either the larva were rejecting Urospora or detrital cover, or positively selecting the cover of Achnantes or detrital matter.

Using the same procedure, comparison of the characteristics of sites selected and not selected by the larvae (the non-selected sites were probably explored by the larvae as they were only 1000 um from the selected site) surprising results were obtained. Detrital texture and detritus abundance inhibited settlement as for the large scale, but Achnantes was now avoided as well by the larvae. In fact, in selecting sites at scales of the order of 1 mm all (additional) living or non living matter attached to the surface (excluding the settlement factor not considered here) was apparently avoided by the larvae.

Finally, since the larvae apparently sought a clean site to settle, it was hypothesized that at scales of less than 0.1 mm the cypris would use microtopographic characteristics of the surface to guide its choice of the exact point of attachment of the antennae, the organs of settlement (Nott and Foster, 1969). This hypothesis was examined by comparing the microtopography of selected sites with that of immediately adjacent unselected sites (Letourneux and Bourget, in press). This was done by making a mould of the site selected and its surroundings. Larvae settled on plastic sheets were embedded in epoxy resin. Then the plastic sheet was peeled off, and using a profilometer a precise transect of the surface across the exact site of attachment of the antennae was obtained. Using fractal dimensions (Mandelbrot 1975, 1982), centers of gravity (using a Bioquant image analyser) as well as a microheterogeneity index (contour length/ linear distance; see Bergeron and Bourget, 1986), the microtopography of the selected sites (300 um) was compared to that of the two adjacent sites. The sites selected by the larvae were significantly more heterogeneous (bearing cavities) than the two adjacent non-selected sites. Further, a close analysis using fractal dimensions indicated that differences between sites become significant when the measuring segment length was (approximately) 40 um or less, thus indicating that the larvae selected (small depressions) microheterogeneous surfaces.

Distribution of Cues in the Two Regions

Does the difference in microhabitat selection between regions correspond to a difference in the distribution of cues? Letourneux and Bourget (in press) examined this question by quantifying substrata characteristics inside and outside of crevices in the Gulf using the SEM and the same variables as above. The substratum outside the crevices was relatively free of biological and inert materials, while that inside crevices supported populations of Achnantes and an unidentified alga (both accounting for the differences between microhabitats). Thus, in the two regions studied, Achnantes appeared to colonize different microhabitats, but the same ones colonized by barnacle larvae.

TOPOGRAPHIC SCALES AND SHIFT OF CUE

A major finding of this study was that there was a shift in (guiding) cues from biological cues (algae) during the broad exploration phase, to a mixture of biological and physical cues during close exploration, to physical cues (microheterogeneity) during final inspection. One question arising from this is whether the larvae actually perceived the biological cue (e.g. Achnantes), as a physical inert structure on the substratum, or by chemicals released by these organisms. For instance, some intertidal insects have been shown to select crevice habitats by perceiving volatile substances emitted by sediment microorganisms (Evans, 1982; 1986). A second question is whether the larvae can actually use a given cue (e.g. Achnantes) in a positive manner at a certain scale and use the same cue in a negative manner at another scale, as our results seem to indicate.

At the large scale (1 m), the need for a biological rather than a physical cue is logical since in a gradient of physical conditions such as in the intertidal zone, physical factors change progressively rather than abruptly and thus could be used directly (as reference marks) for setting abrupt upper limit of settling larvae, at least on the Atlantic coast where there are reportedly no critical tidal levels (Underwood, 1978). There are indications, however, that even on the Pacific coast where critical tidal levels occur (Doty, 1946), that barnacle larvae also use algae as cues (to discriminate between the acceptable and non acceptable zones) for settlement (Strathmann and Branscomb, 1979; Strathmann et al., 1981). The upper limit of intertidal organisms is presumed to be determined by physical factors, usually dessication (Connell, 1975; but see Denley and Underwood, 1979; Underwood and Denley, 1984; Connell, 1985), so it is not surprising that organisms capable of making active choices (accept or reject a settling site) use relatively passive settlers (e.g. plants), ill-equiped to avoid inappropriate sites and thus presumably having an intertidal upper limit set by dessication, as cues. The question of the strength of the association between an active settler and its guiding cue (a passive settler) is posed by the observed distribution of barnacles and Achnantes in the Gulf, where the two species are observed to co-occur in very precise microhabitats, the crevices. Again a study of co-occurrence of the two species is needed to further elaborate this relationship. The alternate hypothesis, the use of different cues in different geographic regions, must also be explored further since some species (e.g. larva of the rotifer *Callotheca* in freshwater environments) appear to be using more than one cue in selecting vegetation onto which to settle in different regions (Wallace, 1977; Wallace and Edmondson, 1986).

Table 1. Comparison of settlement cues used by the larvae of the intertidal barnacle, *Semibalanus balanoides* (Letourneux and Bourget, in press), and that of the subtidal barnacle, *Balanus crenatus* (Hudon et al., 1983). (-) indicates selection against the variable and (+) selection for the variable.

Semibalanus balanoides		*Balanus crenatus*	
Large Scale (metres)			
Urospora wormskjoldii	(-)	% cover of calcareous bases	(+)
Detrital cover	(-)	Position relative to grooves	(+)
Achnantes parvula	(+)	No. of newly metamorphosed individuals	(+)
Detrital aggregation	(+)	No. of larvae settled	(+)
Community characteristics not examined	(?)	No.of barnacles < 3 mm	(+)
		Texture of substratum	(+)
Intermediate Scale (mm)			
Detrital texture	(-)	Abundance of *Synedra* spp.	(-)
Achnantes parvula	(-)	Abundance of *Cocconeis* spp.	(-)
Detrital abundance	(-)	Distribution of detritus	(-)
		Abundance of detritus	(-)
		Silt	(-)
Small Scale (< 0.1 mm)			
Microtopography (<40 um pits)	(+)		

Table 1 compares the settlement cues obtained for *Semibalanus balanoides*) (Letourneux and Bourget, in press) with those of the subtidal species *Balanus crenatus* (Hudon et al., 1983). At intermediate scales (< 1 mm) both species search for a clean site to settle and avoid diatoms and detritus. However, considering the above discussion, at large scales one would expect that subtidal species use other cues than intertidal species. If in Hudon et al. (1983) consistent choices between substrata are indicative of cues used at larger scales, then the presence of adult, spat, larvae and fragments of conspecifics as well as the presence of grooves and texture of the substratum are important cues for *Balanus crenatus* at larger scales.

MODEL OF SETTLEMENT AND SPATIAL AND TEMPORAL SCALES

Settlement refers to all the behavioural events leading to and including attachment. I have summarized in Fig. 7 the main events occurring during the settlement of the intertidal barnacle Semibalanus balanoides, in the light of recent studies by Chabot and Bourget (in prep.) and Letourneux and Bourget (in press). These events are divided in two phases, a planktonic and an epibenthic phase. The first series of events including larval transport and contacts with the substratum,

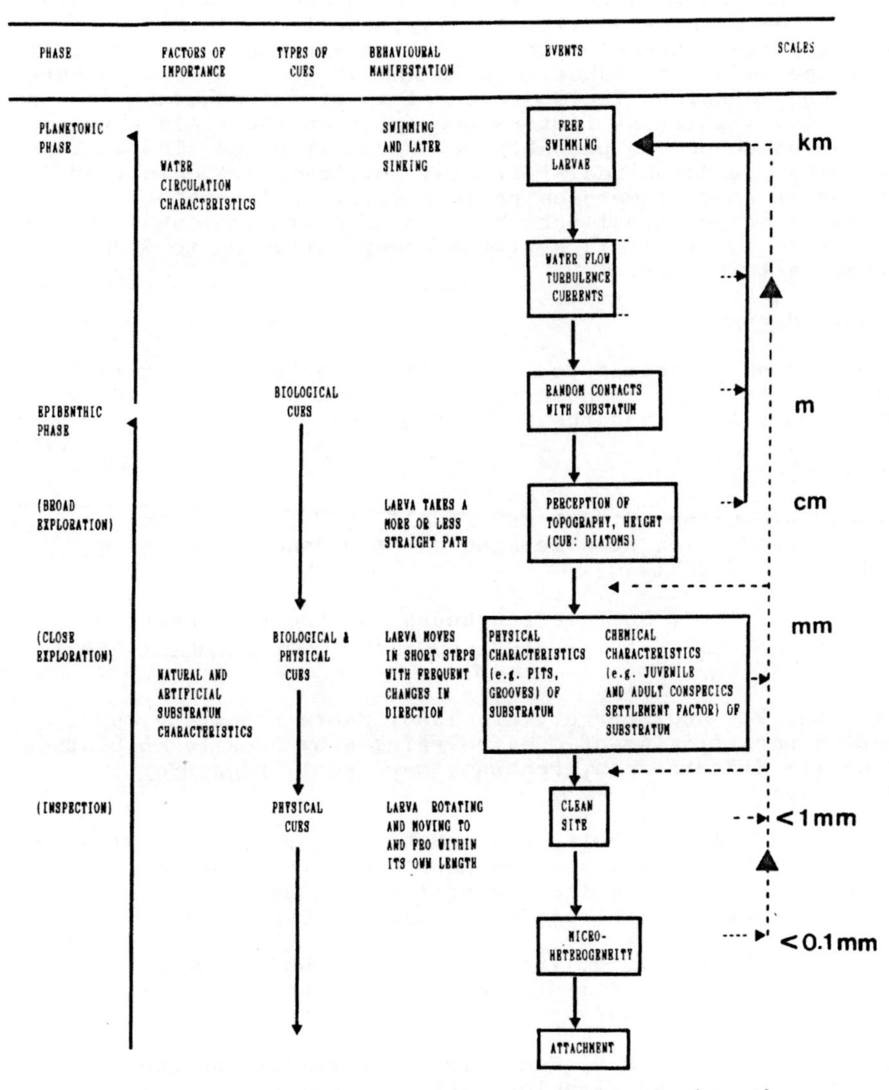

Fig. 7. Summary of the main events occurring during the settlement of the intertidal barnacle Semibalanus balanoides.

are essentially under the control of water flow and turbulence. They are presumed to occur at scales ranging from 1 m to several km. Little information is available concerning this phase. The second phase, the epibenthic phase, immediately preceding settlement, has been separated into three parts by Crisp and Meadows (1963). A broad exploration first takes place, during which the larva explores crudely the site to determine if the overall conditions are acceptable. The scale is in centimeters and the larvae use benthic diatoms on the substratum as cues. If the conditions are adequate then the larva more closely explores the substratum, focusing on its physical and chemical characteristics (grooves, pits and settling factor). If adequate stimuli are preceived, the larva (within the boundaries set by the physical and chemical cues) inspects the surface for a microsite as devoid as possible of diatoms, detritus and silt. Finally, the larva looks for microcavities into which to attach its antennae, presumably to limit the amount of adhesive used and at the same time ensure increased adhesion. The above model is of the sequence of events and settlement factors occurring in the field situation and the sequence can probably be short circuited. The fact that clean, uncolonized artificial substrata are colonized, even after short immersion periods (e.g. in laboratory bioassays or one tidal period in the field), indicates that a clean site in itself is a strong enough stimulus to induce barnacles to settle.

Acknowledgements

I thank John Himmelman for valuable comments and linguistic improvements of the manuscript. This work was supported by NSERC grant A0511 and the FCAR of Québec.

REFERENCES

Barnes, H., Crisp, D. J., and Powell, H.T., 1951, Observations on the orientation of some species of barnacles, J. Anim. Ecol., 20:227-241.

Barnes, H., 1956, Surface roughness and the settlement of Balanus balanoides (L.), Arch. Soc. Zoo. bot. fenn. Vanamo, 10:2.

Bergeron, P., and Bourget, E., 1986, Shore topography and spatial partitioning of crevice refuges by sessile epibenthos in an ice disturbed environment, Mar. Ecol. Prog. Ser., 28:129-145.

Bourget, E., Archambault, D., and P. Bergeron, 1985, Effet des propriétés hivernales sur les peuplements épibenthiques intertidaux dans un milieu subarctique, l'estuaire du Saint-Laurent, Naturaliste can., 112:131-142.

Bousfield, E. L., 1954, The distribution and spawning seasons of barnacles on the Atlantic coast of Canada, Bull. Nat. Mus. Can., No. 132, 112-154.

Bousfield, E. L., 1955, Ecological control of the occurrence of barnacles in the Miramichi Estuary, Bull. Nat. Mus. Can., No. 137:1-69.

Chabot, R., and Bourget, E., in press, The influence of substratum heterogeneity and settled barnacle density on the settlement of cypris larvae. Mar. Biol.

Crisp, D. J., 1955, The behaviour of barnacle cypris in relation to water movement over a surface, J. Exp. Biol., 32:569-590.

Crisp, D. J., 1974, Factors influencing the settlement of marine invertebrate larvae, in: "Chemoreception in Marine Organisms", P. T. Grant, and A. M. Mackie, eds., Academic Press, London.

Crisp, D. J., 1976, Settlement responses in marine organisms, in: "Adaptation to environment: essays on the physiology of marine animals", R. C. Newell, ed., Butterworths, London.

Crisp, D. J., and H. Barnes, 1954, The orientation and distribution of barnacles at settlement with particular reference to surface contour, J. Anim. Ecol. 23:142-162.

Crisp, D. J., and P. S. Meadows, 1963, Adsorbed layers: the stimulus to settlement in barnacles, Proc. Roy. Soc. Lond. (B), 158:364-387.

Connell, J. H. , 1961a, Effects of competition, predation by Thais lapillus, and other factors on natural populations of the barnacle Balanus balanoides, Ecol. Monogr., 31:61-104.

Connell, J. H., 1961b, The influence of interspecific competition and other factors on the distribution of the barnacle Chthamalus stellatus, Ecology, 42:710-723.

Connell, J. H., 1970, A predator-prey system in the marine intertidal region. 1. Balanus glandula and several predatory species of Thais, Ecol. Monogr., 40:49-78.

Connell, J. H., 1975, Some mechanisms producing structure in natural communities: A model and evidence from field experiments, in: "Ecology and evolution of communities", Cody, M. L. and J. M. Diamond (eds). Belknap Press, Cambridge, Mass.

Connell, J. H., 1985, The consequences of variation in initial settlement vs post-settlement mortality in rocky intertidal communities, J. Exp. Mar. Biol. Ecol., 36:269-293.

Daniel, A.; 1955, The primary film as a factor in settlement of marine foulers, J. Madras Univ. 25B:189-200.

Denley, E. J., and A. J. Underwood, 1979, Experiments on factors influencing settlement, survival and growth of two species of barnacles in New South Wales. J. Exp. Mar. Biol. Ecol., 36:269-293.

Doty, M. S., 1946, Critical tide factors that are correlated with the vertical distribution of marine algae and other organisms along the Pacific Coast, Ecology, 27:315-328.

Doyle, R. W., 1976, Analysis of habitat loyalty and habitat preference in the settlement behavior of planktonic marine larvae, Am. Nat., 110:719-730.

Dunbar, M.J., 1979, The relation between oceans, in: "Zoogeography and diversity in plankton", Van der Spoel and Pierrot-Bults, eds., Bringe Scientific Publishers, Utrecht.

El-Sabh, M. I., 1976, Surface circulation pattern in the Gulf of St. Lawrence, J. Fish. Res. Board Can., 33:124-138.

Evans, W. G., 1982, Oscillatoria sp. (Cyanophyta) mat metabolites implicated in habitat selection in Bembidion obtusidens (Coleoptera: carabidae), J. Chem. Ecol., 8:671-678.

Evans, W. G., 1986, Edaphic and allelochemic aspects of intertidal crevices sediments in relation to habitat recognition by Thallassotrechus barbarae (Horn) (Coleoptera: carabidae), J. Exp. Mar. Biol. Ecol., 95:57-66.

Forbes, L., Seward, M. J. B., and Crisp, D. J., 1971, Orientation to light and the shading response in barnacles, in: "Fourth European Marine Biology Symposium", D. J. Crisp, ed., Cambridge University Press, Cambridge.

Feyling-Hanssen, R. W., 1953, The barnacle Balanus balanoides (Linné, 1766) in Spitsbergen, Skr. norsk Polarinstitutt, 98:1-64.

Gaines, S., and Roughgarden, J., 1985, Larval settlement rate: a leading determinant of structure in an ecological community of the marine intertidal zone. Proc. Natl. Acad. Sci. USA., 82:3707-3711.

Gaines, S., Brown, S., and Roughgarden, J., 1985, Spatial variation in larval concentrations as a cause of spatial variation in settlement for the barnacle, Balanus glandula, Oecologia (Berl), 67:267-272.

Gray, J. S., 1974, Animal-sediment relationships, Oceanogr. Mar. Biol. Ann. Rev., 12:223-261.

Gregg, J. H., 1945, Background illumination as a factor in the attachment of barnacle cyprids, Biol. Bull. Mar. Biol. Lab. Woods Hole, 88:44-49.

Hannan, C. A., 1984, Initial settlement of marine invertebrate larvae: the role of passive sinking in a near bottom turbulence flow environment. Ph. D. thesis. Woods Hole Oceanographic Institution and MIT.

Hawkins, S. J., and Hartnoll, R. G. 1983. Settlement patterns of Semibalanus balanoides (L.) in the Isle of Man (1977-981), J. Exp. Mar. Biol. Ecol., 62:271-283.

Hudon, C., Bourget, E., and Legendre, P., 1983, An integrated study of the factors influencing the choice of the settling site of Balanus crenatus cyprid larvae, Can. J. Fish. Aquat. Sci., 40:1186-1194.

Knight-Jones, E. W., Bailey, J. H., and Isaac, M. S., 1971, Choice of algae by larvae of Spirorbis, especially Spirorbis spirorbis, in: "Fourth European Marine Biology Symposium," D.J. Crisp, ed., Cambridge University Press, Cambridge.

Knight-Jones, E. W., and Quasim, S. Z., 1955, Responses of some marine animals to changes in hydrostatic pressure, Nature, Lond., 175:941.

Larman, V. N., and Gabbott, P. A., 1975, Settlement of cyprid larvae of Balanus balanoides and Elminius modestus induced by extracts of adult barnacles and other marine animals, J. Mar. Biol. Assoc. U.K., 55:183-190.

Larman, V. N., Gabbott, P. A., and East, J., 1982, Physico-chemical properties of the settlement factor protein from the barnacle Balanus balanoides, Comp. Biochem. Physiol., 72B:329-338.

Letourneux, F., and Bourget, E., in press, The importance of physical and biological settlement cues used at different spatial scales by the larvae of the barnacle, Semibalanus balanoides, Mar. Biol.

Lewis, J. R., 1964, "The ecology of rocky shores." The English Universities Press Ltd., London.

Mandelbrot, B., 1975, "Les objets fractals: forme, hasard et dimension," Flammarion, Paris.

Mandelbrot, B., 1982, "The fractal geometry of nature," Freeman & Co., San Francisco.

Menge, B. A., 1978, Predation intensity in a rocky intertidal community: relation between predator foraging and environmental harshness, Oecologia (Berl), 34:1-16.

Miller, M. A., Rapean, J. C., and Whedon, W. F., 1948, The role of slime films in the attachment of fouling organisms, Biol. Bull. Mar. Biol. Lab. Woods Hole, 94:143-157.

Moyse, J., and Hui, E., 1981, Avoidance by Balanus balanoides cyprids of settlement on conspecific adults, J. Mar. Biol. Assoc. U.K., 61:449-460.

Newell, R. C., 1979, Biology of intertidal animals. Marine Ecological Surveys Ltd., Faversham, Kent.

Nott, J. A., and Foster, B. A., 1969, On the structure of the antennular attachment organ in the cypris larva of Balanus balanoides (L.), Phil. Trans. R. Soc. (B), 256:115-134.

Petersen, G. H., 1962, The distribution of Balanus balanoides (L.) and Littorina saxatilis, Olivi, var. groenlandica, Mencke in northern west Greenland, Medd. om Gronland, 159:1-47.

Pomerat, C.M., and Reiner E. R., 1942, The influence of surface angle and light on the attachment of barnacles and of the other sedentary organisms, Biol. Bull. Mar. Biol. Lab. Woods Hole, 91:57-65.

Rittschof, D., Branscomb E. S., and J.D. Costlow, 1984, Settlement and behavior in relation to flow and surface in larval barnacles, Balanus amphitrite Darwin, J. Exp. Mar. Biol, 82:131-146.

Smith, F. G. W., 1946, Effect of water currents upon the attachment and growth of barnacles, Biol. Bull. Mar. Biol. Lab. Woods Hole, 90:51-70.

Smith, F. G. W., 1948, Surface illumination and barnacle attachment, Biol. Bull. Mar. Biol. Lab. Woods Hole, 94: 33-39.

Strathmann, R. R., and Branscomb, E. S., 1979, Adequacy of cues to favorable sites used by settling larvae of two intertidal barnacles, in: "Reproductive ecology of marine invertebrates." S. E. Stancyk, ed., The Belle W. Baruch Library in Marine Science No. 9, University of South Carolina Press, Columbia, South Carolina.

Strathmann, R. R., Branscomb, E. S., and Vedder, K., 1981, Fatal errors in set as a cost of dispersal and the influence of intertidal flora on set of barnacles, Oecologia (Berl), 48:13-18.

Underwood, A. J., 1978, A refutation of critical tidal levels as determinants of the structure of intertidal communities on British shores., J. Exp. Mar. Biol. Ecol., 33:261-276.

Underwood, A. J., and Denley, E. J., 1984, Paradigms, explanations, and generalizations in models for the structure of intertidal communities on rocky shores, in: "Ecological communities: Conceptual issues and the evidence," D. R. Strong Jr., D. Dimberloff, L. G. Abele, A. B. Thistle, eds, Princeton University Press, Princeton, New Jersey.

Wallace, R. L., 1977, Adaptive advantages of substrate selection by sessile rotifers, Arch. Hydrobiol. Beih., 8: 53-55.

Wallace, R. L., and Edmondson, W. T., 1986, Mechanism and adaptive significance of substrate selection by a sessile rotifer, Ecology, 67:314-323.

Walker, G., and Yule, A. B., 1984, Temporary adhesion of the barnacle cyprid: the existence of an antennular adhesive secretion, J. Mar. Biol. Assoc. U. K., 64:679-686.

Weiss, G.M., 1947, The effect of illumination and stage of tide on the attachment of barnacle cyprids, Biol. Bull. Mar. Lab. Woods Hole., 93:240-249.

Wethey, D. S., 1984, Spatial pattern in barnacle settlement: day to day changes during the settlement season, J. Mar. Biol. Assoc. U. K., 64:687-698.

Yule, A. B., and Crisp, D. J., 1983, Adhesion of cypris larvae of the barnacle Balanus balanoides, to clean and arthropodin treated surfaces, J. Mar. Biol. Assoc. U. K., 63:261-271.

Yule, A. B., and Walker, G., 1985, Settlement of Balanus balanoides: the effect of cyprid antennular secretion, J. Mar. Biol. Assoc. U. K., 65:707-712.

Analysis of Movement Patterns and Orientation Mechanisms in Intertidal Chitons and Gastropods

Guido Chelazzi, Stefano Focardi (*) and Jean-Louis Deneubourg (**)

University of Florence, Italy
(*) I.N.B.S., Ozzano, Italy
(**) Free University of Brussels, Belgium

INTRODUCTION

Despite their different organization and biology, chitons (Mollusca, Polyplacophora) and gastropods (Mollusca, Gastropoda) share a large number of adaptations to intertidal life, including morpho-functional and behavioural traits. Communication, clustering, aggressiveness and even simple parental cares have been reported in both classes but, as in other animals, the basis of their behavioural adaptation to the intertidal environment is a proper organization of activity in space and time.

The quantity of reports on these topics is impressive, but the number of satisfactory ethological analyses is low. The present contribution is not a comprehensive review of chitons' and gastropods' behavioural ecology (partial reviews may be found in Boyle, 1977; Newell, 1979; Underwood, 1979; Branch, 1981; Hawkins and Hartnoll, 1983), but rather is an attempt to clear up some conceptual and methodological misunderstandings and shortcomings, and to identify the problems in a field that only recently has been opened to quantitative and experimental investigation.

ACTIVITY ORGANIZATION IN TIME AND SPACE

In approaching the study of activity patterns of intertidal chitons and gastropods we must clearly distinguish between occasional, continuous and rhythmic phenomena. Occasional activity includes movements in response to such unpredictable ecological variations as those produced by storms or the sudden appearance of predators and competitors. It also includes the capacity shown by some neritid (Chelazzi and Vannini, 1976, 1980), trochid (Byers and Mitton, 1981; Doering and Phillips, 1983) and littorinid (Gendron, 1977; Hamilton, 1978a) gastropods to regain their optimal zonation upon experimental displacement. A second class of occasional behaviours is given by the avoidance responses elicited in gastropods by contact with, or the approach by potential predators such as sea-stars (Burke, 1964; Feder, 1967; Branch, 1979) or other gastropods (Hoffmann et al., 1978).

Onthogenetic shifting of the population members along the shore produces an evident size-gradient in many gastropods (Branch, 1975b; Underwood, 1979) but the relative importance of individual behaviour (continuous, progressive migration along the sea-land axis) and selective mortality in producing it has not been fully clarified.

Rhythmic activity includes movements related to seasonal, synodic, tidal and diel fluctuations in the shore ecology. Seasonal and synodic movements usually consist in zonal migrations up and down the shore in order to minimize the exposure to stress factors and to optimize the access to resources. Seasonal migrations related to reproduction have been reported in many intertidal gastropods including limpets (Branch, 1975b), throchids (Underwood, 1973) and littorinids (Smith and Newell, 1955; Branch and Branch, 1981). Limpets avoid dehydration and overheating by moving along the sea-land axis in synchrony with seasonal or spring-neap cycles (Breen, 1972; Branch, 1975b, 1981), but the ethological determinism of these long-term rhythmic migrations has yet to be deeply investigated (Hamilton, 1985).

The short-term activity of intertidal chitons and gastropods is organized into temporal units determined by tidal and diel variations of physical and biological factors on the shore. In general, intertidal animals may adopt one of two alternative strategies (Chelazzi and Vannini, 1985). The "isophasic" pattern, typical of such mobile groups as arthropods and vertebrates, consists in rhythmic zonal migrations synchronous and concordant with the tides. These animals perform a dynamic colonization of the intertidal environment, following oscillations in the medium (air or water) to which they are primarily adapted. Some sandy beach whelks, such as *Bullia digitalis* and *B. rhodostoma* (McLachlan et al., 1979; Brown, 1982) adopt this isophasic pattern, performing wide "*Donax* like" migrations. On a shorter scale the isophasic pattern is also shown by some littorinids as the marsh periwinkle *Littorina irrorata* (Hamilton, 1977) and the cerithid *Cerithidaea decollata* climbing on mangrove trunks (Cockroft and Forbes, 1981).

On the contrary, most typical rocky shore chitons and gastropods, due to the high energy cost of their locomotion (Denny, 1980; Horn, 1986) and low speed, adopt an "isospatial" strategy, remaining within a more or less narrow belt along the sea-land axis, with alternating exposure to air and water. The activity of these animals is limited to the time when the condition of their zone is suitable for moving and feeding. Timing of unit phase of activity (u.p.a.) is limited by the morpho-functional organization of the different species, but the very local structure of the environment, including complex relationships with other species is determinant as well (Thain, 1971; Spight, 1982). For this reason a generalization is not possible (Hawkins and Hartnoll, 1983). Among the few chitons studied, the timing of u.p.a. includes tidal and diel components. While the latter seem to predominate in *Nuttalina californica* (Nishi, 1975), a tidal rhythm has been reported in *Mopalia lignosa* (Fulton, 1975). In gastropods the relative importance of diel and tidal components may depend on their zonation on the shore (Zann, 1973a).

Nevertheless, most intertidal chitons and gastropods show combined diel-tidal rhythms of high complexity (Zann, 1973b). As diel and tidal cycles have not the same period, a short-term survey is generally not sufficient to understand the interlocking of the two components in u.p.a. timing (Cook and Cook, 1978) and a satisfactory analysis may require a complete synodic period (about 29 days). Moreover, a marked variation in u.p.a. timing can occur in taxonomically related species, and between different populations of the same species, as in *Patella vulgata* (Hawkins and Hartnoll, 1983). Careful investigation of u.p.a. timing, including the different sources of variability is very important, because the time actually devoted to feeding is a critical parameter in the energy balance of intertidal gastropods and chitons; in addition, u.p.a. definition is necessary for proper scheduling of position sampling for movement analysis.

On the basis of the existing heterogeneous evidence, the feeding excursions performed by rocky shore chitons and gastropods during each u.p.a. are generally classified into three distinct models of increasing complexity: ranging pattern, zonal shuttling and central place foraging.

Ranging pattern. In the first model, sometimes reported as "random pattern", feeding excursions are not orientated toward constant directions and the animals do not return to their previous shelter or to the same shore level. This model has been reported by Underwood (1977) in some australian gastropods and seems to be present in some littorinids as well (McQuaid, 1981; Petraitis, 1982). An example of ranging pattern in chitons has been found with the caribbean *Acanthopleura granulata* (pers. observation) showing highly meandering feeding paths and no long-term preferential rest sites. More generally, this model may be common in species living on non-tidal shores, or when adaptation to the littoral environment is based on the temporal more than spatial organization of activity, linked to a solid morpho-functional specialization to stable microhabitats. Nevertheless, ranging does not mean random: the random walk of these animals must be biased at least by kinetic responses, allowing the long-term stability of their zonal distribution.

Zonal shuttling. The second model, sometimes quoted as "tidal migration", has been described in such gastropods as neritids (Vannini & Chelazzi, 1978; Chelazzi, 1982; Chelazzi et al., 1983c), trochids (Wara & Wright, 1964; Micallef, 1969; Thain, 1971), planaxids (Magnus and Haacker, 1968), and some limpets such as *Acmaea limatula* (Eaton, 1968), *A. pelta* (Craig, 1968), *A. scabra* (Rogers, 1968) which loop along the sea-land axis at each u.p.a. That this shuttling is not a true tidal migration typical of isophasic animals is evident from the fact that - despite their upshore climbing for resting - many intertidal gastropods such as *Nerita textilis* are splashed by waves during high tide. In some species shuttling is paradoxically inverted with respect to the sea level oscillations, with upward movements during low tide, and downward migration prior to high tide, as in *N. polita* (Chelazzi, 1982). Zonal shuttling is due to the separation of feeding and resting zones along the sea-land axis, not simply dependent upon obvious physical factors: predation and competition more often than a simple escape from air or water are probably involved in the evolution of this pattern in such shuttlers as many trochid and neritid gastropods (Thain, 1971) and *Acanthopleura* spp. chitons (Chelazzi et al., 1983a).

Central place foraging. Resting in a definite shelter more or less constant in time and homing to it after each feeding excursion has long since been described in limpets (Davis, 1895) and chitons (Crozier, 1921). Underwood (1979) and Branch (1981) reviewed the occurrence, ecological significance and some operational aspects relative to the so-called "homing behaviour" of limpets, but generalization on the three aspects is again difficult because detailed analysis of movements performed by homer species during each u.p.a., and the modification of individual strategy in time, is often lacking. The sampling of only rest positions is frequent in these studies, since most limpets rest during diurnal low tides, when the shore is accessible for study, while the feeding phase often occurs while the animals are splashed by waves. These sampling limitations produce a fuzzy picture of the spatial behaviour.

Most contributions are concerned with the accuracy of homing performance, scaling the observed behaviour from the "statistical homing" of e.g. *Acmaea digitalis* (Frank, 1964) to the deterministic homing of such species as *Patella depressa* (Cook et al., 1969), *P. vulgata* (Bree, 1959), *P. longicosta* (Branch, 1971), *Collisella scabra* (Hewatt, 1940) and *Notoacmea petterdi* (Creese, 1980). Similar scaling has been observed between congeneric chitons as well (Focardi and Chelazzi, in prep.). But the following aspects of central place foraging are also important: use of natural shelters or active digging of scars, such as in many limpets (Branch, 1981) and the chiton *Acanthopleura gemmata* (Chelazzi et al., 1983a); scar-shell complementarity, as in most limpets, or non-individual scar morphology permitting its use by different conspecifics, as in chitons; scar defense from intruders as in *Acanthopleura gemmata* (Chelazzi et al., 1983b; Chelazzi and Parpagnoli, 1987) or simply abandoned upon intrusion as in *A. brevispinosa* (personal observation).

Moreover, the knowledge of actual movement pattern between two rest events is critical, giving a great deal of information on the spatial relationships between rest site and feeding grounds. It is evident that feeding loops are variously orientated in the different species and in the different populations of the same species: radially distributed around the shelter, as in *Patella vulgata* (Hartnoll and Wright, 1977), or biased along the sea-land axis as in *Acanthopleura gemmata* (Chelazzi et al., 1983a). Moreover, some species are territorial, defending their feeding ground (Stimson, 1970; Branch, 1981), while in others the feeding grounds are not personal. The long-term strategy of homers is very interesting as well, as stressed by Cook and Cook (1981): do the animals intensively exploit a good algal patch or is their feeding conservative, shifting to different places on different excursions? Reiteration in time of individual route recording may reveal interesting aspects of homing pattern, but so far this kind of analysis has been limited to two chitons (Chelazzi et al., 1983a) and two siphonarid limpets (Cook and Cook, 1981). A more detailed analysis of movements may help to interpret the shifting from solitary homing to collective refuging, as observed in many intertidal gastropods and some chitons. This transition often seems intraspecific and dynamic: the mechanism for shifting from solitary to collective resting must be searched for in the amplification of minimal variations in individual behaviour, depending on variations in shore morphology and tidal regime (Chelazzi et al., 1985; Focardi et al., 1985a, 1985b).

An ethological approach to all the above aspects is more necessary than a blind classification of homing performance to understand the homing phenomenon in terms of "central place foraging" theory (Orians and Pearson, 1979). One of the methods which may open some ethological black boxes of the homing phenomenon in chitons and gastropods is LED tracking (or motographic method) (Chelazzi et al., 1983c; 1987). This simple technique permits the integral recording of the spatial strategy of the single animals, provided that the diel component of their u.p.a. is nocturnal. If exported to other species LED tracking may overcome many of the methodological shortcomings indicated by Hamilton (1978b): "absence of data on individuals, lack of multiple position records for individuals, impreciseness of direction and distance measurements, small sample size, short-term observations, and lack of statistical analysis". The last point is a very critical one, as recently stressed by Underwood and Chapman (1985).

But establishing three very distinct models (ranging, zonal and homing) may be misleading. In fact, the zonal pattern is frequently linked to solitary or collective homing (Magnus and Haacker, 1968; Vannini and Chelazzi, 1978) and, conversely, excursions of typical homers are sometime zonally polarized (Chelazzi et al., 1983c), while deterministic homing can alternate with dispersive (ranging) excursions such as observed in the chiton *Acanthopleura granulata* (personal observation).

ANALYSIS OF ORIENTATION MECHANISMS

A frequent outcome of the study on movement patterns is the analysis of underlying orientation mechanisms. Intertidal molluscs have not escaped this trend and some work has been done on mechanisms controlling the behaviours described above. Nevertheless, the quality of information available on chitons and gastropods is disappointing in comparison to that on other groups such as crustaceans (Herrnkind, 1983; Pardi and Ercolini, 1986), with the result that chitons and gastropods make only a timid appearance in recent important reviews on animal orientation (Jander, 1975; Schone, 1984).

Directional orientation. Under the distinct sea-land asymmetry faced by many rocky shore chitons and gastropods, there is probably no pressure to evolve such complex directional mechanisms as the astronomical and magnetic orientation used by some sandy beach arthropods. Some speculations about the existence of "solar orientation" in gastropods (Warburton, 1973) must be discarded on the basis of experiments conducted on neritids (Chelazzi and Vannini, 1976, 1980) and littorinids (Hamilton, 1978a). Also evidence for the use of magnetic field by gastropods (Brown, 1963) and chitons (Ratner, 1976) is very slim. On the contrary, the importance of gravity, light and wave movement in the directional orientation of these groups is evident (Newell, 1979; Creutzberg, 1975; Gendron, 1977) but will not be reviewed here.

The real problem with these intertidal animals lies elsewhere: besides orienting cues, their directional orientation requires a correct modulation of movement polarity to be ecologically adaptive. There is early and recent evidence that zonal movements of molluscs are not based on simple responses to a single cue, but depend on complex regulation of taxes by arrays of releasing stimuli (Chelazzi and Focardi, 1982; Underwood and Chapman, 1985). Some contributions, most on littorinids, demonstrate the dependence of photo- and geotaxis polarity on such external factors as hydration (Gowanloch and Hayes, 1926; Kristensen, 1965) and temperature (Bingham, 1972; Janssen, 1960; Bock and Johnson, 1967), but other physical and biological releasers may be involved as well. Endogenous reversing of taxis is suggested by observation of movements in such shuttlers as *Nerita polita* (Chelazzi, 1982), but while the internal rhythm of general motor activity has been demonstrated in some gastropods (Zann, 1983b; Rohde and Sandland, 1976) the endogenous control of cyclic reversing in photo- and geotaxis polarity is a field open to future research.

Homing mechanisms. Three sets of orienting cues have been suggested to be involved in the homing of chitons and gastropods: internal to the animal (Pieron, 1909), pertaining to the macrosystem (Bohn, 1909) or related to the microsystem onto which the animal moves. Once discarded the importance of idiothetic mechanisms, and of path-integration based on gravity or sun (Cook, 1969), two possible mechanisms of piloting by local information have remained: mnemotaxis, i.e. memorization of micro-landmarks intrinsic to the substrate, and trail following. The mnemotaxis hypothesis (Ohgushi, 1955; Thorpe, 1963; Galbraith, 1965; Jessee, 1968) rises where trail following falls and no direct proof of its validity has been obtained. On the contrary, evidence in support of trail following in solitary (Funke, 1968; Cook et al., 1969; Chelazzi et al., 1987) and collective homers (Lowe and Turner, 1976; Gilly and Swenson, 1978; Trott and Dimock, 1978; Raftery, 1983; Chelazzi et al., 1985) is solid, but incomplete according to some authors.

The following observations are usually reported as disproving the unique importance of trail following in the homing of intertidal chitons and gastropods: i) in some homer species the return and outgoing paths during each feeding loop do not overlap (Beckett, 1968; Cook et al., 1969; Thomas,

1973); ii) some chitons and limpets passively displaced from their scars are nonetheless able to home (Thorne, 1968; Cook, 1969); iii) different methods of trail interruption - including rock chiselling, brushing, or washing with various chemicals - fail sometimes to extinguish the homing performance (Galbraith, 1965; Jessee, 1968; Cook et al., 1969). But the first two counter-proofs fail by admitting that the trail associated information is long lasting and that animals are able to detect a previous, old trail. In fact, there is increasing evidence that this is true in siphonarid limpets (Cook, 1969; 1971) and in neritids (Chelazzi et al., 1985). Concerning the third point, it is interesting what has been observed in the opisthobranch *Onchidium verruculatum* (McFarlane, 1980, 1981) and in the chiton *Acanthopleura gemmata* (Chelazzi et al., 1987): upon reaching the experimental trail interruption these animals perform explorative movements which allow them to re-contact the outward trail and home safely.

Trail associated orienting cues. Although these considerations strongly support the importance of trail following in the homing of chitons and gastropods, the nature of orienting cues contained in their trails and the actual orientation mechanism are still unknown. Cook (1971) quotes three possibilities: radular scrapings on the algal cover; textural discontinuities provided by the mucus; attached or diffusible chemicals. Many authors favour the chemical hypothesis and Funke (1968) admits also that a complex chemical "footprint" is left by *Patella vulgata* in the scar, in order to correctly orientate while resting. Contact chemoreception using pedal receptors has been claimed for *Ilyanassa obsoleta* (Trott and Dimock, 1978), while the cephalic tentacles would be sensitive to trail associated chemicals in littorinids (Peters, 1964; Hall, 1973). Raftery (1983) maintains that chemical information could be important in *Littorina planaxis*, but in his opinion physical properties cannot be ruled out. On the other hand, Bretz and Dimock (1983), on the basis of one of the most complete surveys of trail cues in gastropods, conclude that *Ilyanassa obsoleta* uses mechanical information (textural properties of the mucus) and not such chemicals as proteins. But in a similar analysis, performed on the freshwater snail *Biomphalaria glabrata*, Bousfield et al., (1981) reached the opposite conclusion that "chemical, as opposed to physical factors, play a dominant role in trail following behaviour". The physical hypothesis is indirectly supported by the observation of micro- and macro-patterning of the mucus trail of gastropods (Cook, 1971; Bretz and Dimock, 1983; Stirling and Hamilton, 1986). But chemical content of the trail is even more complex, including such different classes of light and heavy molecules as free aminoacids, lipids, carbohydrates, free proteins and glycoproteins (Wilson, 1968; Hunt, 1970; Grenon and Walker, 1980, Bretz and Dimock, 1983). Future research must proceed both ways but chemical hypothesis seems more attractive upon recognition that the trail associated information is complex, including individual cues and polarity.

Trail individuality. Collective refuger species are expected to release and follow non-individual trails for homing. In fact this is the case for *Ilyanassa obsoleta* (Trott and Dimock, 1978) and *Nerita textilis* (Chelazzi et al., 1985). On the contrary, some mechanisms for own-trail recognition must exist in solitary homers, given the possible interindividual crossing of different feeding paths under natural conditions. Some widely quoted experiments of Funke (1968) show discrimination between personal and conspecific trails in *Patella vulgata* and the same is true for *Onchidium verruculatum* (McFarlane, 1980). Cross-trailing tests in *Acanthopleura gemmata* (Chelazzi et al., 1987) showed that this solitary-homer chiton has a quasi-personal trail: a low trail polymorphysm in the population tested would allow the reduction of inter-individual mistakes in following the outward trail for homing. Nevertheless, other solitary-homers such as *Patella vulgata* (Cook et al., 1969) and *Siphonaria alternata* (Cook, 1971) if transplanted onto conspecific trails do follow them up to the scar. The possibility remains that in these species trail contains both individual- and species-specific information, but in some cases the normal recognition of own trail during the homing could be based on mechanisms external to the trail, such as memorization of the direction taken during the outward journey.

Trail polarity. That trail following is not performed at random concerning polarity has been documented in many gastropods (Stirling and Hamilton, 1986). When central place foraging is combined with zonal shuttling as in Nerita textilis (Chelazzi et al., 1985), correct following may be due to directional mechanisms based on cues external to the trail. Nevertheless, some experiments have ruled out the possibility that polarized trail following is based on such external cues as light or gravity (Cook and Cook, 1975), but the intrinsic mechanisms involved in polarized trail following have not been discovered . Cook (1971) speculates four hypotheses for this capacity in siphonarid limpets: i) chemical macro-gradient along the trail; ii) polarized sequence of different chemicals; iii) polarized physical structure of the trail; iv) differential friction upon retracing the trail in the two opposite directions. Stirling and Hamilton (1986) listed three more: v) chemical micro-gradient; vi) bilateral asymmetry in chemical or physical properties of the trail; vii) differential light reflection in the two opposite directions of the trail. By excluding other mechanisms, the last Authors conclude that a micro-structural polarization of trail is possibly detected by Littorina irrorata. This agrees with the findings of Bretz and Dimock (1983) on Ilyanassa obsoleta, but contrasts those of Gilly and Swenson (1978) who claimed the importance of a chemical macro-gradient as polarity mechanism in Littorina sitkana and L. littorea. Finally, Cook and Cook (1975) obtained the interesting evidence that in Siphonaria alternata trail polarity is lost after retracing.

Trail complexity and multiple use of trail following. That trail following may be used by intertidal gastropods in functional contexts different from homing, such as prey location (Gonor, 1965; Snyder and Snyder, 1971) and mate searching (Hirano and Inaba, 1980) is well known, but in the context of central place foraging it has been usually regarded as a homing mechanism. Recent findings demonstrate that this is an oversimplified view. That homer gastropods and chitons may discriminate between outgoing and return trail has been argued by McFarlane (1981) for Onchidium verruculatum, and by Chelazzi et al., (1987) for the chiton Acanthopleura gemmata. In O. verruculatum displacement tests lead to the conclusion that outward and return trails have different information content. Field observations using LED tracking showed that A. gemmata, besides performing the usual homing-related following of own trail, may reach the feeding place by following its own trail released on the previous night. What is worty is that where the outward and return trails of the previous night do not overlap, the animal counter-follows the return one. This may be based on a different information content of the two branches as in O. verruculatum, or on the intrinsic polarization of the trail.

Whatever the discrimination mechanisms are, the ecological importance of this foodward auto-trailing is evident, allowing the animal to optimize the exploitation of algal grounds. In fact the foodward trail following is not as deterministic as the homeward one. We have hypothesized (Chelazzi et al., 1987) that the amount of foodward trail following may depend on the quality of return trail, which in turn may depend upon the quantity or quality of ingested food. In the same species there is also preliminary evidence for inter-individual trail following in the retrieval of the feeding grounds (Chelazzi et al., 1987), which is tuned at minimal levels thus avoiding local overexploitation of algal patches. In this case communication via trail following may represent a mechanism of ecological regulation for the whole population.

These aspects of chitons' behaviour deserve further work and the study on these particular topics must be extended to intertidal gastropods as well. Nonetheless, they suggest that the trail is a complex eternal memory of multiple ecological significance, which may regulate the different aspects of central place foraging in intertidal molluscs. Besides ants, where trail following play a major role in controlling the ecological relationships of the population, intertidal chitons and gastropods may represent a second study case for this important class of problems in behavioural ecology.

REFERENCES

Beckett, T. W., 1968, Limpet movements. An investigation into some aspects of limpet movements, expecially the homing behaviour, Tane, 14: 43-63.

Bingham, F. O., 1972, The influence of environmental stimuli on the direction of movement of supralittoral gastropod Littorina irrorata, Bull. Mar. Sci., 22: 309-335.

Bock, C. E., and Johnson, R. E., 1967, The role of behaviour in determining the intertidal zonation of Littorina planaxis Philippi 1847 and Littorina scutulata Gould 1849, Veliger, 10: 42-54.

Bohn, G., 1909, De l'orientation chez les patelles, C. R. Acad. Sci., 148: 868-870.

Bousfield, J. D., Tait, A. I., Thomas, J. D., and Towner-Jones, D., 1981, Behavioural studies on the nature of stimuli responsible for triggering mucus trail tracking by Biomphalaria glabrata, Malacol. Rev., 14: 49-64.

Boyle, P. R., 1977, The physiology and behaviour of chitons (Mollusca: Polyplacophora), Oceanogr. Mar. Biol. Ann. Rev., 15: 461-509.

Branch, G. M., 1971, The ecology of Patella L. from the Cape Peninsula, South Africa. 1. Zonation, movements and feeding, Zool. Afr., 6: 1-38.

Branch, G. M., 1975a, Intraspecific competition in Patella cochlear Born, J. Anim. Ecol., 44: 263-282.

Branch, G. M., 1975b, Mechanisms reducing intraspecific competition in Patella spp.: migration, differentiation and territorial behaviour, J. Anim. Ecol., 44: 575-600.

Branch, G. M., 1979, Aggression by limpets against invertebrate predators, Anim. Behav., 27: 408-410.

Branch, G. M., 1981, The biology of limpets: physical factors, energy flow and ecological interactions, Oceanogr. Mar. Biol. Ann. Rev., 19: 235-379.

Branch, G. M., and Branch, M. L., 1981, Experimental analysis of intraspecific competition in an intertidal gastropod, Littorina unifasciata, Aust. J. Mar. Freshwat. Res., 32: 573-589.

Bree, P. J. H., 1959, Homing-gedrag van Patella vulgata L., Kon. Ned. Akad. Wetensch., Versl. Gewone Vergd. Afd. Natuurk., 68: 106-108.

Breen, P. A., 1971, Homing behavior and population regulation in the limpet Acmaea (Collisella) digitalis, Veliger, 14: 177-183.

Breen, P. A., 1972, Seasonal migration and population regulation in the limpet Acmaea (Collisella) digitalis, Veliger, 15: 133-141.

Bretz, D. D., and Dimock, R. V., 1983, Behaviorally important characteristics of the mucous trail of the marine gastropod Ilyanassa obsoleta (Say), J. Exp. Mar. Biol. Ecol., 71: 181-191.

Brown, A. C., 1982, The biology of sandy-beach whelks of the genus Bullia (Nassaridae), Oceanogr. Mar. Biol. Ann. Rev., 20: 309-361.

Brown, F., 1963, How animals respond to magnetism, Discovery, 24: 18-22.

Burke, W. R., 1964, Chemoreception by Tegula funebralis, Veliger, 6 (Suppl.): 17-20.

Byers, B. A., and Mitton, J. B., 1981, Habitat choice in the intertidal

snail Tegula funebralis, Mar. Biol., 65: 149-154.
Chelazzi, G., 1982, Behavioural adaptation of the gastropod Nerita polita L. on different shores at Aldabra Atoll, Proc. R. Soc. Lond. B, 215: 451-467.
Chelazzi, G., Della Santina, P., and Parpagnoli, D., 1987, Trail following in the chiton Acanthopleura gemmata: operational and ecological problems, Mar. Biol., 95: 539-545.
Chelazzi, G., Della Santina, P., and Vannini, M., 1985, Long-lasting substrate marking in the collective homing of the gastropod Nerita textilis, Biol. Bull., 168: 214-221.
Chelazzi, G., Deneubourg, J. L., and Focardi, S., 1984, Cooperative interactions and environmental control in the intertidal clustering of Nerita textilis, (Gastropoda; Prosobranchia), Behaviour, 90: 151-166.
Chelazzi, G., and Focardi, S., 1982, A laboratory study on the short term zonal oscillations of the trochid Monodonta turbinata (Born) (Mollusca: Gastropoda), J. Exp. Mar. Biol. Ecol., 65: 263-273.
Chelazzi, G., Focardi, S., and Deneubourg, J. L., 1983a, A comparative study on the movement patterns of two sympatric tropical chitons (Mollusca: Polyplacophora), Mar. Biol., 74: 115-125.
Chelazzi, G., Focardi, S., Deneubourg, J. L., and Innocenti, R., 1983b, Competition for the home and aggressive behaviour in the chiton Acanthopleura gemmata Blainville (Mollusca: Polyplacophora), Behav. Ecol. Sociobiol., 14: 15-20.
Chelazzi, G., Innocenti, R., and Della Santina, P., 1983c, Zonal migration and trail following of an intertidal gastropod analyzed by LED tracking in the field, Mar. Behav. Physiol., 10: 121-136.
Chelazzi, G., and Parpagnoli, D., 1987, Behavioural responses to crowding modification and home intrusion in Acanthopleura gemmata (Mollusca, Polyplacophora), Ethology, 75: 109-118.
Chelazzi, G., and Vannini, M., 1976, Researches on the coast of Somalia. The shore and the dune of Sar Uanle. 9. Coastward orientation in Nerita textilis Dillwyn (Gastropoda Prosobranchia), Monitore Zool. Ital. (N. S.), 8 (Suppl.): 161-178.
Chelazzi, G., and Vannini, M., 1980, Zonal orientation based on local visual cues in Nerita plicata L. (Mollusca: Gastropoda) at Aldabra Atoll, J. Exp. Mar. Biol. Ecol., 46:147-156.
Chelazzi, G., and Vannini, M., 1985, (Space and time in the behavioural ecology of intertidal animals), S.IT.E. Atti, 5: 689-692.
Cockroft, V. G., and Forbes, A. T., 1981, Tidal activity rhythms in the mangrove snail Cerithidea decollata (Gastropoda, Prosobranchia, Cerithiidae), South African J. of Zoology, 16: 5-9.
Cook, S. B., 1969, Experiments on homing in the limpet Siphonaria normalis, Anim. Behav., 17: 679-682.
Cook, S. B., 1971, A study on homing behaviour in the limpet Siphonaria alternata, Biol. Bull. Mar. Biol. Lab., Woods Hole, 141: 449-457.
Cook, A., Bamford, O. S., Freeman, J. D. B., and Teidman D. J., 1969, A study on the homing habit of the limpet, Anim. Behav., 17: 330-339.
Cook, S. B., and Cook, C. B., 1975, Directionality in the trail-following response of the pulmonate limpet Siphonaria alternata, Mar. Behav. Physiol., 3: 147-155.
Cook, S. B., and Cook, C. B., 1978, Tidal amplitude and activity in the pulmonate limpets Siphonaria normalis (Gould) and Siphonaria alternata (Say), J. Exp. Mar. Biol. Ecol., 35: 119-136.
Cook, S. B., and Cook, C. B., 1981, Activity patterns in Siphonaria populations: heading choice and the effects of size and grazing interval, J. Exp. Mar. Biol. Ecol., 49: 69-79.
Craig, P. C., 1968, The activity pattern and food habits of the limpet Acmaea pelta, Veliger (Suppl.), 11: 13-19.
Creese, R. G., 1980, An analysis of distribution and abundance of populations of the high-shore limpet Notoacmaea petterdi (Tenison-Woods), Oecologia (Berl.), 45: 252-260.

Creutzberg, F., 1975, Orientation in space: animals. 8.1. Invertebrates, in: Marine Ecology, vol. 2, Physiological mechanisms, O. Kinne ed., Wiley and Sons, London.
Crozier, W. J., 1921, Homing behavior in Chiton, Am. Nat., 55: 276-281.
Davis, J. R. A., 1895, The habits of the limpets, Nature, 51: 511-512.
Denny, M., 1980, Locomotion: the cost of gastropod crawling, Science, 208: 1208-1212.
Doering, P. H., and Phillips, D. W., 1983, Manteinance of the shore-level size gradient in the marine snail Tegula funebralis (A. Adams): importance of behavioural responces to light and sea star predators, J. Exp. Mar. Biol. Ecol., 67: 159-173.
Eaton, C. M., 1968, The activity and food of the file limpet Acmaea limatula, Veliger, Suppl., 11: 5-12.
Feder, H. M., 1967, Organisms responsive to predatory sea stars, Sarsia, 29: 371-394.
Focardi, S., Deneubourg, J. L., and Chelazzi, G., 1985a, How shore morphology and orientation mechanisms can affect spatial organization of intertidal molluscs, J. Theor. Biol., 112: 771-782.
Focardi, S., Deneubourg, J. L., and Chelazzi, G., 1985b, The external memory of intertidal molluscs: a theoretical study of trail-following, in: "Mathematics in Biology and Medicine", V. Capasso et al., eds., Springer-Verlag, Berlin Heidelberg.
Frank, P. W. 1964, On home range of limpets, Amer. Nat., 98: 99-104.
Fulton, F. T., 1975, The diet of the chiton Mopalia lignosa (Gould, 1846) (Mollusca, Polyplacophora), Veliger, 18: 34-37.
Funke, W., 1968, Heimfindevermogen und ortstreue bei Patella L. (Gastropoda, Prosobranchia), Oecologia (Berl.) 2: 19-142.
Galbraith, R. T., 1965, Homing behaviour in the limpets Acmaea digitalis and Lottia gigantea, Am. Midl. Nat., 74: 245-246.
Gendron, R. P., 1977, Habitat selection and migratory behaviour of the intertidal gastropod Littorina littorea, J. Anim. Ecol., 46: 79-92.
Gilly, W. F., and Swenson, R. P., 1978, Trail following by Littorina: washout of polarized information and the point of paradox test, Biol. Bull., 155: 439.
Gonor, J. J., 1965, Predator-prey relations between two marine prosobranch gastropods, Veliger, 7: 228-232.
Gowanloch, J. N., and Hayes, F. R., 1926, Contributions to the study of marine gastropods. 1. The physical factors, behaviour and intertidal life of Littorina, Contribs Can. Biol. Fisher. N.S., 3: 135-166.
Grenon, J. F., and Walker, G., 1980, Biochemical and rheological properties of the pedal mucus of the limpet Patella vulgata L., Comp. Biochem. Physiol., 66:451-458.
Hall, J. R., 1973, Intraspecific trail-following in the marsh periwinkle Littorina irrorata (Say), Veliger, 16: 72-75.
Hamilton, P. V., 1977, Daily movements and visual location of plant stems by Littorina irrorata (Mollusca: Gastropoda), Mar. Behav. Physiol., 4: 293-304.
Hamilton, P. V., 1978a, Adaptive visually-mediated movements of Littorina irrorata (Mollusca: Gastropoda) when displaced from their natural habitat, Mar. Behav. Physiol., 5: 255-272.
Hamilton, P. V., 1978b, Intertidal distribution and long-term movements of Littorina irrorata (Mollusca: Gastropoda), Mar. Biol., 46: 49-58.
Hamilton, P. V., 1985, Migratory molluscs, with emphasis on swimming and orientation in the sea hare ,Aplysia, in: "Migration mechanisms and adaptive significance", suppl. to "Contributions to marine sciences".
Hartnoll, R. G., and Wright, J. R., 1977, Foraging movements and homing in the limpet Patella vulgata L., Anim. Behav., 25: 806-810.
Hawkins, S. J., and Hartnoll, R. G., 1983, Grazing of intertidal algae by marine invertebrates, Oceanogr. Mar. Biol. Ann. Rev., 21: 195-282.
Herrnkind, W. F., 1983, Movement patterns and orientation, in: "The biology of Crustacea, vol. 7: Behavior and Ecology", F. J. Vernberg and W. B.

Vernberg eds, Academic Press, New York.

Hewatt, W. G., 1940, Observations on the homing limpet Acmaea scabra Gould, Am. Midl. Nat., 24: 205-208.

Hirano, Y., and Inaba, A., 1980, Siphonaria (pulmonate limpet) survey of Japan. 1. Observations on the behaviour of Siphonaria japonica during breeding season, Publ. Seto Mar. Biol. Lab. 25: 323-334.

Hoffman, D. L., Homan, W. C., Swanson, J., and Weldon, P. J., 1978, Flight responses of three congeneric species of intertidal gastropods (Prosobranchia: Neritidae) to sympatric predatory gastropods from Barbados, Veliger, 21: 293-296.

Horn, P. L., 1986, Energetics of Chiton pelliserpentis (Quoy and Gaimard, 1835) (Mollusca: Polyplacophora) and the importance of mucus in its energy budget, J. Exp. Mar. Biol. Ecol., 101: 119-141.

Hunt, S., 1970, "Polysaccharide-protein complexes in invertebrates", Academic Press, London.

Jander, R., 1975, Ecological aspects of spatial orientation, Annu. Rev. Ecol. Syst., 6: 171-188.

Janssen, C. R., 1960, The influence of temperature on geotaxis and phototaxis in Littorina obtusata (L.), Arch. Neerl. Zool., 13: 500-510.

Jessee, W. F., 1968, Studies of homing behaviour in the limpet Acmaea scabra, Veliger (Suppl.), 11: 52-55.

Kristensen, I. 1965, Habitat of the tidal gastropod Echininus nodulosus, Basteria, 29: 23-25.

Lowe, E. F., and Turner, R. L., 1976, Aggregation and trail-following in juvenile Bursatella leachii pleii, Veliger, 19: 153-155.

Magnus, D. B. E., and Haacker, U., 1968, Zum phanomen der ortsunsteten ruheversammlungen der strandschnecke Planaxis sulcatus (Born) (Mollusca, Prosobranchia), Sarsia, 34: 137-148.

McFarlane, I. D., 1980, Trail-following and trail-searching behaviour in homing of the intertidal gastropod mollusc, Onchidium verruculatum, Mar. Behav. Physiol., 7: 95-108.

McFarlane, I. D., 1981, In the intertidal homing gastropod Onchidium verruculatum (Cuv.) the outward and homeward trails have a different information content, J. Exp. Mar. Biol. Ecol., 51: 207-218.

McLachlan, A., Wooldridge, T., and Van Der Horst, G., 1979, Tidal movements of the macrofauna on an exposed sandy beach in South Africa, J. Zool., 187: 433-442.

McQuaid, C. D., 1981, The establishment and maintenance of vertical size gradient in populations of Littorina africana knysnaensis (Philippi) on an exposed rocky shore, J. Exp. Mar. Biol. Ecol., 54: 77-89.

Micallef, H., 1969, The zonation of certain trochids under an artificial tidal regime, Neth. J. Sea Res., 4: 380-393.

Newell, R. C., 1979, "Biology of intertidal animals", Marine Ecological Surveys, Faversham, Kent.

Nishi, R., 1975, The diet and feeding habits of Nuttalina californica (Reeve, 1847) from two contrasting habitats in central California, Veliger, 18: 28-33.

Ohgushi, R., 1955, Ethological studies on the intertidal limpets. 2. Analytical studies on the homing behaviour of two species of limpets, Jap. J. Ecol., 5: 31-35.

Orians, G. H., and Pearson, N. E., 1979, On the theory of central place foraging, in: "Analysis of ecological systems", Horn, D. J., et alii, eds, Ohio State Univ. Press, Columbus.

Pardi, L., and Ercolini, A., 1986, Zonal recovery mechanisms in talitrid crustaceans, Boll. Zool., 53: 139-160.

Peters, R. L., 1964, Function of the cephalic tentacles in Littorina planaxis Philippi (Gastropoda: Prosobranchiata), Veliger, 7: 143-148.

Petraitis, P. S., 1982, Occurrence of random and directional movements in the periwinkle Littorina littorea (L.), J. Exp. Mar. Biol. Ecol., 59: 207-217.

Pieron, H., 1909, Sens de l' orientation et memoire topographique de la

Patella, C. R. Acad. Sci. (Paris), 148: 530-532.
Raftery R. E., 1983, Littorina trail following: sexual prefence, loss of polarized information, and trail alterations, Veliger, 25: 378-382.
Ratner, S. C., 1976, Kinetic movements in magnetic field of chitons with ferro-magnetic structures, Behav. Biol., 17: 573-578.
Rogers, D. A., 1968, The effect of light and tide on movements of the limpet Acmaea scabra, Veliger (Suppl.), 11: 20-24.
Rohde, K., and Sandland, R., 1976, Factors influencing clustering in the intertidal snail Cerithium moniliferum, Mar. Biol., 30: 203-216.
Schone, H., 1984, "Spatial orientation: the spatial control of behavior in animals and man", Princeton Univ. Press, Princeton.
Smith, J. E.and Newell, G. E., 1955, The dynamics of the zonation of the common periwinkle Littorina littorea (L.) on a stony beach, J. Anim. Ecol., 24: 35-56.
Snyder, N. F. R., and Snyder, H. A., 1971, Pheromone-mediated behavior of Fasciolaria tulipa, Anim. Behav., 19: 257-268.
Spight, T. M., 1982, Risk, reward and the duration of feeding excursions by a marine snail, Veliger, 24: 302-309.
Stimson, J., 1970, Territorial behavior of the owl limpet, Lottia gigantea, Ecology, 51: 113-118.
Stirling, D., and Hamilton, P. V., 1986, Observations on the mechanism of detecting mucous trail polarity in the snail Littorina irrorata, Veliger, 29: 31-37.
Thain, V. M., 1971, Diurnal rhythms in snails and starfish, in:"4th European Marine Biology Symposium", D. J. Crisp ed., Cambridge Univ. Press, Cambridge.
Thomas, R. F., 1973, Homing behaviour and movement rhythms in the pulmonate limpet Siphonaria pectinata L., Proc. Malac. Soc. Lond., 40: 303-311.
Thorne, M. J., 1968, Studies on homing in the chiton Acanthozostera gemmata, Austr. J. Mar. Freshw. Res., 19: 151-160.
Thorpe, W. H., 1963, "Learning and instinct in animals", Methuen, London.
Trott, T. J., and Dimock, R. V., 1978, Intraspecific trail following by the mud snail Ilyanassa obsoleta, Mar. Behav. Physiol., 5: 91-101.
Underwood, A. J., 1973, Studies on the zonation of intertidal prosobranchs (Gastropoda: Prosobranchia) in the region of Heybrook Bay, Plymouth, J. Anim. Ecol., 42: 353-372.
Underwood, A. J., 1977, Movements of intertidal gastropods, J. Exp. Mar. Biol. Ecol., 26: 191-201.
Underwood, A. J., 1979, The ecology of intertidal gastropods, Adv. Mar. Biol., 16: 111-210.
Underwood, A. J., and Chapman, M. G., 1985, Multifactorial analyses of directions of movements of animals, J. Exp. Mar. Biol. Ecol., 91: 17-43.
Vannini, M., and Chelazzi, G., 1978, Field observations on the rhythmic behaviour of Nerita textilis (Gastropoda: Prosobranchia), Mar. Biol., 45: 113-121.
Wara W. M., and Wright, B. B., 1964, The distribution and movements of Tegula funebralis in the intertidal region of Monterey Bay, California (Gastropoda), Veliger (Suppl.), 6: 30-37.
Warburton, K., 1973, Solar orientation in the snail Nerita plicata (Prosobranchia: Neritacea) on a beach near Watamu, Kenya, Mar. Biol., 23: 93-100.
Wilson, R. A., 1968, An investigation into the mucus produced by Limnaea truncatula, the snail host of Fasciola hepatica, Comp. Biochem. Physiol., 24: 629-633.
Zann, L. P., 1973a, Relationships between intertidal zonation and circatidal rhythmicity in littoral gastropods, Mar. Biol., 18: 243-250.
Zann, L. P., 1973b, Interactions of the circadian and circatidal rhythms of the littoral gastropod Melanerita atramentosa Reeve, J. Exp. Mar. Biol. Ecol., 11: 249-261.

Homing Mechanisms of Intertidal Chitons: Field Evidence and the Hypothesis of Trail-Polymorphism

Jean-Louis Deneubourg, Stefano Focardi (*) and Guido Chelazzi (**)

Free University of Brussels, Belgium
(*) I.N.B.S., Ozzano, Italy
(**) University of Florence, Italy

INTRODUCTION

The use of trail-following for homing in intertidal chitons and gastropods is well documented both in the laboratory and in the field (see Chelazzi et al., 1988 for a review). Recent findings suggest that this orientation mechanism may also play an important role in the retrieval of algal patches by these molluscs. This has been observed in A. gemmata where both intra- and inter-individual trail-following occurs during foodward migrations (Chelazzi et al., 1987b). Nevertheless the behavioural mechanisms and cues involved in trail-following are not perfectly clear, and among these the degree of trail-individuality is of particular interest.

Well-controlled analyses of the degree of trail-individuality are scarce. Among the few laboratory experiments are those performed by Funke (1968) which show the capacity to discriminate between personal and conspecific trails in Patella vulgata. However the risk of oversimplification or misinterpretation is high when studying the trail-following mechanism of these species in the laboratory. Gastropods and chitons may give distress responses when handled or put on artificial substrates, and traslocation to unknown natural locations may introduce evident behavioural disturbances (Underwood, 1987). Moreover, additional problems arise when trail-individuality is studied: it is highly probable that the response of the follower to the conspecific trail depends on the trail's 'quality' which in turn may be related to the marker's functional state. In addition the follower's response may be controlled by complex motivational factors: the study of the trail-following in ants has shown that the environment modulates the content of trail information. For instance, humidity, tempera-ture and microscopic substrate features are very important in determining the degree of trail-following (Moser and Blum 1963; Blum and Ross 1965; Torgeson and Akre, 1970). Wilson (1962) has shown that the trail of Solenopsis saevissima has a drastically reduced life time when laid on glass.

These variations in trail effectiveness may produce systematic differences between laboratory and field analyses of trail-following. These practical difficulties induced us to analyse this aspect of the trail-following mechanism, using data relative to the movement pattern of three congeneric (Acanthopleura) intertidal chitons observed in the field. The three species perform looped feeding excursions to a more or less constant rest site and their relatively high population density makes the intersections of inter-individual paths quite common, posing the problem of trail personality.

MATERIALS AND METHODS

The three Acanthopleura spp. (A. gemmata, Indian and Pacific Ocean; A. brevispinosa, Indian Ocean; A. granulata, Caribbean sea) are very similar in morphology and mean size and all three colonize the eulittoral zone on rocky shores of similar composition and structure (limestone cliffs). They are powerful algal grazers, feeding on epilithic and endolithic algae including several green and blue-green algae. At Nimù(Democratic Republic of Somalia) A. gemmata and A. brevispinosa experience large tidal excursions (3.2 m at spring tide), while in the Barbados A. granulata experiences a smaller tidal excursion (1.2 m at spring tide). A. gemmata is almost exclusively found in sheltered areas while the other two species tolerate both exposed and moderately sheltered conditions.

The spatial behaviour of the three species has been analyzed in detail by Chelazzi et al. (1983a) and Focardi and Chelazzi (in prep.). They all perform one feeding loop during each activity phase, but there are important behavioural differences in the homing precision and scar individuality. A. gemmata shows more deterministic homing than A. brevispinosa, while A. granulata often combines ranging movements and homing behaviour. Only A. gemmata shows such complex home-related behaviours as scar digging and its defense against conspecific intruders (Chelazzi et al., 1983 b; 1987a). The feeding migrations of A. gemmata and A. brevispinosa are polarized downward (seaward) while those of A. granulata are relatively scattered around the rest site. The first two species often return of the same algal patch during successive feeding phases.

The data on movement of these three species have the same structure. The sampling method, based on triangulation of fixes from reference stalks, is extensively reported by Chelazzi et al. (1983a). These species rest during diurnal low tides and move about to feed during nocturnal low tides (A. gemmata and A. brevispinosa), or regardless of the tidal regime (A. granulata). Individual positions were sampled twice a day, during diurnal and nocturnal low tides, recording the resting and feeding point for each

Table 1. Number of individuals (N), days of observation and number of recorded excursions in the three studied species

	N	DAYS	EXCURSIONS
A. gemmata	200	20	1370
A. brevispinosa	200	15	1139
A. granulata	150	15	1348

animal. This type of position recording was carried out for at least one semi-lunar month on each species, excluding the neap tide periods.

The outward branch of the feeding loop was estimated by computing the vector connecting the diurnal fix (rest site) to the subsequent nocturnal fix. Movements shorter than 5 cm were discarded. The homeward branch of the feeding loop was estimated by the vector connecting the nocturnal to the next diurnal fix. Detailed analysis of the actual paths followed by the animals (Chelazzi et al., 1987a) shows that the error introduced by this procedure is small for *A. gemmata* and *A. brevispinosa*. Instead, the error introduced for *A. granulata* is harder to calculate becouse this species does not have a definite feeding place which makes the reliability of this procedure lower with respect to the other two species. Table 1 summarizes the general features of the samples.

RESULTS

A. Trail intersection and rest-site exchange

Three parameters have been computed from the data on each species: total number of intersections in the foodward branches of feeding excursions (T), the number single exchanges (E1) and that of double exchange (E2) (Fig. 1A and Table 2A).

The simplest approach to the dynamics of exchanges is to consider that at each trail intersection the probability that a snail will leave its own trail and follow a conspecific one, is null for species which only follow absolutely personal trails, and less or equal to 0.5 for those which follow non-individual trails. In the latter case if the rate of exchange depends on the angle of intersection the value 0.5 is the maximum possible value. The exchange rate:

$$R=(E1+2E2)/2T \qquad (1)$$

for the three species is given in Table 2A. The highest R value is found in *A. granulata* and the lowest in *A. gemmata* but, despite the large differences in movement pattern and spatial behaviour, the observed values are similar.

If an independence of exchanges is hypothesized, then the probability of a double exchange is simply the square of the probability that a chiton will follow a conspecific trail. If the hypothesis of independence of exchanges is not met, i.e., if there is a difference between the observed and the expected number of double exchanges, then a discrimination occurs at the intersection between two conspecific trails, based on internal or external cues. Among the external mechanisms is the possibility that discrimination may be based on the memorization of the broad direction of one's own trail, for instance on the basis of geotactic cues. The smaller the angle between trails, the greater the probability of making a mistake, so following the conspecific trail. This hypothesis may be tested by computing the rate of exchange as function of the minimal angle formed by the intersecting trails. The data plotted in Fig. 2 show that there is no relationship between the exchange rate and angle of incidence in the three species.

The hypothesis that some kind of 'local' effect such as a particular feature of the substrate, could externally bias the exchange rate, was tested, taking into account the level of the shore where intersections occur: it resulted that the exchange rate is independent from the zonal level in all the three species.

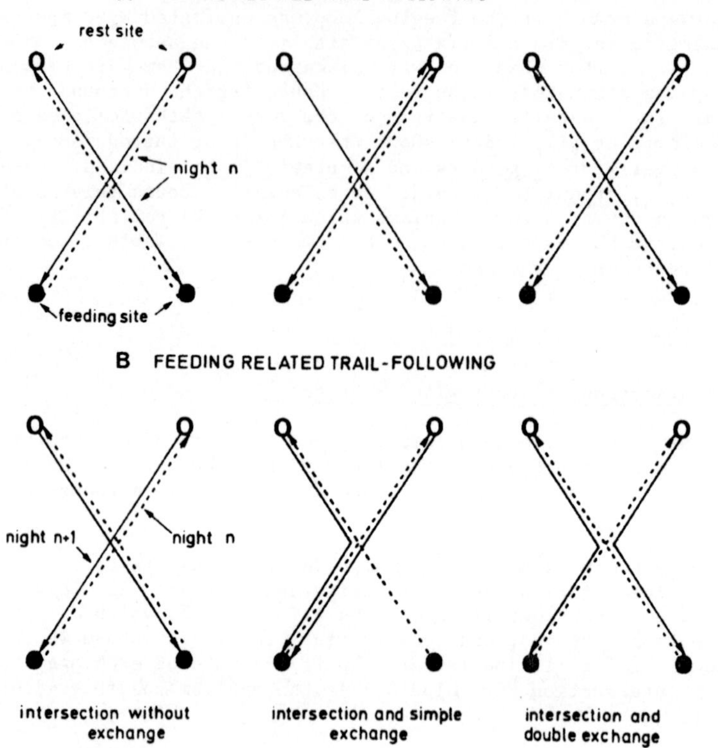

Fig. 1. Scheme of the events considered in the analysis. White ovals, rest site; black ovals, feeding site; solid lines, foodward excursions; dashed lines, homeward excursions

Table 2. The number of intersections (T), single (E1) and double exchanges (E2) and of R is reported for the three species in relation to homing and feeding

	Homing				Feeding			
Species	T	E1	E2	R	T	E1	E2	R
A. gemmata	1108	27	3	.0149	621	81	14	.0878
A. brevispinosa	1489	75	5	.0285	1115	74	7	.0395
A. granulata	1075	67	8	.0386	1324	92	7	.0400

B. Trail intersection and feeding site exchange

Upon observing that feeding ground retrieval may -- at least in *A. gemmata* -- be based on following the homeward trail laid by the animals during the previous feeding loop, a similar analysis was conducted to assess the individuality of trail information involved in the foodward trail-following. The geometrical arrangement of this process is sketched in Fig. 1B and the values of T, E1, E2 and R for the three species are reported in Table 2B.

The exchange rates of *A. granulata* and *A. brevispinosa* are similar, comparable to those obtained for homing, while that of *A. gemmata* is more than 5 times greater, which indicates a large use of conspecific trails during the food search. This effect is even more surprising considering that the trail used by the follower is much older in this situation (about 15-24 hrs) than in the preceding case (3-8 hrs).

As was true for the homing analysis, the angle and point of intersection had no effect on the exchange rate in feeding-related trail following.

Trail polymorphism model

The lack of any evidence that these factors act on the exchange probability, induced us to consider the hypothesis that two chitons exchange trails (and consequently the rest site) when these contain similar information while non-exchanges are partly due to differences detected in the informational content of each trail. To test this hypothesis we developed a mathematical model to assess the existence of differences in the internal features of trails, modulating the exchange probability. The model introduces the hypothesis that the 'quality' of trail released by the animals is polymorphic, in that each animal releases one of a more or less limited number of possible trails.

Let us make the following assumptions about the trail-following mechanism:
1. Within a population of P individuals there are a number of trail types, ranging from 1 (population-specificity and not individual trail-following) to P (strict individual specificity);
2. At each intersection there is a probability (r) that a chiton will follow

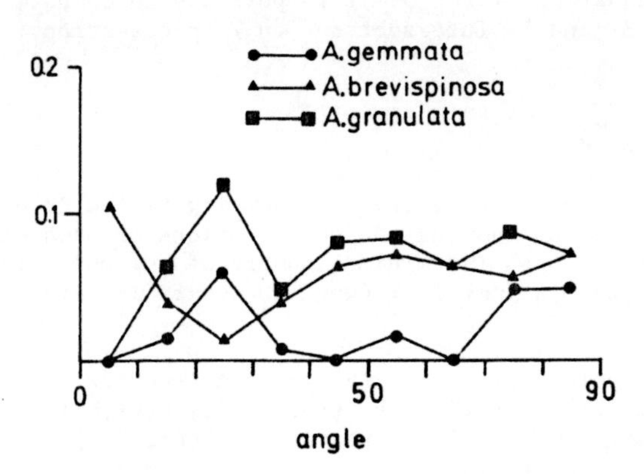

Fig. 2. The exchange rate (ordinates) in function of the minimum angle of intersection (abscissae)

a conspecific trail, if this trail belongs to its own type; otherwise this probability is null. The admissible interval of r values is 0-0.5 thus excluding situations where one's own trail is avoided ($0.5<r\leq 1$).

The probability of M intersections between trails of the same type, in a sample of T crossings, is given by a binomial distribution:

$$Q(M|T) = \frac{T!}{M!\,(T-M)!}\, q^{M}\,(1-q)^{(T-M)} \qquad (2)$$

where q is the probability that the intersecting chitons have the same type of trail, i.e. q is a measure of trail polymorphism.

A first approximation of its value is:

$$q = \sum_{1}^{N} f_i^{2} \qquad (3)$$

where N is the number of trail types and f_i is the fraction of animals with the i-th trail type. For large values of N, q tends to 0, while if N=1 then q is 1.

Remark 1. As the distribution of the f_i is unknown, we can use the value of q as an index of trail polymorphism. For large populations, even or normal distributions of f_i might be conjectured. In the first, simpler, case q = 1/N.

The probability of observing a certain number of simple (E1) and double (E2) exchanges in M intersections of the same type of trails is:

$$P(E1,E2|M) = \frac{M!}{(M-E1-E2)!(E1!E2!)}\, P0^{(M-E1-E2)}\, P1^{E1}\, P2^{E2}\ ; \qquad (4)$$

where:

$$P_0 = (1-r)^2\ ;\ P_1 = 2r(1-r);\ P_2 = r^2\ . \qquad (5)$$

r being the probability that a chiton changes trail when it crosses a trail of the same type.

Associating equations 1 and 3, it is possible to compute the probability of observing E1 and E2 intersections on T intersections:

$$O(E1,E2|T) = \sum_{E1+E2}^{T} P(E1,E2|M)Q(M|T). \qquad (6)$$

In other words, the probability of observing E1 and E2 events in T intersections is given by summing the contributions of each possible number of homomorph intersections. The minimal number of homomorph intersections is cin fact E1+E2 and the maximal is T (when there are no heteromorph intersections).

The numerical solutions of eq. (6) are obtained for each set of values T, E1 and E2 (Table 2), computing the probability O(E1,E2|T) for every couple q and r ($0\leq q\leq 1$; $0\leq r\leq 0.5$). In particular O(E1,E2|T) is computed for the interval E1±10% - E2±10%.

Moreover, it is possible to compute the mean value of E1, E2 as:

$$\langle E1\rangle = 2r(1-r)qT; \quad (7a)$$

$$\langle E2\rangle = r^2 qT. \quad (7b)$$

Finally, the expected value of the total number of exchanges is:

$$\langle E1+2E2\rangle = 2qrT. \quad (8)$$

Figures 3 shows the results relative to the homing related trail-following for the three species. In order to have a clearer presentation, the value N = 1/q has been used (see Remark 1) instead of directly plotting the q values. The peak of the distribution relative to A. gemmata yields about 12 trail types with a value of r = 0.18. On account of the small number of recorded exchanges the precision of this estimate is low. Hovewer, comparing A. gemmata to the other two species, it is clear that this species is characterized by a large degree of trail polymorphism. The maximum value of the distribution of A. brevispinosa is N=4 and r=0.12, while for A. granulata it is N=5 and r=0.2. The latter two species are quite similar and differ sharply from A. gemmata.

The core region of the distributions of probability plotted in Fig. 3 are limited by a probability value of 0.05. The ranges of r values are the same for the three species: 0.06-0.22 for A. brevispinosa, and 0.1-0.3 for A. granulata and A. gemmata. The range of N values is similar for the first two species (2-7 and 3-8, respectively) and quite different for A. gemmata (7-21). This observation suggests that the reduction of the number of exchanges observed in A. gemmata is due to a difference in the degree of polymorphism and not to the value of r.

The corresponding analysis relative to feeding related trail-following is shown in Fig. 4. The differences between A. brevispinosa and A. granulata are always small and their probability distributions are similar to the corresponding homing related trail-following distributions (Fig. 3). The situation of A. gemmata is quite different. While the r value of the spike of its food related trail-following distribution is not particularly different from the value observed for the home related trail-following distribution (0.26 and 0.18, respectively) the number of trail types sharply decreases from 12 to 3.

DISCUSSION

It is generally admitted that intertidal chitons and gastropods which rest in cluster can follow inter-individual trails while solitary homers follow individual trails, which have an adaptative value in maintaining the ownership of a scar (Funke, 1968). The ecological significance of mechanisms reducing the probability of inter-individual trail-following and consequent home-intrusion is evident: reduction of inter-individual competition for the rest site and improved protection of the individual from physical and biological stresses.

Three mechanisms can operate alone or together in reducing inter-individual trail-following: i) spatial separation between the routes of different animals, obtained by territoriality or by concordant polarization (e.g. along the land-sea axis) of the feeding loops during the same activity phase; ii) directional inertia based on kinesthetic orientation mechanisms or various types of taxes, and iii) differentiation in the information content of the trail released by the different animals, combined with their capacity to discriminate between their own and a different trail.

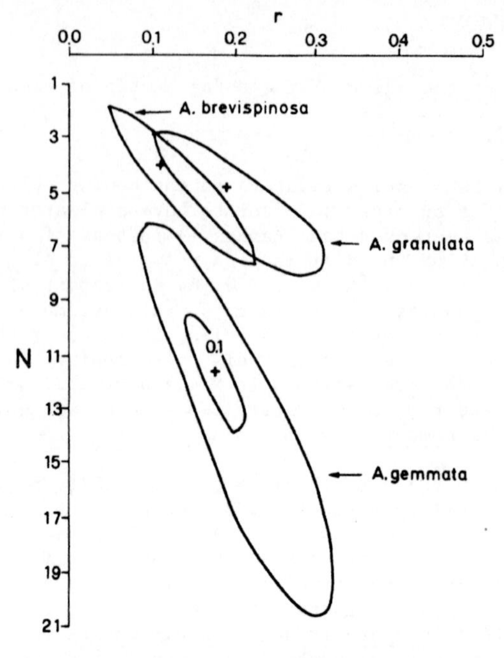

Fig. 3. Homing related trail-following. Probability distribution to observe the experimental values of E1, E2 and T for the three studied species. Abscissae: probability to follow the trail of its own type (r). Ordinates: number of trail types (N). The continous lines connect points with probability 0.05. For A. gemmata is also reported the line connecting points with probability 0.1. Crosses indicate points with the highest probability

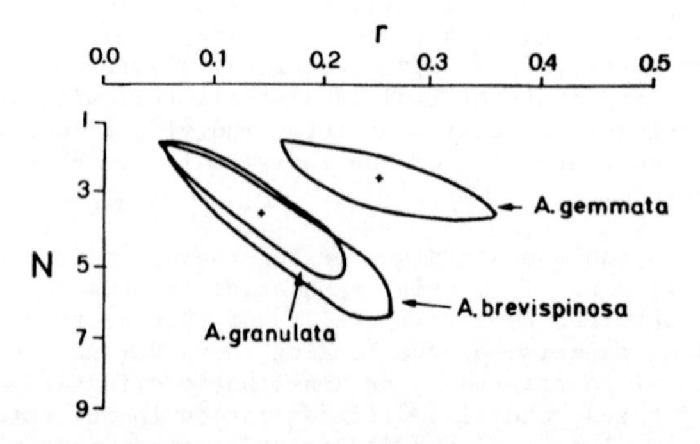

Fig. 4. Feeding related trail-following. Symbols as in Fig. 3

While in many limpets the amount of inter-individual trail-following could be reduced by the development of a personal territory, spatial separation of feeding areas does not occur in the three studied species of chitons. Only A. gemmata shows definite territorial behaviour with hostile interactions, but this is limited to the rest site. On the other hand, a certain degree of interindividual concordance in the direction of the feeding excursions is observed in A. gemmata and A. brevispinosa, while it is completely absent in A. granulata. Nevertheless, the mainly downward oriented feeding loops in A. gemmata and in A. brevispinosa are not rigid enough to cause the separation of individual routes, and crossings are frequent in the three species.

Orientation mechanisms based on external cues such as gravity, light asymmetries and wave movement are widespread in intertidal gastropods, and probably in chitons as well, but our analysis has shown that, if present in the three species, they are not important in discriminating between trails. In fact, whatever the idiothetic or allothetic orienting cue might be, a directional inertia reducing the probability of trail exchange after intersection would produce an inverse relationship between the incidence angle of intersecting and the exchange probability. This does not show up in our data.

Our results show that two conspecific trails are not identical (i.e. r is smaller than 0.5) even when the two trails are supposed to contain similar information. This fact may be explained by non-directional mechanisms. First, there may be a certain time delay in detecting the conspecific trail so that chitons tend to remain on their own route. Second, re-tracing a trail could reduce its information content, as Cook and Cook (1975) have shown in Siphonaria alternata. If the chitons do not reach an intersection simultaneously the r value (here averaged for the whole population) may be reduced to 0.25 in the case of complete destruction of trail-information. The fact that the r values are lower than 0.5 and similar in the three species may reflect the presence of this kind of intrinsic features in their trail following.

The model for the homing-related trail-following revealed a low trail polymorphism in A. brevispinosa and A. granulata which may account for their relatively low home fidelity. On the contrary, the range of q values obtained for A. gemmata (7-21 trail types) suggests that its stronger rest site tenacity depends upon reduction of inter-individual trail-following based on a higher trail polymorphism than in the other two species. Interestingly, the simple count of the exchange rate yielded very small differences between the three species, while the model -- exploiting the information contained in the parameters E1 and E2 -- shows much more evident differences between them.

The fact that A. gemmata does not follow strictly personal trails has to be explained by the cost of developing a large number of morphs concerning trail production and its detection. If the cost of polymorphism increases linearly (or faster) with the number of trail types present in the population, there will be a level above which it is no longer convenient to increase the polymorphism. From inspection of eq. 8 it appears clear that for a given set of T and r values, the total number of exchanges is proportional to 1/N (according to the hypothesis discussed in Remark 1) and in the case of A. gemmata only a small gain is obtained for N values larger than 7.

The most interesting result of the model applied to feeding related trail-following is the strong decrease in the estimate of trail polymorphism of A. gemmata compared to the homing phase. Different physiological hypotheses can be made about the mechanism producing this decrease: i) the homeward and the outward trails may have different informational content;

ii) the age of the trail (longer in the food related trail-following than in the home related trail-following) may modify its composition and part of the information may be lost; iii) the trail may contain both polymorphic and monomorphic cues, but the tracker may vary its response to them according to its motivation.

All these hypotheses are highly speculative, and only the extension of this kind of analysis to other species, combined with controlled experiments, will allow to clarify these physiological aspects. On the other hand, the low degree of trail polymorphism in the feeding related trail-following of algal grazers has important consequences on their ecology, depending upon the dynamics of the exploited resources. At the individual level it may facilitate the discovery of good algal patches, but if the algal film renewal rate is low, it may produce overexploitation and a population crisis. This poses special problems for the feeding related trail-following in intertidal chitons and gastropods, not met in such species as ants where this capacity is basically a recruitment mechanism for food exploitation between genetically related individuals. Interestingly, the ant *Lasius neoniger* reduces food competition using colony-specific trail pheromones (Traniello, 1980).

The application of eq. (6) to the data have shown that the reliability of the results increases quickly with the number of single and double exchanges recorded. The model deserves further mathematical analysis, using simulated and real data, to clearly define its limits of application. In the study of spatial behaviour of intertidal animals this procedure is probably useful for a large number of species. In addition to the conditions listed above, the main requisite for the application of the model is a detailed definition of the movement pattern of the studied species, in order to properly schedule the discrete recording of rest and feeding locations. We have already pointed out that the determination of the rest site is highly reliable for the three species, while the estimate of the feeding location is fairly good for *A. gemmata* and *A. brevispinosa*, but less precise for *A. granulata*.

REFERENCES

Blum, M. S. and Ross, G. N., 1965, Chemical releasers of social behaviour - V. Source, specificity, and properties of the odour trail pheromone of *Tetramonium guineense* (F.) (Formicidae: Myrmicinae), *J. Insect Physiol.*, 11:857-868.

Chelazzi, G., Focardi, S. and Deneubourg, J. L., 1983a, A comparative study on the movement patterns of two sympatric tropical chitons (Mollusca: Polyplacophora), *Mar. Biol.*, 74:115-125.

Chelazzi, G., Focardi, S., Deneubourg, J. L. and Innocenti, R., 1983b, Competition for the home and aggressive behaviour in the chiton *Acanthopleura gemmata* (Mollusca: Polyplacophora), *Behav. Ecol. Sociobiol.*, 14:15-20.

Chelazzi, G., Della Santina, P. and Parpagnoli, D., 1987a, Trail following in the chiton *Acanthopleura gemmata*: operational and ecological problems, *Mar. Biol.*, 95:539-545.

Chelazzi, G. and Parpagnoli D., 1987b, Behavioural responses to crowding modification and home intrusion in *Acanthopleura gemmata* (Mollusca, Polyplacophora), *Ethology*, 75:109-118.

Chelazzi, G., Focardi, S. and Deneubourg, J. L., 1988, The analysis of movement patterns and orientation mechanisms in intertidal chitons and gastropods, *in*: "Behavioural adaptations to intertidal life", G. Chelazzi and M. Vannini, eds., Plenum Press, New York.

Cook, S. B. and Cook, C. B., 1975, Directionality in the trail-following response of the pulmonate limpet *Siphonaria alternata*., *Mar. behav.*

Physiol., 3:147-155.
Funke, W., 1968, Heimfindevermogen und Orstreue bei Patella L. (Gastropoda, Prosobranchia), Oecologia, 2:19-142.
Moser, J. C. and Blum, M. S., 1963, Trail marking substance of the Texas leaf-cutting ant: source and potency, Science, 140:1228.
Torgeson, R. L. and Akre R. D., 1970, Interspecific responses to trail and alarm pheromones by new world army ants, J. Kans. ent. Soc., 43:395-404.
Traniello, J. F., 1980, Colony specificity in the trail pheromone of an ant, Naturwissenschaften, 67:361-362.
Underwood, A. J., 1988, Design and analysis of field experiments on competitive interactions affecting behaviour of intertidal animals, in: "Behavioural adaptations to intertidal life", G. Chelazzi and M. Vannini, eds., Plenum Press, New York.
Wilson, E. O., 1962, Chemical communication among workers of the fire ant Solenopsis saevissima (Fr. Smith), 1. The organization of mass foraging, Anim. Behav., 10:134-164.

Interindividual Variation in Foraging Behaviour within a Temperate and a Tropical Species of Carnivorous Gastropods

Lani West

Hopkins Marine Station
Pacific Grove, California, U.S.A.

ABSTRACT

I studied the feeding behavior of marked individuals of two species of carnivorous snails: Nucella emarginata on the Pacific coast of North America and Thais melones on the Pacific coast of Panama. These snails consumed a variety of invertebrate species such as bivalves, limpets, and polychaetes. Daily observation and mapping of marked snails yielded records of sequential prey eaten by each individual. Both the temperate and tropical species of carnivorous snails were generalized in overall diet but within each population individuals varied in degree of specialization.

Feeding sequences for the 51 temperate and 100 tropical snails that were observed through 5 or more feeding attacks revealed two patterns. First, diets varied from specialized to generalized even among individuals of similar size foraging adjacent to one another in the same habitat. Second, among individuals which exhibited more specialized diets, the prey species of choice was not always the same. These patterns of interindividual variability were not a result of individuals foraging in homogeneous patches of a few prey species since mapped foraging tracks of individual snails intersected and overlapped extensively. Nor were these patterns solely a result of random foraging given the relative abundances of prey species in the habitat.

Shannon-Wiener diversity indices (H') were calculated to obtain the relative specialization within each diet and the proportions of those diets in the population. These analyses revealed that the proportions of specialists choosing rarer foods differed between sites. In the temperate study sites there were more specialists choosing rare foods than in the tropical sites. Further, between tropical sites, proportions of specialists choosing rarer foods differed between sites and seasons. Thus within a population of snails, some individuals may encourage patchiness of particular prey species while others feed less discriminantly. If differences occur in the proportions of these predators and prey species between sites and through time, assessing the impact of a predatory species on its community may require detailed knowledge of interindividual variability in space and time. It is also clear from these studies that factors affecting foraging strategies and their evolution might be most profitably assessed by focussing on the behavior of individuals rather than on population averages.

INTRODUCTION

A general and very basic observation is that members of a given population differ from one another in numerous aspects of their biology. A classic and difficult question in ecology is: What circumstances influence the degree of interindividual variability within a population? An approach to this general problem that I undertook in my research was to study the foraging behavior of individuals of a temperate and a tropical species of carnivorous gastropods. I chose to examine foraging because the effectiveness of an individual's foraging behavior will influence its intake of nutrients and energy, which in turn will affect its growth and reproductive success. Significant variability in foraging behavior among individuals within a population would then provide raw material for natural selection, leading to potentially important consequences for the population. On a community level, changing patterns of individual foraging by predators could have important effects on temporal and spatial patterns of prey distribution and abundance.

The traditional approach to foraging ecology has been to construct a composite picture of an "average" individual's diet by pooling a large number of feeding observations of a number of individuals. Although this type of approach has provided a wealth of information, it is not without its drawbacks, as has been pointed out by Grant (1971) in his work with birds and by Roughgarden (1972, 1979) with lizards. A population that eats a wide variety of prey species could consist of either identical individuals that all eat the same broad diet or it could consist of highly specialized individuals that each eat only a small subset of the prey species eaten by the population as a whole. Theoretically there is a range of situations between these two extremes, but the point here is that an "average" diet may not be a representative diet. So in my work I wanted to examine closely the feeding behavior of a population and document individual diets. The goals were (1) to examine the degree of similarity between diets of individuals living side by side in the same habitat, (2) where inter-individual differences in diet occur, determine whether these differences were due to availability of prey, and (3) examine the proportions of individuals that ate similar diets to determine if snails tended to eat a relatively few number of common diets or many different ones.

METHODS

The temperate studies were carried out at Hopkins Marine Station of Stanford University in Pacific Grove, California, (36°36'N, 121°54'W) U.S.A. Study sites were the eastern shore of Mussel Point (site A), and rocks adjacent to Point Alones (site B); both of which are on the property. The tropical studies were conducted near the Pacific entrance to the Panama Canal (8°45'N, 79°30'W) at the Naos Laboratory of the Smithsonian Tropical Research Institute, Balboa, Panama. Study sites were on the western shore of Naos Island (site A) and the southern shore of Culebra Island (site B).

Two species of snails belonging to the order Neogastropoda, were examined in this study. The temperate *Nucella* (=*Thais*) *emarginata* (see Abbott & Haderlie 1980) may reach 34mm in shell length and the tropical *Thais melones* (see Keen 1971) which bears a much heavier shell, reaches 48mm in length. Both of these snails inhabit the middle and upper intertidal regions of their respective rocky shores. They generally drill through the shells of their prey and consume the soft tissues with their radula and proboscis.

The movements and feeding activities of individually marked snails were followed to record their diets in relation to food availability. In Monterey, California U.S.A., snails were observed once per day during low tide from January 26 through May 18, 1978. On the Pacific coast of Panama snails were observed at least twice daily for three intensive field study periods (2.5 months 1980, 4 months 1982, 4 months 1983). The tropical snail *Thais melones* attacked and consumed prey more quickly than the temperate *Nucella emarginata*. To improve my chances of observing snails actively feeding I searched the study sites twice a day during each outgoing tide when snails were awash in about 0.5 meters of water. I used a mask and snorkle while wading and holding on to the rocks.

Marking of snails and measurement of relative abundance of prey species were conducted at low tide when the study areas were exposed to air. Snails were individually marked *in situ*, using "five-minute" epoxy glue and numbered tags. In the temperate study sites densities of prey were estimated using ten randomly positioned $100cm^2$ quadrats placed over each approximate square meter of study site surface. At site A, $14m^2$ were monitored; at site B, $18m^2$. All potential prey individuals were counted within the quadrats in the field. These measures were made once, two months after the study began. In the tropical study sites densities were estimated from counts of mobile and solitary sessile animals within clear $0.25m^2$ vinyl quadrats (Menge & Lubchenco 1981). At study site A, prey availability was sampled twice (mid Feb. & end of March) during the 2.5 month study period. Six permanent $0.25m^2$ quadrats deliberately located on rock surfaces where snails fed most frequently, were monitored for percentage cover and density of prey. At study site B prey availability was sampled monthly from 9 permanent $0.25m^2$ quadrats (Feb.-May 1982 & June-Sept. 1983) located on rock surfaces where snails fed. Prey availability at site B was also monitored by $0.25m^2$ quadrats placed on other feeding surfaces not included in the 9 permanent quadrats. These sampling periods consisted of two days during each neap tidal cycle totaling 42 quadrats in 1982 and 37 in 1983.

To determine the likelihood that random foraging among available prey would produce the diet observed for each individual, the probability of a given snail eating by chance the observed diet was calculated by computer using the multinomial distribution (Feller 1968, see West 1986a & b for further details):

$$\frac{n!}{k_1!\ k_2!\ \ldots k_r!} \quad p_1^{k_1} p_2^{k_2} p_3^{k_3} \ldots p_r^{k_r}$$

where,

n represents the total number of observed predation events made by one snail.

k_i's are number of times the individual snail fed on prey species i.

p_i $(i=1,\ldots,r)$ represents the relative abundance of prey species i, where $p_1+\ldots+p_r=1$.

I used prey density values for p_i (see methods above) instead of percentage cover values because they were larger and more conservative when used in this test. The density values used were derived from an overall average of the prey available on surfaces where the snails fed at a given site in a given year. For each size class of snails I modified the overall average by subtracting the specific prey species that were not eaten by that particular size class. For example, large snails did not eat serpulid worms so the total number of serpulids were subtracted from the density values of all the quadrats and then percentages were calculated from the remaining density values. This allowed for a broad view of all the microhabitats from which the snails were feeding in a given site and year but took into account only those prey species that were eaten by a given size class of snails. Sizes within prey species were not taken into account because even the smallest snails ate large prey of particular species.

Individual diet diversity was calculated using the Shannon-Wiener index (Krebs 1972) to weight both number and proportions of different species consumed by an individual. A small index value indicates that the individual ate fewer different species or predominantly one species over others. Cluster analyses were calculated using SYSTAT version 3 statistical package (SYSTAT, Inc. Evanston, Ill. USA 1986).

RESULTS

Both the temperate (Nucella emarginata) and tropical snails (Thais melones) were generalized in overall diet (Table 1) but individuals within those populations varied in degree of specialization. Individual diets consisting of five or more feeding attacks showed the following patterns:

Diets varied from specialized to generalized even among individuals of similar size foraging adjacent to one another in the same habitat. For example in the temperate study site A, snail 26 (7th sequence, Fig. 1) ate only one prey species, Mytilus californianus throughout six feeding attacks while snail 72 (11th sequence, Fig. 1) ate three different prey species (Tetraclita, Pollicipes and Balanus). Similarly at the tropical site B in 1983, snail 125 (1st sequence, Fig. 2) ate two species, Siphonaria maura and Lithophaga sp. throughout 11 feeding attacks while snail 242 (4th sequence, Fig. 2) ate five prey species, Acanthina, Nerita, Siphonaria, a cryptic prey, and Ostrea) in six feeding attacks.

Within each study site and among individuals close to the same size, the diets chosen by some individuals differed markedly. For example, snail

Table 1. Prey species eaten by Nucella emarginata and Thais melones in temperate and tropical sites respectively. Corresponding prey symbols are listed for use in Figures 1 & 2.

Temperate Prey	Tropical Prey
B= Balanus glandula	Ac= Acanthina brevidentata
C= Chthamalus spp.	An= Anachis sp.
L= Collisella limatula	Bal= Balanus sp.
S= Collisella scabra	Brc= Brachidontes sp.
M= Mytilus californianus	Chi= chitons
P= Pollicipes polymerus	Chm= Chama sp.
T= Tetraclita rubescens	Cht= Chthamalus sp.
	Crp= crepidula limpet
	Cym= Cymatium sp.
	Fos= Fossarus sp.
	Is= Isognomon sp.
	Lth= Lithophaga sp.
	Mit= Mitra sp.
	Nr= Nerita spp.
	Os= Ostrea spp.
	Ser= serpulids
	Sm= Siphonaria maura
	Tet= Tetraclita panamensis
	Tmel= Thais melones
	Ver= vermetid
	Vit= Vitularia sp.
	X= cryptic prey

64 (Fig 3) ate five Siphonaria limpets and two Ostrea oysters, while snail 65 ate four serpulid polychaetes and one Ostrea.

Some individuals chose the most abundant prey species from the proportions available on the rock surfaces they moved across, while others chose relatively rare prey species. For example snail 7 (Figure 4) ate oysters where oysters were most abundant, while snail 18 ate three vermetids and only one oyster on the same rock surfaces.

None of these patterns appeared to result from biased movements of individual snails among different habitat patches since mapped foraging tracks of individual snails intersected and overlapped extensively. The maps also show that different individuals feeding on the same surfaces can take very different foods. Yet, could the diet patterns simply be a result of random foraging given the relative abundances of prey species in the habitat?

The multinomial distribution allows calculation of the probability of finding for example, 2 mussels and 6 Pollicipes barnacles (snail 27, 6th

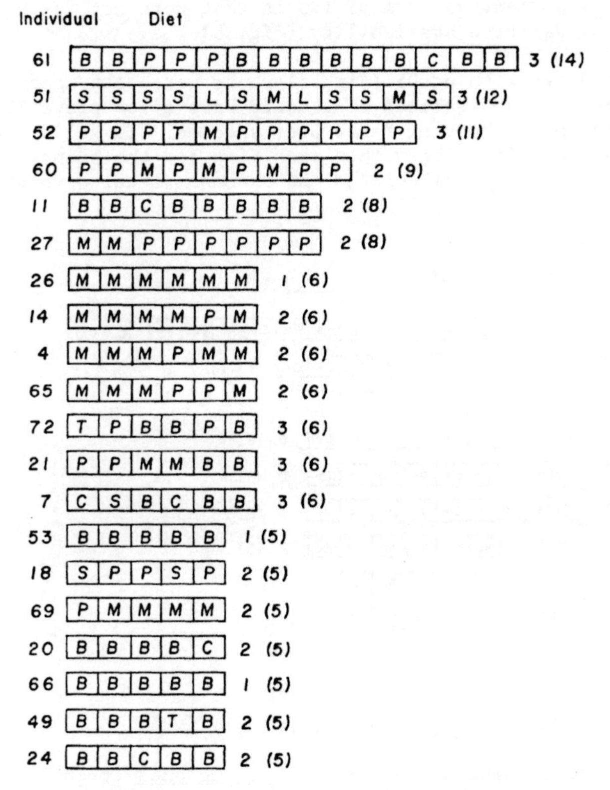

Figure 1. Individual diet sequences of Nucella emarginata (temperate snails) that were observed feeding five or more times at study site A. Snail sizes were from 23mm to 30mm in shell length. Diet symbols are explained in Table 1. The number to the right of each diet sequence is the number of prey species and numbers within parentheses indicate number of observed predation events.

sequence, Fig. 1) given the relative abundance of all potential foods on the rock surfaces where this snail and all the others in its same size class foraged. This technique yields the cumulative probability of each specific diet and all less likely sequences of prey of equal sequence length. Sequences with probabilities < .05 were concluded to depart significantly from the null hypothesis of random foraging. These probability values are conservative because they account only for the number of prey species in the diet, not the specific order in which those prey species occur.

Figure 5 illustrates numbers of diets sorted out by their cumulative diet probabilities. In the temperate sites probabilities of almost all of the diets are <.05, suggesting that snails are not sampling randomly from available prey. At the first tropical site, two thirds of the diets are not different from random foraging from available foods, but one third of those diets are unlikely. At the other tropical site in two different seasons, probabilites of three quarters of the diets are <.05. In general, there was a large percentage of individuals that were not taking prey in proportion to their relative availabilities. The first tropical site is different from the rest and this may be due to a large difference in prey availability. Notice that at site A there were many more oysters available than site B in either the dry or wet seasons (Fig. 6). The snails ate roughly the same proportions of oysters in both cases. At site A it was likely that they would eat oysters but it was unlikely at site B. So the point remains that there was a large percentage of individuals that were not taking prey in proportion to its relative availability (Fig. 5).

Within the diets with p<.05 (Fig. 5) there were both specialized and generalized diets. A generalized diet could have a low probability if the snail was consistently choosing rarer prey species. To get a clearer view of the relative specialization within each diet and the proportions of those diets in the population, I calculated the Shannon-Wiener diversity index

Individual	Diet	
125	Sm Sm Sm Lth Sm Sm Sm Sm Sm Sm Sm	2 (11)
146	Sm X X X Lth Sm Os	4 (7)
198	Sm Sm Bal Sm Sm Os	3 (6)
242	Ac Nr Ac Sm X Os	5 (6)
185	Os Os Os Sm Os Sm	2 (6)
141	Sm Sm X Os Is Is	4 (6)
177	An An Nr An Sm Ac	4 (6)
157	Sm Lth Sm Sm Sm	2 (5)
249	Tm Mit Ac Os Os	4 (5)
147	Os Os Os Os Os	1 (5)
183	Sm Ver Sm Sm Sm	2 (5)
124	Cht Sm Os Sm Cht	3 (5)

Figure 2. Individual diet sequences of Thais melones (tropical snails) that were observed feeding five or more times at tropical site B. These diets represent only the medium size class (20-30mm shell length) snails from 1983. Diet symbols are explained in Table 1. The number to the right of each diet sequence is the number of prey species and numbers within parentheses indicate number of observed predation events.

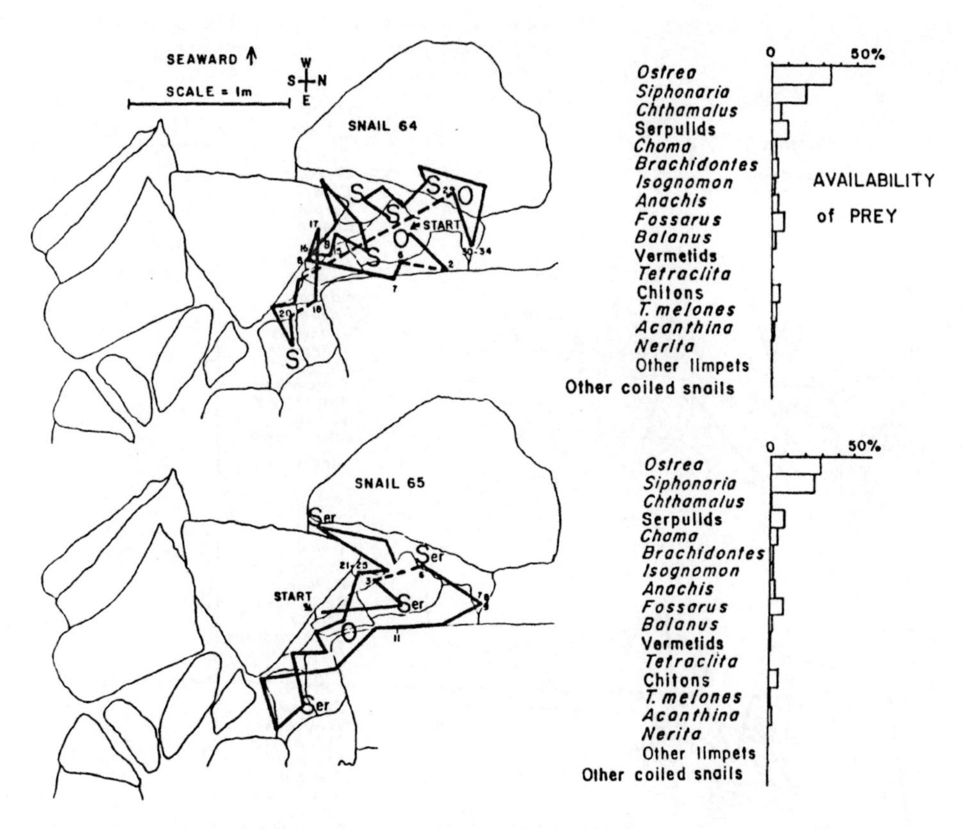

Figure 3. Sections of the net displacement records and sequential feeding attacks made by two small (<20mm) Thais melones in 1980 at tropical site A. Sequential observation periods were numbered (see numbers along the snail path). The number of observation periods that the snail remained at the specific location is also indicated by the numbers. Dotted lines indicate temporary loss of the individual. Letters indicate prey eaten: S = Siphonaria, O = Ostrea, and Ser = serpulid polychaetes. Availability of prey species on the rock surfaces that the snail moved across are summarized at right.

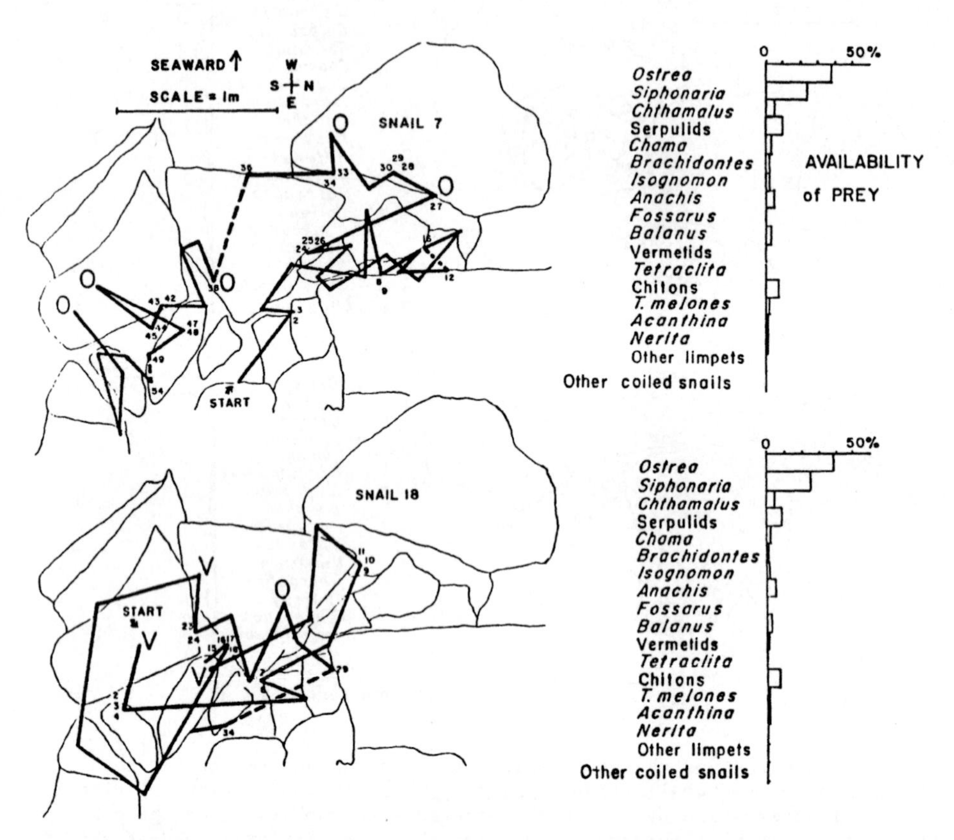

Figure 4. Net displacement record and sequential feeding attacks made by two large (>30mm) Thais melones in 1980 at tropical site A, as in Fig. 5. Letters indicate prey eaten: O = Ostrea and V = vermetid gastropod.

(H') for each individual diet. I then used the observed diversity divided by the maximum diversity possible in each diet to obtain a value for the relative diversity called equitability (H'/Hmax) within each individual diet. When a diet consisted of only one species, it was the most specialized with a value of 0.

Proportions of specialists differed between sites and seasons. Fifteen percent of the temperate snails at site A and 39 percent at site B, were highly specialized (Fig. 7). In the tropical sites, differences in the numbers of specialists between sites and seasons were less dramatic, but at site B, more of the specialized diets contained relatively rare prey. In the temperate sites, all the more specialized diets contained relatively rare prey. Because many individuals are not choosing prey in accordance with the relative abundance of prey species, these differences between study sites and between seasons suggest that a single predatory species may exert different foraging pressures on the invertebrate community in space and time.

I used cluster analysis for an objective way to sort diets into groups based on the proportions of prey species eaten. Cluster analysis measured the difference between every diet and each of the other diets in the data

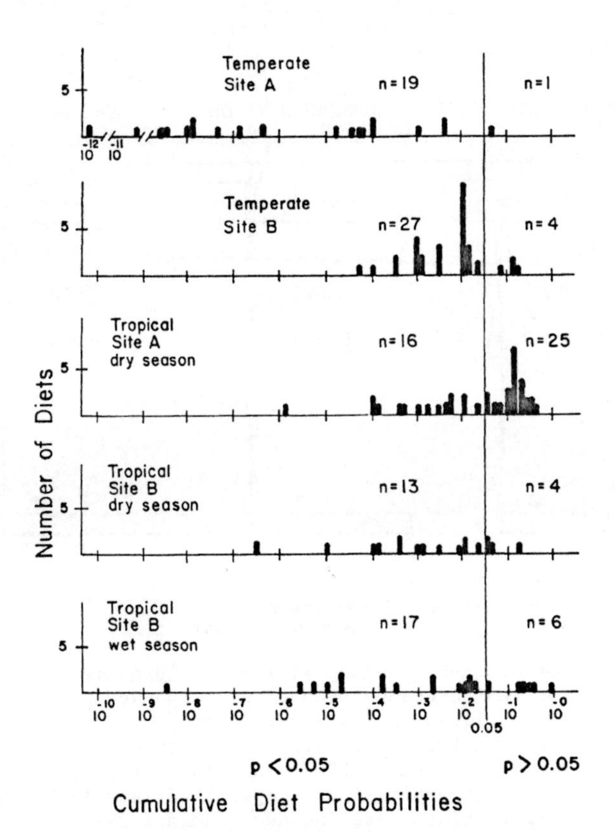

Figure 5. Distributions of cumulative probabilities for each diet. Sequences with probabilities < 0.05 were concluded to depart significantly from the null hypothesis of random foraging.

set. The individuals were then ordered along a scale based on their dissimilarity index. This technique groups snails who's diets were similar; the farther two diets are separated along branches of the dendrogram, the greater their difference in diet. For example, in the temperate studies snail 51 (Fig. 8) ate almost entirely limpets and its diet was the most dissimilar from the rest of the diets. Cluster analysis reveals the numbers of individuals that ate similar diets and whether snails tended to eat a relatively few number of common diets or many different ones.

Temperate diets clustered into four main groups, Pollicipes, Mytilus, Balanus, and the last group made up of mixed diets (Fig. 8). The tropical diets did not cluster as clearly because the diets contained more species and therefore more overlap occurred between individuals' diets (Fig. 9). Six clusters were revealed indicating that the diets emphasized Balanus, serpulids, limpets plus other species, pure limpets, oysters with other species, and pure oysters.

Within the temperate and tropical studies, diets from the same site did not necessarily cluster together. In the temperate sites, clusters of diets that were composed of limpets, Pollicipes barnacles, or mussels originated from site A, while the larger diet clusters containing Balanus barnacles, or mixed diets originated from both sites A and B (Fig. 8). In

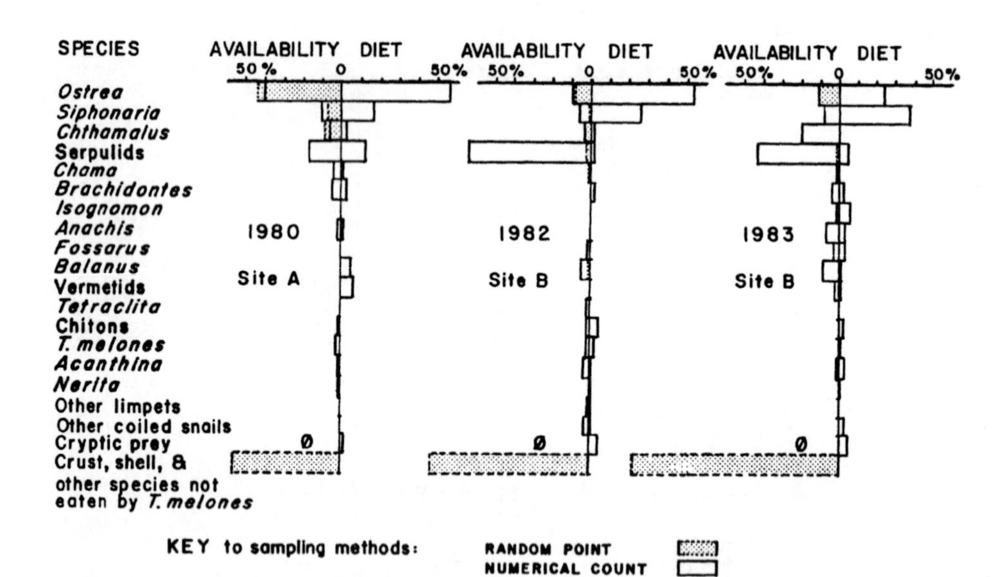

Figure 6. Summary of observed feeding attacks by Thais melones (tropical snails). Solid lines indicate densities of prey changed to numerical percentages while dotted lines indicate percentage cover estimates from 100 random points per $0.25m^2$. Prey availability at site A is a summary of two sampling periods from 6 permanent $0.25m^2$ quadrats. Prey availability at site B in 1982 and 1983 represents four sampling periods each year from 9 permanent quadrats. Both years also include quadrats sampled in additional feeding locations not covered by the permanent quadrats: 42 additional quadrats are included in 1982 and 37 additional quadrats are included in 1983.

the tropical sites clusters of diets that were composed of Balanus and serpulids originated from site A but all the other clusters contained diets that originated from both sites. So within the temperate or tropical snail groups, diets of individuals living in the same study site were not necessarily more similar to each other than diets compared between sites.

DISCUSSION

Within two different species of snails and two contrasting habitats, carnivorous snails showed a notable degree of interindividual variability in feeding behavior. Variability between individuals was not simply a result of snails remaining in different prey patches of food or random foraging among given proportions of prey species in the habitat. Traditionally, variability of this nature has not been examined in marine invertebrate

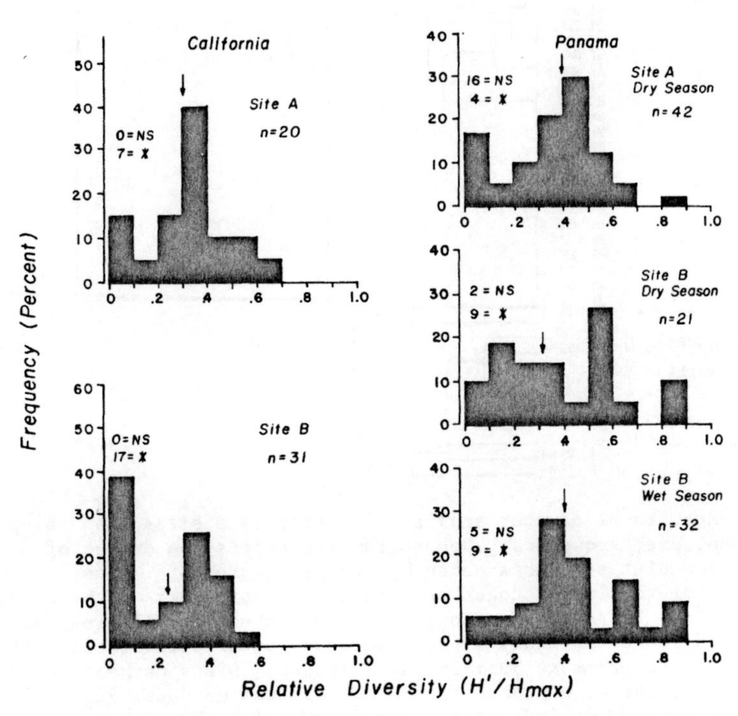

Figure 7. Percentages of individual diet sequences sorted according to the relative diversity within each diet sequence. Relative diversity is calculated from the Shannon-Wiener formula for equitability (Krebs 1972), so that diets with low H'/Hmax contained fewer prey species and were more specialized. Numbers of diets with H'/Hmax < 0.4 and a diet probability of < 0.05 (diet consisted of relatively rare prey species; see methods and Figure 5) are indicated with an asterisk. Numbers of diets with H'/Hmax < 0.4 and a diet probability > 0.05 are indicated with NS. Arrows indicate the mean values.

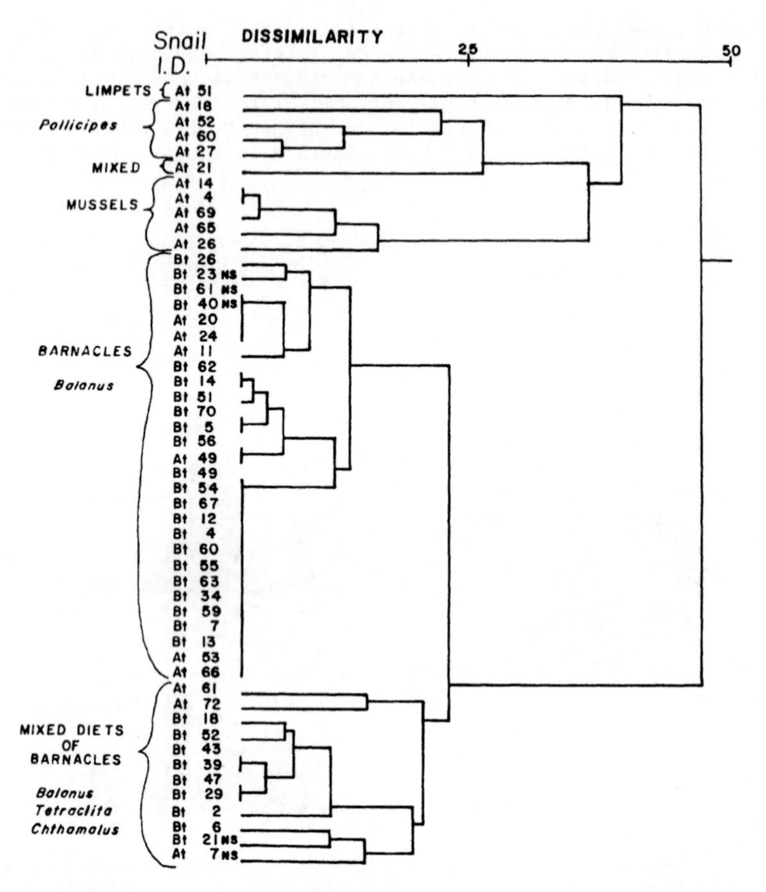

Figure 8. Results of cluster analysis for temperate sites A and B applied to individual diet sequences. Dendrogram represents the degree of dissimilarity between diet sequences eaten by Nucella emarginata. Snails that ate more similar diets cluster together with shorter branches on the dissimilarity scale. Code letters under snail I.D. indicate the following: capital letter A or B indicates study site, lowercase letter t indicates snails were of one size class, 23mm to 30mm in shell length. Diets marked with NS indicate prey sequences with probabilities > 0.05 (see methods; Figure 5), all other diets were < 0.05 indicating that most snails chose rarer prey species.

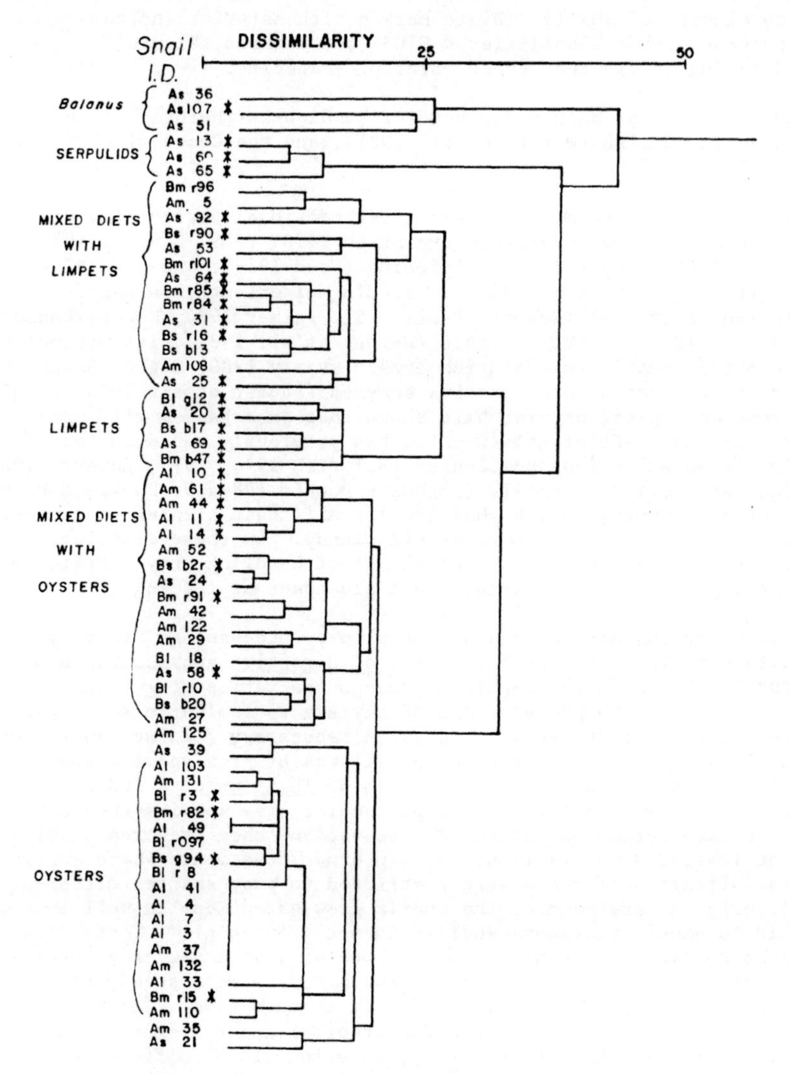

Figure 9. See following page for explanation.

Figure 9 (continued). Cluster analysis applied to individual diet sequences for tropical sites as in Figure 8. Dendrogram represents the degree of dissimilarity between diet sequences eaten by Thais melones in the tropical study sites A (1980) and B (1982). Snails that ate more similar diets cluster together with shorter branches on the dissimilarity scale. Code letters under snail I.D. indicate the following: Capital letter A or B indicates study site, lowercase letters indicate small (<20mm), medium (20-30mm), and large (>30mm) size classes of snails. Diets marked with asterisk indicate prey sequences with probabilities < 0.05 (see methods; Figure 5) indicating snails that chose rarer prey species.

populations, but it has been described for bumblebees (Heinrich 1976, 1979), blue-gill sunfish (Werner et al. 1981), and the Cocos finch (Werner & Sherry 1987).

Some mechanisms that may maintain interindividual variability in populations are: feeding experience (Heinrich 1976, 1979, Laverty 1980, Werner et al. 1981), ingestive conditioning (Wood 1968), and physiological phenomena (Kitting 1980, West 1986b). Feeding experience can modify a consumer's choice of prey (Werner et al. 1981, Palmer 1984, Rowell-Rahier 1984), its techniques of prey capture (Werner et al. 1981), its manipulation of food to obtain nutrients (Heinrich 1979, Laverty 1980, Hughes & Dunkin 1984a) and its production of digestive enzymes (Stuart et al. 1985). In addition some ecological studies have shown that an organism will process a particular food more efficiently when it has previously consumed that food repeatedly: examples include bumblebees (Heinrich 1976, 1979, Laverty 1980), fish (Werner et al. 1981), snails (Hughes & Dunkin 1984a,b, Dunkin & Hughes 1984) and crabs (Cunningham & Hughes 1984). All these studies used energy gained per unit time as indicators of efficiency. In these examples, efficiency is improved in the mechanical act of handling prey. Efficiency may also be improved in the physiological processes of feeding.

Enzyme reactions may enhance or hamper prey processing. In many organisms feeding initiates the synthesis of digestive enzymes (Head & Conover 1983). For example, amphipods change the ratios of gut enzyme activities of amylase to protease and of amylase to laminarinase in response to different foods (Stuart et al. 1985). A laboratory feeding experiment (West 1986b) designed to investigate the effects of prey species and previous feeding experience on shell growth of Thais melones indicated that if snails were allowed to feed on the particular prey species they had previously chosen during one month of observation, they grew reasonably well, or at least did not fair poorly. In other treatments where snails with an established preference were restricted to prey species different from their original preference, the snails grew exceptionally well in a few cases while in other treatments snails did not grow at all. These data suggest that snails continuing to feed on one or just a few prey species may not obtain maximum rates of growth but growth may be more steady than that caused by some specific combinations of prey through time. The examples mentioned above suggest that in both the tropical and temperate species of carnivorous snails in the present study, interindividual differences could be maintained by improved mechanical or chemical handling of prey.

Within the tropical or temperate populations in this study proportions of specialists differed between sites and seasons. Because we know that many individuals were not choosing prey in accordance with the relative abundance of prey species, these differences between study sites and between seasons suggest that a single predatory species may exert different foraging pressures on the invertebrate community in space and time. Through field experiments, Menge (1978) demonstrated that snail phenotype and/or history influenced the feeding rates of two snail phenotypes so that they did not respond in the same way to certain features of the physical environment. In another experimental study, Fairweather et al. (1984) found that for different study sites and for the same study site during different years, the barnacle *Chamaesipho columna* suffered variable rates of predation by the carnivorous snail *Morula marginalba*. Densities of *Morula* were not always as important an influence as the patterns of preference for different prey species by *Morula* and the temporal and spatial variations in the availability of these species. Although my study was not designed to investigate the influences of *Nucella* or *Thais* on the distribution of their prey species, changes in the proportions of individual specialists who prefer the rarer prey species through space and time may be another factor contributing to variation in the effects of predation.

ACKNOWLEDGMENTS

Donald P. Abbott provided inspiration and the suggestion to follow individuals. I thank Bruce Menge and Jane Lubchenco for their encouragement and support throughout this research. Steven Gaines designed the computer program used in calculating diet probabilities and John Lucas provided computer assistance. I thank James Watanabe, Teresa Turner, Terry Farrell, Chris Harrold and Chuck Baxter for tireless and fruitful discussions. I especially thank James Watanabe for assistance in all phases of this research. Support was provided by a Smithsonian short-term fellowship, a Sigma Xi fellowship, and a Smithsonian predoctoral research fellowship to the author. Support was also provided by NSF grant OCE80-19020 to Bruce Menge and Jane Lubchenco and NSF grant OCE-8415609 to Bruce Menge.

LITERATURE CITED

Abbott, D. P. and E. C. Haderlie. 1980. Prosobranchia: marine snails. Pages 230-307 *in*: R. H. Morris, D. P. Abbott and E. C. Haderlie, editors. Intertidal Invertebrates of California. Stanford University Press, Stanford, California, USA.

Cunningham, P. N. and R. N. Hughes. 1984. Learning of predatory skills by shorecrabs *Carcinus maenas* feeding on mussels and dogwhelks. *Marine Ecology Progress Series* 16:21-26.

Dunkin, S. de B. and R. N. Hughes. 1984. Behavioural components of prey-selection by dogwhelks, *Nucella lapillus* (L.), feeding on barnacles, *Semibalanus balanoides* (L.) in the laboratory. *Journal of Experimental Marine Biology and Ecology* 79:91-103.

Fairweather, P. G., A. J. Underwood, and M. J. Moran. 1984. Preliminary investigations of predation by the whelk *Morula marginalba*. *Marine Ecology Progress Series* 17: 143-156.

Feller, W. 1968. An introduction to probability theory and its applications. Volume 1. Third edition. J. Wiley and Sons Incorporated. New York, New York, USA.

Grant, P. R. 1971. Variation in the tarsus length of birds in island and mainland regions. *Evolution* 25:599-614.

Head, E. J. H., and R. J. Conover. 1983. Induction of digestive enzymes in *Calanus hyperboreus*. *Marine Biology Letters* 4:219-231.

Heinrich, B. 1976. The foraging specializations of individual bumblebees. *Ecological Monographs* 46:105-128.

Heinrich, B. 1979. Majoring and minoring by foraging bumblebees, *Bombus vagans*: an experimental analysis. *Ecology* 60:245-255.

Hughes, R. N. and S. de B. Dunkin. 1984a. Behavioural components of prey selection by dogwhelks, *Nucella lapillus* (L.) feeding on mussels, *Mytilus edulis* (L.) in the laboratory. *Journal of Experimental Marine Biology and Ecology* 77:45-68.

Hughes, R. N. and S. de B. Dunkin. 1984b. Effect of dietary history on selection of prey, foraging behaviour among patches of prey, by the dogwhelk *Nucella lapillus* (L.). *Journal of Experimental Marine Biology*

and Ecology 79:159-172.
Keen, A. M. 1971. Sea Shells of Tropical West America. Stanford University Press, Stanford, California, USA.
Kitting, C. L. 1980. Herbivore-plant interactions of individual limpets maintaining a mixed diet of intertidal marine algae. Ecological Monographs 50:527-550.
Krebs, C. J. 1972. Ecology, the Experimental Analysis of Distribution and Abundance. Harper and Row, Publishers, New York, New York, USA. pp.506-508.
Laverty, T. M. 1980. The flower-visiting behaviour of bumble bees: floral complexity and learning. Canadian Journal of Zoology 58:1324-1335.
Menge, B. A. 1978. Predation intensity in a rocky intertidal community. Effect of an algal canopy, wave action and desiccation on predator feeding rates. Oecologia 34:17-35.
Menge, B. A., and J. Lubchenco. 1981. Community organization in temperate and tropical rocky intertidal habitats: prey refuges in relation to consumer pressure gradients. Ecological Monographs 51:429-450.
Palmer, A. R. 1984. Prey selection by Thaidid gastropods: some observational and experimental field tests of foraging models. Oecologia 62:162-172.
Roughgarden, J. 1972. Evolution of niche width. American Naturalist 106:683-718.
Roughgarden, J. 1979. Theory of Population Genetics and Evolutionary Ecology. Macmillan Publishing Company Incorporated, New York, New York, USA. pp.512-513.
Rowell-Rahier, M. 1984. The food plant preferences of Pharatora vitellinae (Coleoptera: Chrysomelinae). B. A laboratory comparison of geographically isolated populations and experiments on conditioning. Oecologia 64:375-380.
Stuart, V., E. J. H. Head and K. H. Mann. 1985. Seasonal changes in the digestive enzyme levels of the amphipod Corophium volutator (Pallas) in relation to diet. Journal of Experimental Marine Biology and Ecology. 88:243-256.
Werner, E. E., G. G. Mittelbach and D. J. Hall. 1981. The role of foraging profitability and experience in habitat use by the bluegill sunfish. Ecology 62:116-125.
Werner, T. K. and T. W. Sherry. 1987. Behavioral niche variation in Pinaroloxias inornata, the "Darwin's finch" of Cocos Island, Costa Rica. Proceedings of the National Accademy of Sciences. USA. 84:5506-5510.
West, L. 1986a. Interindividual variation in prey selection by the snail Nucella (=Thais) emarginata. Ecology 67:798-809.
West, L. 1986b. Prey selection by the tropical marine snail Thais melones: a study of the effects of interindividual variation and foraging experience on growth and gonad development. Doctoral dissertation, Department of Zoology, Oregon State University, Corvallis, Oregon, USA.
Wood, L. 1968. Physiological and ecological aspects of prey selection by the marine gastropod Urosalpinx cinerea (Prosobranchia Muricidae). Malacologia 6:267-320.

Thermal Stress in a High Shore Intertidal Environment: Morphological and Behavioural Adaptations of the Gastropod *Littorina africana*

Christopher D. McQuaid and P. A. Scherman

Rhodes University, Grahamstown
South Africa

INTRODUCTION

Thermal stress, coupled with desiccation, is one of the major physical hazards to be overcome by intertidal organisms. Biogeographic trends are generally temperature mediated (eg. Brown and Jarman, 1978; Bolton, 1985) and, although intertidal organisms generally live well within their physiological limits (Wolcott, 1973; Underwood, 1979), zonation effects are often considered to be linked to the increasing range of temperatures experienced higher up the shore (eg. Sandison, 1967; Sterling, 1982). In cold climates thermal stress is related to problems of very low temperatures and even freezing (resistance to freezing of intertidal invertebrates is reviewed by Murphy, 1983) but in warm temperate, tropical and sub-tropical areas thermal stress takes the form of potential overheating (es. Garrity, 1984). Members of the genus Littorina occupy the highest reaches of the littoral fringe along the coast of southern Africa and are only wet by wave splash or at high spring tides. Consequently they may spend weeks when, especially during the summer, they are subject to high day time temperatures without access to water which could be used for evaporative cooling.

Heat exchange between an animal and its environment may be described by the equation (Porter and Gates, 1969):

$$M + Q_{abs} + C = \varepsilon\sigma T_r^4 + hc\,(Tr - Ta) + E \qquad (1)$$

where M = metabolic heat gain, Qabs = heat absorbed by the animal's surface through radiation, C = heat gained from the substratum by conductance; ε = emissivity of outer surface of the animal; σ = Stefan-Boltzmann constant; Tr = surface temperature of the animal; hc= convection constant; Ta = air temperature and E = heat loss through evaporation of water.

During low tide periods metabolic rates drop to extremely low values and evaporative water loss is minimal (Brown, 1960; McQuaid, 1981). This limits the strategies available for reducing body temperatures in order to avoid overheating. Overheating may be avoided by either reducing heat gain (ie. the left hand side of the equation) or increasing heat loss (the right side). As M (metabolic heat gain) and E (evaporative heat loss)

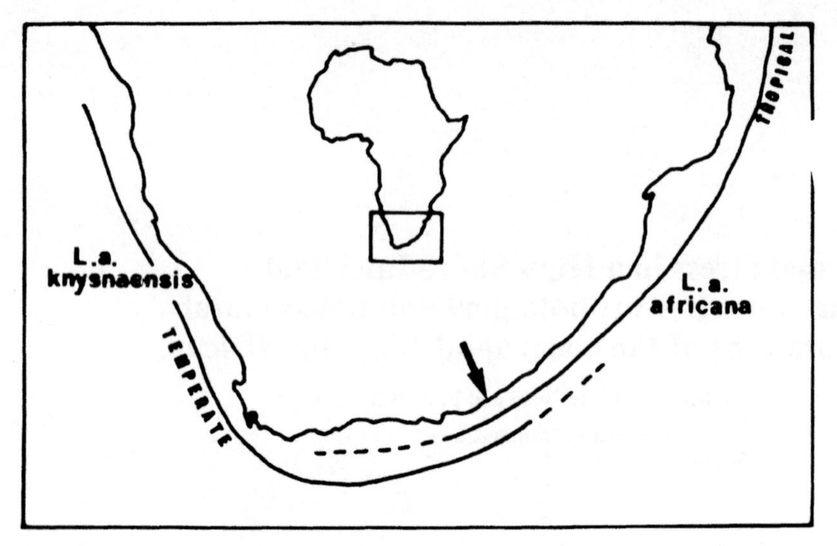

Fig. 1. Distribution of Littorina africana africana and L. a. knysnaensis in southern Africa. Study site arrowed.

approximate to zero in the case of quiescent Littorina the equation may be re-expressed as:

$$Q_{abs} + C = \varepsilon\sigma T_r^4 + h_c (T_r - T_a) \tag{2}$$

Thus the primary options open to Littorina are to decrease radiant heat absorption (Q_{abs}) and conductive heat gain (C) or to increase shell emissivity (epsilon).

Littorina africana africana and L. a. knysnaensis are two sub-species with occasional intermediate forms which exhibit a continuum of shell colour between a pale blue/white morph (L. a. africana) and a dark brown morph (L. a. knysnaensis). There is considerable overlap in distribution but L. a. knysnaensis dominates on the cool west coast of S. Africa and is gradually displaced by L. a. africana along the warm south and east coasts (Hughes, 1979, see Fig, 1). The present study examines the effect of shell colour on thermal relations in these two sub-species and relates this to microhabitat useage and behavioural thermoregulation under natural conditions.

METHODS

Thermal tolerance

Thermal tolerance was examined by deriving LT_{50} values for each subspecies using water baths at temperatures of 30, 35, 40, 45 and 50°C. Small glass bottles, each containing fifteen specimes of one sub-species, were submerged in these baths and one bottle of each sub-species was removed every hour. Condensation inside the bottles was minimal and temperatures were monitored using a thermocouple sealed into a separate bottle. The criterion used for survival was withdrawal of the foot after tactile stimulation. Mortality was checked immediately after removal from the experimental bottles then cool sea water was added to the bottles and mortality checked again 24 hours later.

Effects of shell colour on body temperature

The influence of shell colour on radiant heat uptake was measured by subjecting animals to high light intensities using a 250 V, 200W light bulb directed onto their shells at distances of 7, 14 and 28 cm. This provided light intensities of 1000, 500 and 250 $\mu E.m^{-2}.s^{-1}$ respectively. Light intensities were measured using an integrating quantum radiometer (Li-Cor model LI-188B) and were comparable to values measured in the field. Exposure was continued for 90 minutes and the body temperatures of 20 animals of each sub-species were measured every 30 minutes using a digital thermometer (Fluke model 51) linked to a NI-Cr/Ni-Al thermocouple inserted into the foot. Controls were run simultaneously using the same number of animals of each sub-species after they had been painted black to eliminate differences in shell colour.

This experiment was related to field conditions by painting 20 animals of each sub-species black and replacing them on a horizontal rock surface for 4 hours during an afteroon low tide after which mortality was checked. Control animals were painted with a clear paint. The experiment was carried out twice.

Microhabitat useage and attachment to substratum

Intertidal topography in this area includes many large rocks at the top of the shore with north and south facing surfaces. South facing

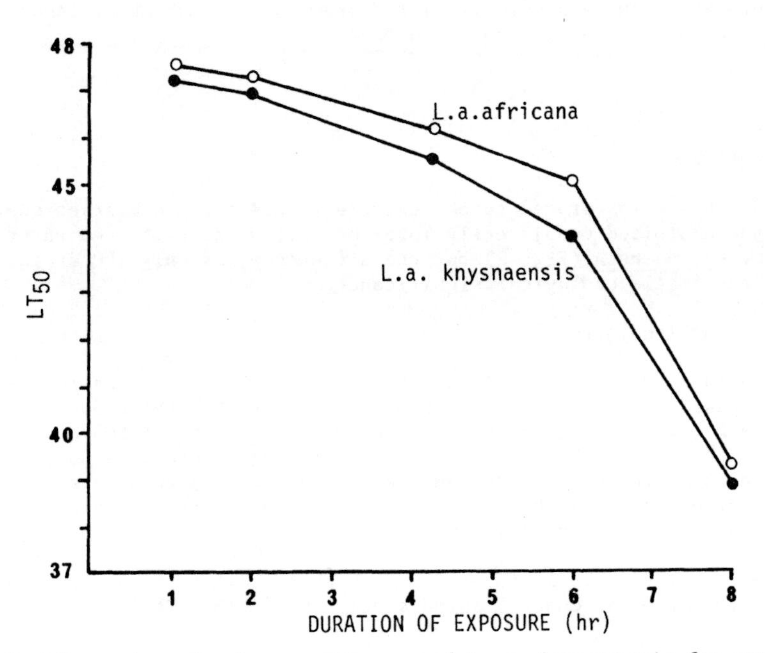

Fig. 2. LT_{50} values for *L. a. africana* and *L. a. kynsnaensis* for various exposure periods.

surfaces are shaded from direct sunlight for most of the day, but experience direct wave action. North facing surfaces are fully insolated but face away from the sea and are sheltered from wave action. The Littorina zone was divided into three vertical sub-zones referred to as A (lowest), B and C. The unconsolidated beach rock of these shores is heavily pitted or pocked by small cup shaped depressions a few centimeters in diameter and depth. This provides two basic habitat types in each zone: open rock and pits. Within these habitats Littorina attach themselves to the rock substratum in three fashions:

a. direct attachment by the foot.

b. foot withdrawn and shell attached to the rock by a film of mucous around the lip

c. animal not in contact with the rock but attached to it, or other animals, by a mucous thread.

The habitat, manner of attachment and shell orientation of all members of each sub-species were recorded during low tide periods in randomly placed 20x20 cm quadrats. Sufficient quadrats were used to provide at least 300 individuals of each sub-species in each zone on north facing (insolated) and south facing (shaded) rock surfaces. Microhabitat availability was quantified in the same quadrats by assesing the percentage cover provided by open rock and pits. Subsequently body temperatures were recorded for insolated animals utilising differente forms of attachment on open rock faces and in pits. Finally size distribution was derived for each sub-species on north and south facing rocks using vernier callipers to measure the maximum shell lenght of all individuals in sufficient 20x20 cm quadrats to provide approximately 300 individuals.

Behavioural responses to rising temperature were examined by inducing foot attachment to glass tubes by submersion in salt water. The tubes, each containing one animal, were then emptied and dried and ten pairs of animals (one of each sub-species per pair) were placed in a glass doored oven in which temperature was raised rapidly (approximately 1°C per minute). The temperature at which each animal shifted from foot to mucous attachment was recorded. The experiment was repeated five times.

RESULTS

Thermal tolerance

Both sub-species proved to be extremely resistant to heat stress. L. a. africana exhibited consistently lower mortality at high temperatures than L. a. knysnaensis (Fig. 2) but the difference was only 1°C at most and probably has little ecological significance.

Effects of shell colour

While there was no significant difference in body temperatures of animals exposed to light intensities of 250 $\mu E.m^{-2}.s^{-1}$ (Fig. 3), body temperatures were significantly lower for L. a. africana at 500 and 1000 $\mu E.m^{-2}.s^{-1}$ (t-test; $P<0.05$ and 0.01 respectively). There were no significant differences in body temperatures of painted aninals at any light intensities, but painted animals were markedly hotter than unpainted individuals ($P<0.01$; Fig. 3).

Animals painted black suffered very high mortality in the field while there was mortality of control animals only on 29/4/87 (Table 1).

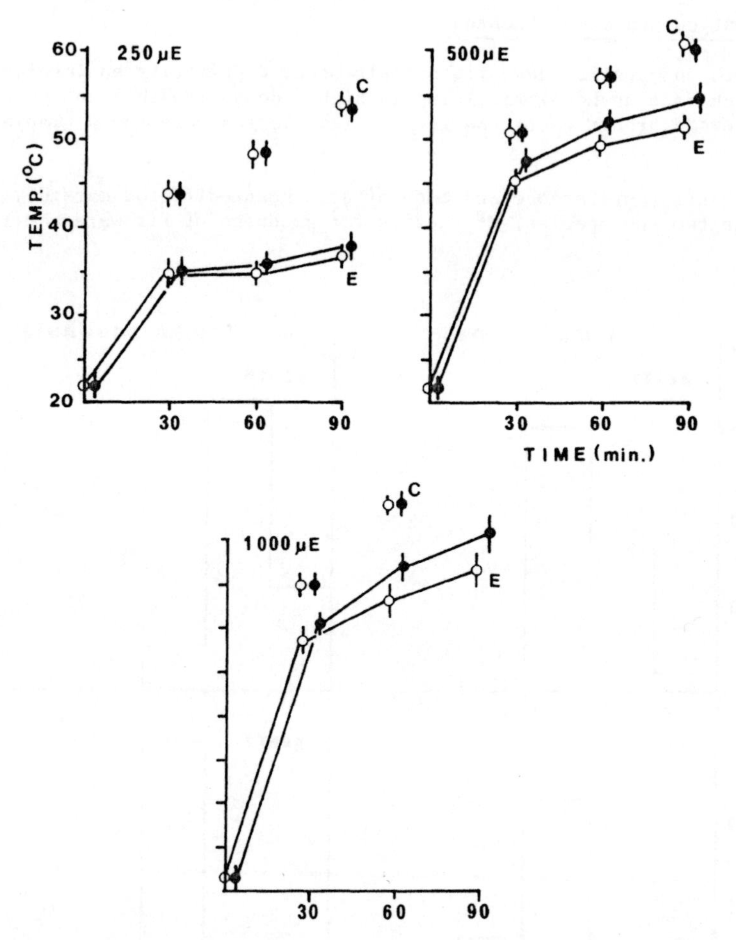

Fig. 3. Body temperatures (±S.D.) of snails at various light intensities. Solid circles L.a.knysnaensis, open circles L.a.africana. E=experimental animals, C=controls

Table 1. Percentage mortality of Littorina on smooth, horizontal surface. Experimental animals painted black, controls painted with clear paint (n=25 in all treatments)

	CONTROL		EXPERIMENTAL	
	L. a. africana	L. a. knysnaensis	L. a. africana	L. a. knysnaensis
29/4/87	0	0	72	64
30/4/87	0	8	80	100

Microhabitat useage and attachment

In both sub-species juvenile animals occured primarily on insolated rocks (which face upshore) resulting in higher densities than on south facing, seaward, rocks where the larger size classes were found (Table 2; Fig. 4).

Microhabitat preference and mode of attachment differed markedly between the two sub-species. X^2 testes for goodness of fit were carried out

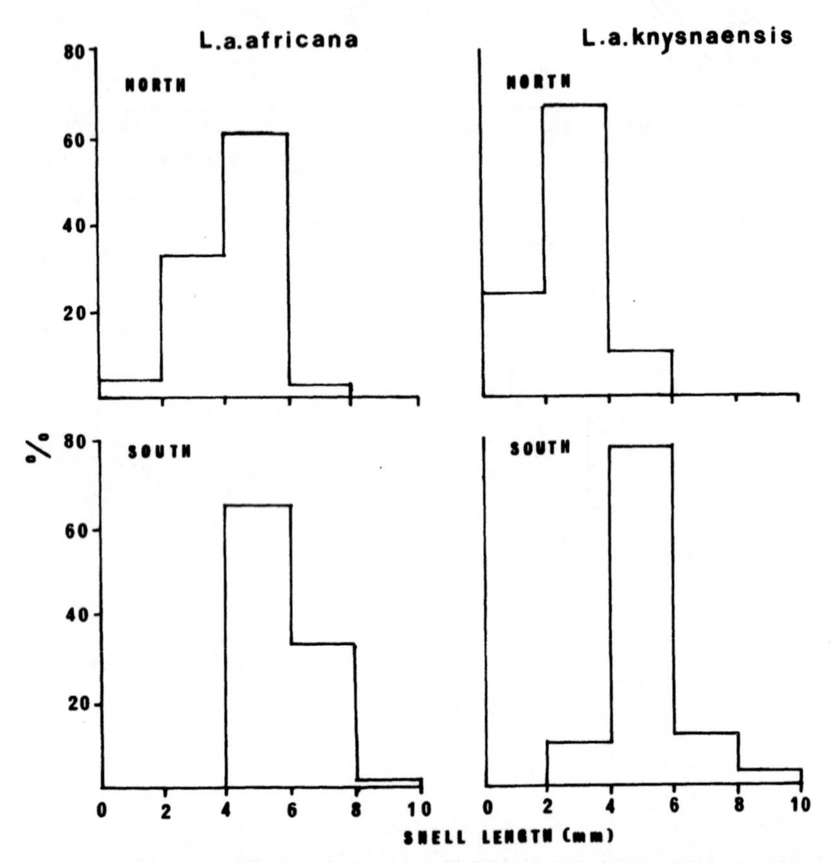

Fig. 4. Size distribution of L. a. africana and L.a.knysnaensis on north and south facing rocks. (n=317, 328 L. a. africana, N,S; 308, 312 L.a. knysnaensis, N,S)

Table 2. Littorina densities. m^{-2} ($\pm$S.D.) on north and south facing rocks. (Number of 20x20 cm quadrats in brackets)

	L. a. africana	L. a. knysnaensis
N -facing(20)	375±23	7575±276
S -facing(20)	195±16	3800±190

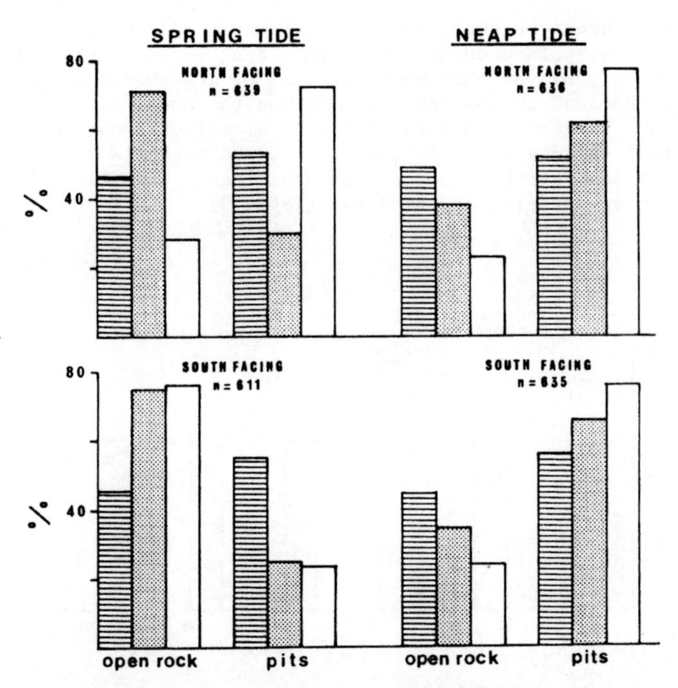

Fig. 5. Habitat availability (% cover - hatched) and useage by L.a.africana (stippled) and L.a.knysnaensis (open) on north and south facing rocks. n values given are totals for both species.

Table 3. Shell mouth orientation of Littorina on north and south facing rocks

	L. a. africana		L. a. knysnaensis	
Shell mouth	N-facing	S-facing	N-facing	S-facing
Up	124	116	132	107
Down	104	100	117	92
Horiz.	97	112	103	112
Total	325	328	352	311

on data for microhabitat availability and useage by each sub-species during spring and neap tides. L. a. africana were found on open surfaces of north or south facing rocks during spring tides, but actively selected pits during neap tides. L. a. knyanaensis also occured primarily in pits during neap and even during spring tides were only abundant on open surfaces on south facing rocks (Fig. 5; $P < 0.01$ in all cases).

Mode of attachment to the substratum was examined during spring low tides and all the animals examined on south facing rocks utilised foot attachment. However, by far the commonest form of attachment on north facing rocks was by withdrawal of the foot and glueing the shell lip to the substratum with a ring of mucous (Fig. 6). L. a. africana used only foot and mucous attachment while a small number of L. a. knysnaensis were attached using mucous threads. Neither sub-species exhibited a significant tendency towards specific orientation of the shell (Table 3).

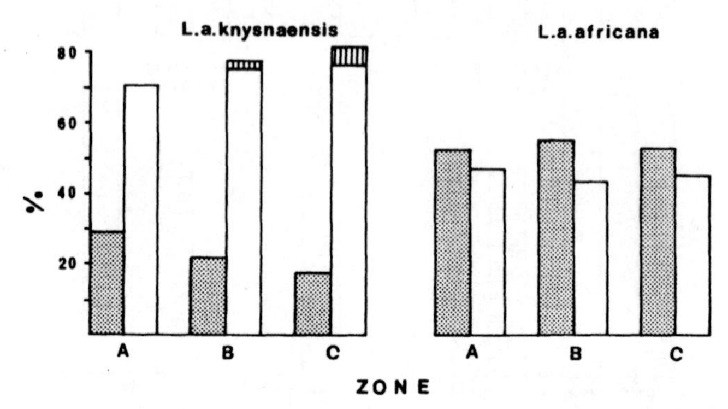

Fig. 6. Mode of attachment of Littorina on north-facing rocks at spring tide. Stippled-foot; opens-mucous; hatched-thread attachment. Note all animals on south-facing rocks used foot attachment.

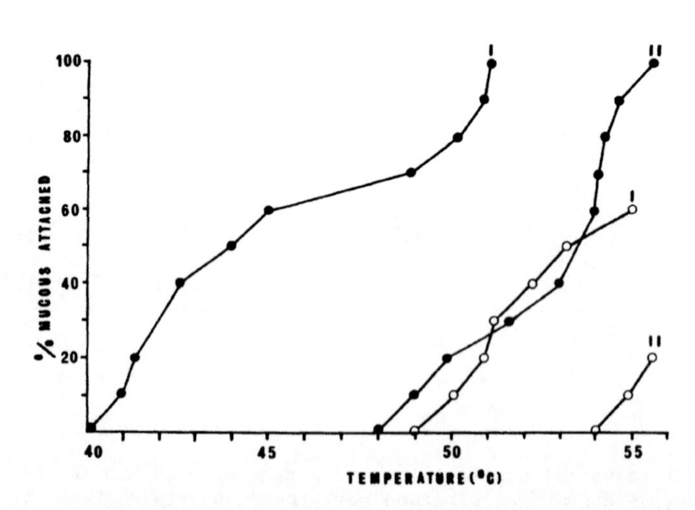

Fig. 7. % of L.a.africana (open circles) and L.a.knysnaensis (closed circles) using mucous attachment. Animals of each sub-species were paired in experiments I and II. All animals were originally foot attached.

When exposed to rapidly rising temperatures both sub-species reverted from foot to mucous attachment. The temperature at which this occured depended on the rate of heating and period of exposure so that exact temperatures have limited significance. However individual L. a. knysnaensis altered their mode of attachment more rapidly than L.a.africana with which they were paired (Wilcoxon matched pairs signed ranks test, $P<0.05$ or 0.01 in all tests). Fig. 7 provides data from two of the five experiments as examples.

Body temperatures in the field

Data on body temperatures of animals using foot attachment in zones A-C on insolated and shaded rock faces are provided in Fig. 8. Values are clearly lower for animals on shaded rocks. Temperatures were very similar in all 3 zones on shaded rocks, but increased higher up the shore on insolated rocks (t-test, $P<0.05$).

Body temperatures of animals using different forms of attachment in zone C on insolated rocks are given in Table 4. These data were collected on a different day from those in Fig. 8 and the values measured are all very high. Animals of both sub-species using foot attachment exhibited significantly higher temperatures than those using mucous attachment (t-test, $P<0.01$). There was no significant difference between mean values for

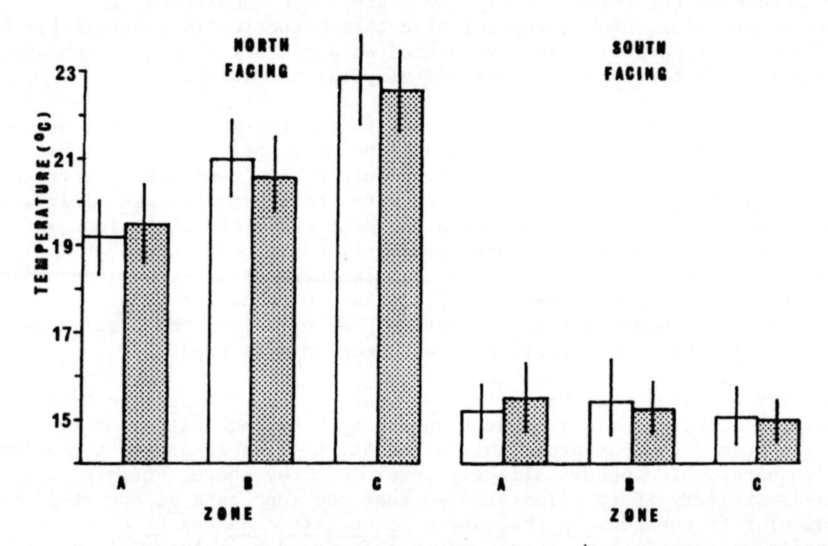

Fig. 8. Body temperatures (means of 20 values ±S.D.) of L.a.africana (stippled) and L.a.knysnaensis (open) in zone A, B and C on north and south facing rocks

Table 4. Body temperatures (±S.D.) of Littorina using foot a mucous attachment in zone C on north-facing rocks 20/3/87. Body temperatures may be much higher in the field e.g. 23/3/87 (n=30 in all cases)

	attachment	L. a. africana	L. a. knysnaensis
20/3	foot	22.71±1.01	22.52±0.92
20/3	mucous	18.21±0.97	20.72±1.07
23/3	mucous	31.60±1.10	34.12±1.02

the two sub-species when foot attachment was used, but with mucous attachment *L. a. knysnaensis* experienced higher body temperatures than *L. a. africana* ($P < 0.05$).

DISCUSSION

High shore gastropos have relatively few possible mechanisms of thermoregulation available to them. In particular, the need to conserve water restricts the use of evaporative cooling and increases the need for behavioural thermoregulation as well as morphological adaptations. Great emphasis nust be placed on a high degree of thermal tolerance and careful microhabitat selection including minimal contact with the substratum if possible. In the present study we see how morphological adaptation ameliorates the need for stringent behavioural control and contributes towards the maintenance of biogeographic trends.

Thermal tolerance as measured in the lab was very high for both sub-species of *Littorina*. Underwood (1979) has pointed out that measurements of lethal thermal limits well in excess of temperatures experienced under normal field conditions may be meaningless as explanations for vertical or biogeographical distribution limits. But vertical and particularly biogeographical limits may be set not by "normal" conditions but by the few days in each year (or perhaps every few years) when conditions are especially stressful. High mortality of both sub-species in a single (rather hot) afternoon if painted black and placed on open rock surfaces emphasises the importance of both shell colour and microhabitat selection.

There were no significant differences in the thermal tolerance of the two sub-species of *Littorina* examined but the pale shell of *L. a. africana* reduced heat gain due to solar heating, resulting in lower body temperatures than for *L. a. knysnaensis* when contact with the substratum was minimised by withdrawal of the foot and mucous attachment. When the whole foot was used for attachment, however, there was considerable heat input by conduction as well as insolation. Body temperatures of *L. a. africana* using foot attachment to the substratum were similar to those for *L. a. knysnaensis*. This suggests that conductive heat gain from the substratum may, at times, override insolation as a source of heat input.

Garrity (1984) found that *Littorina modesta* and *L. aspera* in Panama orientate the shell to minimise heat uptake both by insolation and by conduction from the substratum. This orientation is important to the animals and *L. modesta*, which occurs slightly lower down the shore, suffers increased mortality if re-orientated so that the long axis of the shell is perpendicular to the sun. In the case of *L. a. africana* and *L. a. knysnaensis*, however, there is no such tendency and no evidence that shell orientation inluences mortality. As well as being subject to greater heat input by insolation *L. a. knysnaensis* also exhibits greater sensitivity to rising temperatures and converts from foot to mucous attachment more rapidly. Again this is manifested behaviourally under natural conditions and while both sub-species use foot attachment on shaded (south facing) rocks, a significantly greater proportion of *L. a. knysnaensis* than *L. a. africana* use mucous attachment on isolated rock faces.

The two effects of more rapid heating in direct sunlight and the greater sensitivity of *L. a. knysnaensis* to high body temperatures influence microhabitat selection. Measurements of rock surface temperatures indicate that pits offer a substratum which may be 3 or 4°C cooler as well as providing some shade from the sun and this results in lower body temperatures for animals found in pits. *Littorina africana* occurs in zones bare of macrophytes and presumably grazes off micro-algae. While no measurements were made of food availability, it seems likely that this is minimised in pits by shading and the presence of animals which may both graze and shade the rock themselves. When wet by spray or high spring tides the snails move out onto open surfaces to graze. With prolonged periods of emersion, during neap tides, both sub-species are restricted to pits, but during spring tide periods they remain on open rock surfaces when possible, presumably to maximise the time available for foraging. L. a. africana is capable of this on both north and south facing rocks while it is only possible for L. a. knysnaensis on shaded rocks.

Careful microhabitat selection is most critical during neap tides when the upper parts of the shore are dry for days or even weeks and both sub-species were more rigidly confined to pits during neap than spring tides. This is important as the animals have no way of foretelling how hot conditons will be while they are exposed. The terrestrial snail Theba pisana lives in semi-arid conditions and avoids lethal body temperatures by microhabitat selection and shell orientation. However, it also exhibits an escape response and when temperatures become too high will emerge from its shell and climb up plants etc. to avoid heat re-radiated from the soil (McQuaid et al, 1979). No such escape is possible for high shore Littorina spp. and they remain in their shells even when temperatures go beyond lethal levels. This places even greater importance on the need for careful microhabitat selection, particularly in the case of L. a. knysnaensis which is susceptible to greater heat gain from insolation due to its darker shell.

Considering the biogeogrphy of these two species, they are preyed on by the whelk Nucella dubia (McQuaid, 1985) but there is no evidence of visual predators and so no obvious selection against the more conspicous white morph (cf. Atkinson and Warwick, 1983). Consequently at present there is no information available to account for the replacement of L. a. africana by L. a. knysnaensis towards the west coast of southern Africa. It may, however, be possible to account for the elimination of L. a. knysnaensis towards the semi-tropical east coast. Small Littorina africana are particularly vulnerable to wave action and avoid areas where exposure is great (McQuaid, 1981). Consequently juveniles of both sub-species occur almost exclusively on north facing rocks as they are incapable of withstanding the wave action of south (seaward) facing rocks. This has the effect of forcing the most susceptible size classes into the harshest thermal environment. The biogeographical range of L. a. knysnaensis may therefore be limited by the vulnerability of juveniles to high temperatures during low tide. Vermeij (1973) has described the tendency of high shore and tropical gastropods to decrease heat uptake either through increasing the reflectivity of the shell or the degree of shell sculpturing. L. a. africana decreases its vulnerability to thermal stress, and thus its dependence on the availability of protective microhabitats, by virtue of a paler shell. Kilburn and Rippey (1982) have reported that L. a. africana occurs lower down the sub-tropical shores of Natal (presumably where wave action permits) than where it exists sympatrically with L. a. knysnaensis in the warm temperate Cape. This may be an indication that, while L. a. knysnaensis is excluded, even L. a. africana is thermally constrained in warmer areas. Indeed L. a. africana is in turn replaced by the the highly sculptured Nodilittorina natalensis farther north.

REFERENCES

Atkinson, W.D. and Warwick, T., 1983, The role of selection in the colour polymorphism of Littorina rudis Maton and Littorina arcana Hannaford-Ellis (Prosobranchia: Littorinidae), Biol. J.Linn. Soc., 20:137-151.
Bolton, J.J., 1985, Marine phytogeography of the Benguela upwelling region on the west coast of southern Africa: a temperature dependent

approach, Bot. Mar., 29:251-256.
Brown, A.C., 1960, Desiccation as a factor influencing the vertical distribution of some South African Gastropoda from intertidal rocky shores, Port.Acta.Biol. (B), 7:11-23.
Brown, A.C. and Jarman, N., 1978, Coastal marine habitats, in: "Biogeography and Ecology of southern Africa", M.J.A. Werger, ed., Junk, The Hague.
Garrity, S.D., 1984, Some adaptations of gastropods to physical stress on a tropical rocky shore, Ecology, 65:559-574.
Hughes, R.N., 1979, On the taxonomy of Littorina africana (Mollusca: Gastropoda), Zool.Linn.Soc., 65:111-118.
Kilburn, R. and Rippey, E., 1982, "Sea shells of southern Africa", Macmillan, Johannesburg.
McQuaid, C.D. 1981, The establishment and maintenance of vertical size gradients in population of Littorina africana knysnaensis (Phillips) on an exposed rocky shore, J.Exp.Mar.Biol., 54:77-89.
McQuaid, C.D., 1985, Differential effects of predation by the intertidal whelk Nucella dubia (Kr.) on Littorina africana knysnaensis (Phillipi) and the barnacle Tetraclita serrata Darwin, J.Exp.Mar.Biol.Ecol., 89:97-107.
McQuaid, C.D., Branch, G.M. and Frost, P.G.H., 1979, Aestivation behaviour and thermal relations of the pulmonate Theba pisana in a semi-arid environment, J.Therm.Biol., 4:47-55.
Murphy, D.J., 1983, Freezing resistance in intertidal invertebrates, Ann.Rev.Physiol., 45:289-299.
Porter, W.P. and Gates, D.M., 1969, Thermodynamic equilibria of animals with environment, Ecol.Monogr., 39:227-244.
Sandison, E.E., 1967, Respiratory responses to temperature and temperature tolerance of some intertidal gastropods, J.Exp.Mar.Biol.Ecol., 1:272-281.
Stirling, H.P., 1982, The upper temperature tolerances of prosobranch gastropods of rocky shores at Hong Kong and Dar es Salaam, Tanzania, J.Exp.Mar.Biol.Ecol., 63:133-144.
Underwood, A.J., 1979, The ecology of intertidal gastropods, Adv.Mar.Biol., 16:111-210.
Vermeij, G.J., 1973, Morphological patterns in high-intertidal gastropods: adaptive strategies and their limitations, Mar.Biol. 20:319-346.
Wolcott, T.G., 1973, Physiological ecology and intertidal zonation in limpets (Acmaea): a critical look at "limiting factors". Biol.Bull. Woods Hole, 145:389-422.

Interspecific Behaviour and its Reciprocal Interaction with Evolution, Population Dynamics and Community Structure

George M. Branch and Amos Barkai

University of Cape Town
South Africa

INTRODUCTION

The behaviour of most marine species towards other species is divisible between two distinct phases of the life cycle: larval behaviour, and subsequent behaviour by the adult. Both are constrained to a certain extent by factors intrinsic to each species: for example, morphology and phylogenetic history. Both are also moulded by extrinsic factors, including the physical environment and other organisms. In time, behaviour should evolve in response to these extrinsic factors, so that behavioural responses between species can influence the evolution of species, and the behavioural patterns of species that are tightly interdependent may coevolve. Because of the large number of species that are likely to influence any given species, and the widely different effects that they may have on it, a correspondingly wide array of interspecific behavioural responses can be anticipated. Furthermore, the influence that one species has on another is often modified by prevailing physical conditions or by the presence of a third species. When two or more species interact with a third, their effects on that species may be synergistic or antagonistic, making it difficult to predict the outcome of the interaction, and equally difficult for the target species to develop any single "correct" behavioural response. As a consequence, both variety and flexibility can be anticipated in interspecific behaviour, in contrast to comparatively stereotyped intraspecific behaviour.

The study of larval behaviour is still in its infancy. A great deal is known about settling behaviour, but behaviour during the planktonic stage, particularly in relation to other species, remains largely unexplored. The behaviour of adults is more tractable and better understood. As a working hypothesis, it might be suggested that their behaviour will be most sophisticated and diverse in mobile, active forms, less so in sessile species and very restricted in ectoparasites. Conversely, the behaviour of larvae may be less specialised in mobile species which do not have to rely on the larval stage for habitat or host selection.

Behaviour cannot be considered in isolation. It is both influenced by past evolutionary history and contributes towards future evolution. It plays a role in population dynamics and, in the case of interspecific behaviour, may have a powerful influence on the structure of communities.

There are thus strong links between behaviour and both evolution and ecology. This is, perhaps, most evident in the case of interspecific behaviour (Fig. 1).

This paper forms an introduction to a section of the symposium on "Behavioural adaptations to intertidal life" concerned with interspecific behaviour. In it, we selectively review various kinds of behavioural interactions that occur between species, including larval responses and the behaviour of adults involved in interactions between predators and their prey, and between competitors. The broader implications of behaviour for evolution and ecology are stressed. We conclude the paper by describing two contrasting communities, which we believe are alternate stable states maintained by the behaviour of the participating species.

Within the specific context of the intertidal zone, behaviour is influenced by a particular set of conditions. Physical factors are highly variable, both spatially and temporally, but they are, nevertheless, relatively predictable, being related primarily to tidal and seasonal changes in the environment. Wave action, desiccation, thermal extremes and osmotic stress may all influence behaviour. In addition, intertidal organisms face depredations from both aquatic and terrestrial predators. Another feature of the intertidal zone is that macroalgae and benthic microalgae flourish on hard substrata, whereas they become progressively less important in deeper water as light diminishes. The presence of attached algae greatly modifies both trophic relationships and physical conditions.

INTERSPECIFIC LARVAL BEHAVIOUR

Scheltema (1974) has provided an excellent review of biological interactions that influence the settlement of larvae, and his major points need only be recapitulated here. Many species that live in sediments settle in response to bacteria or other microorganisms in the sediments, and the intensity of settlement changes when the larvae are exposed to different densities or species of bacteria (e.g. Gray, 1967).

The larvae of species that are epibiotic on algae are highly specific in their selection of particular algae, postponing settlement and metamorphosis until the correct alga is available (e.g. Ryland, 1959; Crisp and Williams, 1960; De Silva, 1962; Williams, 1964; Nishihira, 1967). A more recent elaboration of this theme has been the discovery of associations between encrusting coralline algae and grazers. Although some corallines can inhibit settlement of other algae on their surfaces (Johnson and Mann, 1986), many species seem dependent on grazers to prevent fouling by other algae. In one particular case, Steneck (1982) has suggested that the coralline *Clathromorphum circumscriptum* is dependent on the limpet *Acmaea testudinalis* to keep it clear of epiphytes and to prevent overgrowth of other, more productive, algae. He considers that grazing has lead to coevolutionary responses by *Clathromorphum*, which has a multilayered epithallus that protects the meristem, and sunken conceptacles that are relatively safe against limpet grazing. Other encrusting corallines have been shown to be specifically attractive to the settling larvae of grazing molluscs. *Lithothamnion* induces settlement and metamorphosis of the larvae of the chiton *Tonicella lineata* (Barnes and Gonor, 1973), and some encrusting algae appear to release γ-aminobutyric acid, which elicits settlement of the larvae of *Haliotis* (Morse et al., 1979).

In several instances the larvae of slow-moving opisthobranch predators have been shown to settle selectively on species which later form prey (Thompson, 1958, 1962; Hadfield and Karlson, 1969; Harvell, 1984). No one has yet demonstrated that the larvae of prey species avoid settling in areas where their predators occur. Highsmith (1982) has, however, shown that the larvae of the sand dollar *Dendraster excentricus* settle preferentially in sand that contains (or has contained) adults of their own species; a response induced by a chemical cue released by the adults. This has the effect of reducing predation on the newly settled juveniles because the adults exclude *Leptochelia dubia*, a predator on juvenile sand dollars.

Only comparatively recently has it been shown that settling larvae may respond to the presence of other competitors. One of the most important illustrations of this is Grosberg's (1981) demonstration that the larvae of many species avoid settling on plates on which the space-dominating ascidian *Botryllus schlosseri* had already settled. Perhaps the most significant part of Grosberg's work is not that it reveals avoidance of substrata-carrying pre-established competitors, but that the species which showed such avoidance were all species susceptible to overgrowth by *B. schlosseri*. An almost equal number of other species, all of which are not threatened by this species, settled randomly on experimental plates, irrespective of the density of *B. schlosseri*. The importance of this work is that it is the first good evidence that avoidance behaviour has evolved specifically in response to competitive threat. It is notoriously difficult to provide evidence for evolutionary responses because they are often not tractable to experimentation.

Bernstein and Yung (1979) have suggested that the larvae of various epiphytes that occur on the kelp *Macrocystis pyrifera* have evolved settling behaviours which minimise contact between these species and a dominant competitor, *Membranipora membranacea*. During winter, *Membranipora* tends to be concentrated on the uppermost blades of *Macrocystis*, and on plants at the outer (seaward) edge of the kelp bed. At this time of the year the larvae of two other bryozoans, *Hippothoa hyalina* and *Lichenopora buskiana*, and of a polychaete, *Spirorbis spirillum*, settle predominantly at the inner edge of the kelp bed, and on older blades which are not in the canopy. This pattern is established at least partly by the behaviour of the larvae themselves, for even when they have the choice they recruit selectively to older, inner-edge blades. During summer there is an interesting transformation of behaviour patterns: in the warmer water the normal photopositive response of *Membranipora* larvae is abolished, and they settle on an understorey alga, *Pterygophora*. Coincident with the freeing of the *Macrocystis* canopy from domination by *Membranipora*, the settling preferences of *Lichenopora* and *Hoppothoa* are abolished or even reversed, and they now settle readily on the younger, upper, blades of *Macrocystis*. Whether the larvae of the competitively subordinate species respond directly to the presence of *Membranipora* has not been established, but Bernstein and Yung (1979) make out a convincing case that the behavioural patterns displayed (and the seasonal changes in behaviour) have been evolved to minimise the threat of overgrowth which *Membranipora* poses to these species. The complexity of these responses illustrates one of the generalisations touched on in the introduction: that interspecific interactions may often be associated with versatile and flexible behaviour.

As a variation on the theme of how larvae react to competitors, Buss (1981) has shown that the larvae of the arborescent bryozoan *Bugula turrita* settles preferentially amongst established clumps of its own species (although it avoids large groups). Settling in conspecific clumps carries a cost: growth declines with density. Outweighing this disadvantage, however, is the fact that clumps of *Bugula turrita* are seldom overgrown by the encrusting bryozoan *Schizoporella errata*, while solitary colonies are regularly destroyed in this manner. Buss records comparable interspecific behaviour: a closely related species, *B. simplex*, also settles preferentially among medium-sized clumps of B. turrita, and presumably gains a similar advantage from the ssociation.

One of the problems facing workers concerned with larval behaviour is whether the prevalence of newly metamorphosed juveniles in a particular habitat indicates selective settlement, or simply higher post-settlement survival. Woodin (1985) has tackled this question in relation to settling of the larvae of a spionid polychaete, Pseudopolydora kempi. An inhabitant of sheltered sand banks, Pseudopolydora suffers high mortalities from the effects of defaecation by a much larger polychaete, the arenicolid Abarenicola pacifica. Abarenicola deposits faecal castes on the surface, smothering small surface-dwellers. Woodin determined to what extent larvae of Pseudopolydora would establish themselves in cores of sediment which had either (a) an intact Abarenicola, (b) an Abarenicola with an excised tail (which, Woodin argued, would present the same chemical cues to a settling spionid, but would be less active and therefore defaecate less), (c) the scent of an Abarenicola which was left in the core until the experiment began, or (d) no trace of an Abarenicola. Significantly more Pseudopolydora established themselves in the core that had no Abarenicola whereas there were no significant differences between the other treatments. This implies that the larval spionids had selectively settled in the cores that had no hint of Abarenicola. Even the smell of a recently present Abarenicola was sufficient to reduce the rate of settlement.

While it may be coincidence, or a bias on the behalf of researchers, it is worth noting that almost all work on the behaviour of larvae has involved sessile species or sedentary species that have very specific associations with other organisms. These are precisely the kinds of organisms for which sophisticated and elaborate larval behaviour might be predicted: for in the case of mobile species the adults may assume a greater role in locating suitable habitats and food and are not so dependent on larval selectivity for habitat selection, location of prey or mates, and avoidance of competitors.

It is also worth reiterating that all of these behavioural patterns focus on larval settlement and metamorphosis: there has been an assiduous neglect of interspecific behaviour during the planktonic phase.

BEHAVIOUR OF ADULTS

Factors Intrinsic to the Species

To a certain extent innate characteristics, including morphology, constrain not only the behaviour of each species but the future evolutionary pathway it is likely to follow. For instance, Steneck and Watling (1982) have argued for a "functional group approach" when predicting the feeding capabilities of herbivorous molluscs. They suggest that four types of radular structure can be recognised (rhipidoglossan, taenioglossan, docoglossan and polyplacophoran), with a trend towards reduction in the number of teeth and increasing strength of the individual teeth. In parallel with this is an increasing ability to handle relatively tough algae and to excavate the substratum. These trends impose limits on feeding behaviour and what kind of algae can be eaten, and may even dictate the outcome of interactions between species. For instance, patellacean limpets have been shown to outcompete siphonariid limpets (Black, 1979; Creese and Underwood, 1982) and littorinids (Underwood, 1978), possibly because patellacean limpets can excavate deeply into the substratum and remove much of the microalgal growth, while siphonariids and littorinids, having many small teeth in their radulae, can only rasp over the surface

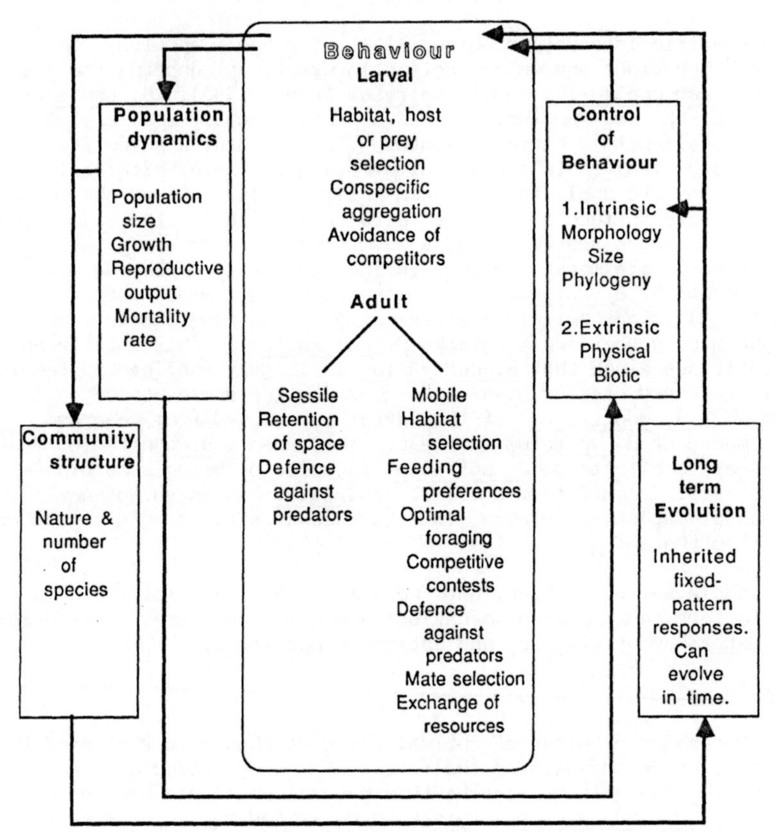

Fig. 1. Flow diagram depicting elements of behaviour in a hypothetical marine animal with both larval and adult phases in the life cycle. Although not exhaustive, the examples of behaviour are intended to indicate that the adults of sessile species are likely to have a more restricted behavioural repertoire than those of mobile species. Both evolved characteristics and extrinsic physical and biotic factors will influence behaviour. Behaviour, in turn, affects population dynamics, both of the species concerned and of other species it interacts with: thus potentially modifying community structure. Both population dynamics and community structure may have an input into the factors affecting short-term behavioural responses and the long-term evolution of behaviour.

of the substratum and remove loosely attached microalgae and filamentous algae. Before falling into the trap of assuming that ecological interactions can automatically be predicted by such morphological considerations, it should be recalled that Siphonaria gigas, the largest of the siphonariids, outcompetes coexisting patellacean limpets (Ortega, 1985), so size may be yet another intrinsic factor that influences behaviour and the outcome of interspecific interactions.

A more specific interaction which illustrates how morphology can influence both behaviour and interspecific interactions concerns the gastropods Busycon contrarium and B. spiratum (Kent, 1983). B. contrarium is the larger of the two species, but has a proportionally smaller foot, and a heavier shell with a narrower mouth. These features allow it to feed on "passive" bivalves (which are slow-moving but have thick shells) because it can use its shell to chip open their shells. Given the choice, however, it prefers to feed on "active" bivalves, which are the major food of B. spiratum. There are two reasons why it seldom does so in the field. Firstly, because of its morphology, it is inefficient at capturing active bivalves. Secondly, B. spiratum (with its large foot, wide mouth and faster movements) is not only more efficient at exploiting the active bivalves, but also aggressively attacks B. contrarium. This interference is sufficiently sustained that B. contrarium is largely confined to feeding on the less preferred passive bivalves. Despite the dominance of B. spiratum over B. contrarium, if the latter is removed from seagrass beds, the numbers of B. spiratum increase. Presumably B. contrarium must have some negative effects on B. spiratum, in spite of being subordinate to it in contests. Apart from this, B. contrarium has an advantage wherever crab predation is intense, for its thicker shell provides greater protection than that of B. spiratum.

This example serves to highlight the limitations that intrinsic morphological factors can impose on behaviour and upon ecological interaction between predators and their prey and between competitors.

Patterns of Foraging

One of the major premises of optimal foraging theory is that predators should select prey which maximise their rate of energy intake per unit of handling time (E/Th). In many cases this general model has been upheld. For example, if Carcinus maenas is given an unlimited choice of mussels of different sizes in the laboratory, it will feed most often on those mussels that give an optimal yield (Elner and Hughes, 1978). It does not, however, exclude suboptimal mussels from its diet, feeding on them less often but roughly in proportion to their abundance. This pattern can be explained if the crabs feed on optimal mussels whenever they encounter them, but initially reject suboptimal mussels and only feed on them if they encounter several suboptimal animals in succession (Elner and Hughes, 1978).

When Carcinus maenas is offered the choice, it selects Littorina rudis rather than L. nigrolineata (Elner and Raffaelli, 1980). This is consistent with the fact that L. rudis yields more energy per unit handling time (because its thinner shell is easier to crush). In theory Carcinus should also select the largest individuals of both species that are available, because their value increases exponentially with size. In fact they preferentially feed on the smallest individuals. Only when account is taken of the success rate of attacks does this become explicable: large littorines may have a higher E/Th, but attacks on them are often unsuccessful. When E/Th is multiplied by success rate it becomes apparent why it is most profitable to attack the smallest littorines. Elner and Raffaelli suggest that the crabs should employ a different method of foraging when feeding on littorines, in comparison with that used against mussels. Mussels have a predictable value, are often encountered, and are clumped so that they can be readily compared. Consequently the crabs feed on optimal-sized individuals. Because the littorines are relatively sparse and isolated, and have an unpredictable value, the crabs appear to attack all individuals they come across, but reject those which do not succumb within a set time. Thus one element of the behaviour of the prey (dispersion pattern) may influence the feeding behaviour of the predator.

To test some of the assumptions of the optimal foraging model under more realistic conditions than are possible in the laboratory, Palmer (1983) maintained three _Nucella_ (=_Thais_) species on prey of different species and sizes, and monitored their growth as an index of the value of the prey items (fast growth being correlated with early reproduction and a high egg production). Subsequently Palmer (1984) showed that those species and sizes of prey yielding high growth rates were indeed the prey items upon which the whelks fed selectively in the field.

Palmer (1984) was also able to show that if the three _Nucella_ species were conditioned by feeding them on prey with a high yield, that they subsequently foraged selectively on prey of high value. If, on the other hand, they were conditioned to prey of low yield, they would feed on all available prey in proportion to their availability. The availability of suitable food can therefore alter the foraging patterns of predators.

Dunkin and Hughes (1984) describe how _Nucella lapillus_ changes both its mode of feeding and the size of barnacles (_Semibalanus balanoides_) on which it feeds, as it gains experience. Inexperienced dogwhelks (accustomed to feeding on mussels) select the largest barnacles available, and attack them by drilling through their shell plates. The largest barnacles yield the greatest flesh weight per unit of handling time, so this response seems explicable on the basis of the profitability of the prey. However, as the dogwhelks gain experience, they selectively opt for medium-sized barnacles. They also switch their mode of feeding, preferring to prize open the opercular valves of barnacles: a process that significantly reduces handling time. One of the benefits of selecting smaller barnacles is that it is possible to prize them open, and experienced dogwhelks prefer to attack barnacles of a size that can be prized open 50% of the time. Thus experience gained by a predator can alter the profitability of its prey.

Despite the benefits of experience and the fact that medium-sized barnacles can be prized open by the dogwhelks while larger individuals cannot, the selection of medium-sized individuals cannot be explained on energetic grounds: larger barnacles still yield the greatest returns. Dunkin and Hughes conclude that one possible reason for selecting medium-sized barnacles is that they can be dealt with more quickly. The quicker a whelk feeds on its prey, the less the chance it will be displaced by another whelk and have its meal stolen (an event Dunkin and Hughes recorded in 12% of the cases when dogwhelks were eating barnacles).

An alternative reason why predators might select prey that can be consumed quickly rather than those most energetically profitable, is that foraging can expose the predator to risks, such as harsh physical conditions or other predators. Menge and Menge (1974) have pursued this idea in examining the feeding preferences of _Acanthina punctulata_, a whelk which occurs intertidally and is exposed to the risk of being washed away by waves if it feeds during high tide. _Acanthina_ shelters in crevices during high tide but forages on the open rock face immediately it is exposed by the receding tide. It feeds preferentially on _Littorina planaxis_ rather than _L. scutulata_, and on either of these littorines rather than barnacles.

These choices reflect the energy gained from these species per unit of handling time. In addition, they restrict the feeding time spent on each prey item, thus reducing the chances that they will still be feeding when the tide rises. Menge and Menge note that *Acanthina* selects *Balanus glandula* ahead of *B. cariosus*, although the two have almost identical *E/Th* values. However, because *B. cariosus* is much larger, it takes 14h to consume, in comparison with 3h for *B. glandula*. Not only will consumption of *B. cariosus* involve greater risk of exposure to wave action, but it also increases the chances of other whelks joining the meal and reducing the value of the prey to the original whelk.

Menge (1978a,b) has attempted to isolate the effects of physical risk on foraging behaviour by comparing the behaviour of *Nucella lapillus* on shores that are exposed to strong wave action with that on sheltered shores. On exposed shores wave action was clearly identified as a source of mortality, and the *Nucella* were confined almost entirely to crevices. On sheltered shores *Nucella* foraged widely over the mid- to low-shore during summer. Although it is potentially vulnerable to desiccation on sheltered shores, in practice it is protected by a canopy of *Ascophyllum* or *Fucus*. Its feeding rate is higher under an algal canopy, higher in cool months than in hot months, higher low on the shore than high on the shore, and is also higher in phenotypes taken from exposed shores than in those from sheltered shores (even when they are transplanted between habitats and measurements made under comparable conditions). There are two particular features of these results worth stressing. Firstly, the existence of different rates of feeding in phenotypes from different habitats raises the possibility of genetic selection of different behaviours under different physical conditions. Secondly, the contrasting behaviours on exposed and sheltered shores have profound implications for the whole community. On exposed shores the impact on *Nucella* on prey species (notably barnacles and mussels) is minimal, except in the immediate vicinity (within 10 cm) of crevices. On sheltered shores they have an important and widespread effect on sessile animals, reducing their cover from almost 100% down to between 0 and 20%. Menge points out that a type of mutualism is involved between the algae and *Nucella*: algal cover enhances the feeding rate of *Nucella* which, in turn, removes organisms such as mussels which would otherwise outcompete fucoid algae.

The interplay between behaviour and risks from physical stress or predation is emphasised in Levings and Garrity's (1983) analysis of diel and tidal foraging movements of two herbivorous gastropods, *Nerita funiculata* and *N. scabricosta*. *N. funiculata* occurs in the midshore, and is restricted to crevices much of the time, by potentially lethal desiccation during day-time low tide and by predatory fish during high tide. Consequently it forages for brief periods during the ebbing and flooding tides, and is confined to within 50 cm of the crevices it occupies. Its impact on comunity structure is correspondingly constrained, and it influences other species only in the immediate vicinity of crevices, where its concentrated grazing bares much of the substratum. *N. scabricosta* is a larger species which occurs high on the shore. It is inactive during high tide, sheltering in crevices or on vertical surfaces, but follows the tide down the shore as it recedes and then back again as it advances. It thus has an effect over much of the intertidal and, *inter alia*, reduces the cover of blue-green algal crusts, restricts settlement and survival of barnacles, and diminishes both the density and mean size of two species of *Littorina*. The differences in the behaviour of *N. funiculata* and *N. scabricosta* seem to be due largely to their different sizes. Individuals of comparable size move similar distances each tidal cycle, but *N. scabricosta* attains a larger size, and it is the larger individuals which roam furthest down the intertidal zone and have the greatest effect on community structure.

Once again, this demonstrates the important effects that behaviour may have on population dynamics and on community structure. There are, however, several factors which can modify behaviour. These combine properties intrinsic to the species itself, such as morphology and size, and extrinsic factors, including the relative availability of different prey items, and risks that feeding may incur, such as exposure to harsh physical conditions, to other predators or to competitors. It is in examining these risks in particular that the unique features of the intertidal zone become apparent: features which supply the selective pressures that shape behaviour.

Behavioural Adaptations of Prey

Many mobile prey species have specific behavioural responses which allow them to escape from predators, including a "running" response, rotational movements that break the animal free from a predator, elevation of the body away from the point of contact (e.g. Bullock, 1953; Clark, 1958; Phillips, 1975; McClintock, 1985), and a more specialised reaction in which toxic or unpalatable parts of the body are erected to cover more vulnerable parts or to prevent attachment (Margolin, 1964; Branch and Branch, 1980). Some species have repugnatorial glands that are discharged on contact with a predator (Young et al., 1986). Sessile species have fewer options open to them, but still respond in ways that may reduce detection or deter predators. Palmer et al. (1982) have, for instance, shown that *Balanus glandula* prolongs withdrawal into its shell after contact with a predator, remaining withdrawn about two to three times longer than it does after contact with a non-predatory species.

Distance chemoreception plays a role in the detection of predators (Phillips, 1975), and oriented movements which reduce the chances of encountering a predator appear to have evolved. For example, Phillips (1975) demonstrated that *Collisella limatula* and *Notoacmea scutum* move downstream if they are on a horizontal surface when they perceive a predatory starfish. If, however, they are on an angled surface, detection of a starfish then reverses their normal photonegative and rheopositive reactions, and they develop a strongly negative geotaxis and move upwards, even if this means moving into the current. At first sight this seems maladaptive, for moving into a current is likely to take the limpet towards the predator. However, considering the unpredictability of turbulent water movements in the intertidal, and the fact that these limpets are zoned above their starfish predators, a net upward movement may be the surest way of escape. The responses of these limpets are species-specific to predators which normally prey upon them. The intensity of their responses (i.e. the distance they move) is also proportional to how close the predator is (Phillips, 1976).

Homing to crevices (Levings and Garrity, 1983), to the undersurface of boulders (Wells, 1980) or to fixed "home scars" (Wells, 1980; Garrity and Levings, 1983) have all been shown to reduce mortality due to predation. The activity rhythms of intertidal animals are tightly linked to the tide and to the day and night. Many of them seem explicable as means of reducing predation. For instance, Bertness et al. (1981) have shown that most tropical intertidal molluscs are inactive and confined to the high-shore, to crevices, or to home scars during high tide, becoming active only for brief periods during the ebbing and flooding tides when they are awash. They interpret this as an adaptation to aquatic predators, which present a threat during high tide, and to desiccation, which is potentially lethal during the day-time low tide. Bertness et al. compare this pattern of activity with that recorded for several temperate species, and conclude that the latter are active when submerged at high tide because the intensity of predation is apparently less on temperate shores. Whether this is generally true remains to be seen. Ortega (1986) has already demonstrated that predation is not always intense in the tropics.

As an alternative to defensive behaviour, prey species can resort to attack. Even a sessile species such as *Crepidula fornicata* effectively repulses *Urosalpinx* by attacking it with its radula. A more complex case has been recorded for two South African limpets, *Patella granatina* and *P. oculus*. When they are small, both species respond to invertebrate predators such as the starfish *Marthasterias glacialis* and the dogwhelk *Nucella dubia* by fleeing. As they get larger (in the region of 40 to 60 mm) they change their behaviour and become aggressive, lifting their shells and crushing them down onto the foot or tubefeet of these predators. The interesting feature of this change in behaviour is that it occurs at different sizes depending on which of the two predators is involved. Comparatively small limpets (>35mm) will attack *Nucella dubia*, but only the largest (>63mm) adopt this behaviour against *Marthasterias glacialis*. Once again, this exemplifies the versatility and flexibility of interspecific behaviour.

Hay et al. (1986) describe how the urchin *Arbacia punctulata* is attracted towards both *Sargassum* and *Gracilaria* during the day, despite the fact that only the latter is consumed in large quantities. By night this chemotactic behaviour alters and they are strongly attracted to *Gracilaria* while their attraction to *Sargassum* decreases. They suggest that the urchins are not necessarily attracted to particular seaweeds because of their value as food. Their attraction to *Gracilaria* both by day and by night may be because it is a desirable food,but possibly they are drawn towards *Sargassum* during the day for a different reason: concealment from visual day-time predators. This hypothesis remains untested, but it does illustrate the potential that predators have for modifying the behaviour of prey species. Elsewhere, *Diadema antillarum* has been shown to shelter in safe sites during the day but to forage openly at night; but the distinction between its day-time and night-time behaviours apparently diminishes in areas where predators are rare (Carpenter, 1984).

A far more controversial issue is whether another urchin, *Strongylocentrotus droebachiensis*, changes its behaviour in the presence of predators such as lobsters and crabs, and whether this change of behaviour has implications for the way in which the urchins affect algae. Bernstein et al. (1981) have argued that when these urchins are scarce, they are compelled to hide in crevices and under boulders to avoid predators. If, however, their densities increase (an increase which Bernstein et al. attribute to a reduction in the intensity of predation) then the urchins can switch their defensive behaviour and form defensive aggregations instead of seeking shelter, and can feed openly and destructively on kelp. Bernstein et al. go further and suggest that the urchins tend to aggregate when a predator is present, so the predators themselves facilitate the formation of aggregations and thus help to "trigger destructive grazing". Clearly, if this is the case, then the switch in defensive behaviour by the urchins is of fundamental importance, for it may be a major contributing factor behind the transformation of healthy kelp beds into "barren" coralline-dominated grounds.

The interpretations of urchin aggregation advanced by Bernstein et al. have been challenged by Vadas et al. (1986). Both in the field and in aquaria they showed that *S. droebachiensis* always moves away from predators rather than aggregating, and flees from recently damaged conspecifics. They agree that urchins do form aggregations, but contend that these are never defensive in function but form in response to algal food. Once they have formed, the aggregations persist as long as there is a nucleus of food available, and do not break up if a predator is introduced. They also showed that crabs and lobsters feed readily on aggregated urchins and, therefore, doubt the efficacy of aggregations as a defence. Some of these differences of opinion may have arisen because urchins tend to gather around surface irregularities when they move away from predators. This is particularly obvious in aquaria, where they accumulate around air pipes and in the corners. Vadas et al. consider that these "two-dimensional associations" are not true aggregations and result largely from the constraints imposed by aquaria. When the urchins were unimpeded in the field, they dispersed away from predators.

The resolution of this argument is an important one for, whereas both groups of authors agree that urchins' grazing is responsible for the massive changes in algal communities documented in several areas, their explanations of how the urchins achieve high numbers and why they aggregate are radically different. The work of Vadas et al. emphasises the need for controlled, realistic, and carefully interpreted experiments to understand the behavioural responses that underpin such ecological phenomena.

Apart from immediate behavioural responses, predators may induce defence mechanisms in their prey. A clear-cut example is the development of spines around the edge of colonies of the bryozoan *Membranipora membranacea* when it is attacked by nudibranchs, notably *Doridella steinbergae* (Harvell, 1984). The development of spines reduces the feeding rate of *Doridella* from 44 zooids per day to 8 per day. Harvell proposes that the two species are coevolved because *Doridella* feeds exclusively on *Membranipora*, its larvae are induced to settle by the presence of the bryozoan, their seasonal appearance is synchronised, and because of the defence mechanism induced in *Membranipora* by nudibranch attack.

Inducible morphological defences provide a link between short-term behavioural responses and longer-term evolutionary changes in morphology and, possibly, also in body size. Vermeij (1978) has argued forcibly that certain morphological features have been evolved to counter predators: features such as thick shells and narrow apertures to gastropod shells. More recently, Palmer (1982) has hypothesised that the repeated trend towards a reduction in the number of parietal plates in the shells of a number of barnacle groups is a response to the evolution of drilling thaid gastropods. Thaids selectively drill at the sutures between plates, and have a higher success rate when they do so (41%) compared with when they drill directly through the parietal plates (12%). Other adaptations the barnacles may have developed are sculpturing and hollow tubes running through the plates, both of which will thicken the plates without a great investment of skeletal material, and will therefore make it more difficult for thaids to drill deeply enough to inject the toxin that relaxes the barnacles' muscles. Palmer (1983) also points out that prey which are less vulnerable to attack because of their morphology or size appear to occur lower on the shore, while more vulnerable species and those which never become large enough to be immune to attack tend to be zoned high on the shore. His argument is based on the premise that marine predators are less of a threat high on the shore; but it does ignore terrestrial predators, notably birds, which are more active in the high-shore. Nevertheless, there are several examples of genera which have thick-shelled species in the low-shore and thin-shelled counterparts higher up the shore (e.g. Elner and Raffaelli, 1980; Branch, 1981, p.329; Palmer, 1983). It remains to be proven that this trend is general.

There are many examples of prey species which attain a size at which they are too large for predators to attack (e.g. Paine, 1966; Palmer, 1983; Hockey and Branch, 1984; Paine et al., 1985). In part, body size is an intrinsic, genetically controlled property of species, and it can influence not only predator-prey relationships but the resilience of communities to disturbance. As an example, Paine et al. (1985) have described how mussels in three different intertidal communities (in Washington State, New Zealand and Chile) react to the experimental removal of a major starfish predator by covering a greater proportion of the substratum and expanding to occupy sections of the shore from which they are normally absent. However, if these manipulated communities are then left to recover, they respond very differently. In Washington and New Zealand the expanded mussel beds retain their domination over the shore for prolonged periods (at least 14 to 16 years), despite the return of starfish to the manipulated areas. In contrast, the beds in Chile rapidly return to their previous condition. Paine et al. ascribe the persistence of the mussels in Washington and New Zealand to their rapid growth to a size which is too large for the starfish to prey on them. In addition, juveniles of both species recruit to byssal threads within the mussel bed, so that the population becomes self-perpetuating. The Chilean mussel, *Perumytilus purpuratus*, is a comparatively small species, and never attains a size at which it is safe from predation. Consequently starfish returning to the manipulated area soon reduce their numbers to normal levels.

Two cautionary notes are, however, necessary. Firstly, body size in invertebrates is extremely plastic and is influenced by environmental conditions. Hockey and Branch (1984) have shown that both growth rate and maximum body size of the limpet *Patella granularis* are altered by food availability. On off-shore islands on the west coast of South Africa, where algal productivity is high because of the run-off of guano from roosting birds (Bosman et al., 1986), *P. granularis* reaches a size of 100mm, well in excess of the largest size that can be attacked by the limpets' major predator, the African black oystercatcher (Hockey and Branch, 1984). This is a major factor contributing to the persistence of limpets on the islands, where predation is intense because of the high density of oystercatchers (up to 35 pairs per km of shoreline). At nearby mainland sites, algal productivity is low, the limpets reach a comparatively small maximum size (c.50mm), and the entire limpet population is open to attack by the oystercatchers. Here the limpets persist only because the oystercatcher density is low (1 to 2 pairs per km).

A second cautionary note about prey attaining a size refuge is that predators may develop behavioural patterns that allow them to circumvent this limitation. For instance, Blankley and Branch (1984) have described how the sub-Antarctic limpet *Nacella delesserti* can reach a size at which it is immune to the attentions of a common starfish, *Anasterias rupicola*, provided the starfish feed as individuals. *Anasterias* can, however, feed co-operatively in groups, and such "cluster feeding" allows it to feed on even the largest limpets, which would otherwise be unattainable.

The vulnerability of a prey species to predation can be altered by its interaction with other species. To exemplify, symbiotic crustacea defend the coral heads in which they live against predatory starfish: a defence initiated by chemical cues from the predator (Glynn, 1980). As a second example, the commensal association between the limpet *Patelloida mufria* and its host, a trochid gastropod, *Austrocochlea constricta*, reduces predation on the limpet. Adults of *Patelloida* are readily consumed by the whelk *Morula marginalba* if they are attached to rocks, but virtually never when they are on a host's shell (Mapstone et al., 1984).

On the other hand, some interactions may increase susceptibility to predation. For instance, aggressive contests between the shrimps *Palaemonetes vulgaris* and *P. floridans* lead to the restriction of *P. vulgaris* to seagrass beds, while *P. floridans* occupies red seaweeds (Coen et al., 1981). In the process, *P. vulgaris* suffers a higher rate of mortality because the more simply structured seagrass provides less shelter from predatory fish. An almost exactly parallel situation exists between P. vulgaris and P. pugio but, in this case, P. pugio is the subordinate species and is displaced from a shelly substratum to mud, where it is preyed upon more easily by fish (Thorp, 1976). A rather different, but comparable, example concerns the interaction between two tube-dwelling polychaetes. One of the species, Axiothella rubrocincta often reduces the densities of another, Pseudopolydora paucibranchiata. At first sight the explanation seems one of direct competition, for both require organic mineral aggregates for feeding and tube construction, and Axiothella only has a negative effect on Pseudopolydora if these aggregates are scarce. In fact, flatfish are the agents responsible for reductions in the numbers of Pseudopolydora, for they prey on this polychaete and their attacks on it are far more successful if organic mineral aggregates are in short supply and the Pseudopolydora have to reach further out of their tubes to search for food (Weinberg, 1979). As Weinberg concludes, "predators may be necessary to effectuate change in community structure initiated by competition".

The behaviour of prey towards their predators is obviously diverse. In many cases it is clearly evolved in response to predators and is highly specific, but it is also influenced by properties intrinsic to the species, by their interactions with other species, and by the extent to which the predators have evolved counter-adaptations.

Behavioural Elements of Competition

In earlier reviews of competition in marine organisms (Branch, 1984, 1985) it has been argued that the outcome of competition between species is related to the mode of competition (exploitation or interference), the nature of the resource competed for (e.g. food or space) and the mobility of the species in question. A review of the literature shows that interference is the most common form of competition in marine organisms (occurring in 80% of competitive interactions) and that, compared with exploitative competition, it is more likely to lead to localised exclusion (being involved in 93.5% of the documented cases of local competitive exclusion).

These facts are relevant to the role of behaviour in competition, for whereas exploitation involves practically no direct behavioural interaction between competitors, behaviour is fundamental to interference. For example, as outlined above, contests between different species of Palaemonetes lead to the localised exclusion of one of the species from preferred habitat (Thorp, 1976; Coen et al., 1981). In similar vein, territorial defence by the fish Embiotica lateralis confines its congener E. jacksoni to deeper water where food is less available (Hixon, 1980). Fighting between different species of mantis shrimps also results in the localised restriction of the subordinate species to a suboptimal habitat (Dingle and Caldwell, 1975). The aggressive reactions of limpets to other herbivores excludes them from the limpets' territories, or even from whole zones of the shore (Stimson, 1970, 1973; Branch, 1975, 1976). Most of these examples concern territoriality or, at least, defence of a general "living area", but non-territorial animals may also have an inhibitory effect on their competitors by interfering with them. For instance, the hermit crab Calcinus obscurus is competitively dominant over Clibanarius albidigitus in fights for shells, so that Clibanarius is largely restricted to inferior shells. Clibanarius is more efficient at gleaning new shells, but Calcinus soon usurps them: in fact Calcinus benefits from association with Clibanarius because the latter increases the retention of new shells in the area and makes them available to Calcinus (Bertness, 1981). Another example is the starfish Pisaster ochraceus, which attacks one of its competitors, Leptasterias hexactis, with its pedicellaria, causing a partial paralysis and a consequent reduction in feeding rate (Menge and Menge, 1974). Interference can also take place by allelopathic control of competitors. As one example, chemicals released by some sponges and ascidians prevent overgrowth by other colonial forms such as bryozoans. Interestingly, this allelopathy is species-specific and has no effect on solitary species which present no competitive threat to the sponges and ascidians (Jackson and Buss, 1975). Such specificity is circumstantial evidence that the allelopathy was evolved specially to deter potential competitors, although there are, of course, other possible explanations for its origin.

These examples suffice to make the point that interference is widespread and that behaviour is fundamental to such competitive interactions. Even subtle differences in behaviour may determine which species wins and under what circumstances. Harger (1968, 1970) has described how the mussel _Mytilus edulis_ tends to crawl clear if covered by sediment, and this gives it an advantage over _M. californianus_ in quiet waters, preventing it from becoming smothered. In mixed clumps of the two species, this behaviour also allows _M. edulis_ to crawl to the outside of clumps, while _M. californianus_ becomes smothered both by the outer layer of _M. edulis_ and by silt that accumulates between the mussels. Thus, in sheltered areas _M. edulis_ is competitively superior to _M. californianus_. On wave-beaten shores, where siltation is not a problem, their competitive abilities are reversed, partly because _M. californianus_ secretes much stronger byssus threads, and partly because it grows faster and to a larger size.

It is equally true that behavioural differences may reduce competition between species. Chelazzi et al. (1983a) outline how two species of chiton, _Acanthopleura brevispinosa_ and _A. gemmata_, occur at fairly similar heights on the shore, have similar diets, and both feed during the nocturnal low tide. However, whereas _A. brevispinosa_ moves randomly during its feeding excursions, _A. gemmata_ moves in an oriented manner down the shore to feed, returning upshore before the tide advances. Thus the two species feed in different and almost non-overlapping zones. Incidentally, while they are feeding the chitons do not react aggressively towards one another, but they do defend their home scars, and will crawl over any intruding chiton and attempt to remove it by "back and forth movements" (Chelazzi et al., 1983b).

The behaviour of interacting competitors can also be ritualised. Different species of symbionts living in heads of the coral _Pocillopora damicornis_, including two fish, a crab and a shrimp, collectively defend their coral head against intruders, but they tolerate each other partly because they have ritualistic signals that indicate their resident status (Lassig, 1977). As another example, the owl limpet _Lottia gigantea_ reacts aggressively by thrusting against other limpets that invade its territories, even to the extent of dislodging them (Stimson, 1970). Wright (1977, 1982) has, however, described how intruding limpets employ a well developed escape response if they come into contact with another limpet while they are on a territory; this response is absent if they are not on another limpet's territory. As a result of this escape behaviour, most territorial contests are ritualised, the mere threat of a territorial animal inducing flight by the intruder. As dislodgement is likely to be fatal to the intruder, and prolonged territorial contests represent a waste of time and energy to the territory holder, ritualisation of the interaction is beneficial to both.

Considerable complexity and flexibility may be evident in behavioural interference between competitors. Wright (1982) notes that whereas a limpet that is intruding on another's territory will flee on contact with the resident, that same animal will react aggressively if it is on its own home scar or territory. This variability in behaviour prompted Wright (1985) to experiment on the proximate factors influencing the responses. Using limpets maintained in aquaria, Wright subjected some limpets to "hostile" treatment (by placing another limpet in contact with them and then pushing that limpet against them to simulate the aggressive response of a territorial animal). Other limpets were subjected to a "benign" treatment (placing a limpet in contact with them and then slowly withdrawing the limpet to simulate retreat of an intruder). The results were clear-cut. Animals treated in a hostile manner became evasive in their response to contact with another limpet; those that received "benign" treatment became aggressive. These behaviours could be reversed if the treatment was reversed, so limpets evidently modulate their behaviour on the basis of previous experience. Wright also showed that benignly trained animals spent more time moving around their territories than those with hostile training: at least hinting that they devote more time to "patrolling" their territories, increasing the chances of encountering an intruder.

This example demonstrates the adaptability of behavioural responses between competitors, a versatility that is well known in "higher" animals, but seldom contemplated for intertidal invertebrates.

Paine and Suchanek (1983) have made the point that the functional role species play in an ecosystem may often be dictated more by their morphology, physiology and behaviour than by their taxonomic affinities. They make a comparison between the mussel *Mytilus californianus* and the solitary ascidian *Pyura praeputialis*, both space-dominating competitors in the intertidal zone. They suggest the domination of these species is achieved because of several properties they share. (1) They are large, fast-growing species. (2) Their larvae recruit to beds of adults, perpetuating the beds and an aggregated pattern of distribution. (3) The adults bind to one another, making it difficult to remove them. These last two points, which are based on behaviour, are particularly interesting because, in combination, they effectively create colonies. Jackson (1977) has previously argued that colonial forms have a number of competitive advantages over solitary species, and nearly always dominate hard substrata in the subtidal; but in the intertidal they are seldom as successful, possibly because interconnectance of individuals and budding preclude a hard case that can counter desiccation. Behaviour can thus powerfully influence the role a species plays.

The outcome of competitive interactions may also be radically modified by the behaviour of the participants. For instance, some species of coral dominate their neighbours by mesenterial digestion of their tissues, but subordinate species may reverse the dominance by developing long sweeper tentacles (Wellington, 1980; Bak et al., 1982; and see Sheppard, 1982 for a review).

The participation of other species can also mute or even reverse competitive domination of one species over another. An obvious case is the role played by predators in reducing the numbers of potentially dominant species. The mussel *Mytilus edulis* is competitively superior to articulated brachipods because it is mobile and, if predators such as fish, starfish, gastropods and crabs are experimentally excluded, will significantly reduce their survivorship by smothering them. On the other hand, brachiopods have a repellent flesh and a low meat weight, so that predators prefer to feed on mussels, reducing the mortality of brachiopods in the process. An interesting twist to this relationship is that in the complete absence of mussels, predators turn their attention to the brachiopods. Thus, whereas predators increase brachiopod survival by reducing competition, mussels reduce predation on the brachiopods by diverting predation (Thayer, 1985).

In another case, Sutherland (1974) has shown that the ascidian Styela is only dominant over the bryozoan Schizoporella if fish are experimentally excluded, but that Schizoporella overgrows Styela if fish are allowed to graze over the substratum. However, Styela is sometimes associated with the hydroid Tubularia, and it then gains sufficient protection from fish to remain dominant over Schizoporella.

Symbionts also have the ability to change the outcome of competition. The xanthid crab Domecia, which lives commensally in coral heads, attacks adjacent coral heads, causing up to 60% mortality, and may tip the balance between their hosts and other corals that pose a competitive threat (Bak et al., 1982). The hydroid Zanclea is frequently associated with colonies of the bryozoan Celloporaria brunnea, which forms a protective calcified layer over it. For its part, Zanclea improves the competitive ability of Celloporaria against other bryozoans and against colonial polychaetes, in addition to repelling nudibranch predators (Osman and Haugsness, 1981).

Even competitors may have the same effect as predators by reducing the numbers of a dominant competitor. To exemplify, the damselfish Eupomacentrus planifrons excludes the urchin Diadema antillarum from its territories because it attacks it aggressively. It is, however, relatively tolerant of another urchin, Echinometra viridis, which can live in the upper branches of corals in the damselfish territories. Removal of damselfish leads to a rapid invasion of Diadema and a corresponding decline of its subordinate competitor, Echinometra (Williams, 1981).

This raises the possibility of indirect mutualisms or commensalisms, in which the positive influence of one species on a second species is not due to its direct effect on that species, but to its negative effect on a third species which is detrimental to the second species. Dethier and Duggins (1984) have demonstrated such a relationship between the chiton Katharina tunicata and three microalgal-eating acmaeid limpets. They removed Katharina from a large section of the shore and doubled its numbers at another site. It would not have been unreasonable to have expected compensatory increases in the numbers of the limpets in the absence of this chiton, on the assumption that it competes with the limpets for food. In fact, the limpets almost completely disappeared from the area lacking Katharina, and doubled their numbers at the site where the density of Katharina had been increased. In addition, they had higher fecundities in the area where Katharina numbers were doubled, compared to control areas. Clearly the chitons had a positive effect on the limpets. Their probable influence on the limpets is an indirect one: they control the growth of macroalgae, which would otherwise outcompete microalgae and monopolise the rock face, reducing or eliminating both the limpets' food and the bare rock they require as a substratum. The fact that the outcome of this experiment was opposite to that that might have been predicted from conventional competition theory, points again to the need for careful experiments to "avoid the pitfalls of assuming the existence of competition" (Dethier and Duggins, 1984, p.215).

Another case in point concerns interactions within a guild of three species of urchins: Strongylocentrotus franciscanus, S. purpuratus and S. droebachiensis (Duggins, 1981). All three occur in similar habitats and eat the same food: competition might be anticipated. In the field, the numbers of S. franciscanus are, however, correlated with those of the other two species, and if the density of S. franciscanus is increased experimentally, this has either a positive effect or no effect on the numbers and reproductive condition of the other species. Conversely, removal of S. franciscanus prefaces declines in the other urchins and their complete disappearance within about 40 days. S. purpuratus and S. droebachiensis appear to derive two benefits from association with

S. franciscanus. First, S. franciscanus is more efficient at trapping drift algae, which can then be used by the other urchins once it is pinned down. Secondly, S. franciscanus, the largest of the three species, is too large for the starfish Pycnopodia helianthoides to feed on. In fact, Pycnopodia avoids areas where S. franciscanus is abundant. S. purpuratus and S. droebachiensis therefore gain protection by associating with S. franciscanus. Following experimental removal of S. franciscanus, Pycnopodia moves in, and both S. purpuratus and S. droebachiensis (which, unlike S. franciscanus, have well developed escape responses to the starfish), decline in numbers as they disperse from the area or are consumed.

It is evident that the interplay of behavioural interactions between competitors, predators and symbionts may have a powerful influence on the outcome of interactions between species and on community structure, and that as more species interact, it becomes more difficult to predict the outcome.

THE EFFECT OF BEHAVIOUR ON COMMUNITY STRUCTURE: A CASE HISTORY

When communities experience similar physical conditions but have a strikingly different biological composition, they present a particular challenge. In this section we summarise a study undertaken by Barkai (1987) on the structure of two such communities, with the aim of showing how biological and, particularly, behavioural interactions influence community structure.

The communities in question are shallow-water systems at adjacent islands on the west coast of South Africa. The two islands, Marcus and Malgas, are situated about 4 km apart, in the mouth of Saldanha Bay (33° 04'S, 17° 57'E). Both are guano islands, and the runoff of guano enhances the productivity of intertidal algae and, thereby, the growth rates and maximum sizes of intertidal grazers (Hockey and Branch, 1984; Bosman et al., 1986; Bosman and Hockey, 1986). Both experience comparable wave action which varies from strong to intense, waves of up to 17m being recorded in the vicinity. Considerable but intermittent disruption of the communities occurs, at intervals of one to three years. In spite of this, the two islands have retained their characteristically different communities for at least 10 years. The two islands also experience similar nutrient levels and similar water temperatures. The waters around Marcus Island are, on average, 2°C warmer than those of Malgas Island but, considering that the daily temperature range can be as much as 8°C at the beginning and end of upwelling cycles, this difference seems trivial. There are therefore no obvious differences in physical conditions that could account for the considerable differences in community structure between the two islands.

Figures 2A and B summarise the differences between the communities. At Marcus Island the benthos is dominated by the black mussel (Choromytilus meridionalis), which covers most of the substratum, forming deep multilayered beds, comprising 87% of the total wet biomass (c. 36,000 g m^{-2}). Associated with the mussel beds is a rich array of animals, the most notable being large numbers of urchins (Parechinus angulosus), three species of holothurians, ophiuroids, crinoids, sediment-dwelling polychaetes and amphipods, and two barnacles, of which the most abundant is Notomegabalanus algicola, which settles mainly on the mussels because primary space is very limited. In addition there are several species of whelks: particularly important are three species of Burnupena. Notable absentees are most species of algae, adults of the ribbed mussel Aulacomya ater, and the rock lobster, Jasus lalandii.

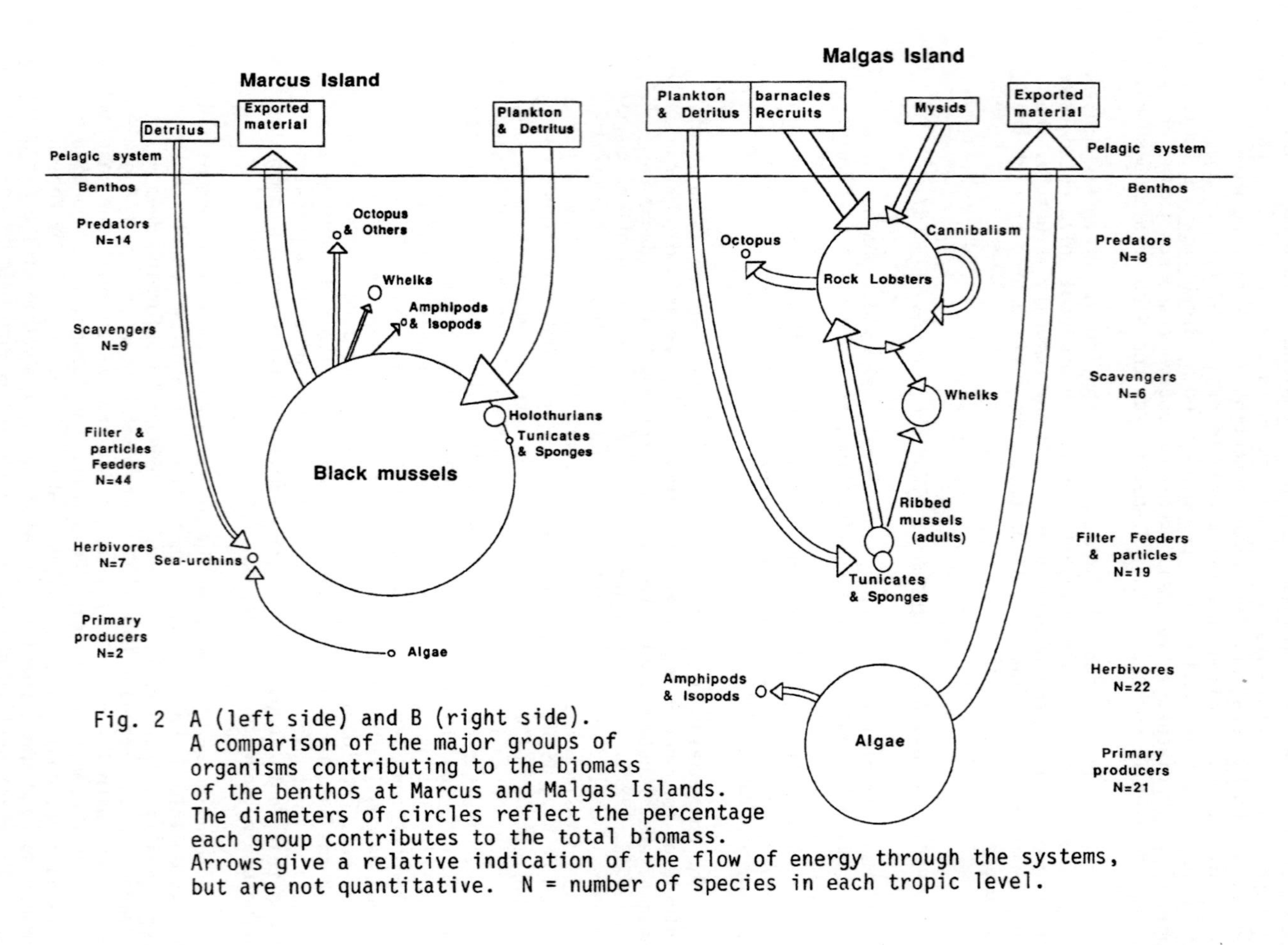

Fig. 2 A (left side) and B (right side).
A comparison of the major groups of organisms contributing to the biomass of the benthos at Marcus and Malgas Islands.
The diameters of circles reflect the percentage each group contributes to the total biomass.
Arrows give a relative indication of the flow of energy through the systems, but are not quantitative. N = number of species in each tropic level.

At Malgas Island the total wet biomass is less (c. 10,000 g m^{-2}), and there are two major contributors to it: the rock lobsters (c. 3,800 g m^{-2}) and various species of algae (collectively c. 4,500 g m^{-2}, if only the understorey species are considered; the canopy of kelp would considerably increase this value). Whereas the flora is relatively rich, the fauna is impoverished, although there are several species of crustaceans that are associated with the algae. Most of the hard substratum is bare, but sponges, solitary ascidians (*Pyura stolonifera*), two species of whelks (*Argobuccinum pustulosum* and *Burnupena papyracea*) and occasional clumps of the ribbed mussel (*Aulacomya ater*) are present. The most notable absentee is the black mussel, *Choromytilus*.

Figures 3 and 4 summarise our ideas on the processes leading to the differences between these communities. Our first hypothesis was that the dense populations of rock lobsters at Malgas Island (Fig. 3) account for the paucity of most species of animals and that, in the absence of grazers and competitors for space, algae flourish (cf Lubchenco and Menge, 1978). This was tested in three ways. Firstly, settlement plates, some protected by cages to prevent predation by rock lobsters and some left unprotected, were installed at both Marcus and Malgas Islands, to monitor which species recruited to the islands, and how they fared, depending on whether they were exposed to predation by rock lobsters or not. Secondly, mussels of both species (*Choromytilus* and *Aulacomya*) were established on some of these plates and allowed time to attach firmly before they were introduced to the two islands. Again, some of these plates were protected by cages and some not. Installing these mussels allowed a test of whether they could survive predation by rock lobsters, but it also provided an opportunity to test if the presence of mussels on the plates influenced recruitment. Finally, the growth rates of those mussels that did survive were monitored, to provide a check on whether any other conditions (such as food availability) were sufficiently different between the islands to explain the sparsity of mussels at Malgas Island.

Almost immediately after the introduction of the plates, rock lobsters descended on the uncaged plates at Malgas and eliminated all the mussels. Throughout the experiment, all the uncaged plates at this island remained almost completely devoid of animal life, although they did support a thin veneer of very young barnacles, which were continually removed by the rock lobsters, and an increasing cover of algae, which eventually dominated these plates. On the caged plates at Malgas, and on both the uncaged and caged plates at Marcus, *Choromytilus* had a low rate of mortality. On the other hand, *Aulacomya* survived only inside the cages at both islands: outside the cages rock lobsters eliminated it at Malgas, whereas at Marcus it was smothered by the black mussels and other, more mobile, species which rapidly moved onto the plates. There seems little doubt that the rock lobsters are responsible for the absence of *Choromytilus* from Malgas.

Monitoring the growth rates of the two species of mussels on the caged plates showed that *Choromytilus* grew at almost identical rates at the two islands, and that *Aulacomya* grew almost 70% faster at Malgas than Marcus. Food availability cannot, therefore, be blamed for the absence of *Choromytilus* and the small numbers of *Aulacomya* at Malgas Island.

Recruitment to the settlement plates was monitored after 2, 4 and 6 months. After 2 months the communities that had developed were distinctly different on the two islands. A Brey-Curtis similarity analysis revealed that the plates fell into two groups, one for each island, and that further subdivision was on the basis of whether or not the plates were protected by cages. After 6 months dense settlements of juvenile mussels had accumulated in all the caged plates at Malgas and, associated with these mussels, an array of other species accumulated. Analysis showed that the recruited

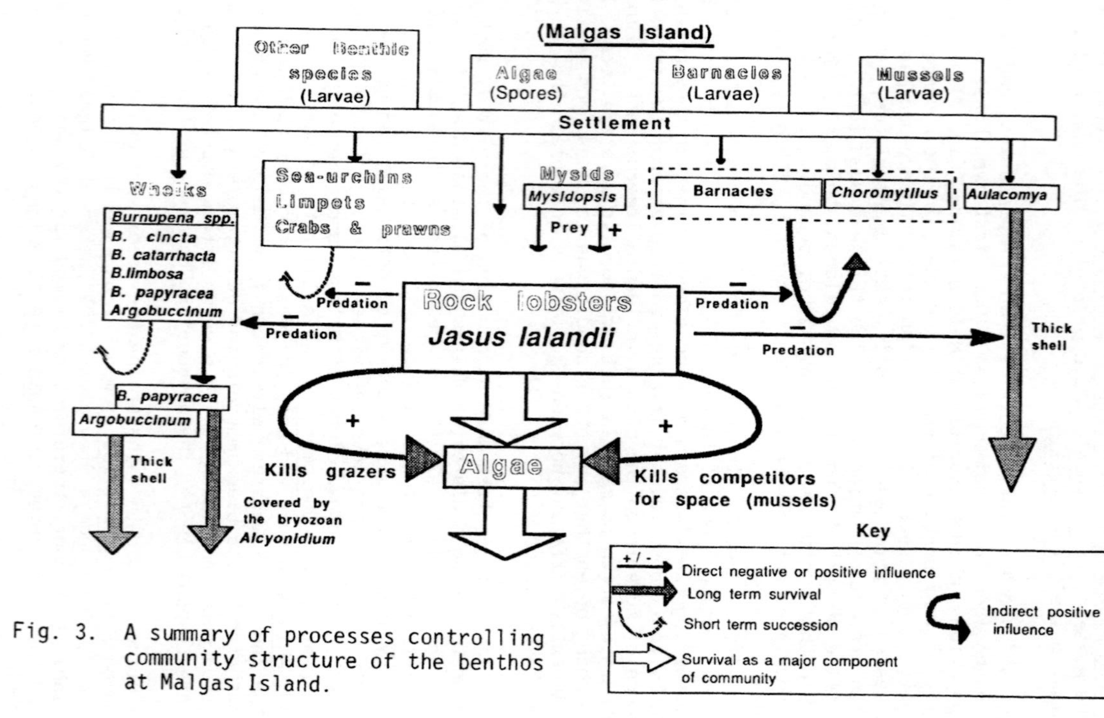

Fig. 3. A summary of processes controlling community structure of the benthos at Malgas Island.

communities on the caged Malgas plates then strongly resembled the caged mussel plates at Marcus; given protection from rock lobsters, the communities had converged. We believe that if it were not for the predatory activities of the rock lobsters, the whole community at Malgas would converge on that now present at Marcus.

Although rock lobsters feed on a wide variety of species, the ribbed mussel Aulacomya is the predominant constituent in its gut contents (Pollock, 1979), and attempts have been made to correlate the growth rate of rock lobsters in different regions with the standing stocks of the benthos (Newman and Pollock, 1974) and, more particularly, with the availability of mussels (Beyers and Goosen, 1987).

Analyses of the feeding behaviour of rock lobsters show that both species of mussel are immune to attack once they have reached a certain "critical size", and that this increases linearly with the size of the rock lobster (Pollock, 1979; Griffiths and Seiderer, 1980). From the rock lobsters' point of view, this places a certain proportion of the mussel population out of reach, and the smaller a rock lobster is, the larger this proportion. Griffiths and Seiderer (1980) calculated that rock lobsters with a carapace length less than 70 mm could only feed on 17% of the biomass of Aulacomya that was present in the area they were considering.

More importantly for the present analysis of the communities present at Marcus and Malgas Islands, Griffiths and Seiderer (1980) compared the relative vulnerabilities of the two species of mussels, Choromytilus and Aulacomya, to attack by rock lobsters. Choromytilus has a shell that is only half as strong as that of Aulacomya, and is more loosely attached to the substratum. Consequently the critical size is higher for Choromytilus than for Aulacomya. When offered the choice, rock lobsters prefer to feed on Choromytilus, selectively attack mussels that are well below the critical size, and concentrate their attack on that portion of the shell that is weakest. They therefore have fairly sophisticated responses when selecting their prey, and give the appearance of foraging optimally when given the opportunity, although we do not yet have enough information to assess if this is the case.

Two other pieces of information help to explain the relative distribution of Choromytilus and Aulacomya at Marcus and Malgas Islands. Firstly, Choromytilus is the faster-growing of the two species, growing two to four times as fast (Griffiths and King, 1979; Griffiths, 1981; Barkai and Branch, in prep.). Secondly, the young recruits of Choromytilus, although settling preferentially amongst other mussels, settle in blankets on exposed surfaces where they are readily accessible to predators. The recruits of Aulacomya appear to select (or survive in) cryptic habitats and, at Malgas Island, are found almost exclusively under boulders, in kelp holdfasts, or hidden beneath large adults of their own species.

Thus the (postulated) differences in settling behaviour, the greater vulnerability of Choromytilus to attack, the higher critical size possessed by Choromytilus, and the selective feeding of rock lobsters on Choromytilus rather than Aulacomya, all combine to explain why Choromytilus should be absent from Malgas Island, while Aulacomya does occur there although, admittedly, in relatively sparse clumps. The size distribution of Aulacomya at Malgas is distinctly bimodal. The bulk of their biomass consists of large individuals which are too large (or perhaps, too uneconomical) to be eaten by the rock lobster, and these are freely exposed on the upper surfaces of rocks. The second mode comprises very small animals that occur in protected habitats.

At Marcus Island, in the absence of rock lobsters, the black mussel

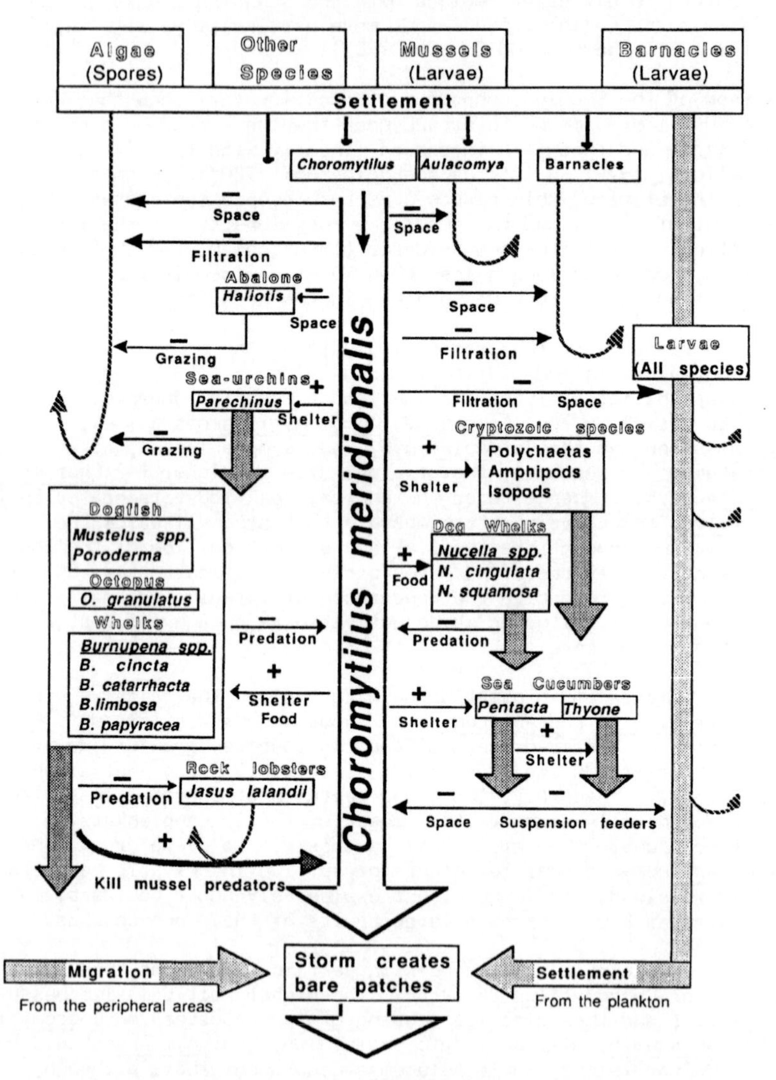

Fig. 4. Summary of the factors controlling community structure of the benthos at Marcus Island.

Choromytilus has established dense beds, but adults of Aulacomya are virtually absent. Recruits and juveniles are found there, but are soon smothered by the faster-growing Choromytilus, and the presence of their dead shells enmeshed in the byssal threads of Choromytilus testify to their fate. Evidently Aulacomya owes its existence at Malgas Island partly to the rock lobsters which prevent Choromytilus from monopolising the area as it does at Marcus. Comparable situations have been well documented in other parts of the world (e.g. Paine, 1974; Lubchenco, 1978; Peterson, 1979).

The relative distributions of the two mussels are therefore due to an interplay of their behaviour, physiology and morphology, and the ways in which these affect their interaction with each other and with rock lobsters.

All of the animal species that achieve a significant biomass at Malgas Island apparently have some defence mechanism preventing their elimination by the dense rock lobster population. The poriferans are probably toxic; the solitary ascidian Pyura stolonifera is encased in a thick, hard tunic; Aulacomya achieves a refuge in size and because it has a thick shell. Experiments have shown that rock lobsters cannot feed on the whelk Argobuccinum pustulosum because its thick shell and strengthened aperture-lip cannot be crushed by the rock lobsters (Barkai and McQuaid, in prep.). The only other macrofaunal species that is common at Malgas is Burnupena papyracea. It is the only representative of the genus at Malgas, although two additional species, B. limbosa and B. cincta occur at Marcus. Tests undertaken with all three species show that rock lobsters select the latter two species in preference to B. papyracea, consuming six to ten B. limbosa or B. cincta for every B. papyracea. In addition, the critical sizes for B. limbosa and B. cincta are much higher than for B. papyracea. Shell strength cannot explain these results: B. cincta has the thickest shell, and the other two have shells of roughly equal thickness. However, B. papyracea almost invariably has a coating of the bryozoan Alcyonidium nodosum on its shell, and this imparts protection. If the bryozoan is experimentally removed, B. papyracea loses its comparative immunity to the rock lobsters (Barkai, 1987; Barkai and McQuaid, in prep.). Interspecific behavioural responses therefore play a major role in this interplay between a predator, a potential prey species, and a symbiont.

Considering that almost all the species contributing significantly to the biomass at Malgas Island are protected from the rock lobsters, the question arises how rock lobsters not only survive there, but achieve a high fecundity and evidently have a high growth rate (Barkai, 1987; Barkai and Branch, in prep.). Examination of their gut contents reveals that newly settled barnacles are the major component for at least six months of the year. Considering the low nutritional value of barnacles, and that they have an average biomass of only 30 g m^{-2} at Malgas, they seem an unlikely source of nutrition. Measurements reveal, however, that the recruits turn over at a high rate, being eliminated by the rock lobsters soon after settlement, and rapidly replaced by a fresh settlement. Their productivity is actually more than enough to meet the energetic needs of the rock lobsters during the six to eight months of peak barnacle recruitment (spring through to autumn). During periods when barnacle recruitment is low (in winter) the rock lobsters feed extensively on planktonic mysids (another unexpected and unrecorded food source), although the mechanism they employ to obtain them is unknown. These findings are of great interest for several reasons. Firstly, they show that despite the selective preferences rock lobsters show for particular species of prey, and even particular sizes of prey (Griffiths and Seiderer, 1980), they can and will make use of unorthodox sources of food never previously even considered as prey. Secondly, prey that are perceived as being "available" in the environment may represent those that can survive predation, rather than those that are suitable prey. This is particularly true in situations like that at Malgas Island, where predation is intense. Thirdly, models of energy flow through ecosystems almost never take into account the considerable turnover that may take place in the form of rapidly consumed recruits: their biomass is small and often undetectable at the scale at which biomass is normally determined for such studies of energy flow.

Returning to Marcus Island, the development of thick mussel beds there seems a consequence of the near absence of rock lobsters. The key question therefore becomes: why are there virtually no rock lobsters there, despite the presence of abundant food in the form of mussels, urchins and whelks? Attempting to answer this question, Barkai (1987) maintained rock lobsters in cages at both islands. No food was made available to the rock lobsters but, despite this, they survived for at least nine months, feeding on organisms that settled on the plates at the bottom of the cages. In the process, they completely prevented any community from developing on the plates, whereas a rich array of species colonised plates in the control cages that lacked rock lobsters. These experiments confirmed that rock lobsters can control community structure by eliminating recruits shortly after they settle, and that such recruits provide sufficient food for the rock lobsters to survive for prolonged periods. They also show that water quality cannot explain their absence from Marcus Island.

In a further experiment, Barkai (1987) transplanted 1000 rock lobsters to Marcus Island from Malgas Island to see what effect they would have on the community. Unexpectedly, they were all killed within one hour of release, the major cause of death being three species of whelks (*Burnupena papyracea*, *B. cincta* and *B. limbosa*) which occur at high densities at Marcus (mean 270 m^{-2}). The whelks almost immediately climbed onto the rock lobsters, which attempted to escape by swimming away, knocking the whelks off in the process. Each time the rock lobsters landed again, they were, however, faced with a fresh assault. Ultimately they were laden down and consumed, the whelks rasping through the soft articulations on the ventral surface of the abdomen. Counts were made of the number of whelks on each rock lobster, and exceeded 300 in most cases. The result of this experiment was all the more unexpected because the whelks usually scavenge for their food and, more particularly, because the rock lobsters normally readily consume two of the three species. The experiment was repeated five times on a smaller scale to confirm the outcome and ensure it was not an artifact of introducing damaged or stressed rock lobsters. Predation of the whelks on the rock lobsters appears to be related to their high densities at Marcus Island: it is difficult to conceive their achieving this feat at lower densities. It is, however, possible that the protection *B. papyracea* derives from its epibiotic bryozoan contributes to this switch of trophic roles. Comparable reversals of competitive status have frequently been described (e.g. Sutherland, 1974; Wellington, 1980; Williams, 1981), but we believe this is the first recorded case of a reversal of a predator/prey relationship, brought about partly by the participation of a symbiont.

This is an important result, for it provides the probable reason why rock lobsters do not occur at Marcus Island. More than that, it suggests a mechanism for the maintenance of the contrasting communities at the two islands. While rock lobsters dominate community structure at Malgas Island and prevent the establishment of all forms of animal life except those resistant to predation, at Marcus Island the whelks occur at such high densities that they preclude rock lobsters from invading the habitat, which otherwise seems suitable for them. In the absence of rock lobsters, *Choromytilus* has developed dense beds and has a strong influence on the associated community. Whether these two communities can be regarded as alternate stable states depends largely on how one defines a "stable" community (see Sutherland, 1974; Paine et al., 1985). What is clear, is that behavioural responses between the species are an integral part of the processes maintaining the two communities, and that flexibility, changes and even reversals of behaviour can all be brought about by changes in the densities of the interacting species, or by the participation of other species.

GENERAL CONCLUSIONS

The interspecific behaviour of intertidal animals is determined partly by factors that are intrinsic to each individual, including its morphology, size, and inherited responses. The evolutionary history of each species may, therefore, place limitations on the kinds of behaviour it is capable of. Behaviour is also strongly influenced by extrinsic factors, including the physical environment and interactions with other species. In the intertidal zone specifically, there is a unique set of physical conditions that influence behaviour. Tidal cycles alternately expose and then submerge the intertidal zone, subjecting intertidal animals to stresses in the form of desiccation, extremes of temperature, changes in salinity, and periodic wave action. These stresses may be potentially lethal, but they are largely predictable on a day-to-day basis. Nevertheless they mould behavioural responses between species and, in many cases, behaviour appears to be constrained by the risk of exposure to adverse physical conditions during particular phases of the tide.

Behaviour is further shaped by interactions with other species. The nature of such interactions are varied, and include escape responses by prey, feeding behaviour of predators and herbivores, and interference competition. Because of the wide range of interactions each species experiences, and the number of species it will encounter, stereotyped interspecific behaviour is unlikely to be appropriate: instead, interspecific behaviour needs to be flexible. Changes in behaviour have been demonstrated in response to changes in body size, to physical conditions, to food availability, to the particular species encountered, and to the participation of additional species in an interaction.

Behaviour plays an important role in influencing population dynamics, and can also shape the nature of communities. At the same time, there are feed-back loops, for both population dynamics (e.g. density or age structure) and community structure (e.g. the presence of other species) also influence what particular behavioural response will be appropriate under any given set of circumstances. Because of this the outcome of behavioural interactions is often difficult to predict, and extrapolation from one situation to another is not easy. It is even more difficult to predict how behavioural responses will influence the population dynamics of the participants or the nature of the community.

Many of these principles are illustrated in this paper, partly by drawing on the literature, and partly by discussing a specific case history, in which two contrasting communities at adjacent islands are compared. The differences between these two communities appear to be dictated by a few key species. At one island predation by rock lobsters controls the nature and abundance of the fauna and flora, while in their absence at the other island, mussel beds proliferate and are associated with a rich fauna and an impoverished flora. The absence of rock lobsters at the second site is due to their consumption by whelks. Whelks are normally preyed upon by rock lobsters and the reversal of this predator/prey relationship is central to understanding the alternate states of these two communities.

REFERENCES

Bak, R.P.M., Termaat, R.M. and Dekker, R., 1982, Complexity of coral interactions: influences of time, location of interaction and epifauna, Mar. Biol., 69:215-222.

Barkai, A., 1987, Biologically induced alternative states in the benthic communities of hard substrata, Ph.D. thesis, University of Cape Town.

Barnes, J.R. and Gonor, J.J., 1973, The larval settling response of the lined chiton, Tonicella lineata, Mar. Biol., 20:259-264.

Bernstein, B.B. and Jung, N., 1979, Selective pressures and coevolution in a kelp canopy community in Southern California, Ecol. Monogr., 49:335-355.

Bernstein, B.B., Williams, B.E., and Mann, K.H., 1981, The role of behavioural responses to predators in modifying urchins (Strongylocentrotus droebachiensis) destructive grazing and seasonal foraging patterns, Mar. Biol., 63:38-49.

Bertness, M.D., 1981, Interference and exploitation competition in some tropical hermit crabs, J. exp. mar. Biol. Ecol., 49:189-202.

Bertness, M.D., Garrity, S.D. and Levings, S.C., 1981, Predation pressure and gastropod foraging: a tropical-temperate comparison, Evolution, 35:995-1007.

Beyers, C.J. de B., and Goosen, P.C., 1987, Variations in the fecundity and size at sexual maturity of female rock lobster Jasus lalandii in the southern Benguela ecosystem, S. Afr. J. Mar. Sci., (in press).

Black, R., 1979, Competition between intertidal limpets: an intrusive niche on a steep resource gradient, J. anim. Ecol., 48:401-411.

Blankley, W.O., and Branch, G.M., 1984, Co-operative prey capture and unusual brooding habits of Anasterias rupicola (Verrill) (Asteroidea) at sub-Antarctic Marion Island, Mar. Ecol. Prog. Ser., 20:171-176.

Bosman, A.L., Du Toit, J.T., Hockey, P.A.R., and Branch, G.M., 1986, A field experiment demonstrating the influence of seabird guano on intertidal primary production, Est. Coastal shelf Sci., 23:283-294.

Bosman, A.L., and Hockey, P.A.R., 1986, Seabird guano as a determinant of rocky intertidal community structure, Mar. Ecol. Prog. Ser., 32:247-257.

Branch, G.M., and Branch, M.L., 1980, Competition between Cellana tramoserica (Gastropoda) and Patiriella exigua (Asteroidea) and their influence on algal standing stocks, J. exp. mar. Biol. Ecol., 48:35-49.

Branch, G.M., 1975, Mechanisms reducing intraspecific competition in Patella species: migration, differentiation and territorial behaviour, J. anim. Ecol., 44:575-600.

Branch, G.M., 1976, Interspecific competition experienced by South African Patella species, J. anim. Ecol., 45:507-529.

Branch, G.M., 1981, The biology of limpets: physical factors, energy flow and ecological interactions, Oceanogr. Mar. Biol. Ann. Rev., 19:235-380.

Branch, G.M., 1984, Competition between marine organisms: ecological and evolutionary implications, Oceanogr. Mar. Biol. Ann. Rev., 22:429-593.

Branch, G.M., 1985, Competition; its role in ecology and evolution in intertidal communities, in: "Species and Speciation", E.S. Vrba, ed., Transvaal Museum Monograph No.4, Transvaal Museum, Pretoria.

Bullock, T.H., 1953, Predator recognition and escape responses of some intertidal gastropods in presence of starfish, Behaviour, 5:130-140.

Buss, L.W., 1981, Group living, competition, and the evolution of cooperation in a sessile invertebrate, Science, N.Y., 213:1012-1014.

Carpenter, R.C., 1984, Predator and population density control of homing behaviour in the Caribbean echinoid Diadema antillarum, Mar. Biol., 82:1/1-108.

Chelazzi, G., Focardi, S., and Deneubourg, J.L., 1983a, A comparative study of the movement patterns of two sympatric tropical chitons (Mollusca: Polyplacophora), Mar. Biol., 74:115-125.

Chelazzi, G., Focardi, S., Deneubourg, J.L., and Innocenti, R., 1983b, Competition for the home and aggressive behaviour in the chiton Acanthopleura gemmata (Blainville) (Mollusca : Polyplacophora), Behav. Ecol. Sociobiol., 14:15-20.

Clark, W., 1958, Escape responses of herbivorous gastropods when stimulated by carnivorous gastropods, Nature (Lond), 181:137-138.

Coen, L.D., Heck, K.L., and Abele, L.G. 1981, Experiments on competition and predation among shrimps of seagrass meadows, Ecology, 62:1484-1493.

Creese, R.G., and Underwood, A.J., 1982, Analysis of inter- and intra-specific competition among intertidal limpets with different methods of feeding, Oecologia (Berl.)., 53:337-346.

Crisp, D.J., and Williams, G.B., 1960, Effect of extracts from fucoids in promoting settlement of epiphytic Polyzoa, Nature (Lond.), 188:1206-1207.

De Silva, P.H.D.H., 1962, Experiments on choice of substrate by Spirorbis larvae (Serpulidae), J. exp. Biol., 39:483-490.

Dethier, M.N., and Duggins, D.O., 1984, An "indirect commensalism" between marine herbivores and the importance of competitive hierarchies, Am. Nat., 124:205-219.

Dingle, H., and Caldwell, R.L., 1975, Distribution, abundance and interspecific agonistic behaviour of two mudflat stomatopods, Oecologia (Berl.), 20:167-178.

Duggins, D.O., 1981, Interspecific facilitation in a guild of benthic marine herbivores, Oecologia(Berl.), 48:157-163.

Dunkin, S. de B., and Hughes, R.N., 1984, Behavioural components of prey-selection by dogwhelks, Nucella lapillus (L), feeding on barnacles, Semibalanus balanoides (L), in the laboratory, J. exp. mar. Biol. Ecol., 79:91-103.

Elner, R.W., and Hughes, R.N., 1978, Energy maximisation in the diet of the shore crab Carcinus maenas, J. anim. Ecol., 47:103-116.

Elner, R.W., and Raffaelli, D.G., 1980, Interactions between two marine snails, Littorina rudis Maton and Littorina nigrolineata Gray, a predator, Carcinus maenas (L), and a parasite, Microphallus similis Jägerskiold, J. exp. mar. Biol. Ecol., 43:151-160.

Garrity, S.D., and Levings, S.C., 1983, Homing to scars as a defence against predators in the pulmonate limpet Siphonaria gigas, Mar. Biol., 72:319-324.

Glynn, P.W., 1980, Defence by symbiotic Crustacea of host corals elicited by chemical cues from predator, Oecologia (Berl.), 47:287-290.

Gray, J.S., 1967, Substrate selection by the archiannelid Protodrilus rubropharyngeus, Helgol. wiss. Meeresunters, 15:253-269.

Griffiths, C.L. and King, J.A., 1979, Energetic costs of growth and gonad output in the ribbed mussel Aulacomya ater, Mar. Biol., 53:217-222.

Griffiths, C.L., and Seiderer, J.L., 1980, Rock-lobsters and mussels - limitations and preferences in a predator-prey interaction, J. exp. mar. Biol. Ecol., 44:95-109.

Griffiths, R.J., 1981, Population dynamics and growth of the bivalve Choromytilus meridionalis (Kr) at different tide levels, Est. Coastal Shelf Sci., 12:101-118.

Grosberg, R., 1981, Competitive ability influences habitat choice in marine invertebrates, Nature (Lond.), 290:700-702.

Hadfield, M.G., and Karlson, R.H., 1969, Externally induced metamorphosis in a marine gastropod, Am. Zool., 9:1122.

Harger, J.R.E., 1968, The role of behavioural traits in influencing the distribution of two species of sea mussel, Mytilus edulis and Mytilus californianus, Verliger, 11:45-49.

Harger, J.R.E., 1970, The effect of wave impact on some aspects of the biology of sea mussels, Veliger, 12:401-414.

Harvell, C.D., 1984, Predator-induced defense in a marine bryozoan, Science, N.Y., 224:1357-1359.

Hay, M.E., Lee, R.R., Guieb, R.A., and Bennett, N.M., 1986, Food preference and chemotaxis in the sea urchin Arbacia puntulata (Lamarck) Philippi, J. exp. mar. Biol. Ecol., 96:147-153.

Highsmith, R.C., 1982, Induced settlement and metamorphosis of sand dollar (Dendraster excentricus) larvae in predator-free sites: adult sand dollar beds, Ecology, 63:329-337.

Hixon, M.A., 1980, Competitive interactions between California reef fishes of the genus Embiotica, Ecology, 61:918-931.

Hockey, P.A.R., and Branch, G.M., 1984, Oystercatchers and limpets: impact and implications. A preliminary assessment, Ardea, 72:199-206.

Jackson, J.B.C., 1977, Competition on marine hard substrata: the adaptive significance of solitary and colonial strategies, Am. Nat., 111:743-767.

Jackson, J.B.C., and Buss, L., 1975, Allelopathy and spatial competition among coral reef invertebrates, Proc. Nat. Acad. Sci. USA, 72:5160-5163.

Johnson, C.R., and Mann, K.H., 1986, The crustose coralline alga, Phymatolithon Foslie, inhibits the overgrowth of seaweeds without relying on herbivores, J. exp. mar. Biol. Ecol., 96:127-146.

Kent, B.W., 1983, Patterns of coexistence in whelks, J. exp. mar. Biol. Ecol., 66:257-283.

Lassig, B.R., 1977, Communication and coexistence in a coral community, Mar. Biol., 42:85-92.

Levings, S.C., and Garrity, S.D., 1983, Diel and tidal movement of two co-occurring neritid snails: differences in grazing patterns on a tropical rocky shore, J. exp. mar. Biol. Ecol., 67:261-278.

Lubchenco, J., 1978, Plant species diversity in a marine intertidal community: importance of herbivore food preferences and algal competitive abilities, Am. Nat., 112:23-39.

Lubchenco, J., and Menge, B.A., 1978, Community development and persistence in a low rocky intertidal zone, Ecol. Monogr., 48:67-93.

Mapstone, B.D., Underwood, A.J., and Creese, R.G., 1984, Experimental analyses of the commensal relation between intertidal gastropods Patelloida mufria and the trochid Austrocochlea constricta, Mar. Ecol. Prog. Ser., 17:85-100.

Margolin, A.S., 1964, The mantle response of Diodora aspersa, Anim. Behav., 12:187-194.

McClintock, J.B., 1985, Avoidance behaviour and escape responses of the sub-Antarctic limpet Nacella edgari (Powell) (Mollusca : Gastropoda) to the sea star Anasterias perrieri (Smith) (Echinodermata : Asteroidea), Polar Biol., 4:95-98.

Menge, B.A., 1978a, Predator intensity in a rocky intertidal community. Relation between predator foraging activity and environmental harshness, Oecologia (Berl.), 34:1-16.

Menge, B.A., 1978b, Predator intensity in a rocky intertidal community. Effect of an algal canopy, wave action and desiccation on predator feeding rates, Oecologia (Berl.), 34:17-35.

Menge, B.A., and Menge, J.L., 1974, Prey selection and foraging period of the predaceous rocky intertidal snail Acanthina punctulata, Oecologia (Berl.), 17:293-316.

Morse, D.E., Hooker, N., Duncan, H., and Jensen, L., 1979, γ-aminobutyric acid, a neurotransmitter, induces planktonic abalone larvae to settle and begin metamorphosis, Science (N.Y.), 204:407-410.

Newman, G.G., and Pollock, D.E., 1974, Growth of the rock lobster Jasus lalandii and its relationship to benthos, Mar. Biol., 24:339-346.

Nishihira, M., 1967, Observations on the selection of algal substrata by hydrozoan larvae, Sertularella miurensis in nature, Bull. Mar. Biol. Stn Asamushi, 13:34-48.

Ortega, S., 1985, Competitive interactions among tropical intertidal limpets, J. exp. mar. Biol. Ecol., 90:11-25.

Ortega, S., 1986, Fish predation on gastropods on the Pacific coast of Costa Rica, J. exp. mar. Biol. Ecol., 97:181-191.
Osman, R.W., and Haugsness, J.A., 1981, Mutualism among sessile invertebrates: a mediator of competition and predation, Science, N.Y., 211: 846-848.
Paine, R.T., 1966, Food web complexity and species diversity, Am. Nat., 100:65-75.
Paine, R.T., 1974, Intertidal community structure. Experimental studies on the relationship between a dominant competitor and its principle predator, Oecologia (Berl.), 15:93-120.
Paine, R.T., Castilla, J.C., and Cancino, J., 1985, Perturbation and recovery patterns of starfish-dominated intertidal assemblages in Chile, New Zealand, and Washington State, Am. Nat., 125:679-691.
Paine, R.T., and Suchanek, T.H., 1983, Convergence of ecological processes between independently evolved competitive dominants: a tunicate-mussel comparison, Evolution, 37:821-831.
Palmer, A.R., 1982, Predation and parallel evolution: recurrent parietal plate reduction in balanomorph barnacles, Paleobiology, 8:31-44.
Palmer, A.R., 1983, Growth rate as a measure of food value in thaidid gastropods: assumptions and implications for prey morphology and distribution, J. exp. mar. Biol. Ecol., 73:95-124.
Palmer, A.R., 1984, Prey selection in thaidid gastropods: some observational and experimental field tests of foraging models, Oecologia (Berl.), 62:162-172.
Palmer, A.R., Szymanska, J., and Thomas, L., 1982, Prolonged withdrawal: a possible predator evasion behaviour in Balanus glandula (Crustacea: Cirripedia), Mar. Biol. 67:51-55.
Peterson, C.H., 1979, The importance of predation and competition in organising the intertidal epifaunal communities of Barnegat Inlet, New Jersey, Oecologia (Berl.), 39:1-24.
Phillips, D.W., 1975, Distance chemoreception triggered avoidance behaviour of the limpets Acmaea (Collisella) limatula and Acmaea (Notoacmea) scutum to the predatory starfish Pisaster ochraceus, J. exp. Zool., 191:199-209.
Phillips, D.W., 1976, The effect of species-specific avoidance response to predatory starfish on the intertidal distribution of two gastropods, Oecologia (Berl.), 23:83-94.
Pollock, D.E., 1979, Predator-prey relationships between the rock lobster Jasus lalandii and the mussel Aulacomya ater at Robben Island on the Cape west coast of Africa, Mar. Biol., 52:347-356.
Ryland, J.S., 1959, Experiments on the selection of algal substrates by polyzoan larvae, J. exp. Biol., 36:613-631.
Scheltema, R.S., 1974, Biological interactions determining larval settlement of marine invertebrates, Thalass. Jugoslav., 10:263-298.
Sheppard, C.R.C., 1982, Coral populations on reef slopes and their major controls, Mar. Ecol. Prog. Ser., 7:83-115.
Steneck, R.S., 1982, A limpet-coralline algal association: adaptations and defenses between a selective herbivore and its prey, Ecology, 63:507-522.
Steneck, R.S., and Watling, L., 1982, Feeding capabilities and limitation of herbivorous molluscs: a functional group approach, Mar. Biol., 68:299-319.
Stimson, J., 1970, Territorial behaviour of the owl limpet, Lottia gigantea, Ecology, 51:113-118.
Stimson, J., 1973, The role of the territory in the ecology of the intertidal limpet, Lottia gigantea (Gray), Ecology, 54:1020-1030.
Sutherland, J.P., 1974, Multiple stable points in natural communities, Am. Nat., 108:859-873.
Thayer, C.W., 1985, Brachiopods versus mussels: competition, predation, and palatability, Science, N.Y., 228:1527-1528.

Thompson, T.E., 1958, The natural history, embryology, larval biology and post-larval development of Adalaria proxima (Alder and Hancock) (Gastropoda, Opisthobranchia), Phil. Trans. roy. Soc. Lond. Ser. B, Biol. Sci., 242:1-58.

Thompson, T.E., 1962, Studies on the ontogeny of Tritonia hombergi Cuvier (Gastropoda, Opisthobranchia), Phil. Trans. roy. Soc. Lond. Ser. B. Biol. Sci., 245:171-218.

Thorp, J.H., 1976, Interference competition as a mechanism of coexistence between two sympatric species of grass shrimp Palaemonetes (Decapoda: Palaemonidae), J. exp. mar. Biol. Ecol., 25:19-35.

Underwood, A.J., 1978, An experimental evaluation of competition between three species of intertidal prosobranch gastropods, Oecologia (Berl.), 33:185-202.

Vadas, R.L., Elner, R.W., Garwood, P.E., and Babb, I.G., 1986, Experimental evaluation of aggregation behaviour in the sea urchin Strongylocentrotus droebachiensis. A reinterpretation, Mar. Biol., 90:433-448.

Vermeij, G.J., 1978, "Biogeography and Adaptation Patterns of Marine Life", Harvard University press, Cambridge Massachusetts, 332pp.

Weinberg, J.R., 1979, Ecological determinants of spionid distributions within dense patches of deposit-feeding polychaetes Axiothella rubrocincta, Mar. Ecol. Prog. Ser., 13:301-314.

Wellington, W., 1980, Reversal of digestive interactions between Pacific reef corals: mediation by sweeper tentacles, Oecologia (Berl.), 47: 340-343.

Wells, R.A., 1980, Activity pattern as a mechanism of predator avoidance in two species of acmaeid limpet, J. exp. mar. Biol. Ecol., 48:151-168.

Williams, A.H., 1981, An analysis of competitive interactions in a patchy back-reef environment, Ecology, 62:1107-1120.

Williams, G.B., 1964, The effect of extracts of Fucus serratus in promoting the settlement of Spirorbis borealis (Polychaeta), J. mar. biol. Assoc. U.K., 44:397-414.

Woodin, S.A., 1985, Effects of defecation by arenicolid polychaete adults on spionid polychaete juveniles in field experiments: selective settlement or differential mortality? J. exp. mar. Biol. Ecol., 87:119-132.

Wright, W.G., 1977, Avoidance and escape: two responses of intertidal limpets to the presence of the Territorial Owl Limpet Lottia gigantea, West. Soc. Naturalists, 1977:50.

Wright, W.G., 1982, Ritualised behaviour in a territorial limpet, J. exp. mar. Biol. Ecol., 60:245-251.

Wright, W.G., 1985, "The Behavioural Ecology of the Limpet Lottia gigantea: Interaction between Territoriality, Demography and Protandric Hermaphroditism", Ph.D. thesis. University of California, San Diego, 223pp.

Young, C.M., Greenwood, P.G., and Powell, C.J., 1986, The ecological role of defensive secretions in the intertidal pulmonate Onchidella borealis, Biol. Bull., 171:391-404.

Influence of the Presence of Congeneric Species on the Behavioural Presences of *Hydrobia* Species

Richard S. K. Barnes

University of Cambridge, U.K.

INTRODUCTION AND REVIEW

"The small prosobranchs of the genus Hydrobia ... belong undoubtedly to the quantitatively most important species of the invertebrate macrofauna on flats in estuaries and shallow lagoons everywhere along the coasts of Northwestern Europe." (Muus. 1967). To this can be added that they are equally important members of the macrofauna of fully marine, but sheltered, localities, and around the margins of the Arctic Ocean and Mediterranean/Black/Caspian Sea systems. Three species occur in northwest Europe - H. ulvae (Pennant), which is by far the commonest, most tolerant and most widespread of the three, H. ventrosa (Montagu), and H. neglecta Muus, the rarest and most specialized in its digestive repertoire (Fenchel, 1975a; Hylleberg, 1976; Lassen & Hylleberg, 1978; Cherrill & James, 1985).

The majority of the localities which have been investigated support only one of the three species, and to some extent this can be interpreted as being consequent on different habitat preferences and environmental tolerance ranges (Muus, 1967; and see Lassen & Hylleberg, 1978). Least controversial here is their differential occupation of tidal and non-tidal habitats. H. ulvae is the only species widely to occur in the marine intertidal zone. It is also the only one to remain active when uncovered by water and to climb above the water level in laboratory aquaria (pers. obs; and see Barnes, 1981a, b). In contrast, H. ventrosa and H. neglecta are largely restricted to non-tidal coastal lagoons and equivalent habitats, although H. ulvae occurs in these as well as on the shore (Muus, 1967; Bishop, 1976; etc.) The position with respect to the other major variable generally considered to affect their distribution patterns, salinity (see Muus, 1967; Lassen & Hylleberg, 1978; Hylleberg, 1986), is more problematic. Work on Danish populations (Muus, 1967; Fenchel, 1975a; Hylleberg, 1975; 1986) has suggested that the three species tend to occur in different segments of the ambient salinity gradient; with H. ventrosa being found in the least saline regions (< 20°/oo), and H. ulvae in the most fully marine (> 10°/oo). This accords with the experimentally determined salinity optima of Danish material: H. ventrosa 10-20°/oo; H. neglecta 25°/oo; and H. ulvae 25-30°/oo. In East Anglia, U.K., however, all three inhabit full strength sea water (35°/oo), and H. ulvae occurs in salinities as low as those normally tolerated by H. ventrosa (Cherrill & James, 1985; Barnes, 1987a; and see McMillan, 1948). Barnes (1987a) was unable to find any statistically significant differences between the salinities of the East Anglian lagoons in which the various species occurred. Even in Denmark, the apparent salinity boundaries demarkating their ranges vary from site to site (Muus, 1967), and all three show peak reproductive activity at the same salinity (c. 20°/oo) (Lassen & Clark, 1979) (see also Fish & Fish, 1977; 1981). What does appear certain is that there are many lagoons in which any of the three could survive successfully: the main problem here may well be gaining access to these often largely landlocked habitats (Fenchel, 1975a; Barnes, 1987b).

In some regions, co-occurrence of two or all three species has been reported. Thus of the 76 East Anglian sites containing *Hydrobia* surveyed by Cherrill & James (1985), 10 possessed two species and a further one all three; considering the coastal lagoons of East Anglia, 25% of the localities support more than one *Hydrobia* species (data in Cherrill & James, 1985, and Barnes, 1987a). Equivalently, 45% of the 85 stations in the Danish Limfjord sampled by Fenchel (1975a) contained at least two species; and extensive co-existence has been reported from the Kattegat and Little-Belt-Sea shores of Denmark (Hylleberg, 1986). In both East Anglia and the Limfjord, the joint presence of *H. ulvae* and *H. neglecta* was the least frequently observed combination.

Field observations and laboratory experiments (Fenchel, 1975b; Fenchel & Kofoed, 1976; Cherrill & James, 1987b) indicate that where more than one species do occur together in the same habitat, they compete and evidence for three potential effects of such interspecific competition exists. In stable mixed populations, Fenchel (1975b) and Fenchel & Kofoed (1976) suggested that character displacement with regard to body size had occurred, thereby permitting *H. ulvae* and *H. ventrosa* to co-exist by utilizing different parts of a common resource spectrum (probably food). Character displacement, however, is notoriously easier to postulate than it is to demonstrate conclusively (Connell, 1980; Levinton, 1982); Hylleberg (1986) failed to find any evidence for the phenomenon in other mixed Danish populations, and Cherrill & James (1987a) found no evidence for it in East Anglia. Secondly, many mixed populations appear far from stable. At Broad Water, Norfolk, UK, for example, what in 1982/83 was an abundant *H. ventrosa* population with a few individuals of *H. ulvae* (Cherrill & James, 1985) is now a locality for numerous *H. ulvae* and relatively rare individuals of *H. ventrosa* (pers. obs.). Fenchel (1975a) concluded that the distributions of the *Hydrobia* species in the Limfjord were heavily influenced by dispersal and colonization of new localities, and by interspecific competition between the original occupants and later arrivals, with, under most circumstances, the species most likely to displace another being *H. ulvae* (and see Christiansen & Fenchel, 1977). Broad Water perhaps provides an example of such competitive displacement (but see Cherrill & James, 1987b). Competitive displacement is also known in other mud snails, eg in interactions between *Cerithidea californica* and *Ilyanassa obsoleta* (Race, 1982). Thirdly, mixed populations do not in fact occur in a number of shared habitats. In most of the East Anglian lagoons with more than one species, for example, the different species are either found in different regions of the shared lagoon or occupy different microhabitats (pers. obs.). Behavioural selection of different microhabitats within a shared system is therefore another potential mechanism for minimizing the effect of interspecific competition, and this amongst other aspects of their behavioural ecology is currently under investigation in the Department of Zoology at Cambridge. This paper addresses the question of whether the behavioural habitat preferences of one species are influenced by the presence of another.

MATERIALS AND METHODS

Material of H. ulvae was obtained from the sandy bed of Spiral Marsh Creek, Scolt Head Island, Norfolk (British Grid ref. TF804467); that of H. ventrosa from the surface of mud and from submerged Ruppia cirrhosa in 'Lagoon 0', Bawdsey, Suffolk (TM359409); and that of H. neglecta from filamentous green algae floating on the surface or attached around the margins of 'Lagoon 6', Shingle Street, Suffolk (TM373437) (see Barnes, 1987a for details of the two lagoonal sites). At each locality, the species named was the only Hydrobia present.

The experimental procedures used were essentially those of Barnes & Greenwood (1978) and Barnes (1979) in their study of the sediment preferences of H. ulvae, and here only major departures from those procedures will be specified. Each species was first separately offered a choice between two natural sediment types: a mud obtained from the sides of a salt-marsh creek on Scolt Head Island, and a sand from the bed of a similar creek on the same island. A total of 150 snails was used in each replicate and they had available to them two mud quadrants and two sand quadrants, all overlain by some 5mm of local sea water (31-33°/oo). The number of Hydrobia on the two sediment types was then counted after 24 hours. (150 snails were chosen as the experimental number so that if all individuals moved into the one sediment type, the density there - 20,000m^{-2} - would lie well within the range of densities displayed in the field.

In a second set of experiments, the three possible combinations of different pairings of species were investigated by placing 150 specimens of each of two species in the choice chambers at the same time (40,000m^{-2} is still within the density range of all three species), and again counting the individuals of the different species in the two sediment types after a period of 24 or 48 hours.

After a pilot run in July 1986, two series of these two experiments were conducted in September 1986, in each case using freshly collected Hydrobia and sediments. In the first series, the two experiments were carried out sequentially, whilst in the second series, begun after an interval of 10 days, they were run concurrently.

RESULTS

The reactions of the three species varied with the precise characteristics of the 'mud' and 'sand' used in the two experimental series and hence the two are treated separately below.

(i) The First Experimental Series

When tested separately, all three species clearly preferred the mud, which was visibly rich in motile algae, to the sand (Fig. 1A) (individual $\chi^2 > 25$; $p < 0.001$).

When tested in pairs, however, some significant departures from these preferences were displayed (Fig. 1B), especially in respect of H. neglecta. Regardless of the species co-occurring with them, both H. ulvae and H. ventrosa maintained a significant preference for the mud (ulvae $\chi^2 200$; $p < 0.001$. ventrosa $\chi^2 > 70$; $p < 0.001$). In the presence of either H. ulvae or H. ventrosa, however, the proportion of H. neglecta on the sand increased significantly ($\chi^2 > 60$; $p < 0.001$), significantly changing its apparent preference ($\chi^2 > 250$; $p < 0.001$).

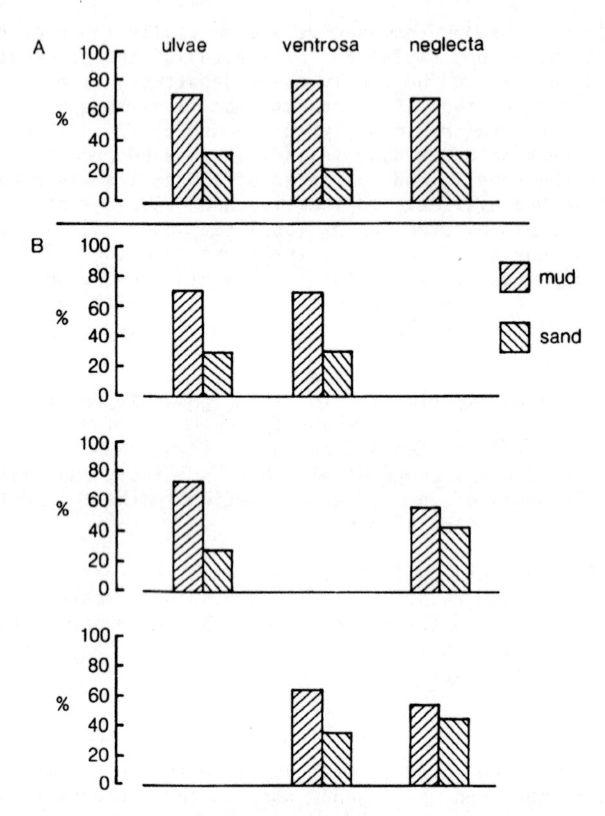

Fig. 1 The first series of experiments: the proportions of the three Hydrobia species in the two test sediments; (A) when tested separately (N = 550 or 600 for each species), and (B) in the three possible pairings (N = 1050 or 1200 for each species).

(ii) The Second Experimental Series

The mud used in this series lacked the visibly obvious motile algae, and although *H. ulvae* again displayed a marked preference for the mud (χ^2 18; $p < 0.001$), the other two species did not discriminate between the sediments provided ($\chi^2 < 1$; $p > 0.25$) (Fig.2A).

When tested in pairs, however, although *H. neglecta* did not change its apparent indifference ($\chi^2 < 3.7$; $p > 0.05$), both *H. ulvae* and *H. ventrosa* did exhibit significant alterations of their preferences (Fig. 2B). When tested against either of the other two, *H. ulvae* increased its disproportional occupation of the mud ($\chi^2 > 30$; $p < 0.001$) and *H. ventrosa* likewise displayed a consistent alteration to its distribution ($\chi^2 > 25$; $p < 0.001$) now showing a preference for the mud when in company with either *H. ulvae* or *H. neglecta* ($\chi^2 > 30$; $p < 0.001$).

(iii) Common Features of the Two Series

a) In the presence of either *H. ventrosa* or *H. neglecta*, *H. ulvae* always maintained, and in some cases increased, its preference for the mud.

b) In the presence of *H. ulvae* or *H. ventrosa*, *H. neglecta* displayed a relative increase in its utilization of the sand, either directly (as in the first series), when the other two maintained their preference for mud, or (as in the second series), via an increased preference for the mud on the part of *H. ulvae* or *H. ventrosa*. In either event, *H. neglecta* always formed the majority of the individual snails on the sand, whilst the other species were the more numerous on the mud.

c) In the presence of *H. ulvae* or *H. neglecta*, *H. ventrosa* either maintained (albeit in some cases with a significant reduction) or changed to a preference for the mud.

DISCUSSION

It should be emphasised strongly that these experiments were in no way designed to investigate the sediment preferences of the three *Hydrobia* species. The precise sediments used were entirely arbitrary: all that was required by the experimental rationale was that each species, when alone, should display an identifiable reaction to the test sediments that could conceivably be modified in the presence of another species of the same genus. The preferences or reactions displayed cannot therefore be taken to indicate any general behavioural preference for sediments of any particular type, and indeed it is entirely possible that different reactions would be elicited when offered other sediments describable as 'mud' or 'sand'.

If, however, a given *Hydrobia* species is uninfluenced by the presence of another (the null hypothesis), the logic of the experimental design is such that either (i) the proportion of that species in a given test sediment should not vary dependent on the presence or absence of another *Hydrobia* species, or (ii) any variation from (i) above should result solely from changed total *Hydrobia* density, in which case the two tested species should vary together to the same extent. Any departure from these two predictions must indicate some form of interspecific effect. Further, the extremely short time-span of the experiments renders it improbable that exploitative competition could be the cause of any departure observed; rather, direct modifications of behaviour elicited by the presence of the congener (or behavioural 'interference competition' which amounts to the same thing) can be implicated.

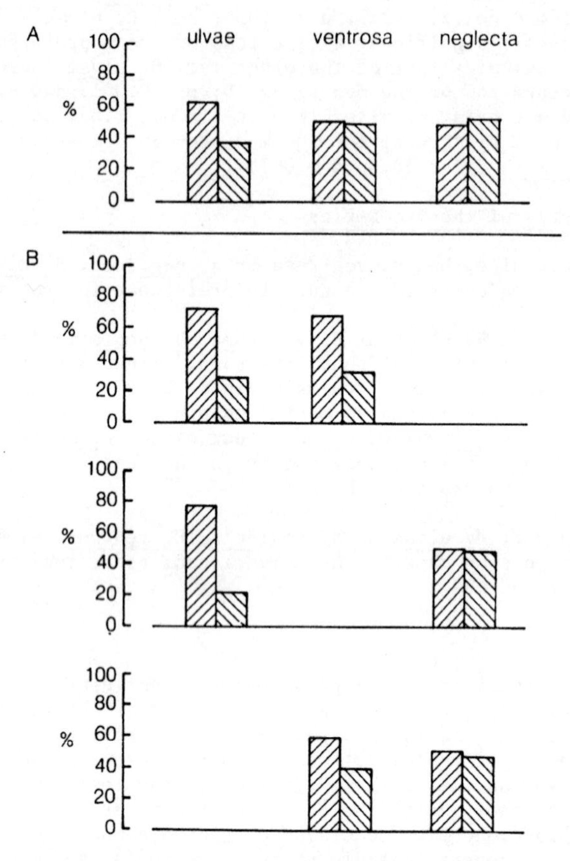

Fig. 2 The second series of experiments: the proportions of the three Hydrobia species in the two test sediments (see Fig 1); (A) when tested separately (N = 300 for each species), and (B) in the three possible pairings (N = 900 for each species).

On this basis, with the exception of the joint presence of H. ulvae and H. ventrosa, pairings of the various species resulted in significant changes to the apparent behavioural preferences of all three species, and these changes were such as to result in any two tested species tending to predominate on different substrata. Modification of habitat preference in the presence of congeners did, therefore, occur in these experiments, and hence presumably could equally occur in the field when two species become sympatric. Should the shared habitat possess a variety of (in this case) substrata, then co-existence would clearly be facilitated by such changes in behavioural preferences. Several field studies have indicated that different Hydrobia species may occupy different regions within any shared site (see the Introduction to this paper and Fenchel 1975a; Lassen & Hylleberg, 1978; etc.). Nevertheless, the degree of co-existence permitted by this mechanism may well be rather limited, especially, of course, in relatively uniform habitats. H. ulvae, for example, was the species least affected by the presence of others in this study, and it is also the most widespread, most tolerant, and, it has been argued (Fenchel, 1975b; Fenchel & Kofoed, 1976; Lassen, 1979) under most environmental conditions the competitively superior species. The preference for mud described here does appear to be a general phenomenon (Barnes & Greenwood, 1978), but it also thrives on sand, both intertidally (Barnes, 1979) and in lagoons (Barnes, 1987a). Except in the most dilute of brackish waters, no modification of habitat preferences would appear likely to remove a congener from the range of habitats and microhabitats which can be colonized successfully by H. ulvae. It may purely be its presumed poor abilities of lagoonal colonization (Cherrill & James, 1987b; Barnes, 1987b) which prevent H. ulvae from occupying more individual lagoonal sites than it does. Since, in these experiments, H. neglecta did not show lesser behavioural displacement in the presence of H. ulvae than in the presence of H. ventrosa, they do not shed any light on why H. ulvae + H. neglecta should be a rare combination, and H. ventrosa + H. neglecta a relatively common one.

In contrast to the other pairings, in the presence of H. ulvae, H. ventrosa diverged not away from the sediment type preferred by its congener but towards it (Fig. 2). This is scarcely in accord with Fenchel's (1975b) finding that at high salinities the competitive advantage lies with H. ulvae, although it may be significant that Cherrill & James (1987b) have shown that in a mixed population of these two species in East Anglia the intensity of intraspecific competition exceeds that interspecifically, rendering competitive exclusion unlikely. The somewhat unexpected behaviour of H. ventrosa might find an explanation in the activities of H. ulvae somehow making some specific source of sedimentary food more available to H. ventrosa, but at the moment all here is purely conjectural.

SUMMARY

The behavioural preferences displayed experimentally by Hydrobia ulvae, H. ventrosa and H. neglecta (Mollusca: Gastropoda) when separately offerred a choice between two sediments are modified, sometimes highly, in the presence of a congener. This divergence in apparent habitat preferences may facilitate co-existence of Hydrobia species in at least some habitat types. Under the experimental conditions, the preferences of H. ulvae were modified least, and those of H. neglecta most.

ACKNOWLEDGEMENTS

The author is deeply grateful to the Nature Conservancy Council for permission to work and stay in the Scolt Head NNR, and to Mr A.G.C. Barnes, Mr D.K.A.Barnes, Mr R.A.Cadwalladr and Ms D.D.Kohn for assistance in collecting the Hydrobia and in counting out the experimental animals.

REFERENCES

Barnes, R.S.K., 1979, Intrapopulation variation in Hydrobia sediment preferences. Est. coast. mar. Sci., 9, 231-234.

Barnes, R.S.K., 1981a, An experimental study of the pattern and significance of the climbing behaviour of Hydrobia ulvae. J. mar. biol. Ass., UK, 61, 285-299.

Barnes, R.S.K., 1981b, Factors affecting climbing in the coastal gastropod Hydrobia ulvae. J. mar. biol. Ass, UK, 61, 301-306.

Barnes, R.S.K., 1987a, Coastal lagoons of East Anglia, UK. J. coast. Res., in press.

Barnes, R.S.K., 1987b, The faunas of land-locked lagoons: random differences and the problems of dispersal and persistence. In prep.

Barnes, R.S.K. & Greenwood, J.G., 1978, The response of the intertidal gastropod Hydrobia ulvae (Pennant) to sediments of differing particle size. J. exp. mar. Biol. Ecol., 31, 43-54.

Bishop, M.J., 1976, Hydrobia neglecta in the British Isles. J. moll. Stud., 42, 319-326

Cherrill, A.J. & James, R., 1985, The distribution and habitat preferences of four species of Hydrobiidae in East Anglia. J. Conch., 32, 123-133

Cherrill, A.J. & James, R., 1987a, Character displacement in Hydrobia. Oecologia, Berl., 71, 618-623

Cherrill, A.J. & James, R., 1987b, Evidence for competition between mudsnails (Hydrobiidae) in the field; an experiment and review. Hydrobiol., in press.

Connell, J.H., 1980, Diversity and the coevolution of competitors, or the ghost of competition past. Oikos, 35, 131-138.

Christiansen, F,B, & Fenchel, T., 1977, Theories of Populations in Biological Communities. Springer, Berlin.

Fenchel, T., 1975a, Factors determining the distribution patterns of mud snails (Hydrobiidae). Oecologia, Berl., 20, 1-17.

Fenchel, T., 1975b, Character displacement and coexistence in mud snails (Hydrobiidae). Oecologia, Berl., 20, 19-32.

Fenchel, T. & Kofoed, L.H., 1976, Evidence for exploitative interspecific competition in mud snails (Hydrobiidae). Oikos, 27, 367-376.

Fish, J.D. & Fish, S., 1977, The effects of temperature and salinity on the embryonic development of Hydrobia ulvae (Pennant). J. mar. biol. Ass., UK, 57, 213-218.

Fish, J.D. & Fish, S., 1981, The early life-cycle stages of Hydrobia ventrosa and Hydrobia neglecta with obervations on Potamopyrgus jenkinsi. J. moll. Stud, 47, 89-98.

Hylleberg, J., 1975, The effect of salinity and temperature on egestion in mud snails (Gastropoda: Hydrobiidae). Oecologia, Berl., 21, 279-289.

Hylleberg, J., 1976, Resource partitioning on basis of hydrolytic enzymes in deposit-feeding mud snails (Hydrobiidae). II Studies in niche overlap. Oecologia, Berl., 23, 115-125.

Hylleberg, J., 1986, Distribution of hydrobiid snails in relation to salinity, with emphasis on shell size and co-existence of the species. Ophelia, Suppl. 4, 85-100.

Lassen, H.H., 1979, Reproductive effort in Danish mudsnails (Hydrobiidae). Oecologia, Berl., 40, 365-369.

Lassen, H.H. & Clark, M.E., 1979, Comparative fecundity in three Danish mudsnails (Hydrobiidae). Ophelia, 18, 171-178.

Lassen, H.H. & Hylleberg, J., 1978, Tolerance to abiotic factors in mudsnails (Hydrobiidae). Natura jutl., 20, 243-250

Levinton, J.S., 1982, The body size - prey size hypothesis: the adequacy of body size as a vehicle for character displacement. Ecology, 63, 869-872.

McMillan, N.F., 1948, Possible biological races in Hydrobia ulvae (Pennant) and their varying resistance to lowered salinity. J. Conch., 23, 14-16.

Muus, B.J., 1967, The fauna of Danish estuaries and lagoons. Medded. fra Danm. fisk. - og Havunders., 5, 1-316.

Race, M.S., 1982, Competitive displacement and predation between introduced and native mud snails. Oecologia, Berl., 54, 337-347.

Optimal Foraging in the Intertidal Environment: Evidence and Constraints

Roger N. Hughes

University College of North Wales
Bangor, U.K.

ABSTRACT

Dietary selectivity, a prerequisite of Optimal Foraging, has been demonstrated in diverse animals feeding intertidally. Moreover under laboratory conditions, crabs, dogwhelks, asteroids and fish select prey which tend to maximize the rate of food intake, and this also has been shown to occur on the shore in dogwhelks and in a wading bird. In other cases, constraints including the tidal limitation of time available for foraging and the risks associated with competitors and predators make animals forage more opportunistically, blurring the underlying propensity to feed selectively. But they do not operate continuously or with constant intensity and it would be premature to assume that energy maximization is restricted to brief or unusual conditions, having little relevance in nature. Controlled experimentation on the shore is needed to show the relative importance of energy maximization and foraging constraints in shaping natural diets.

INTRODUCTION

Natural diets are seldom random samples of available food, but are often biased in favour of certain items. Selectivity could result from simple mechanical relationships between trophic apparatus and food, or from complex behavioural responses of the forager to large sets of information. In either case it is relevant to ask whether or not the selective feeding enhances genetic fitness.

This question was first broached theoretically by MacArthur & Pianka (1966) and Emlen (1966a), who independently applied economics models to foraging behaviour, initiating what has since become known as Optimal Foraging Theory (reviewed in the marine context by Hughes, 1980). A subset, Optimal Diet Theory (ODT), is concerned specifically with food selection. In its basic form, ODT proposes that selective feeding increases fitness by maximising the net rate of energy gain, so maximising the potential for all vital processes in addition to maintenance. Using this Energy Maximization Premise (EMP), models predict diets from the quality and abundance of available food. Positive tests of these predictions could in principle verify the EMP, establishing it as a general law of animal behaviour (although this is refuted by Pierce & Ollason (1987)). Inevitably, however, the fruitful simplicity of the models is achieved at some expense

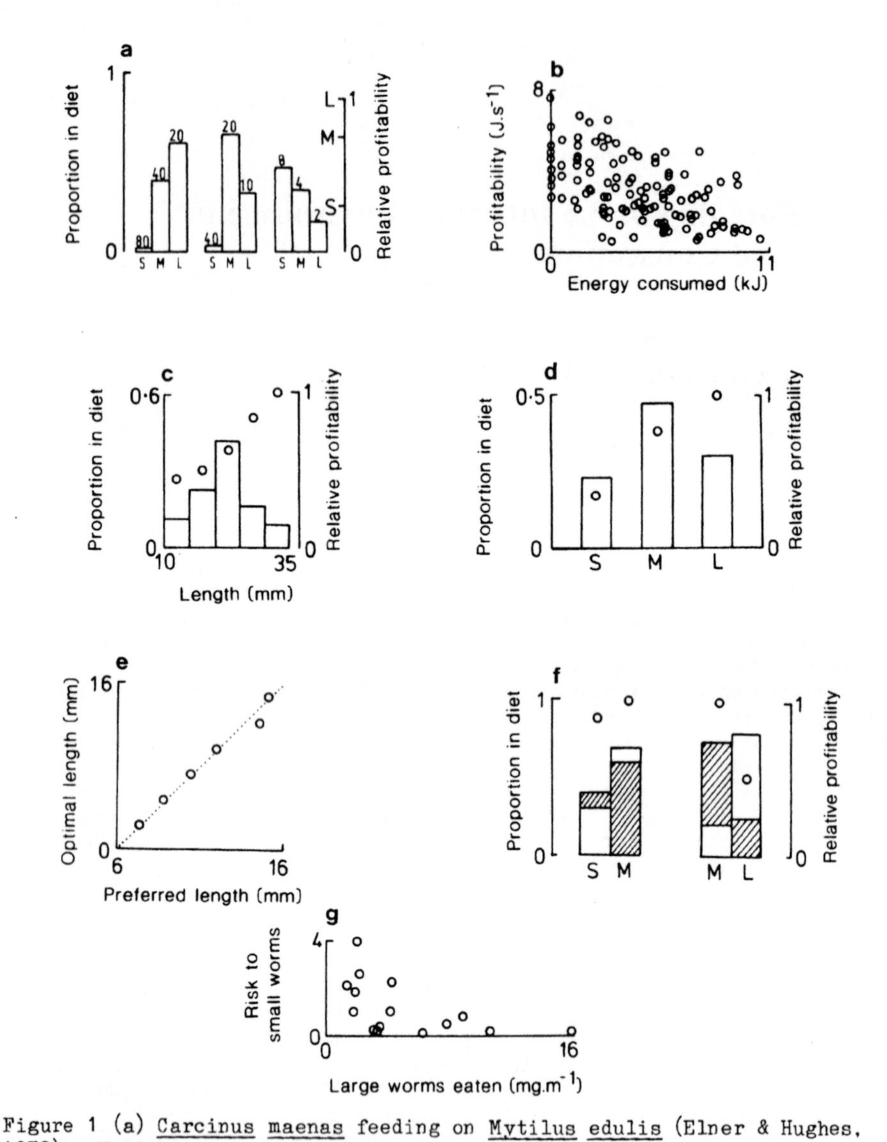

Figure 1 (a) Carcinus maenas feeding on Mytilus edulis (Elner & Hughes, 1978). Numbers presented as shown.
(b) Liocarcinus puber feeding on Carcinua meanas (ap Rheinallt, 1986).
(c) Nucella lapillus feeding on Mytilus edulis (Hughes & Dunkin, 1984a).
(d) Asterias rubens feeding on Mytilus edulis (ap Rheinallt, 1982).
(e) Spinachia spinachia feeding on Neomysis integer (Kislalioglu & Gibson, 1976a). (f) Lipophrys pholis feeding on Carcinus meanas (ap Rheinallt, 1982). Prey were presented in pairs. Shaded = satiated fish, unshaded = starved fish.
(g) Tringa totanus feeding on Nereis diversicolor (Goss-Custard, 1977a). Risk = (No. eaten per m of search path per minute x 1000)/ number per m^2.
In all figures s = small, M = medium, L = large.

of reality. For example. it is assumed that the time available for foraging is unlimited, that foraging imposes no risks to survival, and that there are no competitive interactions. These are precisely the conditions likely to be faced by intertidal foragers. Energy maximization therefore should be regarded as a central tendency influencing feeding behaviour, but which is likely to be compromised by other factors. This conceptual framework continues as it has in the past, to inspire productive experimental observations of foraging behaviour.

SELECTIVE FEEDING

Size selection

Most predators encounter prey of different sizes and frequently feed disproportionately on some of them. When prey are all of the same species, size is often the key factor influencing selection. Larger prey may yield more flesh, but they are also likely to take longer to catch. subdue and ingest. Consequently, profitability (yield of flesh/handling time) will be a function of prey size. Moreover, larger (or smaller) prey may be more apt to escape or to successfully resist attack, so decreasing their average profitability. Consequently, expected maximum profitability will occur at a particular size of prey, the optimal size. ODT predicts that optimally sized prey should always be accepted, but progressively more suboptimal sizes should also be accepted as the availability of better items declines. Experimental tests of the predictions therefore must determine the relationship between profitability and prey size, comparing selection with the expected optimal size and the availability of prey. These relationships are reviewed below for several important types of intertidal predator.

Crabs

Crabs often show pronounced size selection when feeding on molluscs, for example they select mussels close to the optimal size (Elner & Hughes, 1978; Elner & Jamieson, 1979; Hughes & Seed, 1981; Du Preez, 1984; ap Rheinallt & Hughes, 1985). In accordance with theory, suboptimal sizes become acceptable when the optimal mussels are scarce (Fig.1a). Preferred sizes of mussel. however, do not always coincide exactly with the predicted optima (Elner & Hughes 1978; Hughes & Seed, 1981), and no matter how plentiful are the optimal mussels, a small number of suboptimal sizes are taken in proportion to their relative abundance (Elner & Hughes, 1978). The latter phenomenon of "partial preference" (Charnov, 1976) is open to a variety of interpretation: for example the crab misjudges prey-size, the crab samples different sized mussels to assess the range available, or the crab ingests a proportion of suboptimal prey to compensate for the time spent identifying this category as a whole (Hughes, 1979).

Ways of discriminating among such possibilities have not been devised, but perhaps the most likely explanation lies in the way crabs perceive the abundances of different sized mussels. These are not randomly mixed in nature, but tend to be sorted topologically within the clump, so that a crab is likely to encounter size categories in "runs" (Ameyaw-Akumfi & Hughes, 1987). Consequently, average rates of encounter estimated from prolonged foraging experience are less appropriate than moving averages based on more immediate experience (Elner & Hughes, 1978) and this will cause the acceptability of suboptimal mussels to fluctuate as the crab forages through the clump.

How do crabs select mussels? The simplest mechanism would be for crabs to attack all mussels encountered and to reject those that do not yield after a certain amount of handling. Small mussels might tend to slip from the chelae, while large mussels might not yield to pressure before the crab loses interest. Optimally sized mussels remain securely in the grip of the chelae and yield within a relatively short handling time. This optimal size depends on the size and strength of the chelae, so increasing with crab size. The maximum time for which a crab will try to break a mussel before rejecting it is determined by hunger. satiated crabs persisting for a shorter time than starving ones. Consequently, crabs will consume a wider range of mussel sizes when hungry. Thus, Liocarcinus puber appear to lose a proportion of small Mytilus edulis owing to limited chelal dexterity, while large mussels are rejected only after attempts to open them have failed (ap Rheinallt & Hughes, 1985; ap Rheinallt, 1986).

Transducers embedded in dummy mussels (Elner, 1978) reveal the handling procedure adopted by Carcinus maenas prior to rejection of the prey (Fig. 2). Initially, a large force is applied, which will break any shell that is sufficiently small or is intrinsically weak. If the shell does not yield, the mussel is shifted in position and pulses of smaller forces are applied which possibly inflict cumulative microfractures (Boulding & Labarbera, 1986), or eventually bear upon a plane of weakness. This procedure allows the crab to sustain its attack without risking damage to the chela. The persistence time increases with hunger: starved C.maenas indiscriminately attack the first mussels encountered, persisting with them if large but quickly becoming more selective after one or two have been ingested (Jubb et al., 1983; Ameyaw-Akumfi & Hughes, 1987). Rejection is also stimulated by concurrent detection of other mussels (Jubb et al,. 1983; ap Rheinallt, 1986). This hunger mediated mechanism could also account for size selection by crabs feeding upon gastropods. Compared with mussels, the vulnerability of littorinids and thaidids is a less predictable function of size once the shell is larger than a certain threshold. When feeding on these prey, C.maenas attack all snails encountered, rejecting those that do not yield flesh after a certain time (Hughes & Elner, 1979; Elner & Raffaelli, 1980).

Size selection by crabs may, however, involve factors beyond the simple mechanism described above. C.maenas frequently appear to reject, rather than lose, small mussels (Elner & Hughes, 1978; Jubb et al., 1983).

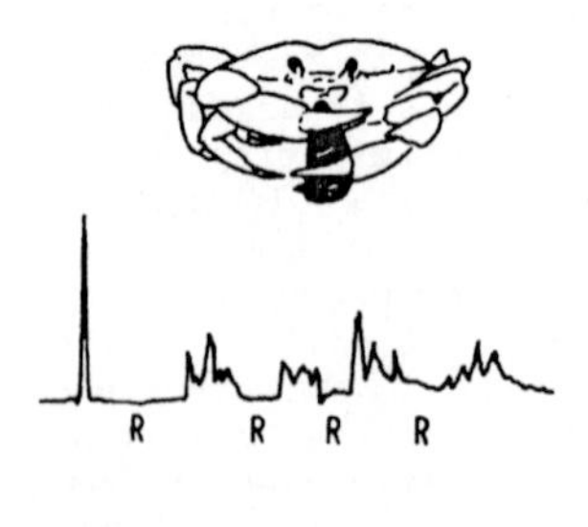

Figure 2. Carcinus maenas crushing Mytilus edulis. Upper figure shows typical feeding position, force being applied to the narrower, umbonal end of the mussel. Lower figure shows the output of a strain gauge measuring the force applied to the mussel by the crab (Elner, 1978). R = interval when mussel is reorientated.

Relative strengths of stimuli from the mussel held in the chelae and from others touched by the walking legs may play a part in the decision to reject (Jubb et al., 1983). When crabs feed on mobile prey, visual stimuli influence prey selection in addition to any mechanical factors. Movement is a stronger stimulus than size to Liocarcinus puber preying upon juvenile C.maenas, but smaller prey escape from grasp more frequently (ap Rheinallt, 1986). With these prey, handling is not hindered by mechanical resistance and profitability is not significantly correlated with prey size, but ingestion may be retarded by the accumulation of exoskeletal fragments in the cardiac stomach, and as a result profitability declines with increasing amount of food consumed(Fig. 1b).

Of course, size selection is not inevitable, and would not be expected where prey are scarce or require considerable effort for their capture. Callinectes sapidus foraging for infaunal clams cannot readily judge size before the prey are exposed by the relatively laborious process of digging, and under these circumstances the crabs persist with all items encountered (Blundon & Kennedy, 1982). Any size selection would occur only by differential availability, for example if larger clams were buried beyond the reach of the crabs.

Gastropods

Size selection by predatory gastropods is known mainly from muricids and naticids, both of which have specialized apparatus for drilling the calcareous shells of their prey. Drilling is a lengthy process even for small prey, so profitability may be expected to increase with yield of flesh, hence prey size, upto the point of satiation. This appears to be the case for muricids feeding undisturbed in the laboratory (Dunkin & Hughes, 1984; Hughes & Dunkin, 1984a). However in normal circumstances, larger prey may be devalued by the robbing of flesh by conspecific predators (Emlen, 1966b), whereas the value of smaller prey may be increased by the use of toxin to shorten penetration time (Palmer, 1982). Prey are not attacked indiscriminately: many items may be encountered before one is chosen (J.L.Menge, 1974; Palmer, 1984) and different categories of prey may be taken disproportionately to their relative abundance (West, 1986).

Individual items are crawled over for up to 2h before rejection or acceptance (Hughes & Dunkin, 1984a). It is not known what stimuli the snails use to select their prey, but penetration sites are not haphazard and may shift as a result of experience, suggesting that the preliminary inspection could involve prospecting for a suitable point of entry (Dunkin & Hughes, 1984; Hughes & Dunkin, 1984a). Also, dislodgement by waves can threaten snails handling their prey (J.L.Menge, 1974) and the possibility of gaining a firm grip may influence rejection or acceptance. In this way, size selection could have a mechanical basis. In contrast to crabs, however, it is unlikely that muricids would routinely adopt a "giving-up-time" rule for selecting their prey. Penetration of the prey's shell by drilling is a more predictable process than crushing, but it generally takes a long time (J.L.Menge, 1984). perhaps even several days with large mussels (Hughes & Dunkin, 1984a). Premature abandonment therefore would be uneconomical.

In the laboratory,Nucella lapillus prefer barnacles or mussels somewhat smaller than those seeming to be the most profitable from the average flesh content and handling time (Fig. 1c) perhaps so avoiding a higher risk of intrusion by kleptoparasites, associated with larger items (Emlen, 1966b). Strong evidence of selection according to profitability has been documented, however, for three other species of Nucella (=Thais) feeding on barnacles and mussels (Palmer, 1984). As in the case of crabs, the Nucella spp. fed unselectively when starved, becoming more selective after maintenance on higher quality prey. Selection was not strongly influenced by ingestive conditioning (Wood, 1968; Hall et al., 1982), nor was it frequency dependent (Murdoch, 1969), but appeared to be based on the profitabilities of available prey and on recent feeding history (see also J.L.Menge, 1974). Whether the latter involved some expectation of prey quality and availability, as envisaged by Charnov's (1976) model, or acted merely through hunger level is not known.

Naticids also select prey according to size and although the attack behaviour of some, particularly temperate species, seems unresponsive to experience (Edwards & Huebner, 1977; Kitchell et al., 1981), it appears to be more flexible in others (Hughes, 1985; Ansell & Morton, 1987).

Asteroids

Forcipulate asteroids such as Asterias, Leptasterias, Pisaster, Heliaster and Pycnopodia are formidable predators, which by virtue of their eversible stomach and suckered tube-feet, are able both to subdue large individual prey and to ingest multiple small prey, even when these are firmly attached to rocks. Asteroids such as Luidia and Astropecten, unable to evert their stomach and lacking suckers on their tube feet, tend to feed on items not firmly attached to hard substrata. Size selection has been recorded in both groups (Kim, 1969; Hancock, 1974; Doi, 1976; Ribi et al, 1977; Anger et al., 1977; Hylleberg et al., 1978).

Asterias rubens foraging for M.edulis in aquaria, select intermediate sized mussels, even though larger items potentially yield more flesh per unit handling time (Fig. 1d). Small mussels are rejected, particularly when larger prey are contacted during the handling process. Medium and large prey are attacked as encountered, but the latter with less success. As mussels become larger, an increasing proportion are abandoned after about an hour, amounting to some 5 - 50% of the normal handling time and occurring approximately when the stomach would be everted (ap Rheinallt, 1982). A.rubens, therefore, appear to reject small mussels, but to accept all larger ones abandoning those that do not yield after some effort has been made to force the valves apart. This mechanism of size selection is analogous to the one proposed for crabs. Indirect evidence of it is available for Pisaster ochraceus, whose natural diet encompasses a minimum and a maximum size of Mytilus californianus, both of which increase with larger Pisaster (Paine, 1976). In several forcipulates, however, it has been noticed that many prey items may be contacted before one is attacked (Menge, 1972; Birkeland, 1974), so that a more complicated mechanism of prey selection, perhaps involving factors additional to size, could be operating.

Luidia clathrata ingests whole infaunal prey, and although capable of ingesting relatively large items, it preferentially takes smaller ones (McClintock & Lawrence, 1981). Similarly, L.sarsi selects smaller ophiuroids (Fenchel, 1965) and Astropecten irregularis selects smaller infaunal bivalves (Christensen, 1970). Although larger prey may escape more frequently (Christensen, 1970), the differential ingestion of smaller items could result from active selection, whereby the stomach is packed more efficiently, so maximizing the contained volume of digestible material (Fenchel, 1965; McClintock & Lawrence, 1981).

Fish

For many species of inshore fish, prey size is correlated with predator size (Kislalioglu & Gibson, 1975). Fifteen-spined sticklebacks, Spinachia spinachia, prefer mysids of a size maximizing the ratio of yield to handling time (Fig. 1e). This corresponds to a size beyond which handling time rises sharply, owing to limited capacity of the jaws to hold the prey. Hungry fish take prey above and below the optimal size, but selectivity increases towards satiation. Handling time lengthens with decreasing hunger, interacting with prey size in such a way that suboptimal prey become even less profitable relative to the optimum.

It appears that S.spinachia use visual recognition stimuli to evaluate prey, rather than the mechanical, trial and error apparently used by crabs, dogwhelks and starfish. Presumably, suboptimal prey generate weaker stimuli, while responsiveness of the fish declines as hunger abates, with the result that suboptimal prey are ignored. In addition to size, recognition stimuli include movement, colour and shape. Size and movement are the most important, with movement being marginally in first rank (Kislalioglu & Gibson, 1976b). As well as attracting the predator's attention, movement may indicate vulnerability and, together with size, will be an important measure of expected profitability.

Blennies, Lipophrys pholis, feeding on juvenile crabs, C.maenas, respond strongly to movement and size, the former being the more evocative (ap Rheinallt 1982). When presented freely, smaller prey are more active and hence are attacked more frequently than larger items, but when movement is restricted, larger prey are preferred by hungry blennies. Decreasing hunger lessens the motivation to attack, especially with larger prey (Fig. 1f), and lengthens handling time. Consequently fish close to satiation abandon larger prey prematurely. Because of the longer time required to subdue and manipulate larger items, their profitability is less than that of intermediate sized prey (Fig. 1f). As with other predators, therefore, blennies tend to select items closer to the experimentally predicted optimum when partially satiated.

Birds

Birds respond mainly to visual cues and often select prey according to size. Oystercatchers, Haematopus spp., feed extensively on molluscs, using their powerful bill to break or prise open the shell. H.ostralegus avoid small cockles, Cerastoderma edule, unless larger ones are scarce (O'Connor & Brown, 1977; Sutherland, 1982), but use information additional to size when selecting mussels (Durrell & Goss-Custard, 1984). In the latter case, individual birds learn to attack mussels by one of three methods: stabbing the bill between the slightly gaping valves to sever the adductor muscle, or smashing the shell by dorsal or ventral hammering. Stabbers select mussels that are gaping, dorsal hammerers tap the shell and apparently ascertain its vulnerability, selecting those that are relatively thin and eroded, while ventral hammerers seem to use visual cues associated with shell thinness, perhaps the brown colouration and lack of fouling characteristic of fast-growing, hence thin-shelled mussels. These methods of evaluation are more accurate than ones based on size alone because the amount of flesh in a mussel of a given shell length is so variable. Nevertheless mistakes are made, especially by ventral hammerers, and a proportion of mussels are abandoned if they do not yield to attack.

Redshank, Tringa totanus, select larger individuals of the polychaete, Nereis diversicolor, taking more smaller when fewer larger items are available. As predicted by ODT when evaluation is instantaneous, feeding on the less profitable items depends not on their own availability, but only on that of better items (Fig. 1g). Apparently the birds use size of the burrow entrance, or movement of the worm as recognition stimuli.

Species selection

Different prey species in addition to possible differences in size, may require different attack methods, or have different nutritional qualities, perhaps leading to a preference hierarchy such as that shown by Morula marginalba feeding among a range of intertidal prey (Fairweather et al., 1984). As with any category ranked in profitability, ODT predicts that the acceptance of suboptimal species depends on the availability of more profitable prey, and consequently predation may appear to be frequency dependent (Hubbard et al, 1982). Redshank, T.totanus, prefer the amphipod, Corophium volutator, to the polychaete, N.diversicolor, possibly because the former, although smaller, yields some scarce nutrient (Goss-Custard, 1977b). As predicted by ODT, predation intensity on Nereis, even when abundant, is inversely proportional to the availability of Corophium (Fig.3).

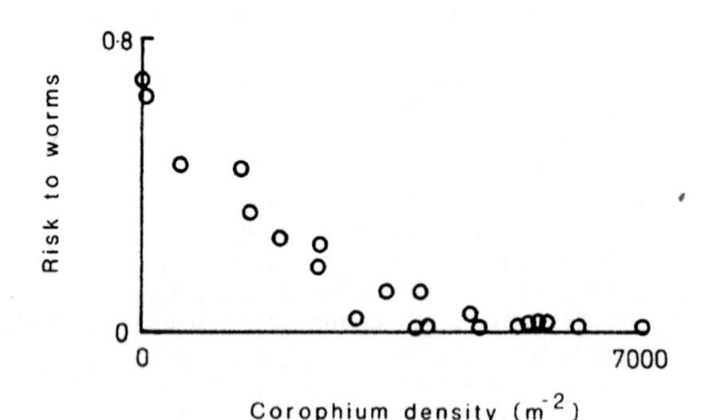

Figure 3. Tringa totanus feeding on mixture of Nereis diversicolor and Corophium volutator (Goss-Custard, 1977b). Risk = (number eaten per m of search path per min x 1000)/ number per m^2.

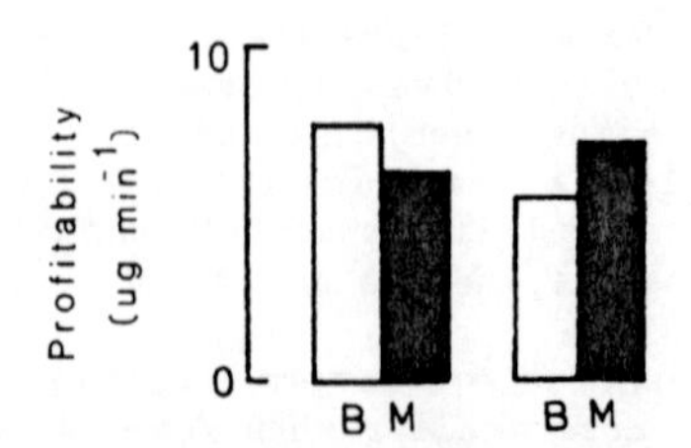

Figure 4. Nucella lapillus feeding on Semibalanus balanoides (B) and M. edulis (M) (Hughes & Dunkin, 1984b). Shaded = dogwhelks maintained on mussels, unshaded = dogwhelks maintained on barnacles.

Selection of specific prey tends to be enhanced by experience, during which 'ingestive conditioning' or the development of 'search images' takes place. Through mechanisms such as associative learning or improved attack efficiency, ingestive conditioning and search images may raise the profitability of specific prey. Predation intensity will be frequency dependent, since the experience and reinforcement required to develop and sustain ingestive conditioning or search images are possible only when the specific prey are relatively abundant.

Neither process has been recorded in crustaceans, although C.maenas quickly reduce handling time by as much as 60% on gaining experience with bivalve and gastropod prey, thereby showing potential for frequency dependent predation (Cunningham & Hughes, 1984).

Ingestive conditioning is well known in gastropods (Wood, 1968; Hall et al., 1982). N.lapillus feed on barnacles and mussels, but develop a strong preference for one or the other after prolonged monospecific diets (Hughes & Dunkin, 1984b). As a result of experience, handling time is reduced to such an extent that the profitabilities of large barnacles and small mussels can be reversibly transposed (Fig. 4). In the case of Acanthina spirata, preferences strengthened by experience cause frequency dependent predation in the laboratory (Murdoch, 1969).

The effect of experience on the foraging behaviour of asteroids has received little attention, although ingestive conditioning has been reported for some species (Sloan, 1980).

Frequency dependent predation caused by learned modifications of foraging behaviour, including search-image formation, is well known among fish (Dill, 1983) and birds (Clarke, 1969), but there is little information on species foraging intertidally.

THE ENERGY MAXIMIZATION PREMISE

Evidence, as presented above, suggests that the selective feeding observed in a wide range of predators tends to maximize the net rate of energy intake while foraging. Only a few studies (eg Mittelbach, 1983) have tested the central assumption that energy maximization enhances genetic fitness. Even these few were not direct tests, since they used growth as an index of fitness, under the reasonable assumption that larger size increases survivorship and fecundity. Currently the most rigorous test for any animal has been that made by Palmer (1983, 1984) on dogwhelks. Nucella spp. were found to grow faster on a diet of their preferred prey, medium sized Balanus glandula, than on diets of the less preferred items (Fig. 5a). Moreover, in N.emarginata the faster growth resulted in earlier sexual maturity, and in both N.emarginata and N.canaliculata there was some evidence of enhanced fecundity. Similarly in another study (Garton, 1986), scope for growth (absorbed energy - respired and excreted energy) was significantly greater for small Thais haemastoma maintained on their preferred prey, small oysters, than for those maintained on larger oysters, although there was scarcely any such difference for large T.haemastoma (Fig. 5b). Under natural circumstances, growth rate and final asymptotic size of Morula marginalba are strongly correlated with the quality of available prey (Moran et al., 1984).

CONSTRAINTS ON FORAGING

Limited foraging time and risks of mortality

To a greater or lesser extent, tides influence the behaviour of all animals foraging on the shore. Food may be available only at certain stages of the tide and movement may be curtailed either by the need to reduce desiccation or to avoid dislodgement by waves (Moran, 1985). On shores where the substratum usually dries out at low tide, thaidids search for prey only when submerged, although they often remain in the open, motionless on top of their prey, when the tide recedes (Fairweather & Underwood, 1983; Palmer, 1984; West. 1986). The reverse may be true, however, on shores persistently buffeted by strong waves. In such a habitat, *Acanthina punctulata* searches for prey only during low tide, remaining fixed to its prey when subsequently awash (J. L. Menge, 1974).

In addition to facing environmental hazards, intertidal foragers may themselves fall prey to larger animals and the two kinds of risk can interact, constraining foraging time yet further. On the Pacific coast of Panama, the muricid *Purpura pansa* does not forage on rising tides, when it would run the risk of being eaten by puffer fish, and forages only briefly on ebbing tides during the day, when desiccation is severe. Prolonged foraging by *P.pansa* in this habitat is confined to nocturnal low tides (Garrity & Levings, 1981).

Constraints on foraging time and concurrent risks of mortality may have important effects on diet selection (McNair, 1979; Lucas, 1983). *Acanthina punctulata* prefers littorinids to barnacles, in accordance with their profitability. On the falling tide, *A.punctulata* feeds selectively, but towards the return of the tide it more readily accepts prey as they are encountered. Moreover, whereas in the laboratory *A.punctulata* will commence searching immediately after a meal, on the shore it usually waits until the next low tide. This snail therefore appears to compromize energy maximization by minimizing the risk of mortality from wave action (J. L. Menge, 1974).

Possibly, differences among shores may generate natural selection for different, genetically controlled traits in foraging behaviour. *Nucella lapillus* from shores exposed to heavy wave action consume more prey per unit time than those from calmer shores, when tested in both habitats (B. A. Menge, 1978). This behavioural difference is correlated with a difference in shell architecture, and genetic control of the behaviour, although highly probable, remains to be proven, since feeding rate could be determined morphologically. It will be valuable to investigate any intraspecific differences in foraging behaviour which may be genetically controlled and therefore potentially responsive to natural selection, since this could shed light on the relative contribution of Optimal Foraging to genetic fitness. For example, on calmer shores, where searching is less frequently curtailed by the environment, lower predation rates by individual *N.lapillus* may accompany more selective feeding and a greater tendency to maximize the net rate of energy intake. On exposed shores, higher individual predation rates may involve less selective feeding and a greater tendency to minimize foraging time, at the expense of energy maximization.

Whereas some predators, such as muricids, are able to continue handling prey *in situ* throughout the tidal cycle or, such as *Pisaster ochraceus*, are able to carry prey back to subtidal refuges, others must terminate their consumption of prey at a certain phase of the tide. This may affect prey selection, since if only a short time is left to eat prey, any item requiring a long handling time must be abandoned before its consumption is complete. *C.maenas* attacking intertidal mussels may be unable to detach some individuals, which must be left behind as the tide recedes. Mussel size will therefore be less closely correlated with profitability and feeding may be expected to be less selective when time is constrained in this way. Data, however, are lacking and experiments are needed.

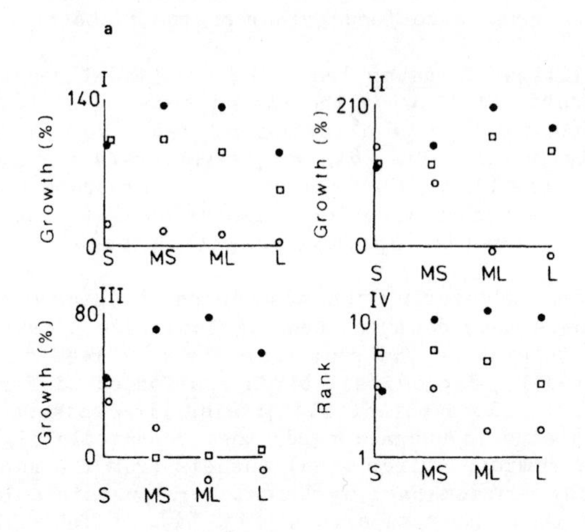

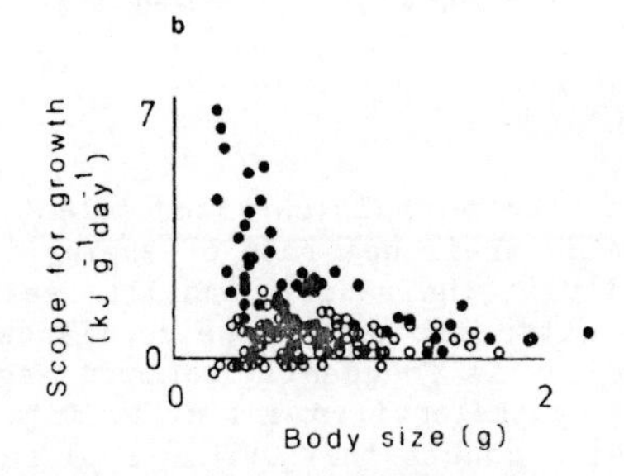

Figure 5. (a) Growth performance (% change in body mass during 1 month) of dogwhelks fed on different prey. (Palmer, 1983). ● = *Balanus glandula*, ○ = *Semibalanus cariosus*, □ = *Mytilus edulis*. S = small, MS = medium small, ML = medium large, L = large. I = *Nucella emarginata*, II = *N.canaliculata*, III = *N.lamellosa*, IV = preference rank shown by all *Nucella* spp. as a function of prey size and species. (b) scope for growth of *Thais haemastoma* as a function of body size (Garton, 1986). ● = small oyster, ○ = large oyster diet.

Competition

Partly as a result of energy subsidies generated by the tides and waves, intertidal habitats generally are extremely productive. Consequently they are the focus of intense feeding activity, with ensuing competition among foragers. Competition strongly influences foraging behaviour, tending to lower the net rate of energy intake. The probability of this event is increased by the synchronizing effect of the tides: residents emerge and immigrants converge to feed either on the flood or the ebb.

In suitable localities, C.maenas leave their subtidal resting places and move onshore in hoards, following the rising tide. Over 100 crabs per tide have been recorded visiting 0.1 m^2 of mussel bed (Dare et al., 1983). The crabs range widely in size (Fig. 6a) and larger crabs are aggressively dominant over smaller ones (Fig. 6b). As a result, most crabs are forced to feed more opportunistically than they do in isolation, and size selection no longer appears to be based on 'optimal prey' (Fig. 6c).

As the tide recedes, uncovering the mussels and causing crabs to leave the shore, oystercatchers move down to feed. A dominance hierarchy allows the top ranking birds to exclude the rest from the best feeding stations (Ens & Goss-Custard, 1984). Subdominant birds are forced to forage in suboptimal places and they lose potential foraging time because unlike the dominant birds whose status is unchallenged, they repeatedly fight among themselves. Moreover, dominant birds steal mussels from the subdominants (Fig. 7a). Consequently, subdominant oystercatchers sustain a lower net rate of energy intake than their superiors (Fig. 7b). Effects of this competition on prey selection have not been reported, but are likely to depress selectivity among subdominant birds, who have less time for foraging and face the risk of losing many of the items they have chosen.

RELEVANCE TO COMMUNITY ECOLOGY

There is no doubt that most animals can feed selectively, that selective feeding tends to maximize their net rate of energy intake or to optimize the nutrient quality of their diet, and that selective feeding can promote their growth, fecundity, and hence genetic fitness. There is also no doubt that selective feeding is frequently compromized, or even prevented, by constraints imposed by the environment or by interactions with other animals. This may raise doubts that ODT is applicable beyond the simplified conditions of the laboratory. Such doubts are clearly unfounded in cases where Optimal Diet selection has been shown to occur in the field, such as the Nucella spp. studied by Palmer (1984). But they may be more serious when the field data are less decisive.

Earlier in this review it was shown that Optimal Diet selection by crabs has been demonstrated many times in laboratories, and it was also shown that the foraging behaviour of crabs in the field is constrained by the tides and by intraspecific competition. Such constraints tend to favour opportunistic at the expense of selective feeding and this may give the impression that crabs do not adopt Optimal Foraging behaviour on natural mussel beds (Dare et al. 1983). Opposite conclusions drawn from the field and laboratory studies reflect the two levels of approach involved, one focussing on populations, the other on individuals. Losses of mussels observed in the field result from the integrated foraging activities of many crabs, which are not only foraging under varying degrees of constraint, but are of different sizes, each having its own preferred size of prey. Field data, on their own, might suggest that crabs indiscriminately consume all mussels small enough to be opened, but could not indicate the

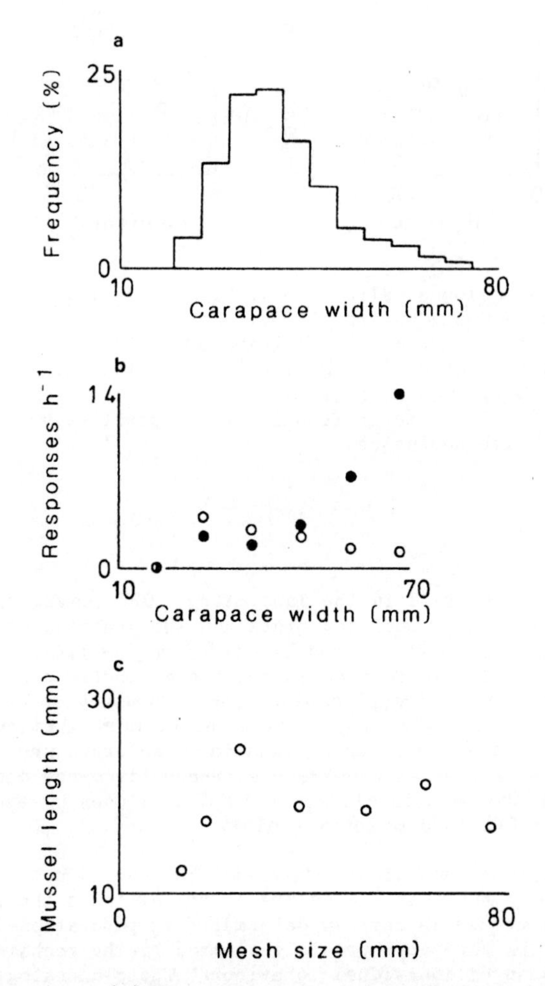

Figure 6. (a) Size-frequency distribution of crabs migrating onshore to feed on mussels (Dare et al, 1983).

(b) Interactions among crabs feeding on mussels at high tide, monitored by video (Cunningham, 1983). • = aggressive response, o = fleeing response.

(c) Mean size of mussels taken from clumps enclosed within cages of increasing mesh size. Sizes of mussels eaten do not change systematically as the size-range of feeding crabs increases (Cunningham, 1983).

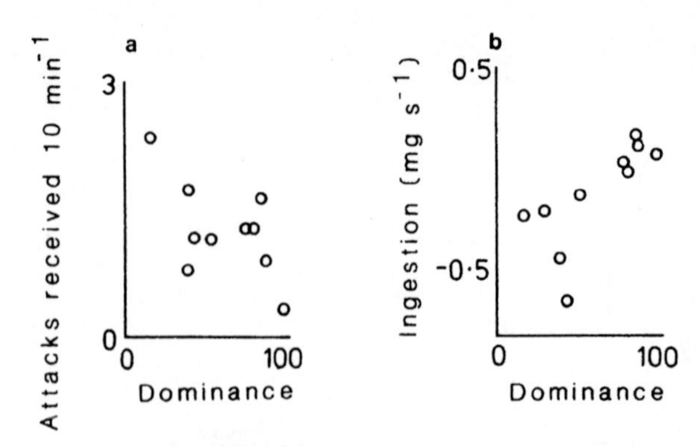

Figure 7. Haematopus ostralegus feeding on Mytilus edulis (Ens & Goss-Custard, 1983).
(a) The frequency of being attacked, during which prey is stolen, is negatively correlated with dominance (% of encounters won).
(b) The rate of food intake is positively correlated with dominance.

degree of selectivity observed in the laboratory. ODT however, predicts both the selective feeding of isolated crabs and the opportunistic feeding of those constrained by competition and limited foraging time. Superficial lack of evidence for selective feeding in the field, therefore, does not necessarily show that ODT is inapplicable. The laboratory observations, and certain field data, strongly suggest that energy maximization is a central tendency, toward which foraging behaviour converges when released from constraints. The fact that constraints frequently cause diets to fall short of the optimum does not invalidate the EMP. It does however, question the relevance of ODT to community ecology.

ODT helps to explain individual behaviour, but should not be expected to cope with the hugely multivariate nature of predation at the population level. Community structure is largely determined by population-level events and therefore is more appropriately studied by the techniques of population ecology than of individual behaviour. The old analogy of Boyle's Law and the movement of individual gas molecules comes to mind.

Nevertheless, rules governing individual behaviour may have effects that penetrate to the population level and hence influence community structure. This has been strikingly demonstrated by Fairweather et al. (1984), who showed how a preference hierarchy can cause predators to influence community structure in different ways, depending on the availability of preferred prey. Morula marginalba severely reduces the density of low-ranking prey, such as the barnacle Chamaesipho columna, only when higher ranking prey are scarce, otherwise it ignores them (Fairweather, 1985). Effects such as this could confound predator exclusion experiments that do not take into account the relative abundances of prey within the predator's preference hierarchy.

Frequency dependent predation has a long pedigree in theoretical ecology as a potential stabilizing force in the community (Murdoch & Oaten, 1975). Experimental evidence of such an effect, however, is almost totally lacking. Earlier in this review it was shown that some intertidal foragers are known to select prey in a frequency dependent manner, but whether this ever stabilizes the prey populations by preventing their local dominance or extinction remains questionable. Spatial heterogeneity, causing local differences in population status, probably has a greater stabilizing influence than frequency dependent predation (but see Lawton, 1987) and may cause individuals within a predator population to adopt different diets. A wide range of dietary specialization and generalization can occur among Nucella emarginata foraging within a local habitat, even though individual diets appear to be highly consistent when observed over several sequential attacks (West, 1986).

Applications of ODT based solely on energy maximization are severely limited in scope. Natural constraints need to be taken into account, as do details of individual faraging history and demographic provenance. To do this, it will be necessary to examine the foraging behaviour of experimental subjects under a wider range of conditions (Stickle & Bayne, 1987) and to combine laboratory and field experimentation.

ACKNOWLEDGEMENTS

I am most grateful to my wife, Helen, for typing the manuscript and for helping to initiate my studies of feeding behaviour.

REFERENCES

Ameyaw-Akumfi, C. & Hughes, R.N. 1987. Behaviour of Carcinus maenas (L.) feeding on large Mytilus edulis L. How do they assess the optimal diet? Mar.Ecol.Progr.Ser. in press.

Anger, K., Rogal, U., Schreiver, G. & Valentin, C. 1977. In-situ investigations on the echinoderm Asterias rubens as a predator of soft-bottom communities in the western Baltic Sea. Helgolander wiss.Meeresuntersuchungen 29: 439 - 459.

Ansell, A.D. & Morton, B. 1987. Alternative predation tactics of a tropical naticid gastropod. J.Exp.Mar.Biol.Ecol. in press.

ap Rheinallt, T. 1982. The foraging behaviour of some marine predators. Ph.D. thesis, University of Wales, Bangor.

ap Rheinallt, T. 1986. Size selection by the crab Liocarcinus puber feeding on mussels Mytilus edulis and on shore crabs Carcinus maenas: the importance of mechanical factors. Mar.Ecol.Progr.Ser. 25: 63 - 70.

Birkeland, C. 1974. Interactions between a sea pen and seven of its predators. Ecol.Monogr. 44: 211 - 232.

Blundon, J.A. & Kennedy, V.S. 1982. Mechanical and behavioural aspects of blue crab, Callinectes sapidus (Rathbun), predation on Chesapeake Bay bivalves. J.Exp.Mar.Biol.Ecol. 65: 47 - 65.

Boulding, E.G. & Labarbera, M. 1986. Fatique damage: repeated loading enables crabs to open larger bivalves. Biol.Bull. 171: 538 - 547.

Charnov, E.L. 1976. Optimal foraging: attack strategy of a mantid. Amer. Nat. 110: 141 - 151.

Christiansen, M.E. 1970. Feeding biology of the sea-star Astropecten irregularis. Ophelia 8: 1 - 134.

Clarke, B.C. 1969. The evidence for apostatic selection. Heredity 24: 347 - 352.

Cunningham, P.N. 1983. Predatory activities of shorecrab populations. Ph.D. thesis. University of Wales, Bangor.

Cunningham, P.N. & Hughes. R.N. 1984. Learning of predatory skills by shorecrabs Carcinus maenas feeding on mussels and dogwhelks. Mar. Ecol.Progr.Ser. 16: 21 - 26.

Dare, P.J., Davies, G. & Edwards, D.B. 1983. Predation on juvenile Pacific oysters (Crassostrea gigas Thunberg) and mussels (Mytilus edulis L.) by shore crabs (Carcinus maenas (L.)). Fish.Res.Tech. Rep.No.73, MAFF., Lowestoft.

Dill, L.M. 1983. Adaptive flexibility in the foraging behaviour of fishes. Can.J.Fish.Aquat.Sci. 40: 398 - 408.

Doi, T. 1976. Some aspects of the feeding ecology of the sea stars, genus Astropecten. Publ.Amakusa Mar.Biol.Lab. 4: 1 - 19.

Du Preez, H.H. 1984. Molluscan predation by Ovalipes punctatus (De Haan) (Crustacea: Brachyura: Portunidae). J.Exp.Mar.Biol.Ecol. 84: 55 - 71.

Dunkin, S. de B. & Hughes, R.N. 1984. Behavioural components of prey selection by dogwhelks, Nucella lapillus (L.), feeding on barnacles, Semibalanus balanoides (L.), in the laboratory. J.Exp.Mar.Biol. Ecol. 79: 91 - 103.

Durrell, S.E.A. le V.dit & Goss-Custard, J.D. 1984. Prey selection within a size-class of mussels, Mytilus edulis, by oystercatchers, Haematopus ostralegus. Anim.Behav. 32: 1197 - 1203.

Edwards, D.C. & Huebner, J.D. 1977. Feeding and growth rates of Polinices duplicatus preying on Mya arenaria at Barnstable Harbor, Massachusetts. Ecology 58: 1218 - 1236.

Elner, R.W. 1978. The mechanics of predation by the shore crab, Carcinus maenas (L.), on the edible mussel, Mytilus edulis L. Oecologia (Berlin) 36: 333 - 344.

Elner R.W. & Hughes, R.N. 1978. Energy maximization in the diet of the shore crab, Carcinus maenas. J.Anim.Ecol. 47: 103 - 116.

Elner, R.W. & Jamieson, G.S. 1979. Predation of sea scallops, Placopecten magellanicus, by the rock crab, Cancer irroratus, and the American lobster, Homarus americanus. J.Fish.Res.Bd.Can. 36: 537 - 543.

Elner, R.W. & Raffaelli, D.G. 1980. Interactions between two marine snails, Littorina rudis Maton and Littorina nigrolineata Gray, a predator Carcinus maenas (L.), and a parasite Microphallus similus Jagerskiold. J.Exp.Mar.Biol.Ecol. 43: 151 - 160.

Emlen, J.M. 1966a. The role of time and energy in food preference. Amer. Nat. 100: 611 - 617.

Emlen, J.M. 1966b. Time, energy and risk in two species of carnivorous gastropods. Ph.D. dissertation, University of Washington, Seattle.

Ens, B.J. & Goss-Custard, J.D. 1984. Interference among oystercatchers Haematopus ostralegus, feeding on mussels Mytilus edulis in the Exe estuary. J.Anim.Ecol. 53: 217 - 231.

Fairweather, P.G. 1985. Differential predation on alternative prey, and the survival of rocky intertidal organisms in NSW. J.Exp.Mar.Biol. Ecol. 89: 135 - 156.

Fairweather, P.G. & Underwood, A.J. 1983. The apparent diet of predators and biases due to different handling times of their prey. Oecologia (Berlin) 56: 169 - 179.

Fairweather, P.G., Underwood, A.J. & Moran, M.J. 1984. Preliminary investigations of predation by the whelk Morula marginalba. Mar.Ecol.Progr.Ser. 17: 143 - 156.

Fenchel, T. 1965. Feeding biology of the sea-star Luidia sarsi Duben and Koren. Ophelia 2: 223 - 236.

Garrity, S.D. & Levings, S.C. 1981. A predator-prey interaction between two physically and biologically constrained tropical rocky shore gastropods: direct, indirect and community effects. Ecol.Mongr. 51: 269 - 286.

Garton, D.W. 1986. Effects of prey size on the energy budget of a predatory gastropod, Thais haemastoma canaliculata (Gray). J.Exp.Mar. Biol.Ecol. 98: 21 - 23.

Goss-Custard, J.D. 1977a. Optimal foraging and the size selection of worms by redshank Tringa totanus. Anim.Behav. 25: 10 - 29.

Goss-Custard, J.D. 1977b. The energetics of prey selection by redshank, Tringa totanus (L.), in relation to prey density. J.Anim.Ecol. 46: 1 - 19.

Hall, S.J., Todd, C.D. & Gordon. A.D. 1982. The influence of ingestive conditioning on the prey species selection in Aeolidia papillosa (Mollusca: Nudibranchia). J.Anim.Ecol. 51: 907 - 921.

Hancock, D.A. 1974. Some aspects of the biology of the sunstar Crossaster papposus (L.). Ophelia 13: 1 - 30.

Hughes, R.N. 1979. Optimal diets under the energy maximization premise: the effects of recognition time and learning. Amer.Nat. 113: 209 - 221.

Hughes, R.N. 1980. Optimal foraging theory in the marine context. Oceanogr.Mar.Biol.Ann.Rev. 18: 423 - 481.

Hughes, R.N. 1985. Predatory behaviour of Natica unifasciata feeding intertidally on gastropods. J.Moll.Stud. 51: 331 - 335.

Hughes, R.N. & Dunkin, S.de B. 1984a. Behavioural components of prey selection by dogwhelks, Nucella lapillus (L.), feeding on mussels, Mytilus edulis L., in the laboratory. J.Exp.Mar.Biol.Ecol. 79: 45 - 68.

Hughes, R.N. & Dunkin, S de B. 1984b. Effect of dietary history on selection of prey, and foraging behaviour among patches of prey, by the dogwhelk, Nucella lapillus (L.). J.Exp.Mar.Biol.Ecol. 79: 158 - 170.

Hughes, R.N. & Elner, R.W. 1979. Tactics of a predator, Carcinus maenas, and morphological responses of the prey, Nucella lapillus. J.Anim. Ecol. 48: 65 - 78.

Hughes, R.N. & Seed, R. 1981. Size selection of mussels by the blue crab Callinectes sapidus: energy maximizer or time minimizer? Mar. Ecol. Progr.Ser. 6: 83 - 89.

Hubbard, S.F., Cook, R.M., Glover, J.G. & Greenwood, J.J.D. 1982. Apostatic selection as an optimal foraging strategy. J.Anim.Ecol. 51: 625 - 633.

Hylleberg, J., Brock, V. & Jorgensen, F. 1978. Production of sublittoral cockles, Cardium edule L., with emphasis on predation by flounders and seastars. Natura Jutl. 20: 183 - 191.

Jubb, C.A., Hughes, R.N. & ap Rheinallt, T. 1983. Behavioural mechanisms of size selection by crabs, Carcinus maenas (L.), feeding on mussels, Mytilus edulis L. J.Exp.Mar.Biol.Ecol. 66: 81 - 87.

Kim, Y.S. 1969. Selective feeding on several bivalve molluscs by starfish, Asterias amurensis Luken, Bull.Fac.Fish.Hokkaido Univ. 19: 244 - 249.

Kitchell, J.A., Boggs, C.H., Kitchell, J.E. & Rice, J.A. 1981. Prey selection by naticid gastropods: experimental tests and applications to the fossil record. Paleobiology 7: 533 - 552.

Kislalioglu, M. & Gibson, R.N. 1975. Field and laboratory observations on prey-size selection in Spinachia spinachia (L.). Proc. 9th Europ. Mar.Biol.Symp. pp.29 - 41.

Kislalioglu, M. & Gibson, R.N. 1976a. Prey 'handling time' and its importance in food selection by the 15-spined stickleback Spinachia spinachia (L.). J.Exp.Mar.Biol.Ecol. 25: 151 - 158.

Kislalioglu, M. & Gibson, R.N. 1976b. Some factors governing prey selection by the 15-spined stickleback, Spinachia spinachia (L.). J.Exp.Mar.Biol.Ecol. 25: 159 - 169.

Lawton, J.H. 1987. Population dynamics in a patchy world. Nature, in press.

Lucas, J.R. 1983. The role of time constraints and variable prey encounter in optimal diet choice. Amer.Nat. 122: 191 - 209.

MacArthur, R.H. & Pianka, E.R. 1966. On optimal use of a patchy environment. Amer.Nat. 100: 603 - 609.

McLintock, J.B. & Lawrence, J.M. 1981. An optimization study on the feeding behaviour of Luidia clathrata Say (Echinodermata: Asteroidia). Mar.Behav.Physiol. 7: 263 - 275.

Menge, B.A. 1978. Predation intensity in a rocky intertidal community. Oecologia (Berlin) 34: 17 - 35.
Menge, J.L. 1974. Prey selection and foraging period of the predaceous rocky intertidal snail, Acanthina punctulata. Oecologia (Berlin) 17: 293 - 316.
Mittelbach, G.G. 1983. Optimal foraging and growth in bluegills. Oecologia (Berlin) 59: 157 - 162.
Moran, M.J. 1985. The timing and significance of shellboring and foraging behaviour of the predatory intertidal gastropod Morula marginalba Blainville (Muricidae). J.Exp.Mar.Biol.Ecol. 93: 103 - 114.
Moran, M.J., Fairweather, P.G. & Underwood, A.J. 1984. Growth and mortality of the predatory intertidal whelk Morula marginalba Blainville (Muricidae): the effects of different species of prey. J.Exp.Mar. Biol.Ecol. 75: 1 - 17.
Murdoch, W.W. 1969. Switching in general predators: experiments on predator specificity and stability of prey population. Ecol.Mongr. 39: 335 - 354.
Murdoch, W.W. & Oaten, A. 1975. Predation and population stability. Adv. Ecol.Res. 9: 1 - 131.
O'Connor, R.J. & Brown, R.A. 1977. Prey depletion and foraging strategy in the oystercatcher Haematopus ostralegus Oecologia (Berlin) 27: 75 - 92.
Paine, R.T. 1976. Size limited predation: an observational and experimental approach with the Mytilus - Pisaster interaction. Ecology 57: 858 - 873.
Palmer, A.R. 1982. Predation and parallel evolution: recurrent parietal plate reduction in balanomorph barnacles. Paleobiology 8: 31 - 44.
Palmer, A.R. 1983. Growth rate as a measure of food value in thaidid gastropods: assumptions and implications for prey morphology and distribution. J.Exp.Mar.Biol.Ecol. 73: 95 - 124.
Palmer, A.R. 1984. Prey selection by thaidid gastropods: some observational and experimental field tests of foraging models. Oecologia (Berlin) 62: 162 - 172.
Pierce, G.J. & Ollason, J.G. 1987. Eight reasons why optimal foraging theory is a complete waste of time. Oikos 49: 111 - 118.
Ribi, G., Scharer, R. & Ochsner, P. 1977. Stomach contents and size frequency distributions of 2 coexisting seastar species, Astropecten aurancianus and A.bispinsosus, with reference to competition. Mar.Biol. 43: 181 - 185.
Sloan, N.A. 1980. Aspects of the feeding biology of asteroids. Oceanogr. Mar.Biol.Ann.Rev. 18: 57 - 124.
Stickle, W.B. & Bayne, B.L. 1987. Energetics of the muricid gastropod, Thais (Nucella) lapillus. J.Exp.Mar.Biol.Ecol. in press.
Sutherland, W.J. 1982. Do oystercatchers select the most profitable cockles? Anim.Behav. 30: 857 - 861.
West, L. 1986. Interindividual variation in prey selection by the snail Nucella (=Thais) emarginata. Ecology 67: 798 - 809.
Wood, L. 1968. Physiological and ecological aspects of prey selection by the marine gastropod Urosalpinx cinerea (Prosobranchia, Muricidae). Malacologia 6: 267 - 320.

Sexual Difference in Resource Use in Hermit Crabs; Consequences and Causes

Peter A. Abrams

University of Minnesota
Minneapolis, U.S.A.

INTRODUCTION

Most ecological theory assumes that a species consists of identical individuals, or of individuals whose properties are determined by age or size. In many species, however, there are significant differences in the ecological roles of males and females. Such between-sex differences are potentially important, both in determining the nature of population regulation within a species, and in determining the nature of interactions between species. Although ecologists have devoted large amounts of effort to quantifying the differences in resource use between species (e.g. Schoener, 1974), there has been relatively little quantification of differences in resource use between the sexes. The present paper is an attempt to quantify sex-related differences in resource use in two intertidal hermit crab species, and to analyze the causes and consequences of these differences.

Hermit crabs, because of their unique form of resource use, provide excellent subjects for studying population regulation and competition. There is abundant evidence that many hermit crab species compete for empty gastropod shells, both within and between species (Vance, 1972; Spight, 1977; Kellogg, 1977; Hazlett, 1981; Abrams, 1980, 1981a,b; Bertness, 1981a,b,c). Previous studies concerned with intraspecific competition have generally tried to determine whether the supply of empty gastropod shells limited the population size of the hermit crabs (Vance, 1972; Spight, 1977). One study treated the effect of shell exchanges on the relationship between crab and shell size (Scully, 1982). Studies of interspecific competition have tried to determine whether different crab species affect the quality of shells occupied by other crab species (Bertness, 1981a,b,c; Abrams, 1980, 1981a, 1987a,b). The relative amounts of inter- and intra-specific competition have been estimated for several different species assemblages of hermit crabs (Abrams, 1980, 1981b, 1987a,b; Abrams et al. 1986).

Although sexual dimorphism in hermit crab size, and relative use of differently sized shells, has been noted previously (e.g. Fotheringham, 1976; Bertness 1981c, 1982; Blackstone, 1985; Blackstone and Joslyn, 1984), between-sex differences in resource use have seldom been quantified systematically, and their implications for the nature of competition for shells have not been explored. Bertness (1981c) is the most complete treatment of sexual differences in shell use by hermit crabs. He studied three species on the Pacific coast of Panama, and noted that males were more successful than females in obtaining shells in shell fights, and appeared to find and occupy empty shells more rapidly than females in the laboratory. He also showed that males have larger average and maximum sizes in these three species, and attributed the differences between males and females to sexual selection having acted to increase the competitive effort of males.

The remaining sections of this paper: (1) review previous work on competition for shells between *Pagurus hirsutiusculus* and *Pagurus granosimanus*, the two most common intertidal hermit crab species in the Greater Puget Sound Basin, Washington, U.S.A.; (2) quantify differences between males and females in shell utilization and shell-related behaviors; (3) discuss the implications of these differences for the processes of inter- and intra-specific competition; (4) discuss the possible causes of the observed dimorphism in resource use. The final section will also review the available data on sexual dimorphism in other hermit crab species.

EVIDENCE FOR COMPETITION FOR SHELLS WITHIN AND BETWEEN *PAGURUS GRANOSIMANUS* AND *PAGURUS HIRSUTIUSCULUS*

In 1972, Richard Vance demonstrated that addition of empty snail shells increased the population size of *Pagurus hirsutiusculus* on an isolated rocky reef in the San Juan Islands. This remains the most direct demonstration of shell limitation of a hermit crab population. At a second site in the San Juan Islands, Spight (1977) showed that the population size of *Pagurus granosimanus* was correlated with the death rate (during the preceding months) of the most common large species of snail at that site. Vance (1972), Nyblade (1974), and Abrams (1987a) have all noted that these two hermit crab species overlap in their use of habitats and both species and sizes of shells. Vance (1972) and Abrams (1987a) show that the majority of individuals of both species inhabitat shells that are significantly smaller than the preferred size. There have been repeated demonstrations that occupancy of small shells reduces hermit crab fitness (Hazlett, 1981), including demonstrations by Vance (1972) and Nyblade (1974) for the species studied here. Abrams (1987a) showed experimentally that shells intially obtained by individuals of one species would often be obtained by members of the other species of hermit crab if the intial occupant were removed, and the shell replaced in its original locations. Thus, there is strong evidence for both inter- and intra-specific competition for shells.

Abrams (1987a,b) quantified the differences in shell utilization between *P. hirsutiusculus* and *P. granosimanus*. The species differed in habitat use (*P. hirsutiusculus* had a higher upper limit and mean height in the intertidal zone), size distribution of shells used (most of the small shells were occupied by *P. hirsutiusculus* and most of the large shells by *P. granosimanus*), and relative use of different shell species (of the two most common species of small gastropod, *P. granosimanus* preferred *Littorina scutulata* and *P. hirsutiusculus* preferred *Littorina sitkana*). Field experiments suggested that *P. granosimanus* occupied empty shells approximately four times as rapidly as *P. hirsutiusculus*. As a result of these differences, it was estimated that each species experienced approximately an order of magnitude more competition from members of its own species than from members of the other species at sites in the San Juan Archipelago (Abrams, 1987a).

The work reviewed above gathered data on sexual differences in shell utilization in both species of hermit crab, but did not use these differences in estimating the relative magnitudes of inter- and intra-specific competition. Because male hermit crabs can mate with more than a single female (personal observations, Hazlett, this volume), it is possible that the reproductive output of a species is relatively insensitive to the number of males. If there were sufficient differences between male and female shell use, this could alter conclusions about the relative magnitudes of inter- and intra-spedific competition. Competitive effects upon males would have very little effect on the reproductive output of the population. The results presented below suggest that there are significant differences between the sexes in shell and habitat use in both *P. hirsutiusculus* and *P. granosimanus*, but these differences result in relatively minor changes in the previous estimates of relative magnitudes of inter- and intra-specific competition between the species.

Abrams (1987a) should be consulted for additional details about collecting methods and the natural history of these two hermit crab species.

EVIDENCE OF SEXUAL DIFFERENCES IN RESOURCE UTILIZATION: *PAGURUS HIRSUTIUSCULUS* AND *PAGURUS GRANOSIMANUS*

For most species of hermit crab, the probability of an individual using a shell is determined by the shape of the shell (usually determined by the species of gastropod that produced it), the size of the shell, and the type of habitat in which the shell first becomes available (Abrams, 1980, 1987a,b). Differences in these three resource dimensions largely determine the amount of competition for shells between two species or two groups within a species. Differences in relative ability to find newly available shells and to force exchanges of shells (via shell fighting) can also affect the probability of an individual obtaining a given shell, and thus, the magnitude of competitive effects (Bertness, 1981a; Abrams, 1980, 1981a). Various types of damage (e.g. holes in the shell, epiphytic growth on the shell) also affect the probability of use by crabs, but neither different species of hermit crab nor groups within species have been shown to differ significantly in their relative preference for damaged and undamaged shells.

There is a close correlation between crab size and shell size (e.g. Abrams et al. 1986; Hazlett, 1981), and the distribution of body sizes within two different groups provides a rough indication of the amount of overlap in use of differently sized shells. Table 1 presents body size distributions of male and female *P. hirsutiusculus* and *P. granosimanus* from two different sets of sites. Both species at both sites have heavily female biased total sex ratios. For both species at both locations, the sex ratio in small size classes is approximately equal, there is a large excess of females in intermediate size classes, and an excess of males in the largest size classes. It should be noted that this pattern can be consistent with very little or no difference in average male and female sizes, as in the case of the San Juan sample of *Pagurus granosimanus*, or a larger average size for females, as in the case of Outer Coast *P. granosimanus*. The between-sex difference in mean sizes is larger, and the overlap in size distributions less, for *P. hiusutiusculus* than *P. gransosimanus*. The abrupt change from a female biased to a male biased sex ratio in the largest size classes suggests that males are willing to inhabit shells that females consider unsuitable. This conclusion is supported by shell selection experiments (reported below), in which large male *P. hirsutiusculus* were much more likely to move into a larger shell than were large females.

If there were no difference in the relative shell-size occupied by male and female crabs, the above body-size distributions would directly quantify differences in use of different shell sizes. Linear regression analysis

Table 1
Size Distributions of *Pagurus hirsutiusculus* and *Pagurus granosimanus*

	P. hirsutiusculus				*P. granosimanus*			
Shield Length	San Juan		Outer Coast		San Juan		Outer Coast	
	M	F	M	F	M	F	M	F
<1.5mm	18	18	21	28	2	2	3	3
1.5-2	61	154	64	130	21	17	18	22
2.0-2.5	90	212	113	337	17	27	59	65
2.5-3	59	72	168	448	15	21	73	83
3-3.5	49	6	230	271	21	19	86	148
3.5-4	37	2	161	191	17	26	109	208
4-4.5	10	0	74	120	19	45	99	422
4.5-5	6	0	41	42	20	37	78	221
5-5.5	1	0	13	2	12	8	58	147
5.5-6	0	0	10	0	11	4	27	92
6-6.5	1	0	0	0	10	0	31	37
≥6.5	2	0	0	0	4	0	14	8
Total	314	464	895	1558	169	206	665	1358
Average M	2.63 mm		3.22 mm		3.66 mm		4.01 mm	
Average F	2.34 mm		2.93 mm		3.64 mm		4.19 mm	
Overlap	.674		.797		.793		.833	

Crab sizes are shield lengths. Overlaps are calculated using Schoener's (1970) proportional similarity.

was used to determine the relationship between crab size (measured by shield length) and shell size (greatest linear dimension) for males and females of each species of hermit crab at the San Juan sites. The analysis was done for several of the most commonly occupied shell species. There was no significant difference (t-test, $p = 0.05$) between either regression slopes or intercepts for male and female *P. granosimanus* occupying the same shell species (for *Littorina scutulata*, *Calliostoma ligatum*, *Searlesia dira*, and *Nucella lamellosa*). Crab size-shell size relationships of males and females which had selected shells from a large array of empty shells in the laboratory also did not differ significantly from each other. When the linear dimensions of different gastropod species were converted to a common size measurement (as described in Abrams, 1987a), the overlap in distributions of occupied shells for males and females was 0.789, virtually identical to the overlap in body sizes.

There were significant differences in shell size-crab size relationships between male and female *P. hirsutiusculus*. Regression lines for all commonly occupied shell species (*Littorina sitkana*, *Littorina scutulata*, and *Nucella emarginata*) had lower slopes for males than for females. For *Littorina sitkana*, shell size was related to body size by:

(shell length in mm) = 2.16 + 2.10 (crab shield length) ($n = 102$) for males, and

(shell length in mm) = 0.79 + 2.82 (crab shield length) ($n = 93$) for females.

Male and female regression lines crossed at a small crab size (1.9mm). Thus, most males occupied smaller shells than similarly sized females, and this difference became more pronounced as crab size increased. The

Table 2
Shell Species Use by Male and Female Hermit Crabs, San Juan Island

A. Small *P. granosimanus* (<3.5 mm)	M	F
Littorina sitkana	186	201
L. scutulata	55	63
B. Large *P. granosimanus* (>4.0 mm)	**M**	**F**
Searlesia dira	211	351
Nucella emarginata	189	328
Calliostoma ligatum	59	101
C. Small *P. hirsutiusculus* (<2.3 mm)	**M**	**F**
Littorina sitkana	374	436
L. scutulata	88	233

Figures give numbers of crabs collected in each shell species from collections reported in Abrams (1987a).

difference was also present in crabs that had been allowed to select shells from a large array of empty shells in the laboratory; these regression lines for *P. hirsutiusculus* in selected *Littorina sitkana* shells were:

(shell length) = 1.81 + 2.50 (crab shield length) (n = 52) for males

(shell length) = 1.55 + 2.95 (crab shield length) (n = 67) for females

Data regarding shell occupancy at Outer Coast sites showed the same pattern. Regression lines for males and females occupying *L. sitkana* shells were:

(shell length) = 3.62 + 1.56 (crab shield length) (n = 204) for males

(shell length) = 2.21 + 2.16 (crab shield length) (n = 351) for females

Using the complete set of San Juan data, shell sizes were converted to a common scale (as described in Abrams (1987a)), and the frequency distribution of shell sizes occupied was calculated for both sexes. The overlap in shell size distributions was 0.770, significantly larger than the overlap in body size distributions.

The shell occupancy data for both crab species show the same qualitative pattern as the size overlap data; equal numbers of males and females in the smallest shell sizes, with a large excess of females in intermediate size classes and an excess of males in the largest size classes. The numerical dominance of male *P. hirsutiusculus* in the largest shell size classes is less pronounced than their relative abundance in the largest body size classes.

The amount of competition between two groups of crabs is also affected by the overlap in the distributions of shell species used by similarly sized members of each group. Table 2 compares the shell species occupied by similarly sized males and females for both *P. hirsutiusculus* and *P. granosimanus* collected at the San Juan Island sites. There is no significant difference between male and female *P. granosimanus* in their shell species utilization. Male *P. hirsutiusculus* use a significantly greater proportion of *Littorina sitkana*, and a significantly lower proportion of *Littorina scutulata* shells than do females. The greater utilization of *L. sitkana* by males is not an indirect result of differences in habitat use by males and females (see below), because the same pattern may be seen in shell species utilization in a single tidepool; for crabs in the 1.8mm-2.3mm (shield length) size range, males had a greater ratio of *L. sitkana* to *L. scutulata* occupied than did females in 13 of 17 tidepools surveyed, with equal ratios in two of the remaining four pools. Shell selection experiments showed that both sexes preferred *L. sitkana* to *L. scutulata* (Abrams 1987a). Males collected in *L. scutulata* were more likely to switch to *L. sitkana* when offered large numbers of empty *L. sitkana* than were females; 28 of 30 males in such experiments switched shells, compared to 21 of 30 females.

Male *P. hirsutiusculus* differed from females in having a greater tendency to exchange shells when offered a large variety of empty shells in the laboratory. Shell selection experiments were carried out for 14 sets of 20 individuals from 10 different microsites. The proportion of males moving into a new shell exceeded the proportion of females in 10 of 14 cases; 69% of all males switched shells compared to 51% of females. This difference is more pronounced when the analysis is restricted to large individuals (greater than 3.3mm shield). In this group, 78% of males switched shells compared to 24% of females.

I also examined the possibility that shell fighting might affect the distribution of shells between males and females. Groups of twenty individuals of each sex occupying marked, recently selected shells were released in the tidepools from which they had previously been collected at Deadman Bay, San Juan Island. The experiment was done for both *P. hirsutiusculus* and *P. granosimanus*. Tidepools were chosen to have both species present in large numbers and to have sex ratios of both species with the less abundant sex making up at least 40% of the population. All marked shells were collected three days after release, and the species and sex of the occupant determined. Recovery rates of marked shells ranged from 40% to 75%. None of the shells was recovered containing an individual of a different sex of the same species, suggesting that shell exhanges forced by shell fighting were not an important aspect of competition. More extensive experiments on shell fighting were reported previously (Abrams, 1987a).

Previous work on interspecific resource partitioning between hermit crab species (Abrams 1980, 1981a,b, 1987a,b) has generally found that differences in habitat use have the greatest effect in reducing the probability that an individual of one species will deprive an individual of a second species of a shell. Sex-related differences in habitat use among hermit crabs have apparently not been reported previously. Differences in habitat use between the sexes of *P. granosimanus* and *P. hirsutiusculus* were investigated by determining sex ratios in a variety of microhabitats on San Juan Island. Crabs were collected at low tide (as described in Abrams 1987a) from a variety of microhabitats having high population densities of hermit crabs: in tidepools, under rocks, and in beds of the alga *Fucus distichus*. Each sex could be described by a distribution over microhabitats, and the overlap of these distributions was determined using a measure of proportional similarity (Schoener, 1970). Results for several sampling periods are given in Table 3. Both *P. hirsutiusculus* and *P. granosimanus* have significant between-sex differences in habitat utilization. Relatively high tidepools usually contained an excess of male *P. hirsutiusculus*, and microhabitats that did not contain standing water at low tide (i.e. everything other than tidepools) generally had an excess of females. The overlap of males and females was recalculated for the smallest and largest 50% of the individuals of each sex; these size-specific overlaps did not differ significantly from each other.

Table 3
Habitat Overlap Figures for Male and Female
Hermit Crabs at San Juan Island Sites

Period	*P. hirsutiusculus*	*P. granosimanus*
Summer 1983	.572	--
Summer 1984	.588	.776
Summer 1985	.536	.759
large individuals	.542	.629
small individuals	.533	.833
Spring 1985	.642	.741

Overlaps were calculated as described in text.

Although some microhabitats contained primarily a single sex of *P. granosimanus*, there were no obvious environmental factors that distinguished male and female microhabitats. Sexual differences in habitat use were greater for large (the largest 50% of the population) than for small (smallest 50%) *P. granosimanus*. Repeated samples of several microsites during the summers of three different years did not reveal any reversals of the relative abundances of the two sexes. Samples of a predominantly male *P. hirsutiusculus* pool high in the intertidal zone during the spring of 1985 revealed an approximately 1:1 sex ratio. Because of the lack of midday low tides in nonsummer months, high pools may be safer habitats at those times. As Table 3 shows, however, the difference between male and female habitat distributions did not disappear during the spring. There were no obvious seasonal shifts in the overlap in habitat distributions of male and female *P. granosimanus*.

Some differences in habitat distributions could occur if size- or sex-specific mortality factors differed between microhabitats and if crabs very seldom moved between microhabitats. Although well over 90% of the individuals of each of these species do not move between microsites on a given day, over 50% of marked individuals of both crab species returned to their original pool within 24 hours of being moved to a different pool 10-15 meters away (unpublished observations). Thus, it appears that active habitat selection is involved in producing the observed distributions.

Before considering the consequences of these differences in resource use for the process of competition, I will review the less extensive data available for a third intertidal hermit crab that has some overlap in geographical distribution with the two species discussed above. *Pagurus samuelis* occurs sympatrically with the two species discussed above at locations on the outer coast of the Olympic Peninsula, but is absent from the San Juan Archipelago (Nyblade, 1974; Abrams, 1987b). Less extensive observations were made on sexual differences in resource use in this species at several locations on the Olympic Peninsula of Washington and Vancouver Island. The pooled data for these sites are discussed below. The size distribution of *P. samuelis* is similar in form to that of *P. granosimanus*. There are approximately equal numbers of males and females less than 2 mm in size (43 males and 47 females). There is a large excess of females in the intermediate size classes (988 females and 699 males), reaching a maximum ratio of 2.15 : 1 in the 4.67-5mm shield length size class. Of the 45 largest individuals, 33 are male. The overlap in size distributions is

0.9114. The total sex ratio is 775 males:1047 females. The mean female size is larger than mean male size by 0.02 mm. Unlike *P. hirsutiusculus*, male *P. samuelis* consistently occupy slightly larger shells relative to their body size than do females, and this difference is greatest for relatively small individuals. Regression lines for *P. samuelis* occupying *Littorina sitkana* shells from a single site are as follows:

For males: Shell length = 3.13 + 2.14(shield length) n = 229

For females: Shell length = 1.95 + 2.38(shield length) n = 277

Both slopes and intercepts are different at the 0.05 level (t- test), and the lines predict that males will occupy larger shells for all body sizes less than 4.91mm shield length, which includes over 90% of all individuals. The same pattern was found when samples from single tidepools were analyzed, so the above result is not a consequence of differences in habitat use. Only a small amount of sex ratio data are available for individual tidepools for *P. samuelis*, but samples from five very high intertidal pools at Mukkaw Bay, Washington State, USA all had male biased sex ratios. Given the female bias for the species as a whole, this suggests some sexual differences in habitat preferences.

CONSEQUENCES OF MALE-FEMALE RESOURCE PARTITIONING FOR INTRA- AND INTER-SPECIFIC COMPETITION

Evidence of multiple matings by individual male crabs suggests that the hermit crab species studied thus far are characterized by female demographic dominance; the egg production is insensitive to the number of males (or their shell quality). More studies must be done to confirm this, but if it is true, it suggests that competitive effects are best measured by considering only effects on the females in the population.

The data on resource partitioning reviewed above suggests that a female hermit crab of *P. hirsutiusculus* or *P. granosimanus* will experience larger competitive effects from females of its own species than from males of its species. The magnitude of the difference between male and female effects on females may be estimated as follows, using *P. hirsutiusculus* as an example. (1) Because competitive effects are being measured as effects upon reproductive output, size classes of females should be weighted by the relative reproductive output of an individual in each size class. Because reproductive output is proportional to mass, it is proportional to the cube of shield length. (2) Considering males and females of equivalent sizes, the probability of a male depriving a female of a shell relative to the probability that another female will do so is roughly proportional to overlap in habitat distributions, overlap in species of shell used at a given microsite, and to the relative shell acquisition ability of males.

The methods for estimating relative intensities of competition are discussed in Abrams (1987a). The results of reanalyzing the data in Abrams (1987a) for San Juan Island sites are: (1) Female shell supply for *P. hirsutiusculus* is affected 3.6 times more by other females than by males. This results from intersexual differences in habitat use, shell preference, and the high female:male ratio in size classes with the largest number of large-bodied females. This figure (3.6) takes into account the larger shell acquisition rate of males.

(2) Female shell supply for *P. granosimanus* is affected 2.5 times more by other females than by males within its species. This figure results from habitat differences and from the female biased sex ratio in size classes with many females.

(3) The effect of *P. granosimanus* on *P. hirsutiusculus* is slightly larger than that estimated considering both sexes of the *P. hirsutiusculus* population (Abrams, 1987a, p. 242). Effects of *P. granosimanus* on the

first three size classes of *P. hirsutiusculus* (see Table 11, p. 242, Abrams, 1987a) increased by 28%, 23%, and 15% respectively. These three size classes contain over 80% of the females. This does not, however, change the conclusion that the intraspecific competitive effects experienced by *P. hirsutiusculus* are much larger than the interspecific effects.
(4) The effect of *P. hirsutiusculus* on *P. granosimanus* is not significantly different from that estimated considering both sexes of *P. granosimanus*. None of the values of the size-specific competitive effects calculated in Abrams, 1987a (Table 11) is changed by more than 5% if male *P. granosimanus* are removed from the calculations.

CAUSES OF OBSERVED PATTERNS OF RESOURCE PARTITIONING

Three factors seem to be the most likely candidates for the causes of sexual dimorphism in resource use. The first is the differential availability of energy for growth. Males do not have to produce eggs, and can therefore devote more energy to growth. Bertness (1981a) found that males of a Pacific Panamanian hermit crab grew approximately 3 times as fast as females in the laboratory when both had access to an effectively unlimited number of shells and equal access to food. He suggested that this might explain the sexual dimorphism of all 3 common intertidal species he studied in Panama. This idea will be referred to as the growth hypothesis.

The second hypothesis is the sexual selection hypothesis (Bertness, 1981c). The basic idea is that fitness increases more rapidly with size for males than for females because of intrasexual competition between males to obtain matings. Female egg production is roughly proportional to body mass in a number of species of hermit crab (Nyblade, 1974). The benefits of larger size for males would be expected to be greater than those for females because: (a) maximum reproductive output is greater in males than in females, due to the ability of a male to fertilize more than one female; and (b) larger size in males increases both the number of potential mates and the expected success of the individual in intrasexual competition over possession of females. A male must be larger than a female to mate with her; this is true of all pairs of all species that I have observed engaging in copulatory or precopulatory behavior, and Bertness (1981c) reports similar observations. In many species, the male drags the female around by the shell for a significant period of time before the female is ready to mate (Hazlett, 1966); this generally requires that the male be larger. The males in precopulatory pairs of *P. hirsutiusculus* that were collected in tidepools on San Juan Island, Washington were significantly larger than the average size of males in the same tidepool. (The males in 25 precopulatory pairs collected in four different tidepools averaged 1.62 times larger in shield (anterior cephalothorax) length than the average sized male in their respective tidepools.) I have observed males fighting over the possession of a female in several hermit crab species, and in all takeovers that I have observed (15 cases for *P. hirsutiusculus*), the successful male was larger than the other male. These considerations suggest that for sufficiently large crab sizes, male reproductive output will increase more rapidly with size than female reproductive output.

The third hypothesis is intersexual character displacement as a result of competition for shells. Selander (1966) proposed that some cases of sexual dimorphism in hawks might be the result of divergence of the sexes to avoid competition. A recent mathematical analysis by Slatkin (1984) suggests that such displacement can occur by a mechanism very similar to displacement between species. If this is true, sexual dimorphism might occur even in the absence of any inherent differences in growth potential and in the absence of sexual selection.

Table 4
Size Distributions and Sex Ratios for Subtidal Hermit Crabs from San Juan Archipelago

Species	No. M/No. F	Av. M/Av. F	Overlap	No. M/No. F largest 10%
Elassochirus tenuimanus	203/231	5.60/5.27	.777	30/13
Labidochirus splendescens	29/39	6.05/5.90	.557	5/1
Paguristes turgidus	41/26	11.60/9.60	.754	6/1
Pagurus ochotensis	107/104	8.15/9.75	.758	11/10
Pagurus kennerlyi	155/129	7.05/5.93	.701	25/4
Pagurus dalli	219/219	4.06/3.82	.776	33/11
Pagurus capillatus	393/204	5.50/5.25	.661	37/14
Pagurus beringanus	274/322	5.59/5.10	.780	47/13
Pagurus aleuticus	46/26	4.48/5.19	.796	4/3

Data are from collections reported in Nyblade (1974) and Abrams et al. (1986).

A fourth possible reason for sexual dimorphism is that the ecological roles of shells are different for male and female crabs. However, shells perform much the same function in male and female crabs. These functions include protection from predators, cannibalism, and dessication (Hazlett, 1981). For females (but not males), the shell must protect the eggs, which are carried outside the body, on the pleopods, often for extended periods of time. However, there are a number of species (see below) in which size and species preferences for different shells has been shown to be the same for male and female hermit crabs. Thus, this hypothesis will not be considered below.

These three hypotheses are not mutually exclusive alternatives, and it is likely that all three affect resource use in many hermit crab species. In addition, these hypotheses about causes do not provide a complete explanation of magnitudes of sexual dimorphism. The differences in food supply and risk of mortality in different habitats place constraints on the magnitude of the intersexual difference in habitat use. Similarly, size-specific mortality factors will influence size dimorphism. Nevertheless, it is of interest to review the available information to determine whether one of these three causal explanations accounts for any of the observed resource use patterns.

Evaluating hypotheses about causes of observed resource utilization patterns requires comparative data. The three species discussed thus far do not provide a sufficient basis for this. Information about sexual differences in size in a group of subtidal hermit crab species and differences in size at several locations in a terrestrial hermit crab species are reviewed below.

Table 4 presents sex ratios and summary statistics for size distributions for nine of the hermit crab species occurring subtidally in the vicinity of San Juan Island, Washington, USA (see Abrams et al. 1986). Most of the samples upon which this table is based were obtained by dredging (Nyblade, 1974; Abrams et al., 1986), and small-scale habitat segregation could not be accurately assessed. It is possible that the large excess of males in *P. capillatus* results from sampling a predominantly male habitat is a species with sex-biased habitat segregation. Almost all of the *P. capillatus* were obtained from a single location which was near the reported (Hart, 1982) upper limit of the depth distribution of this species. All of the *Paguristes turgidus* obtained at five of the six sites dredged in 1983 were males (the sixth site had an approximately equal sex ratio); there is no doubt that this species exhibits pronounced sexual differences in habitat use. *Pagurus ochotensis* and *P. aleuticus* appear unusual in having a larger average size for females than males. In each case, this results from an excess of males recorded for the smaller size classes. This may be an error, because crabs were classified as male if they lacked gonopores on the third pair of limbs; it is possible that external sexual expression had not yet developed. However, both species lack the male bias in larger size classes seen in the other species. It would be interesting to determine whether there are corresponding differences in mating systems that might reduce the presumed large male mating advantage.

P. kennerlyi, P. beringanus, E. tenuimanus, P. dalli, and *Paguristes turgidus* have body size patterns similar to the intertidal species; equal sex ratios in the smallest size classes, female bias in intermediate classes, and male bias in the largest size classes. Sufficient data on crab-size, shell-size relationships is available for *P. dalli, E. tenuimanus, P. kennerlyi*, and *P. beringanus* to show that similarly sized males and females occupy similarly sized shells. Thus, shell-size overlaps are similar to body-size overlaps.

Unlike the intertidal species, there are four cases of significantly male biased sex ratios and two cases of 1:1 sex ratios out of the nine subtidal species surveyed. None of the species represented by sample sizes greater than 100 individuals has a sex ratio as strongly female biased as that of all three intertidal species discussed above. As a result, the competitive effect of males on females is larger than for the three intertidal species discussed above.

The data on subtidal species is especially useful in evaluating the passive growth difference model for sexual dimorphism. If this were the main factor affecting the amount of dimorphism, then dimorphism should be greatest in the species with the greatest reproductive output. Nyblade (1974) measured the reproductive output of all of the species listed in Table 4 except for *Pagurus aleuticus*, and for the three intertidal species discussed above. Using weight of eggs produced per year as a measure of reproductive effort (data from Nyblade, 1974, p. 155), one finds a correlation coefficient between (mean male size)/(mean female size) and reproductive output of .009. Thus, the passive growth model fails as a predictor of the amount of sexual size dimorphism.

Sexual selection and intersexual character displacement are both expected to have larger effects in species that have high population densities. This would argue for greater dimorphism among intertidal species because of the generally lower population densities in the subtidal zone (Abrams et al., 1986). The ratio of mean male size to mean female size is actually greater in most subtidal species than in the intertidal ones. The ratio of mean sizes is, however, a poor measure of sexual dimorphism in resource use strategies. The more biased sex ratio of intertidal species suggests that the male-female differences in this group have a greater effect on survival. In any case, however, the observations do no distinguish the relative importances of sexual selection and character displacement in causing differences in shell use.

Table 5
Size Distributions and Sex Ratios for *Coenobita compressus*

Site	No. M/No. F	Av. M/Av. F	Overlap	No. M/No. F largest 10%
Secas #1	117/98	10.09/9.24	.685	20/2
Secas#2	34/45	4.63/5.24	.773	2/6
Venado Beach	47/35	13.72/11.01	.529	8/1
Bahia Honda	36/48	8.06/8.33	.701	5/3
Coiba	39/23	13.74/11.52	.630	5/1
Jicaron	204/218	10.08/9.65	.904	25/18
Boyscout Beach	173/238	12.70/12.64	.882	25/15

Data are from collections described in Abrams (1978)

Observations of a single species in which males and females are known to have identical shell-size and shell-species preferences can be used to show the effect of demographic variables on dimorphism, independent of differences in the 3 causal factors. The terrestrial hermit crab, *Coenobita compressus*, has apparently identical male and female shell preferences and use (Abrams, 1978) on the Pacific coast of Panama. Sexual dimorphism at several different sites is described in Table 5. There is considerable variation in size overlaps, ratios of mean sizes, and overall sex ratios. There are two sites at which females are slightly larger than males in mean size; both of these sites have a female biased sex ratio, while the majority of sites have male biased sex ratios. The male bias was unexpected because sexual selection is expected to operate on *Coenobita compressus*, and this would normally result in greater male mortality than if there were no competition for matings. It may be that females at some sites incur heavy mortality when they enter the water to release their eggs.

SUMMARY

Differences between male and female hermit crabs in shell and habitat use, and shell-related behaviors have been shown for *Pagurus hirsutiusculus* and *Pagurus granosimanus*. Less extensive data on a third intertidal species, *Pagurus samuelis,* suggests that it also exhibits between-sex differences. These differences result in females with high reproductive output competing primarily with other females within each species. Taking male-female differences into account increases a previous estimate of the competitive effect of *P. granosimanus* on *P. hirsutiusculus*, but the effect remains much smaller than the effect of *P. hirsutiusculus* on its own shell supply. Three potential causes of intersexual differences are discussed, and data from other hermit crab species are reviewed.

ACKNOWLEDGMENTS

This work was supported in part by NSF grant BSR-8306998 to the author. Carl Nyblade contributed data, and Janet Pelley and Sallie Sheldon helped with field and laboratory work. Janet Pelley also assisted with last minute manuscript preparations, for which I am very grateful.

LITERATURE CITED

Abrams, P.A. 1978. Shell selection and utilization in a terrestrial hermit crab, Coenobita compressus. Oecologia 34:239-253.

Abrams, P.A. 1980. Resource partitioning and interspecific competition in a tropical hermit crab community. Oecologia 46:365-379.

Abrams, P.A. 1981a. Shell fighting and competition between two hermit crab species in Panama. Oecologia 51:84-90.

Abrams, P.A. 1981b. Alternative methods of measuring competition applied to two Australian hermit crabs. Oecologia 51:233-240.

Abrams, P.A. 1981c. Competition in an Indo-Pacific hermit crab community. Oecologia 51:241-249.

Abrams, P.A. 1987a. An analysis of competitive interactions between three hermit crab species. Oecologia, 72:233-247.

Abrams, P.A. 1987b. Resource partitioning and competition for shells between intertidal hermit crabs on the outer coast of Washington State. Oecologia, 72:248-258.

Abrams, P.A., C.F. Nyblade, and S. Sheldon. 1986. Resource partitioning and competition for shells in a subtidal hermit crab species assemblage. Oecologia 69:429-445.

Bertness, M.D. 1981a. Competitive dynamics of a tropical hermit crab assemblage. Ecology 62:751-761.

Bertness, M.D. 1981b. Pattern and plasticity in tropical hermit crab growth and reproduction. Am. Nat. 117:754-773.

Bertness, M.D. 1981c. Interference, exploitation, and sexual components of competition in a tropical hermit crab assemblage. J. Exp. Mar. Biol. Ecol. 49:189-202.

Bertness, M.D. 1982. Shell utilization, predation pressure, and thermal stress in Panamanian hermit crabs: an interoceanic comparison. J. Exp. Mar. Biol. Ecol. 64:159-187

Blackstone, N.W. 1985. The effects of shell size and shape on growth and form in the hermit crab Pagurus longicarpus. Biol. Bull. 168:75-90.

Blackstone, N.W., and Joslyn, A. R. 1984. Utilization and preference for the introduced gastropod Littorina littorea by the hermit crab Pagurus longicarpus (Say). J. exp. mar. Biol. & Ecol. 80:1-9.

Fotheringham,N. 1976. Population consequences of shell utilization by hermit crabs. Ecology 57:570-578.

Hart, J.F.L. 1982. Crabs and their relatives of British Columbia. Hand book, British Columbia Provincial Musuem. Victoria, B.C.

Hazlett, B.A. 1966. Social behavior of the Paguridae and Diogenidae of Curacao. Stud. Fauna Curacao 23:1-143.

Hazlett, B.A. 1981. The behavioral ecology of hermit crabs. Ann. Rev. Ecol. Syst. 12:1-22.

Kellogg,C.W. 1977. Coexistence in a hermit crab species ensemble. Biol. Bull. 153:133-144.

Nyblade, C.F. 1974. Coexistence in sympatric hermit crabs. Ph.D. thesis. University of Washington. Seattle.

Schoener, T.W. 1970. Nonsynchronous spatial overlap of lizards in patchy habitats. Ecology 51:408-418.

Schoener, T.W. 1974. Resource partitioning in ecological communities. Science 185:27-39.

Scully, E.P. 1982. The behavioral ecology of competition and resource utilization among hermit crabs. In: Studies in adaptation: the behavior of higher crustacea. Ed. by S. Rebach and D. Dunham. New York, J. Wiley.

Selander, R. K. 1966. Sexual dimorphism and differential niche utilization in birds. Condor, 68:113-151.

Slatkin, M. 1984. Ecological causes of sexual dimorphism. Evolution 38:622-630.

Spight, T. 1977. Availability and use of shells by intertidal hermit crabs. _Biol. Bull._ 152:120-133.
Vance, R.R. 1972. Competition and mechanism of coexistence in three sympatric species of intertidal hermit crabs. _Ecology_ 53:1062-1074.

Stabilizing Processes in Bird-Prey Interactions on Rocky Shores

Philip A. R. Hockey and Alison L. Bosman

University of Cape Town
South Africa

INTRODUCTION

In recent years, rocky shores have become the focus of many ecologists endeavouring to understand the processes leading to patterns in community organization (Paine, 1974; Menge, 1976; Menge and Lubchenco, 1981; Underwood et al., 1983; Dethier, 1984; Connell, 1985; Hartnoll and Hawkins, 1985; Paine et al., 1985; Marsh, 1986). The nature and frequency of disturbance, either physical (Paine and Levin, 1981; Jara and Moreno, 1984; Ebeling et al., 1985), chemical (Bosman and Hockey, 1986), biological (Johnson et al., in press) or a combination of these factors (Menge, 1976; Dethier, 1984), have been a central point in the debate on the causes of pattern. Predatory processes have received much attention and have been found to be both stabilizing (Connell, 1961; Paine, 1974; Hughes, 1985) and potentially destabilizing (Underwood et al., 1983; Katz, 1985; Johnson et al., in press), although the definition of stability or persistence within communities is scale-dependent (Connell and Sousa, 1983) and therefore varies with the objectives of an individual study. Many of the classical demonstrations of the impact of predators are derived from manipulations of predator populations (e.g. Paine et al., 1985). Although the effects of such manipulations are often dramatic, evidence for the frequency, or even occurrence of such phenomena in nature is lacking (Connell, 1985), and analysis of the stability of predator-prey relationships in natural communities has remained an unexplored aspect of population interactions (Katz, 1985).

Earlier studies of predation concentrated almost exclusively on the predatory behaviour of invertebrates (e.g. Menge, 1976, 1978; Menge and Sutherland, 1976). A few studies document dramatic predation intensity by vertebrate predators, including birds (Feare, 1966, 1977), fish (Cook, 1981) and mammals (Simenstad et al., 1978), and the potential importance of such mobile predators is highlighted by Edwards et al., (1982). They note that effects of such predators are "hard to demonstrate and easy to overlook or misinterpret, but assuming them to be unimportant may perpetuate oversimplified models of community interactions".

Birds are widespread and frequently abundant components of rocky intertidal communities, with high mobility and energy requirements relative to invertebrate and fish predators. However, few attempts have been made to integrate studies of avian and invertebrate ecology in a littoral environment, and the extent to which birds impose selective forces on rocky shore invertebrates is therefore largely unknown (Feare and Summers, 1985). Some studies report heavy predation by birds on some prey species (e.g. Frank, 1982; Hockey and Branch, 1984) suggesting that birds may sometimes play a very important role in community dynamics.

This study aims to elucidate the relationships between a suite of sympatric avian predators and their respective prey populations on rocky mainland shores and offshore islands of the southwestern Cape, South Africa. An assumption is made that there is temporal long-term stability in this system over a geographical range spanning at least 100 km of coast. Some historical evidence exists for the avian communities having been similar over the past 300 years (Hockey, 1987), and analyses of prehistoric middens of indigenous peoples dating back 10,000 years show that there have been no significant changes in dominant invertebrate species present on the shore during this period (Parkington, 1976).

Rocky intertidal invertebrate communities of the southwestern Cape typically are dominated in the lowshore by mussels, and in the mid- and high-shore by the limpet *Patella granularis*. Mats of foliose algae are uncommon in the mid- and high-shore on mainland coasts, but are prevalent at islands (Bosman and Hockey, 1986).

The rocky shore avifauna of the area may be divided into three categories, based on foraging behaviour and dispersion patterns - two categories are of individual resident species, one is of a group of migrants. Of the dominant resident species, African Black Oystercatchers *Haematopus moquini* are territorial and specialized in their foraging (on limpets and mussels) in contrast to Kelp Gulls *Larus dominicanus* which are itinerant and feed opportunistically. Several smaller migratory species are present during the austral summer, notably Turnstones *Arenaria interpres*, Sanderlings *Calidris alba* and Curlew Sandpipers *C. ferruginea*. These migrant species prey predominantly on the infauna of mussel and algal beds.

Given the suggestion of long-term, large-scale community persistence, this study aims to examine interactions between these different avian predators and their prey, with special reference to explaining processes which might lead to stability in the relationships.

STUDY AREA AND METHODS

Avian community structure and diet

The study was concentrated at small (9-46 ha) granitic islands in Saldanha Bay, southwestern Cape, South Africa, and along mainland rocky shores between Saldanha Bay and Cape Columbine to the north (Fig. 1). All shorebirds along the 1.6 km shore of Marcus Island were counted, at low tide, at least once a month between May 1983 and January 1985.

Diet of African Black Oystercatchers was determined by a) direct observation, and b) collections of shells from 'middens' where adults feed their chicks (Hockey and Underhill, 1984). Diet of Kelp Gulls was determined from a) regurgitated pellets, b) collections of shells broken on 'anvils' and c) analysis of stomach contents. Monthly collections of pellets were made at fixed locations at Marcus Island between March 1983 and August 1984. Forty-three pellets were analysed in detail to determine species composition and relative abundance of prey species. Undamaged mussel valves recovered from pellets were measured (n = 5 325). Kelp Gulls drop large mussels onto anvils to break them open (Siegfried, 1977), and shells dropped at four anvils on Marcus Island, ranging in size from 21m^2 to 79m^2 and totalling

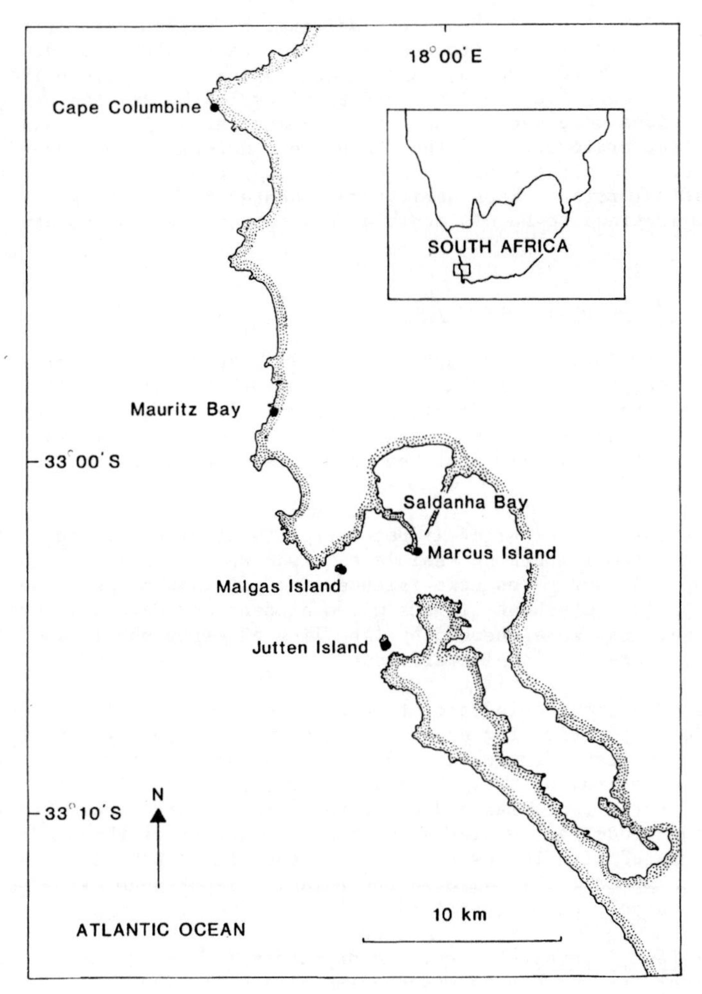

Fig. 1. Map of the study area

$217m^2$, were collected monthly between March 1983 and December 1984. All material was cleared from each anvil at each visit. All shells were identified to species and measured when possible: shells too badly damaged to be measured were assumed to have the same size distribution as measurable shells. The flesh energy content of shells on each anvil at each visit was calculated from equations in Griffiths (1981), and was standardized to kJ per $100m^2$ of anvil per day. The total area of the island used as anvils was estimated using line transects. Mean monthly energy values determined from the four study anvils were extrapolated to the whole island to provide an estimate of mussel-flesh energy obtained by gulls per day from anvil-dropped mussels. Using a DEE for Kelp Gulls of 800kJ (Walsberg, 1983), the number of gulls (per day) which could be supported from this food source alone was calculated. Diet of Turnstones (n = 26) was determined from stomach analyses.

Spatial utilization of the intertidal area by shorebirds for feeding was determined from 247h of instantaneous activity scans (Altmann, 1974), at intervals of five or 10 minutes, made between January and May in 1983 and 1984. Scans were made over full or half tidal cycles by day and night: night observations were made using a 4.7 X image-intensifier with a laser. Foraging habitat preferences of the birds were determined by recording their activities in eight shore habitats ranging from the high- water mark to the infratidal fringe. Each habitat represented an assemblage sufficiently distinct to be discernible from the observation point.

Effects of birds on prey populations

Limpets. Population demography of *Patella granularis* was determined at Jutten Island (high oystercatcher density) and on the mainland at Cape Columbine (low oystercatcher density) (Hockey, 1983). Demography of limpets in areas accessible and inaccessible to oystercatchers was compared, and was related to prey size preferences of oystercatchers (as determined from shells in middens).

To determine the extent of limpet mortality attributable to oystercatchers, 500 limpets accessible to birds were labelled (using white epoxy) at each of four sites (two islands, two mainland sites). The subsequent survival of these limpets was monitored at monthly intervals. (Rates of label loss were checked by labelling 50 empty shells and glueing them onto the shore: no labels were lost.)

Mussels. The inevitable rate of natural mortality resulting from intraspecific competition for space in the mussel *Choromytilus meridionalis* was determined by recording packing densities at different sizes and relating these to growth rate. Predation rates, and prey size preferences of three predators, two avian and one invertebrate, were determined in the field, and then modelled cumulatively. The consumption of the predator array was compared with losses due to intraspecific competition (Griffiths and Hockey, in press). This allowed the relative importance of birds as predators of mussels to be quantified.

Small infaunal invertebrates. The predatory impact of small migrant waders on algal infauna was assessed using bird exclusion cages. Two $1m^2$ cages were placed in the midshore at Marcus Island where beds of *Enteromorpha* and *Ulva* spp. occur naturally. The cages stood 150mm above the substratum, with a frame of 12mm galvanized steel rods and a 40 x 20mm diamond steel mesh. The legs of the cage were drilled approximately 300mm into the granite substratum. The area within each cage was cleared of algae and heat-dried. Twenty-five 100 X 100mm high intensity polystyrene plates were glued to the rock as an artificial substratum for algal settlement and growth (McQuaid, 1981; Bosman and Hockey, 1986). The periphery of the cleared area was delimited with copper-based anti-fouling paint as an additional precaution against intrusion by large grazing herbivores. An uncaged control was established in an identical manner adjacent to each cage. The experiment was set up six months before the first plates were removed, to allow development of a mature algal community comparable with that on the surrounding rock. Five settlement plates were removed from each cage and each control in January, March, April, May, June and November, and algal infauna was compared between the two treatments. To eliminate possible bias resulting from differential algal growth inside and outside cages, all algae were removed from the plates, oven-dried, and infauna counts were expressed for each plate as numbers per 2g AFDW of algae. Differences in mean invertebrate numbers between cages and controls were compared using Mann-Whitney 'U' tests.

Table 1. Comparison of bird densities (per km of shore) at Marcus Island (this study) and on 33.8 km of rocky shore between Saldanha Bay and Cape Columbine (data from Underhill and Cooper, 1984). Densities of Turnstones and migrant waders at Marcus Island refer to summer counts (October to March, n = 10 counts), and of oystercatchers and gulls refer to the whole year (n = 21 counts). Data from Underhil and Cooper (1984) refer to a single midsummer count

Species	Marcus Island mean	S.D.	Mainland count
African Black Oystercatcher	61.3±	9.2	1.7
Kelp Gull	129.2±	207.4	8.3
Turnstone	71.2 ±	25.7	26.8
Other migrant waders	23.5 ±	11.6	10.1

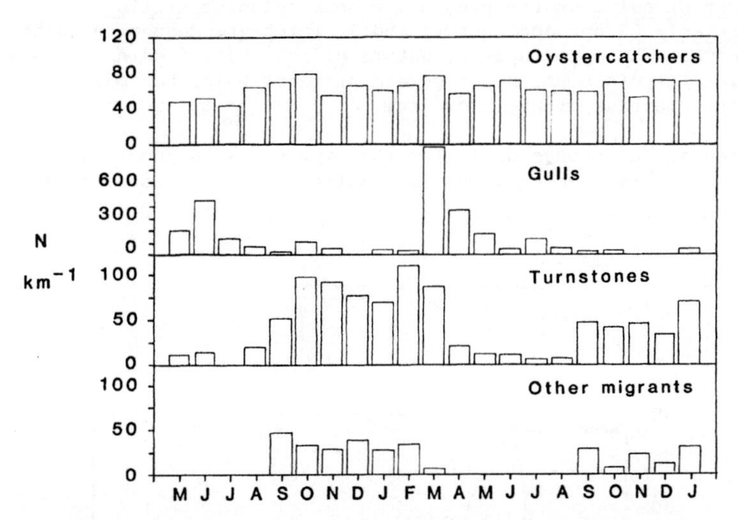

Fig. 2. Seasonal fluctuations in shorebird densities (n per km) at Marcus Island

RESULTS

Avian communities

Oystercatchers , Kelp Gulls and small waders all were markedly more common at Marcus Island than on the nearby mainland (Table 1). At Marcus Island, oystercatcher density fluctuated little throughout the study period, whereas density of Turnstones and other migrant waders fluctuated in a seasonably predictable manner (Fig. 2). Kelp Gull densities fluctuated dramatically, but with less evidence of a seasonal pattern (Fig. 2).

Avian diets

African Black Oystercatchers. Diet varied within pairs, with females taking proportionally more mussels than males, and males taking more limpets. *Patella granularis* comprised 17-31% (by numbers) of prey items of males and 1-5% of female prey items (direct observations, n = 2 pairs). Corresponding values for mussels were 27-68% (males) and 38-87% (females). The mollusc component of the diet ranged from 72% to 90% (2 pairs). Larger samples of molluscs from middens (n = 29 348 shells) at three other islands indicated that *P. granularis* made up 24-36%, and mussels 52-66% of molluscan prey remains: no other dietary component comprised more than 11% of midden contents at these islands (Hockey and Underhill, 1984).

Kelp Gulls. A minimum of 29 prey species was identified from pellets (n = 43) and stomach analyses (n = 13) of Kelp gulls at Marcus Island. Mussels were found in 85% of stomachs and in 77% of pellets. No other class of prey occurred in more than 30% of stomachs or 10% of pellets. Of 8 546 shells recovered from anvils, 98.8% were large mussels. Kelp Gulls fed on mussels in two different ways, either swallowing small whole mussels (recovered from pellets) or dropping large mussels (recovered at anvils), resulting in a bimodal pattern of prey size selection, compared to the unimodal pattern of prey size selection by oystercatchers (Fig. 3). Of mussels on anvils, 99.8% were *Choromytilus meridionalis*: 89.3% of mussels recovered from pellets were either *C. meridionalis* or the presumed recent colonist *Mytilus galloprovincialis* (Grant and Cherry, 1985), which are morphologically similar species. The opportunistic nature of Kelp Gull feeding renders a quantitative diet description per se almost impossible, but mussels clearly are the most important single prey item.

Turnstones. Turnstones dominated the migrant wader population (Fig. 2), and the diet of this species alone is considered here. A total of 4 009 prey

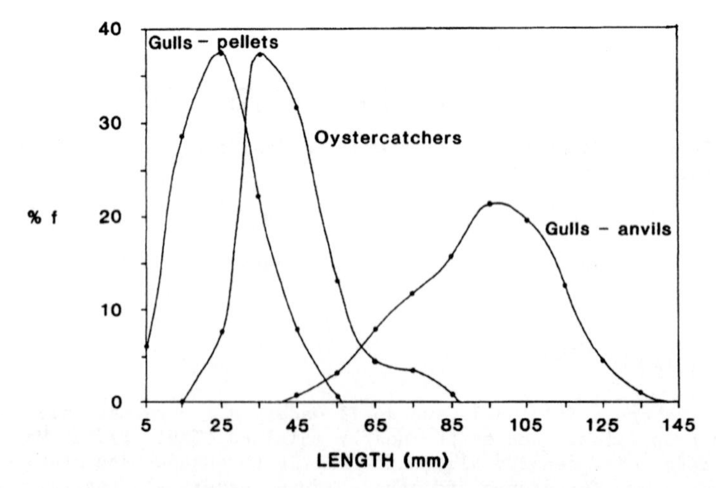

Fig. 3. Size selection of mussels by Kelp Gulls and African Black Oystercatchers

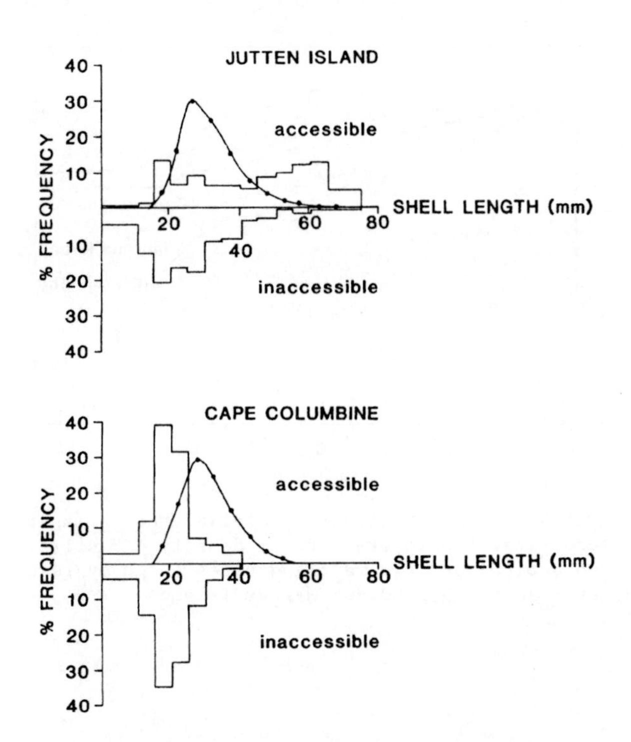

Fig. 4. Comparative size structures of populations of *Patella granularis* respectively accessible and inaccessible to oystercatchers at Jutten Island and Cape Columbine (redrawn from Hockey and Branch, 1984)

items was recovered from 26 stomachs: 14.7% of these items were terrestrial in origin (mainly ticks, *Ornithodorus capensis*, which Turnstones obtain at breeding colonies of seabirds). Of the intertidally-derived component, the six numerically dominant classes of prey were insects (dipteran larvae) 27.2%; grastropods 23.6%; isopods 19.9%; polychaetes 17.3%; amphipods 3.6%; and, pelecypods 2.4% (= 94% of intertidally-derived prey). All six prey classes were represented in the infauna of algal beds where bird exclusion experiments were sited.

Impact on prey populations

Oystercatchers and limpets. There were major differences in the demography of *P. granularis* populations between Jutten Island and Cape Columbine, and between sites at Jutten Island accessible and inaccessible to oystercatchers (Fig. 4). Limpets at Jutten Island grow larger than at Cape Columbine, and consequently a proportion of the population had a physical refuge in size from predation by oystercatchers (Fig. 4). The slower-growing limpets at Cape Columbine (Bosman and Hockey, in prep. a,b) have no such refuge. At Jutten Island (high oystercatcher density) there was clear depletion of those limpet size classes preferentially preyed on by oystercatchers. In addition to demographic differences between island and

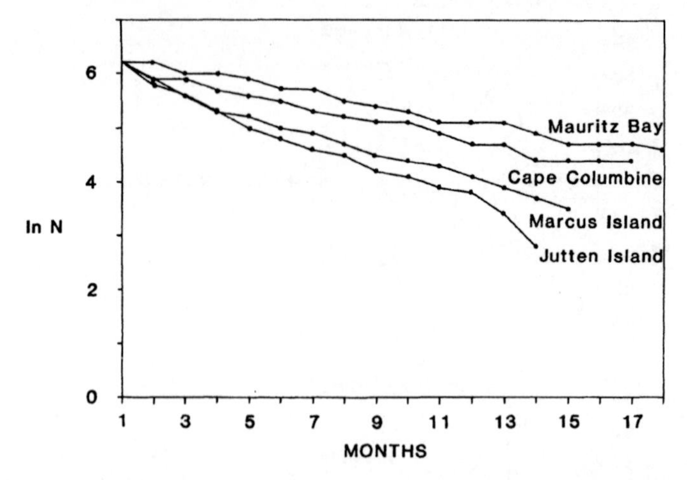

Fig. 5. Comparative mortality rates of Patella granularis at two mainland sites (low oystercatcher density - Mauritz Bay and Cape Columbine) and two mainland sites (high oystercatcher density - Jutten Island and Marcus Island)

mainland sites, densities of limpets in accessible sites were consistently lower at islands than on the mainland (see Hockey and Branch, (1984) for primary data source). Differential mortality rates between island and mainland sites further supported the assertion that oystercatchers are the most significant predator of P. granularis > 15mm in length (Figs 4,5). The probability of a limpet surviving one year at Jutten Island was 7%, at Marcus Island was 8%, and at both Mauritz Bay and Cape Columbine was 27% (Fig. 5). The mortality rate at sites at Marcus Island which were inaccessible to oystercatchers was identical to the mortality rate at accessible sites at Cape Columbine (Bosman and Hockey in prep. b). These calculations are size-independent: at two of three islands, limpets > 50mm in length had a significantly lower mortality than individuals < 50mm. At a mainland site, this trend was reversed but not significant (Bosman and Hockey, in prep. b). Hockey and Branch (1983) provide further evidence for the strong selection pressure exerted by oystercatchers on P. granularis, and show phenotypic differences between populations subjected to different intensities of predation.

Despite the high mortality rate and relatively low density of P. granularis at Jutten Island, female gonadial release at Jutten Island (ca. 60 wet g m^{-2} y^{-1}) was an order of magnitude greater than at Cape Columbine (ca. 6 wet g m^{-2} y^{-1}) (Hockey and Branch 1984). This was due mostly to the generally larger sizes of limpet at Jutten Island, with ca 66% of gonad output at this site being accounted for by individuals which had a refuge in size from oystercatcher predation.

Kelp Gulls, oystercatchers and mussels. Large mussels, which can be broken by dropping on anvils, are an important food source for Kelp Gulls at Marcus Island (Fig. 6). Figure 6 contains two sources of bias with respect to counts of gulls and calculations based on anvil mussels. Firstly, counts of gulls refer only to those birds counted on the shore at low tide and do not reflect the variable numbers of birds in the interior of the island or loafing offshore. Secondly, during severe storms, very large numbers of mussels are washed ashore broken and gulls can remove flesh from these mussels in situ without resorting to the use of anvils. For much of the year, storm-washed mussels apparently could supply the energy needs of the entire gull population (Fig. 6). However, these mussels are scavenged rather than depredated, and the primary source of disturbance causing their loss to the (subtidal) mussel population is physical (wave action) rather than predatory. The availability of such mussels to the gulls is dependent on wave action (Fig. 7) and therefore, this source of food, although readily detectable, is temporally unpredictable. At certain times of year, particularly during the summer months when storms are infrequent, small mussels assume greater importance in the diet (Fig. 6). Griffiths and Hockey

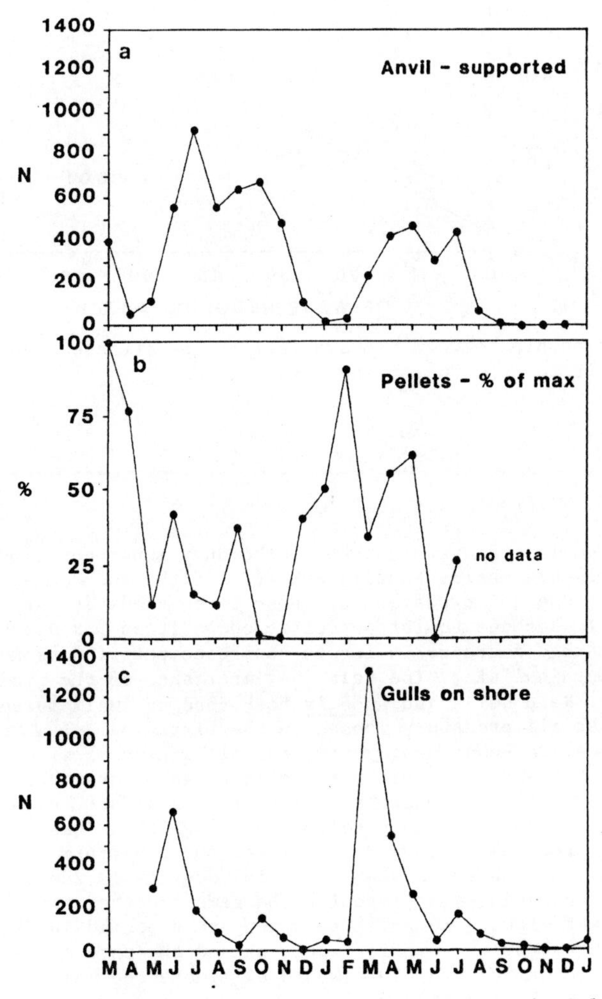

Fig. 6. Comparison of a) calculated number of Kelp Gulls supported by anvil-dropped mussels alone, b) relative seasonal abundance of mussels in Kelp Gull pellets, and c) numbers of Kelp Gulls counted on the shore, at Marcus Island

(in press) estimated that Kelp Gulls derived 25% of their annual energy requirements from small mussels. (This is probably an overestimate and their model therefore exaggerates the predatory impact of the gulls - see below.) Oystercatchers were assumed to derive 50% of their annual energy needs from mussels (after Hockey and Underhill, 1984).

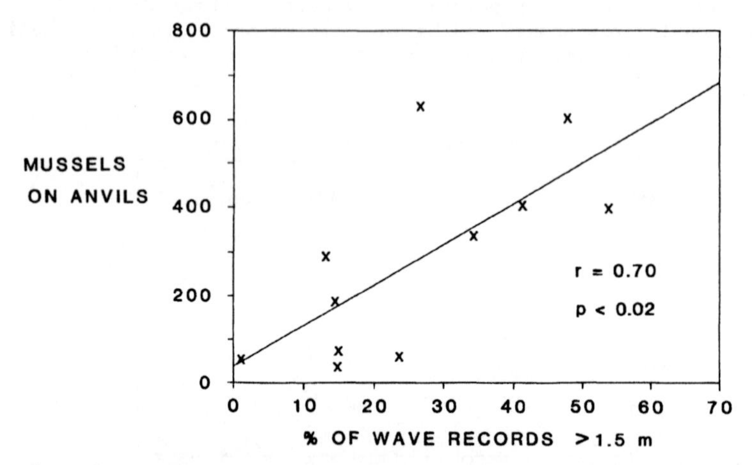

Fig. 7. Relationship between the availability of stranded subtidal mussels to Kelp Gulls, and wave action

Apart from gulls and oystercatchers, the only important predators of intertidal mussels are whelks _Nucella_ spp. (Griffiths and Hockey, in press). Losses of mussels due to predation by these three predators are compared with inevitable losses due to intraspecific competition for space in Fig. 8A. There is a marked decrease in losses due to competition as mussels approach their terminal size. The relative importance of the predators also varies over time. Kelp Gulls and _Nucella_ both feed on small mussels and account for almost all predatory losses in the first year of a mussel cohort's existence. In subsequent years, mussels achieve a size refuge from these predators, but enter the preferred size range of oystercatchers which become the major predator of mussels from year 2 onwards (Fig. 8A). When mussels are small, losses due to predators are less than losses due to competition, but from year 4 onwards, oystercatchers perform all the thinning necessary to achieve a stable packing density in the mussel population. The consequences of imposing the same predator array at the same density on a slower growing mussel population are depicted in Fig. 8B. From year 4 onwards, predator consumption would exceed the rate of natural mortality and, eventually, predators would eliminate the mussel population.

Migrant shorebirds and algal infauna. Peak numbers of small shorebirds are present during the austral summer, and Turnstones are present for longer and in greater numbers than any other species (Fig. 2). Using data for the

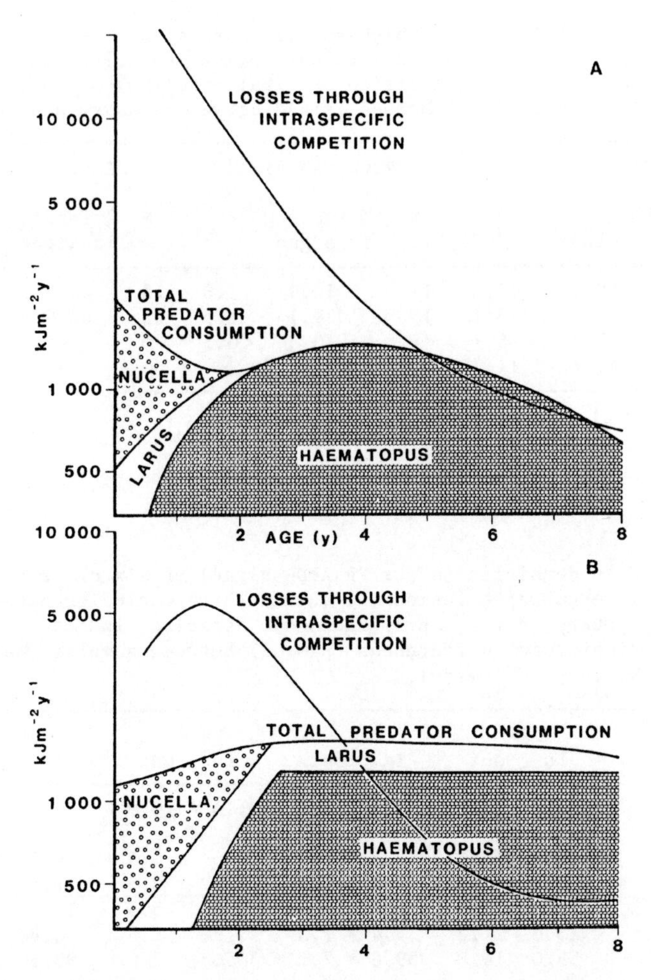

Fig. 8. Model of the comparative rates of mortality due to predation and intraspecific competition in a fast-growing intertidal mussel population at Saldanha (8A), and the consequences of imposing the same predator array on a slower-growing mussel population (8B). Reproduced from Griffiths and Hockey, (in press)

most common migrants, Turnstones, Sanderlings and Curlew Sandpipers, it is clear that the potentially greatest predatory impact on algal infauna occurs in February, when numbers of Turnstones and Sanderlings are maximal and birds are feeding intensively in algal mats (Table 2, Fig. 2). Significant differences were found between invertebrate densities within and outside cages in January and March, but not in April or June (Table 3, cf. Fig. 2). In March, following the month of greatest predation intensity in algal mats, there were significant differences in abundances of four of the six invertebrate classes (Table 3). (There was an overall tendency for dipteran larvae, an important food of Turnstones, to be greater outside than within cages, suggesting that adult dipterans do not readily fly through the mesh of cages in order to lay eggs.)

Table 2. Numbers, utilization of high and midshore algal mats (% of total feeding activity) and 'impact coefficient' (of birds on algal infauna, Il = N.% utilization of algal mats/100) of three species of migrant shorebirds at Marcus Island between January and April

	TURNSTONES			CURLEW SANDPIPER			SANDERLING			
	N	% feeding in algae	Il	N	% feeding in algae	Il	N	% feeding in algae	Il	Il
Jan	102	19.1	19.5	11	14.9	1.6	18	44.2	8.0	29.1
Feb	164	26.3	43.1	16	9.1	1.5	23	66.9	15.4	60.0
Mar	129	5.2	6.7	4	0.3	<0.1	0	-	0	6.7
Apr	28	41.7	11.7	0	-	-	0	-	0	11.7

Table 3. Mean densities (n per 2g AFDW algae) of six classes of invertebrates inside and outside bird exclusion cages in January, March, April and June. Asterisks indicate significant differences ($p<0.05$) between samples (Mann-Whitney "U" tests)

	JAN		MAR		APR		JUN	
	in	out	in	out	in	out	in	out
Dipteran larvae	9.2	14.2	1.6	4.0	1.0	* 10.0	32.0	24.0
Isopods	1.6	0.0	3.4	* 1.3	2.0	0.5	4.4	12.5
Polychaetes	6.0	* 1.0	7.2	* 0.2	1.8	2.5	1.4	* 4.5
Gastropods	13.4	* 1.4	6.0	4.8	5.6	5.5	16.1	3.0
Amphipods	23.8	21.2	5.4	* 1.0	0.9	1.5	0.8	5.0
Bivalves	55.0	14.8	59.8	* 7.1	30.0	7.0	93.6	25.0

Of the six classes of invertebrates, all except gastropods were more abundant outside cages in June than in March. In June, three of the six invertebrate classes were more abundant outside cages than within them. Although this difference was significant only for polychaetes, it suggests that cages were not 'preferred habitats' for these invertebrates and that results are not biased for this reason.

DISCUSSION

Shorebirds in the southwestern Cape interact as predators with at least three of the spatially dominant intertidal assemblages of the region, namely mussel beds, areas dominated by limpets, and the infauna of algal mats. These interactions occur simultaneously over extensive areas of shore, and yet mechanistically are qualitatively different with respect to the population dynamics of the prey. In all three cases, however, the processes involved apparently promote stability rather than instability in the predator-prey relationship. In the case of oystercatchers and limpets, predator density is inversely correlated with prey density (limpets are more abundant where oystercatchers are rare), but positively correlated with prey productivity, which, ultimately, is a function of primary production. Primary production at islands is much higher than on the mainland (Bosman and Hockey, 1986), and limpet growth rates are correspondingly fast (Bosman and Hockey, in prep. a). On any physically heterogeneous shore, a certain proportion of prey will have physical refuges from predators (Branch, 1985; Marsh, 1986), and it has been suggested that these achieve greater importance in tropical than in temperate regions due to the high mobility and year-round foraging of predators in tropical regions (Menge and Lubchenco, 1981). In the situation described here, the stabilizing effects of such refuges to prey populations are supplemented by an additional physical refuge in size which is a consequence of the combination of fast limpet growth rate and high tenacity (Branch and Marsh, 1978). At sub-Antarctic islands, where oystercatchers are absent, adhesion forces of the dominant *Nacella* limpet species are low, and these limpets are an important food source for the (relatively small) populations of Kelp Gulls, but achieve refuges by extending their distribution into deep water (Blankley and Branch, 1985; Branch, 1985). The *Nacella* limpets in the Straits of Magellan, where oystercatchers are fairly common, adhere strongly to the rock (pers. obs). In contrast to the sub-Antarctic situation there is no evidence that Kelp Gulls ever eat large limpets in either the Straits of Magellan or South Africa. In instances where oystercatchers prey on smaller limpets, such as *Collisella digitalis* (Hahn, 1985), the lack of a size refuge results in larger limpets being restricted to spatial refuges: no such pattern exists with respect to *P. granularis* and many large limpets occur in sites accessible to predators.

African Black Oystercatchers mostly are territorial throughout the year, and the predatory pressure they exert will, except for the period during which they are feeding chicks, be fairly consistent. *Patella granularis* populations experience mortality from sources other than predation, notably from intraspecific competition and 'bulldozing' of juveniles by adults (Hockey and Branch, unpubl.), and physical stresses such as desiccation (Branch 1981). However, there is strong evidence that oystercatchers are by far the most important predators of these limpets, and are capable of imposing high mortalities above the background level. Density-dependent predation and frequency-dependent predation are both thought to be mechanisms promoting stable equilibria (Katz, 1985), but the highly specific oystercatcher/*P. granularis* interaction could perhaps be better described as 'production-dependent', with the key to stability being the high reproductive output of individuals which escape predation by virtue of a refuge in size.

If island densities of oystercatchers and their concomitant high predation pressure were to be imposed on mainland limpet populations which lack such size refuges it is easy to conceive that the predator-prey relationship may become inherently unstable, as happens in areas where man is a major intertidal predator (e.g. Hockey and Bosman, 1986).

Not all oystercatchers which feed on limpets are territorial throughout the year. Frank (1982) observed the effects of seasonal (winter) foraging by flocks of non-territorial oystercatchers *H. bachmani* on limpets. He found that numbers of *Notoacmaea persona* > 20mm in length fell by two orders of magnitude, and *N. scutum* by one order of magnitude during the winter. However, he presented no data to show whether this severe depredation by oystercatchers persisted on a year to year basis, or whether his observations represent a spatially episodic predation event. Lewis and Bowman (1975) deduced that oystercatchers *Haematopus ostralegus* were the major predator of post-juvenile *P. vulgata*. They noted that because the birds operate in small flocks, their influence is both local and 'somewhat random'. At one study site, 81% of limpets present disappeared in two months following the seasonal reappearance of the birds.

Thus, oystercatchers have been shown to have two types of effect on limpet prey populations. The predation intensity of a resident, territorial species is attuned to the life-history characteristics of its prey in a manner that leads to long-term, large-scale stability, whereas seasonally flocking species can have a potentially destabilizing, but spatially restricted, impact on their prey populations.

Mussels experience losses from a more diverse predator array than do limpets. The quantification of post-settling juvenile mortality in limpets is difficult, but in mussels is relatively easy (Griffiths and Hockey, in press). However, little work has been done on the impact of birds on mussel populations of rocky shores. Marsh (1986) presents some evidence, albeit not conclusive, that shorebirds may affect the rate of mussel recruitment. However, the effects that he was able to demonstrate and to attribute to birds were few, and referred only to 'small' mussels at 'intermediate' densities. He did not consider the longer-term consequences of natural mortality as mussels grow. His argument that shorebirds prevent mussels becoming established on smooth, exposed substrata in the high intertidal is based on the observation that mussels in this zone occur only in crevices 'inaccessible to birds': we suggest that such crevices are more important in providing refuges from desiccation than from predation.

The arguments presented here for an apparent lack of impact of avian predators on mussel populations relate to established mussel beds, and thus contain an inherent assumption of successful recruitment. The role played by recruitment variability in determining stability properties of intertidal communities is poorly understood, due to problems in quantifying such events (Choat, 1982; Hawkins et al., 1983; Connell, 1985; Hartnoll and Hawkins, 1985; Johnson and Mann, in press). Zwarts and Drent (1981) studied a population of *Mytilus edulis* in which there was a gradual reduction in densities of mussels of sizes preferred by oystercatchers, following persistent poor recruitment, due to heavy predation on first-year mussels by an increasing population of Herring Gulls *Larus argentatus*. In accord with the tenets of optimal foraging theory, oystercatchers responded to this resource decrease by increasing their dispersion on the mussel beds and by prey-switching. Similar stabilizing processes have been well documented in invertebrate predator-prey systems (Hughes, 1985). Based on the model of Griffiths and Hockey (in press), we suggest that, given a scenario of predictable or 'adequate' settlement and recruitment, the subsequent predation pressure imposed by natural predators is not, in itself, destabilizing. However, it is probably the factor or factors influencing settlement intensity which ultimately control the dynamics of mussel communities.

The predation of algal infauna by small shorebirds may be likened functionally to the observations of flocking oystercatchers made by Lewis and Bowman (1975) and Frank (1982). The shorebirds have a major depletion effect on their prey populations, but only for a short period of the year: the prey are thus afforded a temporal refuge from predation. Most studies of prey depletion by shorebirds have been undertaken on soft substrata (Schneider, 1978; Schneider and Harrington, 1981; Quammen, 1984), and we found qualitatively similar effects operating on hard substrata. Many Arctic migrants leave the breeding grounds well before the winter collapse of prey stocks (e.g. Pitelka, 1959), but Schneider and Harrington (1981) found that substantial prey reductions occurred at migratory stopover sites. They suggest that, for a highly mobile group such as birds, migration is a mechanism to obtain an almost continuous food supply when seasonal pulses of productivity follow predictable latitudinal sequences. Persistent unidirectional prey depletion, even allowing for equalization of prey numbers (Schneider, 1978), inevitably tends towards disequilibrium. However, it is the very mobility of these predators which provides the mechanism for long-term stability in the face of short-term instability. High P/B ratios of algal infauna enable invertebrate numbers to build up prior to the next period of intense predation pressure.

Predation of algal infauna by birds does not affect the nature of the primary space occupier in these communities, namely the algae. By contrast, the pulsed predation of limpets by oystercatchers observed by Frank (1982), led to colonization of the denuded areas by foliose algae: the time for which this new community persisted is not known. The persistent mats of foliose algae at offshore islands in the southwestern Cape are a consequence of a) consistent predation by oystercatchers on limpets (Hockey and Branch, 1984) and b) high algal production resulting from nutrient enrichment by seabird guano (Bosman and Hockey, 1986; Bosman et al., 1986). Thus, predation by oystercatchers, in this instance, indirectly facilitates the feeding of small shorebirds (Hockey and Branch, 1984).

Shorebirds have been demonstrated on several occasions to exert selective forces on morphology, colour polymorphism and behaviour of their prey (Hartwick, 1981; Hockey and Branch, 1983; Sorenson, 1984; Branch, 1985; Mercurio et al., 1985; Hockey et al., in press). Birds other than shorebirds have also been shown to have regulating effects on either prey abundance or demography (Askenmo et al., 1977; Holmes et al., 1979; Gradwohl and Greenberg, 1982; Cooper et al., 1984; Howard and Lowe, 1984). However, almost without exception, studies of terrestrial species have failed to identify the functional nature, and particularly the consequences, of the observed predation. The study of predatory birds in habitats such as forests (Holmes et al., 1979; Gradwohl and Greenberg, 1982) inevitably are beset with interpretive problems because of the number of unknowns and limited knowledge of the life-history traits of so many of the component species. The functional roles played by shorebirds both as predators and nutrient-recyclers (Bosman and Hockey, 1986; Bosman et al., 1986), although not invariably easy to detect (Botton, 1984), appear to be worthy of more attention from intertidal ecologists than they have received in the past. Fundamental to this is that birds be recognized as an integral component of intertidal systems (Feare, 1966), which by virtue of their mobility and high energy demand, have the potential to be used in studies of a range of biologically stabilizing and destabilizing processes.

SUMMARY

Birds are an integral and important component of rocky intertidal communities and interact with their prey populations in a number of stabilizing and potentially destabilizing manners. These processes may vary according to life-history parameters and morphologies of the prey, variations in primary and secondary production rates, and the spatio-temporal incidence of predation. It is difficult, if not impossible, to generalize about the processes and consequences of predation by shorebirds across the range of species involved, although prey refuges (of varying types) are likely to be a common factor underpinning such interactions. Three quite different stabilizing predator-prey relationships were identified in sympatric avian predator communities in the southwestern Cape, South Africa. All three relationships, although involving different prey species and predator behaviour patterns, were characterized by the escape of some prey individuals by virtue of a refuge in time, space or size. The reproductive contributions of such prey lead to community persistence, assuming that settlement is successful. The variability in avian predator-prey interactions on rocky shores renders birds as a group potentially of great value in elucidating differences in population processes of species at lower trophic levels, and thereby improving our understanding of patterns and processes in intertidal communities.

ACKNOWLEDGEMENTS

This study was funded by the South African National Committee for Oceanographic Research and the Foundation for Research Development of the South African Council for Scientific and Industrial Research; and the University of Cape Town. We thank the Director of the Sea Fisheries Research Institute, Marine Development Branch, for permission to work on and for providing transport to islands under his control. We also thank the many colleagues who helped with various aspects of this study, especially those ornithologists who failed to appreciate fully the need for measuring thousands of limpets, mussels and other marginally-mobile creatures of the shore.

REFERENCES

Altmann, J., 1974, Observational study of behaviour: sampling methods, Behaviour, 49:227-265.

Askenmo, C., von Bromssen, A., Ekman, J. and Jansson, C., 1977, Impact of some wintering birds on spider abundance in spruce, Oikos, 28:90-94.

Blankley, W.O. and Branch, G.M., Ecology of the limpet Nacella delesserti at Marion Island in the Sub-Antarctic southern ocean, J. Exp. mar. Biol. Ecol., 92:259-281.

Bosman, A.L. and Hockey, P.A.R., 1986, Seabird guano as a determinant of rocky intertidal community structure, Mar. Ecol. - Prog. Ser., 32:247-257.

Bosman, A.L. and Hockey, P.A.R., in prep a, Importance of variation in primary production in determining the structure and function of rocky intertidal algal and herbivore communities.

Bosman, A.L. and Hockey, P.A.R., in prep b, Life-history parameters of populations of the limpet Patella granularis: the dominant roles of food supply and predation pressure.

Bosman, A.L., Hockey, P.A.R. and Siegfried, W.R., in press, The influence of coastal upwelling on the functional structure of rocky intertidal communities, Oecologia.

Bosman, A.L., du Toit, J.T., Hockey, P.A.R. and Branch, G.M., 1986, A field experiment demonstrating the influence of seabird guano on intertidal primary production, Estuar. coast. Shelf Sci., 23:283-294.

Botton, M.L., 1984, Effects of Laughing Gull and shorebird predation on the intertidal fauna at Cape May, New Jersey, Estuar. coast. Shelf Sci., 18:209-220.

Branch, G.M., 1981, The biology of limpets: physical factors, energy flow and ecological interactions, Oceanogr. Mar. Biol. Anm. Rev., 19:235-380.

Branch, G.M., 1985, The impact of predation by Kelp Gulls Larus dominicanus on the sub-Antarctic limpet Nacella delesserti, Polar Biol., 4:171-177.

Branch, G.M. and Marsh, A.C., 1978, Tenacity and shell shape in six Patella species: adaptive features, J. exp. mar. Biol. Ecol., 34:111-130.

Choat, J.H., 1982, Fish feeding and the structure of benthic communities in temperate waters, Ann. Rev. Ecol. Syst., 13:423-449.

Connell, J.H., 1961, Effects of competition, predation by Thais lapillus and other factors on natural populations of the barnacle Balanus balanoides, Ecol. Monogr., 31:61-104.

Connell, J.H., 1985, Variation and persistence of rocky shore populations, in: 'The Ecology of Rocky Coasts', P.G. Moore and R. Seed, eds, Hodder and Stoughton, London.

Connell, J.H. and Sousa, W.P., 1983, On the evidence needed to judge ecological stability or persistence, Amer. Nat., 121:789-824.

Cook, S.B., 1981, Fish predation on pulmonate limpets, Veliger, 22:380-381.

Cooper, S.D., Winkler, D.W. and Lenz, P.H., 1984, The effect of grebe predation on a brine shrimp population, J. Anim. Ecol., 53:51-64.

Dethier, M.N., 1984, Disturbance and recovery in intertidal pools: maintenance of mosaic patterns, Ecol. Monogr., 54:99-118.

Ebeling, A.W., Laur, D.R. and Rowley, R.J., 1985, Severe storm disturbances and reversal of community structure in a southern California Kelp forest, Mar. Biol., 84:287-294.

Edwards, D.C., Conover, D.O. and Sutter, 1982, Mobile predators and the structure of marine intertidal communities, Ecology, 63:1175-1180.

Feare, C.J., 1966, The winter feeding of the Purple Sandpiper, Brit. Birds, 59:165-179.

Feare, C.J., 1967, The effect of predation by shorebirds on a population of dogwhelks Thais lapillus, Ibis, 109:474.

Feare, C.J. and Summers, R.W., 1985, Birds as predators on rocky shores, in: "The Ecology of Rocky Coasts", P.G. Moore and R. Seed, eds, Hodder and Stoughton, London.

Frank, P.W. 1982, Effects of winter feeding on limpets by Black Oystercatchers, Haematopus bachmani, Ecology, 63:1352-1362.

Gradwohl, J. and Greenberg, R., 1982, The effect of a single species of avian predator on the arthropods of aerial leaf litter, Ecology, 63:581-583.

Grant, S.W. and Cherry, M.I., 1985, Mytilus galloprovincialis Lmk. in southern Africa, J. exp. mar. Biol. Ecol., 90:179-181.

Griffiths, C.L. and Hockey, P.A.R., in press, A model describing the interactive roles of predation, competition and tidal elevation in structuring mussel populations, S. Afr. J. Mar. Sci.

Griffiths, R.J., 1981, Population dynamics and growth of the bivalve Choromytilus meridionalis (Kr.) at different tidal levels, Estuar. coast. Shelf Sci., 12:101-118.

Hahn, T.P., 1985, Effects of predation by Black Oystercatchers (Haematopus bachmani) Audubon) on intertidal limpets (Gastropoda, Patellacea), Unpubl. MSc Thesis, Stanford University.

Hartnoll, R.G. and Hawkins, S.J., 1985, Patchiness and fluctuations on moderately exposed rocky shores, Ophelia, 24:53-63.

Hartwick, E.B., 1981, Size gradients and shell polymorphism of limpets with consideration of the role of predation, Veliger, 23:254-264.

Hawkins, S.J., Southward, A.J. and Barrett, R.L., 1983, Population structure of Patella vulgata L. during succession on rocky shores in southwest England, Oceanol. Acta, 1983:103-107.

Hockey, P.A.R., 1983, The distribution, population size, movements and conservation of the African Black Oystercatcher Haematopus moquini, Biol. Conserv., 25:233-262.

Hockey, P.A.R., 1987, Saldanha Bay: a history of exploitation, Afr. Wildl., 41:79-82.

Hockey, P.A.R. and Bosman, A.L., 1986. Man as an intertidal predator in Transkei: disturbance, community convergence, and management of a natural food resource, Oikos, 46:3-14.

Hockey, P.A.R. and Branch, G.M., 1983, Do oystercatchers influence limpet shell shape?, Veliger, 26:139-141.

Hockey, P.A.R. and Branch, G.M., 1984, Oystercatchers and limpets: impacts and implications. A preliminary assessment, Ardea, 72:199-206.

Hockey, P.A.R. and Underhill, L.G., 1984, Diet of the African Black Oystercatcher on rocky shores: temporal, spatial and sex-related variation, S. Afr. J. Zool., 19:1-11.

Hockey, P.A.R., Bosman, A.L. and Ryan, P.G., in press, The maintenance of polymorphism and cryptic mimesis in the limpet Scurria variabilis by two species of Cinclodes (Aves: Furnariinae) in central Chile, Veliger.

Holmes, R.T., Schultz, J.C. and Nothnagle, P., 1979, Bird predation on forest insects: an enclosure experiment, Science, 206:462-463.

Howard, R.K. and Lowe, K.W., 1984, Predation by birds as a factor influencing the demography of an intertidal shrimp, J. exp. mar. Biol. Ecol., 74:35-52.

Hughes, R.N., 1985, Rocky shore communities: catalysts to understanding predation, in: P.G. Moore and R. Seed. eds, "The Ecology of Rocky Coasts", Hodder and Stoughton, London.

Jara, H.F. and Moreno, C.A., 1984, Herbivory and structure in a midlittoral rocky community: a case in southern Chile, Ecology, 65:28-38.

Johnson, C.R. and Mann, K.H., in press, Diversity, patterns of adaptation and the stability of kelp (Laminaria longieruris) beds on the Atlantic coast of Nova Scotia, Ecol. Monogr.

Katz, C.H., 1985, A non-equilibrium marine predator-prey interaction, Ecology, 66:1426-1438.

Lewis, J.R. and Bowman, R.S., 1975, Local habitat-induced variations in the population dynamics of Patella vulgata L., J. exp. mar. Biol. Ecol., 17:165-203.

Marsh, C.P., 1986, Rocky intertidal community organization: the impact of avian predators on mussel recruitment, Ecology, 67:771-786.

McQuaid, C.D., 1981, The establishment and maintenance of vertical size gradients in populations of Littorina africana knysnaensis (Philippi) on an exposed rocky shore, J. exp. mar. Biol. Ecol., 54:77-89.

Menge, B.A. 1976, Organization of the New England rocky intertidal community, role of predation, competition and environmental heterogeneity, Ecol. Monogr., 46:355-393.

Menge, B.A., 1978, Predation intensity in a rocky intertidal community. Effect of an algal canopy, wave action and desiccation on predator feeding, Oecologia, 34:17-35.

Menge, B.A. and Lubchenco, J., 1981. Community organization in temperate and tropical intertidal habitats: prey refuges in relation to consumer pressure gradients, Ecol. Monogr., 51:429-450.

Menge, B.A. and Sutherland, J.P., Species diversity gradients: synthesis of the roles of predation, competition and temporal heterogeneity, Amer. Nat., 110:351-369.

Mercurio, K.S., Palmer, A.R. and Lowell, R.B., 1985, Predator-mediated microhabitat partitioning by two species of visually cryptic, intertidal limpets, Ecology, 66:1417-1425.

Paine, R.T., 1974, Intertidal community structure: experimental studies on the relationship between a dominant competitor and its principal predator, Oecologia, 15:93-120.

Paine, R.T. and Levin, S.A., 1981, Intertidal landscapes: disturbance and the dynamics of pattern, Ecol. Monogr., 51:145-178.

Paine, R.T., Castilla, J.C. and Cancino, J., 1985. Perturbation and recovery patterns of starfish dominated intertidal assemblages in Chile, New Zealand and Washington State, Amer. Nat., 125:679-681.

Parkington, J.E., 1976, Coastal settlement between the mouths of the Berg and Olifants Rivers, Cape Province, S. Afr. archaeol. Bull., 31:127-140.

Pitelka, F.A., 1959, Numbers, breeding schedule and territoriality in Pectoral Sandpipers of northern Alaska, Condor, 61:233-264.

Quammen, M.L., 1984, Predation by shorebirds, fish and crabs on invertebrates in intertidal mudflats: an experimental approach, Ecology, 65:529-537.

Schneider, D.C., 1978, Equalization of prey numbers by migratory shorebirds, Nature, 271:353-354.

Schneider, D.C. and Harrington, B.A., 1981, Timing of shorebird migration in relation to prey depletion, Auk, 98:801-811.

Siegfried, W.R., 1977, Mussel-dropping behaviour of Kelp Gulls, S. Afr. J. Sci., 73:337-341.

Simenstad, C.A., Estes, J.A. and Kenyon, W.K., 1978, Aleuts, sea otters and alternate stable-state communities, Science, 200:403-411.

Sorenson, F., 1984, Designer limpets and their avian consumers, Amer. Malac. Bull., 2:80.

Underhill, L.G. and Cooper, J., 1984, Counts of waterbirds on the coastline of southern Africa, 1976-1984, Unpubl. Report, Western Cape Wader Study Group and Percy FitzPatrick Institute of African Ornithology, Cape Town.

Underwood, A.J., Denley, E.J. and Moran, M.J., 1983, Experimental analyses of the structure and dynamics of mid-shore rocky intertidal communities in New South Wales, Oecologia, 56:202-219.

Walsberg, G.E., 1983, Avian ecological energetics, in: "Avian Biology, Vol. 7" eds, D.S. Farner, J.R. King and K.C. Parkes, Academic Press, London.

Zwarts, L. and Drent, R.H., 1981, Prey depletion and the regulation of predator density: Oystercatchers (Haematopus ostralegus) feeding on mussels Mytilus edulis, in: N.V. Jones and W.J. Wolff, eds, "Feeding and Survival Strategies of Estuarine Organisms", Plenum Press, New York and London.

Behavioural Plasticity as an Adaptation to a Variable Environment

Brian A. Hazlett

University of Michigan
U.S.A.

INTRODUCTION

As a class of phenotypic characteristics, the outstanding feature of behavior is its flexibility in interfacing organisms with an environment which is variable in time and space. When and where an organism executes any behavioral act is crucial in determining the functionality of the behavior patterns. To the degree that an environment is variable, flexibility in behavioral execution is critical and this is especially the case when the timing or extent of the variability is unpredictable. Limitations on flexibility are to be expected in organisms only to the extent that they face an environment that is predictable.

As a type of environment, the intertidal zone may well be the epitome of variability. Virtually every aspect of the abiotic environment is variable in both predictable and unpredictable ways. The tidal flux which defines the zone imposes a variation in light levels and spectral composition, turbidity, temperature, nutrient availability, availability of respiratory gases, density of the surrounding fluid medium, and the very category of environment an organism faces. The latter refers to the alteration between aquatic and terrestrial environment that a single location in the intertidal affords. Clearly it would be rare when only one abiotic factor varied, rather multiple factors will usually vary in concert (Alderdice, 1972), and often in non-linear ways. Driven by variability in these abiotic features, the biotic environment to which any species is exposed (prey, predators, mutualists, competitors, etc.) is also variable in time and space. The fact that wave action is an important physical force in the intertidal, adds another abiotic feature that is highly variable. Many aspects of morphology, physiology, and behavior can be viewed as adaptations to wave action and we should keep in mind that organisms must face extreme variability in the magnitude of that physical force.

Let me consider the two aspects of variability, temporal and spatial, in more detail. Any individual must of course face both aspects during its lifetime. The temporal variability imposed upon an intertidal organism by tidal variation is depicted in the logo for this symposium, but in an idealized manner. The temporal pattern of regular variation in water level superimposed on a lunar cycle illustrates very predictable variability. But this "simple" pattern is complicated by at least two major factors in most locations. Seasonal patterns can add another level of predictable variation while storms and other climatic events add variation in both the timing and magnitude of changes in water level which are not easily predicted by organisms. Even when the seasonal occurence of storm-related variation in water level can fall into the predictable category, the timing and magnitude of the events which drastically alter the intertidal organism's environment are unpredictable. If all of the above complexities (which are faced by individual organisms in one locale) were not enough, the variation faced by individuals in different populations of a species adds even more uncertainty to the environment(s) encountered by (the individuals making up) a species. Local topographic and hydrographic features add idiosyncratic features to the above patterns in both their predictable an unpredictable aspects.

Thus we are led to a consideration of spatial variability. Let me first mention the rather extreme case of sandy beaches. While perhaps the most common type of intertidal zone, at least in tropical and sub-tropical areas of the world, sandy beaches are a formidable place from the viewpoint of environmental predictability. Not only do all comments on temporal variability mentioned above apply, but the physical location of the intertidal zone can move quite readily. In some cases the physical shifts are small and regular while in other case the beach "moves" great distances. Even without disturbances (storms), beaches move considearbly as a result of current patterns and physical features nearby (river inputs, rocky promontories, etc.). All these factors combine to create a high degree of spatio-temporal variability in environmental factors.

The intertidal of rocky shores are less readily altered over time but still present a high degree of environmental variability from location to location. Slight differences in the orientation of shores, on both a macro- and micro-scale, lead to differences in abiotic and biotic features on an almost centimeter-by-centimer level. To illustrate this, consider the variability in the temporal pattern of feeding and shelter-seeking in the gastropod *Morula marginalba*. Moran (1985) found that the snails' responses to tidal variation (feed when submersion allows, seek shelter when exposure will be prolonged) varied markedly between locations depending upon position within the intertidal *and* degree of exposure of the locaton to sunlight. This is quite logical given the added stress that direct sunlight during a period of immersion would add, but points out the variability in behavior patterns to be expected given environmental variability.

Given this picture of the intertidal as a "sea" of variability, what can organisms do to cope and flourish in such an environment? Life histories vary among organisms such that one "approach" could be the classic r-selected organism which produces many, many offspring and survives evolutionarily by the chance that a few make it. As a base line, the morphology and physiology of an intertidal organism must be robust to the extremes in physical features presented in the intertidal. Developmental plasticity which allows some degree of optimality in a variety of situations may also aid in fitting animals to an unpredictable environment. Flexibility in life history parameters (Hines, 1986) would also be favored in unpredictable environments. But I contend, that it is the flexibility afforded by behavior patterns which really allows animals to succeed in the intertidal.

The behavior patterns of primary importance will vary widely from species to species. And much of the considerations of any one of us must be with the particular behavior patterns that are shown by one type or taxon in which we are interested. But the real "adaptation" for intertidal life is the preservation (non-loss) of flexibility in behavior patterns which allows adjustment to a hostile environment. Particular acts are simply specific examples of the phenotypic fit to environmental conditions. The overall attribute which has been selected for (and is therefore the adaptation to the environment in the strict sense) is the flexibility itself which behavior affords as a spatial-temporal modifier of organism-environment fit, both on a day-to-day basis or during particularly critical choice points in the organism's life history.

To go beyond the generality I just mentioned, I want to briefly examine several types of behaviors, in each case mindful of the general theme of flexibility as the adaptation. These types of behavior will be: larval behaviors, movements and migrations, and learning about the microhabitat.

LARVAL BEHAVIOR PATTERNS

Planktonic forms are by definition affected extensively by abiotic forces and the complexity of the patterns of these forces can lead one to invoke an almost random process in the description of the distribution of larvae (Jackson, 1986). The larvae of intertidal forms exhibit considerable differences in the extent to which larval behavior is believed to influence distribution patterns. The responses of some crustacean larvae to abiotic features appear to increase the probability that those larvae will be somewhat near suitable habitat when they metamorphose (Cronin and Forward, 1986) while larvae of some polychaetes (Levin, 1986) and echinoderms (Banse, 1986) appear to be passively dispersed. This dispersal can, at least in the case of teleplanic larvae, carry individuals such great distances (Scheltema, 1986) that uncertainty concerning the details of environmental features is virtually guaranteed.

Whatever mechanisms get larvae near suitable habitat for settlement, those larvae must utilize some environmental cue(s) to select a specific site for settlement (or avoid particular sites (Woodin, 1986)) or face the demographic consequences of random settlement. A rather common behavior which is not limited to intertidal forms but certainly occurs there is to respond to chemicals from conspecific individuals (gregarious settlement). By settling out where others have settled before (and survived), the larva is using what ethologists might call a sign stimulus. The extreme variability in abiotic factors which occurs in the intertidal makes utilization of stimuli associated directly with those factors a risky type of response. By responding to the smell or taste of conspecifics, a larva is utilizing the existence of those animals as a sampling advice, signalling that an environmental location has, at least in the past, been suitable habitat for that kind of animal. While there may be additional reasons for settling near conspecifics (particularily for sessile forms), for all species the individuals are effectively taking an average of the environmental situation by responding to just one cue---conspecifics (Highsmith, 1982; Burke, 1986).

Of course, not all species appear to utilize such a mechanism and in the extreme case it has been suggested that a nearly random pattern of initial settlement occurs (Bell and Westoby, 1986). This initial pattern is followed by within habitat movements to more appropriate places according to the Bell-Westoby model.

MOVEMENTS: SHORT- AND LONG-TERM

One obvious way an organism can succeed when faced with a variable environment is to move in such a way that the less-favorable aspects of the environment are avoided. The movements can be as simple as the muscular contractions which lead to a (sessile) barnacle closing the plates of the capitulum and thus creating a form which is resistent to the effects of exposure. More complex movements can involve some form of locomotion up and down the intertidal, maintaining the individual animals in favourable habitat. Patterns which are behaviorally even more comlex would include learned orientation towards particular refuge sites such as the movement to crevices in rocks by the grapsid crab *Pachygrapsus transversus* (Abele et al., 1986) or to particular burrows by individuals of the ocypodid crab *Dotilla sulcata* (Fishelson, 1983). We can predict that intertidal organisms that orient to hiding places must do so more quickly and accurately than similar animals in other marine habitats since the physical dangers of being exposed at low tide add reinforcement to the other pressures for finding hiding places, e.g. predator avoidance. In some cases, the response shown by an individual is variable; the hermit crab *Clibanarius vittatus* may either move offshore as the tide receeds, bury itself in the mud/sand and thus avoid some aspects of terrestriality, or just sit on the surface of the mudflat and wait until the tide comes back (Hazlett, 1981b).

An alternative mechanism controlling short-term movements, one which does not necessarily involve past experience, is the trail following shown by several species of chiton (Chelazzi et al., 1983). By following cues which result from earlier movements, an animal can accurately orient to a specific location within the intertidal witout learning being necessarily involved. However, this mechanism potentially restricts the animal to a given path of movement which may not be the quickest means of return to a safe location. Another alternative mechanism is having another animal execute the movement. The intertidal snail *Ilyanassa obsoleta* changes its behavior when infected with the trematode *Gynaecotyla adunca* and moves further up the shore than is normal for the snail. Having been moved by its host, the cercariae of the trematode can then more readily be transmitted to its next host, semiterrestrial crustaceans (Curtis, 1987).

Whatever the complexity of the short-term movement which keeps the animal in a more favorable portion of its environment, two general timing mechanisms can line up the phase of tidal cycle and the movement. Endogenous tidal and lunar rhythms may help control the timing of the movements (Neuman, 1981) and in that case the flexibility of the timing is often set by early experience (Forward et al., 1986). Once set, the rhythm persists. By this means, the unpredictability of cycles from location to location is met by the one-time setting of the rhythm while the regular variation in tidal height is met by the rhythm itself. The type of input serving as zeitgeber for the rhythm can be water movement, temperature changes, light level changes, or hydrostatic pressure changes (Neuman, 1981).

The second class of timing mechanisms is to simply react to immediate cues associated with changes in water level caused by tidal movement. Obviously in the field individuals will modify the exact timing of behavior by attention to both immediate input and biological rhythms. The rhythm involved can be in the sensitivity of the animal to external input such as light level, thus amplifying the animal's readiness to react to direct cues from tides (Neuman, 1981).

Movements over any distance can of course be expensive, both in terms of energetics and exposure to predators. One alternative to long movements can be shorter movements which result in the formation of aggregations. The aggregation behavior of many intertidal forms can serve, among other things, to retain water between members of the group and otherwise buffer the extent of environmental stress (Reese, 1969). As shown by the work of Snyder-Conn (1980, 1981) on hermit crab aggregations, the regular variation in aggregation formation can again vary in its extent and details of timing from location to location.

While the above examples are organisms that regularly live in the interdidal zone on a day-to-day basis, additional short-term movements into the intertidal are made by animals that normally live elsewhere. Onshore movements for reproductive purposes, usually linked closely with seasonal and lunar rhythms, illustrate another type of behavioral adaptation for (temporary) intertidal life. The reproductive movements of _Limulus polyphemus_ are a good example of this and one where general movement towards the intertidal may be timed via an endogenous rhythm but actual movement into the (high) intertidal is timed via specific cues from the immediate environment (Barlow et al., 1986). Since _L. polyphemus_ lays eggs only during the highest tides of a lunar cycle and since the times of those highest tides at any location are influenced by unpredictable weather conditions (Barlow et al., 1986) the horseshoe crab must utilize input from more exact cues.

The other type of movement relating to the intertidal zone is long-range movement or migration. Seasonal off-shore migrations have been reported for a number of species including several species of hermit crabs (Fotheringham, 1975; Rebach, 1981). These movements result in the avoidance of the intertidal zone during particularily harsh conditions.

HERMIT CRAB LEARNING AND SHELL UTILIZATION

The last category of behavioral adaptations to intertidal life which I wish to mention (and which leads into some specific examples I wish to present) is the class of adaptations which I find most interesting---learning to match behavior patterns with the particular details of the piece of the intertidal in which an individual finds itself. I have already alluded to this mechanism twice above: grapsid crabs learning the location of refuge sites in the local environment and the setting of endogenous rhythms by local cues. The latter is not learning in the classic sense but it does represent a modification of (the timing of) behavior by past input. But I wish to illustrate this category mostly with the examples I know best---hermit crabs and one aspect of their microhabitat, shells. While I will make use of work on this system to illustrate the role of past experiece on behavior I do not believe this system to be unusual. While I can not prove it, I believe that I found evidence of the multiple roles of past experience with these crustaceans only because I have looked for it, not because they are especially flexible in their behavior. Of course, as pointed out by Reese (1969), occupation of gastropod shells (i.e. being a hermit crab) is, in part, an adaptation to a harsh environment---affording not only protection from both terrestrial and aquatic predators but from the environmental extremes characteristic of the intertidal.

For the vast majority of crustaceans which we call hermit crabs, the behavior patterns by which individuals utilize empty gastropod shells as shelter are critical to the crab's existence. Orientation to shells, selection of a shell to enter, entry of a shell, and subsequent living in the shell are all behaviors which directly affect the Darwinian fitness of an individual crab. Failure to find or efficiently enter a shell at all can rapidly and markedly increase the chances of predation while selection of the wrong kind or size of shell can affect both the risk of the crab being eaten and its reproductive opportunities.

When a hermit crab glaucothoe settles out of the plankton and metamorphosis dictates occupation of a shelter (=gastropod shell), the exact array of shells available for potential occupation can be both diverse in kind and variable from location to location. While there are exceptions, most species of crabs can be found in a wide array of species of snail shells (Hazlett, 1981a). I have shown some years ago that the initial entry of a crab into a shell is strongly influenced by past experience (Hazlett, 1971; Hazlett and Provenzano, 1965). If an individual crab has had no experience handling objects, its first several attempts at shell entry are poorly oriented and acts occur out of sequence. Such mistakes disappear from the crab's repertoire after a few entries.

However, the aspect of shell-related behavior I want to focus upon is that of shell selection. The kind of shell a crab will enter when given a choice among shell types not only varies from species to species, but, within limits, among the individuals of a species. Ghilchrist (1985) has shown that the shell type(s) an individual experiences early in life can influence the preferred kind of shell a crab chooses. Given the highly variable array of shells to which a young crab could be exposed upon metamorphosis, this is a quite reasonable situation. If we were to imagine a crab which was genetically programmed to search until it found just the "right" species of empty gastropod shell, it is easy to imagine that crab using up much energy in such a search, perhaps never finding that "right" shell, and failing to partition time and energy to reproduction. Such an imaginary crab would have little chance (evolutionarily) when in competition with a crab with a flexible shell preference. Flexibility in preference would also be advantageous when new shell types become available, such as the human-mediated introduction of shell species. Blackstone and Joslyn (1984), Wicksten (1977), and Drapkin (1963) have all shown that crabs can rapidly take advantage of new resource types as result of crabs' behavioral plasticity.

The species of shell to which a hermit crab will be exposed (and thus have to live in) can vary greatly. The species of gastropods present in an area varies depending upon many factors, some as simple as the types of rock and the height of the rocks. The latter affects submersion time which influences the species of gastropods found in the area (McGinnes and Underwood, 1986), which in turn affects the shell resources available to a hermit crab. Crabs of one species will be found in different shell species depending, in part, upon the shells present (Bertness, 1980). Other environmental factors also influence shell choice. Predation pressure can influence not only the species of shells crabs occupy in the field (Bertness, 1982) but also their preference in the laboratory. Individuals of _Pagurus pollicaris_ from field populations that were exposed to more frequent predation from octopi, initially preferred hydroid-colonized shells when given a choice but after some time in the laboratory, no longer showed such a preference (Brooks and Mariscal, 1985). Similar changes in crab responses to cnidarian symbionts have been reported for _Dardanus arrosor_ and the cue triggering a change in crab behavior is olfactory detection of cephalopods (Balasch and Mengual, 1974).

Even within a species of shell, the size of shell preferred by a given size of crab can be influenced by past experience. Abrams (1978) reported that different populations of the land hermit crab _Coenobita compressus_ differ in the size of shells they prefer. Scully (1979, 1983) has shown that in populations of _Pagurus longicarpus_, individuals which have had more limited access to empty shells in the field, select larger shells when placed in a free choice situation, than individuals which have had access to some empty shells in the field. Childress (1972) has suggested that crabs usually select shells which are actually larger than immediately optimal with the result that growth can occur and a shell provide the best fit on the average for the period of time an individual crab is likely to inhabit that shell. The observations of Scully strongly suggest that past experience of individuals influences what the crab's estimate of finding a larger shell in the future is. This in turn affects its immediate choice of shells when allowed free choice of shell size.

Under most circumstances, hermit crabs investigate an empty shell with their appendages before attempting shell entry. Information about the size and condition of the shell is presumably obtained by insertion and movement of the chelipeds, ambulatory legs, or both types of appendages. It is well established that hermit crabs, like many crustaceans, can show compensatory plasticity (Reese, 1983) and vary the appendages utilized in shell investigation if appendage loss occurs. This plasticity could be either "hardwired" or the result of individual experience. However, the finding of Scully (1986) that intact individuals from different populations of *Pagurus longicarpus* utilize different patterns of appendage use suggests an experiential role since these populations are probably not genetically distinct (Scully, 1983).

The flexibility in shell selection ability can not be limited to young stages. As crabs grow they must find larger shells to occupy but as a result of the distribution and demographics of snails, they must continually occupy new kinds of shells. As the maximum size of one species of shell becomes too small for larger crabs, new types of shell must be occupied if positive growth is to continue. That positive growth is adaptive can be inferred both from the wide-spread relationship (positive exponential function) between female size and number of eggs and by the fact that when provided with larger shells, female hermit crabs partition energy to growth rather than immediate reproduction (Bertness, 1981). Data we have recently collected show that the probability of a male obtaining matings is also strongly influenced by crab size (Figure 1). Continual switching of shell types as crabs grow means crabs must constantly modify the details of their shell-related behavior patterns as new species are handled, entered, exchanged, and generally utilized.

Shell preferences of hermit crabs have also been demonstrated to shift, even as adults (Elwood et al., 1979). Medium-sized individuals of *Pagurus bernhardus* altered their shell species preference based upon recent past experience with shells. The learned species preference held even when shell weight was adjusted such that crabs had to choose between correct weight and recently experienced species of shell.

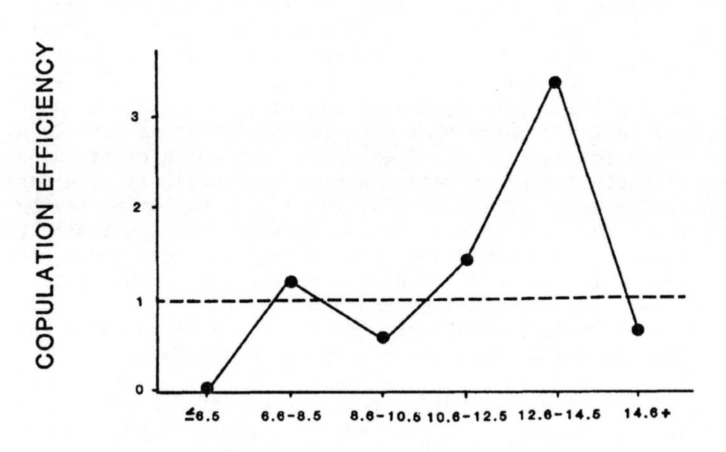

Fig. 1. Relationship between size of male *Calcinus tibicen* and proportion of all copulations obtained by males in a size class divided by the proportion of males in that size class.

When crabs first experience a new shell type, behavior patterns can appear sub-optimal. The intertidal Caribbean species Clibanarius antillensis normally occupies Cerithium shells when very young, but as they grow they eventually occupy other species, including Nerita. Laboratory observations have indicated that crabs which have had experience only with Cerithium shells are initially unable to select the proper size Nerita shell (Hazlett, 1987a). Orientation to the globus form (vs. the high spiral of Cerithium) is both slow in execution and highly variable in space (i.e. "incorrect") and occupation of shells one chosen is erratic. Individuals which would enter Cerithium shells in a few seconds would hold Nerita shell for several minutes before attempting an entry and quite often changed into yet another shell within minutes after occupying a Nerita. The crabs behaved as if they were unable to treat the new shell shape properly as a shell object. After some days of experience with such shells, entry was smooth, quick, and efficient.

The behavior patterns shown by hermit crabs during shell exchanges are also influenced in several ways by past experience. The initial shell exchanges of young crabs are quite different from later exchanges by the same individuals. Initiating crabs attempt to execute shell rapping patterns from improper orientations and often fail to respond properly to tactile signals from the non-initiating crab (Hazlett and Provenzano, 1965). Only after experience both as initiator and non-initiator do the behaviors seen during a shell exchange resemble those of larger crabs. Even after the initial exchange experiences have shaped the general sequence of acts, details of exchange are influenced further by past experience. Crabs communicate by visual displays prior to the execution of specific patterns related to shell exchange. Older individuals of Pagurus bernhardus utilize a variant of the war-of-attrition strategy (Maynard Smith and Parker, 1976) of communication by visual displays, whereas younger crabs appear to follow a different strategy (Hazlett, 1982). This war-of-attrition strategy is the more logical one in terms of short-term payoff but my results suggest that crabs need experience in interactions before they utilize that set of behavior patterns.

Additional effects of past experience can be seen in that crabs inexperienced with shell exchanges with any particular type of shell tend to follow the aggressive model of resource exchange while experienced animals follow the negotiations model of resource exchange (Hazlett, 1987a). Young Clibanarius antillensis which had had limited experience in shell exchanges executed many more raps than experienced animals (mean of 66 compared to a mean of 24 raps per interaction), suggesting that the information on shell size transmitted during rapping (Hazlett, 1987b) is not as readily utilized by crabs with limited experience. Each shell type which crabs encounter probably has slightly different relationships between shell size parameters and fundamental frequency (Field et al., 1987) and the latter may be used by crabs during shell exchange behavior as an indicator of shell size (Hazlett, 1987b). As a result, crabs would not be able to accurately judge shell size during exchanges until they had had some experience with that type of shell. Thus as animals grow and must learn how to efficiently handle and select new types of shells, they must also learn how to exchange these types of shells in the most efficient manner, via negotiations.

However, flexibility in behavior patterns has its limits and this can have very serious consequences on the animal's fitness. For example, consider two intertidal areas on the north coast of Jamaica. An area to the east of the dock at the Discovery Bay Marine Laboratory has populations of the hermit crab Calcinus tibicen which occupy a variety of shell species including numbers of large individuals in Nerita and Cittarium shells. Just 25 meters away, to the west, the large individuals of C. tibicen are primarily in Leucozonia and Cymatium shells. In both habitats, shell

Table 1. Relative frequencies of successful copulations and shell species occupation by male Calcinus tibicen

	Percent of Shells Occupied by Males	Percent of Copulations Observed
Leucozonia nassa	28.4	49.1
Cittarium pica	24.1	12.2
Thais deltoidea	11.4	24.5
Nerita sp.	9.7	0
Cymatium pileare	7.3	12.2
Bursa thomae	4.8	0
Astraea tecta	4.1	0
Tegula fasciata	0.8	1.7
Others (10 species)	6.2	0

occupation simply reflects shell availability. For large males in particular, this has extremely important consequences. Based upon laboratory observations, individuals occupying Nerita or Cittarium shells have very low probability of obtaining matings while similar sized males in Cymatium shells obtain many matings (Table 1). It would appear that the few males in good shells in the east site in particular would obtain many matings while those inhabiting Nerita shells would obtain almost none. (Needless to say, crabs in Cymatium shells will not exchange shells with crabs in Nerita or Cittarium shells.) But the crabs are caught in a bind: Cittarium shells are the only kind available for very large crabs. Thus a male's reproductive success increases as he increases in size, but falls suddenly if the only shell of adequate size is the "wrong" species (Figure 1).

This is where the limits on flexibility become critical. The shapes of these "bad" shells simply do not allow proper execution of the precopulatory patterns. The male can not hold the female in the opposed position and rotate her smoothly (Hazlett, 1966) because of weight distribution of the shell (Nerita) and/or the angle of the aperture relative to the columellar angle (Cittarium). The latter species is shaped such that even normal locomotion is difficult, much less smooth execution of pre-copulatory movements. From our observations, almost none of the individuals in these shells had overcome the difficulties of these shells, perhaps pointing to behavioral or even morphological limitations. I have argued that flexibility is an important adaptation in an unpredictable environment, and it is, but watching the clumsy attempts at mating by hermit crabs in the wrong kind of shell indicates to me that there clearly are limits.

EFFECTS OF PAST EXPERIENCE: A FRAMEWORK

While I have considered in the above section, several examples of how past experience can influence the behavior of intertidal organisms, I would like to briefly consider a broader framework. In discussing this framework, I should first comment on the question of the relative contributions of past experience and genetic constraints in the development of behavior. In my consideration of hermit crab behavior mentioned above, I have clearly emphasized the number of instances where the experiments or observations have indicated an influence by past experience in shaping the behavior of hermit crabs. I have done this, in part, because experiments on past experience are much easier to do than genetic analyses of behavior and thus such studies are available in the literature. However, another motivation for emphasizing the role of past experience in shaping behavior patterns which are immediately functional (and therefore adaptive in a broad sense) is my wish to point out the need for study of both types of input---genetical and experiential---and their interaction.

My plea for a balance of studies and interpretations is hardly meant to advocate any disregard of the roles of genetics. Clearly the genome guides development in such a way that animals with different genotypes behave differently. This is just as true for behavior patterns which are strongly shaped by past experience as it is for patterns which are influenced relatively little by past experience. However, the demonstration that different genotypes behave differently when all individuals are exposed to the same environment should not be taken as evidence that genes alone are somehow responsible for the patterns observed; It would be just as foolish of me to claim that past experience alone is responsible for the shell-utilization behaviors of any individual hermit crab. The demonstration of some role of genotypic differences or the demonstration of some role of experiential input are just the beginning steps in gaining an understanding of the behavior of any population of animals. The latter is, of course, just the beginning information for an understing of how behavior has evolved, i.e. what are the behavioral adptations which intertidal organisms exhibit?

When I observe an adult Calcinus tibicen entering a gastropod shell smoothly and quickly, do I conclude that shell entry behavior has evolved? I see a functional behavior pattern which results in the individual crab avoiding predation and obtaining a shelter which will allow optimal growth and reproductive output. But I would argue that the pattern which we can see is not the adaptation in the strict sense. Rather the functional behavior pattern is the result of a number of different processes and capabilities. The crab learned how to handle objects as a result of manipulating objects in its environment, to enter shells in general as a result of shell entry experience, and to select appropriate sized shells of a particular species following experience with that kind of shell. The crab also had in its repertoire, as a naive post-glaucothoe, the elements of shell entry behavior. Thus the functional pattern of shell entry is a result of a series of adaptations: the elements of shell entry, the tendency to handle objects, the ability to learn from past experience in handling objects, and the maintenance of flexibility in details of shell entry and shell selection. Appropriate shell entry behavior, which increases Darwinian fitness, is the result of all of these adaptations. It is emergent outcome of processes which occur at a variety of levels (Schneirla, 1972). These adaptations in turn can influence a variety of behavior patterns, not just shell entry or shell selection. The ability of the young crab to be influenced by general manipulative experience is not an adaptation for shell entry (alone) but rather a learning capability which influences many patterns. I can predict that crabs which manipulated objects early in life will not only enter shells more efficiently but will forage more efficiently and go through precopulatory behavior more effectively.

Having placed my past comments in such a framework, I wish to complicate the analysis of the more tractable aspects of the development of behavior patterns, the roles of past experience. The examples of experiential input which I have given and those that each of you could add are, for the most part, what I will call simple, direct-link experiential situations. A grapsid crab learns where crevices are in a rock by direct experience with shelter on that rock. A hermit crab develops a preference for a type of shell after experience with that type of shell. The behavioral category influenced is the same in the past and future situations. Such direct effects can be complicated, as illustrated by the memory system of the honey bee. Apis millifera quickly learns the odor, color, and shape of a flower and in later responding to objects it requires appropriate input in all three parameters (Gould, 1984).

However, in addition, influences can be of a less direct, more general nature. When young hermit crabs are given the opportunity to handle objects in general (but not gastropod shells), they make less mistakes in entering gastropod shells (Hazlett, 1971). The specific sequence of behavior used to enter shells (and only used to enter shells) is executed correctly the first time *if* the larvae had other objects to handle. Thus generalizations of experience can effect very specific sets of behavior patterns, patterns not obviously related (in form or function) to the experiential situation encountered earlier. Extra-dimensional training and cross-modal transfer (Thomas, 1969) are examples of a general category of influences I will call diffuse effects. Extra-dimensional training refers to the situation where an animal is conditioned to discriminate along one dimension or aspect within a modality but its behavioral responses are tested along a different dimension or aspect of the same modality (Thomas, 1970). In cross-modal transfer, animals conditioned to respond to a greater intensity of input in one modality (for example) would then respond similarly to greater intensity of input in another modality. Learning a discrimination in one modality can also be influenced by the presence of unrelated stimuli with which the animal has had some past experience (Wigal et al., 1984).

In some cases we may wish to say the diffuse effects are obvious. When early exposure to environmental variability sets the pattern of tidal rhythms, we easily accept that all aspects of an animal's behavior are influenced. Yet what we have in that case is one kind of input influencing every aspect of an animal's later behavior. The embryonic crab may have been influenced early by a pattern of change in salinity (Forward et al., 1986) but later responses to light , water pressure, temperature, etc. are all influenced by that earlier experience.

What I am suggesting is a much more complex gestalt of how behavior develops. We can not simply expose an experimental group to a specific environmental input and later see if it responds to that specific situation differently. Or rather if we do that, we can not conclude much in general about the role or nature of past experience for that kind of animal, or even about the role of past experience for the particular kind of behavior in that kind of animal in which we have an interest.

To turn this argument on its side, I am interested in how the behavior of animals evolves. Knowing that a behavior pattern is good for an animal in some way (appears to increase its fitness) suggests that the pattern is adaptive (in the evolutionary sense). But that tells me nothing about how the pattern is controlled, nothing about the developmental mechanisms (direct, diffuse, etc.) which lead to the pattern. And until I can understand those developmental mechanisms, my guesses about how evolution has acted are just that---guesses. We must understand the pattern of ontogeny of behavior if we are to suggest what the adaptations are for any particular species or group.

As a final point, I want to mention an obvious ommision in my talk thus far. I have premised this talk on the idea that variability and predictability in an environment (the intertidal) should be associated with flexibility in behavior. Stated evolutionarily, in an unpredictable evironment organisms will be favored which are able to adapt (behaviorally) to that unpredictability. Yet I have simply implied that intertidal organisms are flexible. Without saying it, I have implied that they are more flexible than similar organisms which inhabit more predictable environments. I should now present data that allow comparisons between related forms that live in the intertidal and, for example, the deeper benthic zones. However, what would I compare---the data are not available. Thus I would end with a plea for a rather ambitious program. Complete studies of the relative importance of experiential input throughout the life

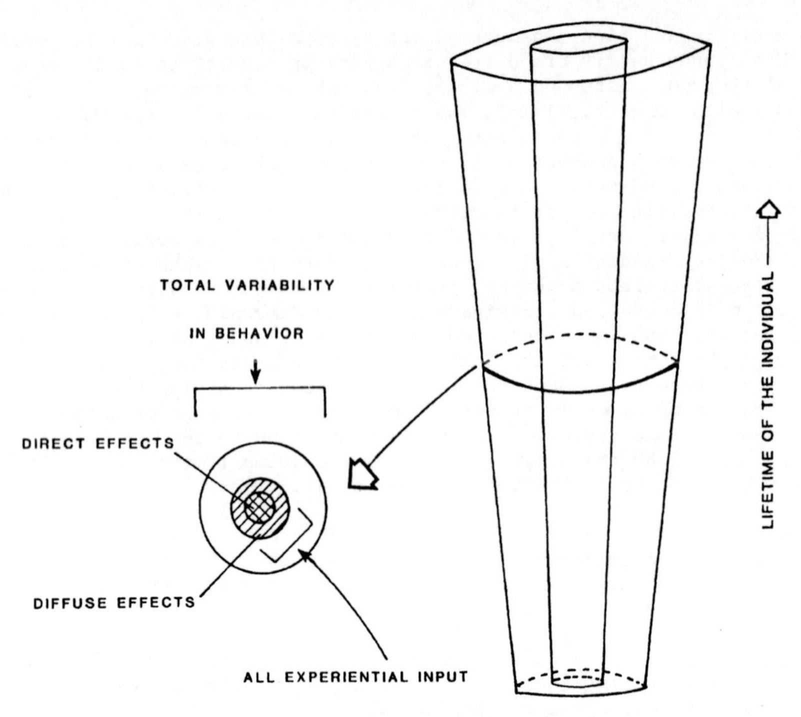

Fig. 2. Hypothetical information solid representing the proportion of variability in behavior attributable to the effects of past experience

of organisms in several types of habitats which vary in environmental predictability are needed. Without such data, we will continue to research particular cases or examples of behavioral adaptations to particular situations without testing the bigger picture.

Studies involving all life stages are essential for making comparisons between animals living in different habitats in part, because differences in the ecology of individual species preclude the short-cut of looking for the relative roles of past experience or flexibility at any one stage. While the pattern of greater-flexibility-in-young-stages may be a common one, in other cases the young stages may be less flexible. Scapini (1986) has shown that the orientation behavior of individuals of Talitrus saltator from Mediterranean population is initially influenced very little by experience but that modulation of the early orientational behavior by experience occurs later.

During its lifetime, an individual may use one means of coping with variability at one stage (gregarious settlement as a post-larva), a second means shortly after that (setting of tidal rhythms by local zeitgebers), and yet another mechanism throughout its life (response to immediate cues associated with tidal movement). To make comparisons between habitats (intertidal vs. subtidal) we would need data on the role of past experience in all phases of a series of species. Lacking the hard data to make such a comparison, all I can do is suggest a methodology.

The variability in the execution of acts in time and space can be measured/represented as uncertainty. When uncertainty of the form or timing of behavior patterns is reduced by (knowledge of) past experience, this can be quantified via the use of information theory (Hazlett and Estabrook, 1974). To state it differently, knowing that past experience can influence the execution or timing of acts is the same as an increase in predictive ability of the biologist. How much the predictive ability is increased can be measured by use of information theory methods. At any given stage in the life history of a species, we can consider the results of many experiments and observations and estimate the total role of past experience in the behavior of the animal by summing the percent uncertainty reduced across all categories of behavior patterns. If we now look at all the life history stages and consider the role of past experience throughout the life of an average individual we could construct an "information solid" (Figure 2). The total influence of past experience is then estimated as a "volume" of information. The absolute and relative volumes of species could then be compared.

Such estimates of the relative roles of all types of past experience throughout the life of different species, living in different habitats would form the basis of testing the general hypothesis that environmental unpredictability favors (via natural selection) greater flexibility. It is possible that such a hypothesis will be falsified and that a multitude of specific behavioral adaptations to intertidal life, which do not necessarily involve influences of past experience, will prove to be a better explanation of the patterns of behavior seen in intertidal animals. However, I feel we do need a framework for testing the alternative that the adaptation of importance for organisms in the intertidal is flexibility itself.

ACKNOWLEDGEMENTS

I wish to thank Linda Baron and Catherine Bach for their comments on this manuscript. Portions of this work were supported by a grant from the Psychobiology Program of the National Science Foundation (BNS-8604092).

REFERENCES

Abele, L. G., Campanella, P.J., and Salmon, M., 1986, Natural history and social organization of the semiterrestrial grapsid crab *Pachygrapsus transversus*, *J. Exp. Mar. Biol. Ecol.*, 104:153-170.

Abrams, P., 1978, Shell selection and utilization in a terrestrial hermit crab, *Coenobita compressus* (H. Milne Edwards), *Oecologia*, 34:239-253.

Alderdice, D. F., 1972, Factor combinations, Responses of marine poikilotherms to environmental factors acting in concert, *in*: "Marine Ecology", Vol. 1, O. Kinne (ed.), Wiley-Interscience, London.

Balasch, J., and Mengual, V., 1974, The behavior of *Dardanus arrosor* in association with *Calliactis parasitica* in artificial habitat, *Mar. Behav. Physiol.*, 2:251-260.

Banse, K., 1986, Vertical distribution and horizontal transport of planktonic larvae of echinoderms and benthic polychaetes in an open coastal sea, *Bull. Mar. Sci.*, 39:162-175.

Barlow, R. B. Jr., Powers, M. K., Howard, H., and Kass, L., 1986, Migration of *Limulus* for mating: relation to lunar phase, tide height, and sunlight, *Biol. Bull.*, 171:310-329.

Bell, J. D., and Westoby, M., 1986, Variation in seagrass height and density over a wide spatial scale: effects on common fish and decapods, *J. Exp. Mar. Biol. Ecol.*, 104:275-295.

Bertness, M. D., 1980, Shell preference and utilization patterns in littoral hermit crabs of the Bay of Panama, J. Exp. Mar. Biol.Ecol., 48:1-16.
Bertness, M. D., 1981, The influence of shell type on hermit crab growth rate and clutch size (Decapoda, Anomura), Crustaceana, 40:197-205.
Bertness, M. D., 1982, Shell utilization, predation pressure, and termal stress in Panamanian hermit crabs: an interoceanic comparison, J. Exp. Mar. Biol. Ecol., 64:159-187.
Blackstone, N. W., and Joslyn, A. R., 1984, Utilization and preference for the introduced gastropod Littorina littorea (L) by the hermit crab Pagurus longicarpus (Say) at Guilford, Connecticut, J. Exp. Mar. Biol. Ecol., 80:1-9.
Brooks, W. R., and Mariscal, R. N., 1985, Shell entry and shell selection of hydroid-colonized shells by three species of hermit crabs from the northern Gulf of Mexico, Biol. Bull., 168:1-17.
Burke, R. D., 1986, Pheromones and the gregarious settlement of marine invertebrate larvae, Bull. Mar. Sci., 39:323-331.
Chelazzi, G., Focardi, S., Deneubourg, J. L., and Innocenti, R., 1983, Competition for the home and aggressive behavior in the chiton Acanthopleura gemmata (Blainville) (Mollusca: Polyplacophora), Behav. Ecol. Sociobiol., 14:15-20.
Childress, J. R., 1972, Behavioral ecology and fitness theory in a tropical hermit crab, Ecol., 53:960-964.
Cronin, T. W., and Forward, R.B. Jr., 1986, Vertical migration cycles of crab larvae and their role in larval dispersal, Bull. Mar. Sci., 39:192-201.
Curtis, L. A., 1987, Vertical distribution of an estuarine snail altered by a parasite, Science, 235:1509-1511.
Drapkin, E. I., 1963, Effect of Rapana bezear Linne (Mollusca, Muricidae) on the Black Sea fauna, Dokl. Akad. Nauk SSR, 151:700-703.
Elwood, R. W., McClean, A. and Webb, L., 1979, The development of shell preferences by the hermit crab Pagurus bernhardus, Anim. Behav., 27:940-946.
Field, L. H., Evans, A., and MacMillan, D. L., 1987, Sound production and stridulatory structures in hermit crabs of the genus Trizopagurus, J. mar. biol. Ass. U. K., 67:89-110.
Fishelson, L., 1983, Population ecology and biology of Dotilla sulcata (Crustacea, Ocypodidae) typical for sandy beaches of the Red Sea, in: "Sandy Beaches as Ecosystems", A. McLachlan and T. Erasmus (eds.), Junk, Hague.
Forward, R. B. Jr., Douglass, J., and Kenney, B. E., 1986, Entrainment of the larval release rhythm of the crab Rhithropanopeus harrisii (Brachyura; Xanthidae) by cycles in salinity change, Mar. Biol., 90:537-544.
Fotheringham, N., 1975, Structure of seasonal migrations of the littoral hermit crab Clibanarius vittatus (Bosc), J. Exp. Mar. Biol. Ecol., 18:47-53.
Ghilchrist, S. L., 1985, Ecology of juvenile hermit crab shell use: Field and laboratory comparisons, Am. Zool., 25(4):60A.
Gould, J. L., 1984, Natural history of honey bee learning, in: "The Biology of Learning", P. Marler and H. S. Terrace (eds.), Berlin, Spinger-Verlag.
Hazlett, B. A., 1966, Social behavior of the Paguridae and Diogenidae of Curacao, Stud. Fauna Curacao, 23:1-143.
Hazlett, B. A., 1971, Influence of rearing conditions on initial shell entering behavior of a hermit crab (Decapoda, Paguridea), Crustaceana, 20:167-170.
Hazlett, B. A., 1981a, The behavioral ecology of hermit crabs, Ann. Rev. Ecol. Syst., 12:1-22.
Hazlett, B. A., 1981b, Daily movements of the hermit crab Clibanarius

vittatus, Bull. Mar. Sci., 31:177-183.
Hazlett, B. A., 1982, Resource value and communication strategy in the hermit crab Pagurus bernhardus (L), Anim. Behav., 30:135-139.
Hazlett, B. A., 1987a, Hermit crab shell exchange as a model system, Bull. Mar. Sci., 44:in press.
Hazlett, B. A., 1987b, Information transfer during shell exchange in the hermit crab Clibnarius antillensis, Anim. Behav., 35:218-226.
Hazlett, B. A., and Estabrook, G., 1974, Examination of agonistic behavior by character analysis. I. The spider crab Microphrys bicornutus, Behaviour, 48:131-144.
Hazlett, B. A., and Provenzano, A. J. Jr., 1965, Development of behavior in laboratory reared hermit crabs, Bull. Mar. Sci., 15:616-633.
Highsmith, R. C., 1982, Induced settlement and metamorphosis of sand dollar (Dendraster excentrius) larvae in predator-free sites: adult sand dollar beds, Ecol., 63:329-337.
Hines, A. H., 1986, Larval problems and perspectives in life histories of marine invertebrates, Bull. Mar. Sci., 39:506-525.
Jackson, G. A., 1986, Interaction of physical and biological processes in the settlement of planktonic larvae, Bull. Mar. Sci., 39:202-212.
Levin, L. A., 1986, The influence of tides on larval availability in shallow waters overlying a mudflat, Bull. Mar. Sci., 39:224-233.
Maynard Smith, J., and Parker, G. A., 1976, The logic of asymmetric contests, Anim. Behav., 24:159-175.
McGuinness, K. A., and Underwood, A. J., 1986, Habitat structure and the nature of communities on intertidal boulders, J. Exp. Mar. Biol. Ecol., 104:97-123.
Moran, M. J., 1985, The timing and significance of sheltering and foraging behaviour of the predatory intertidal gastropod Morula marginalba Blainville (Muricidae), J. Exp. Mar. Biol. Ecol., 93:103-114.
Neumann, D., 1981, Tidal and lunar rhythms, in: "Handbook of Behavioral Neurobiology", Vol. 4, Biological Rhythms, J. Aschoff ed., Plenum Press, N. Y.
Rebach, S., 1981, Use of multiple cues in short range migrations of Crustacea, Am. Midl. Nat., 105:168-180.
Reese, E. S., 1969, Behavioral adaptations of intertidal hermit crabs, Am. Zool., 9:343-355.
Reese, E. S., 1983, Evolution, neurothology, and behavioral adaptations of crustacean appendages, in: "Studies in Adaptations, The Behavior of Higher Crustacea", S. Rebach and D. W. Dunham, eds., Wiley Interscience, N. Y.
Scapini, F., 1986, Inheritance of solar direction finding in sandhoppers, 4, Variation in the accuracy of orientation with age, Monitore zool. ital. (N.S.), 20:53-61.
Scheltema, R. S., 1986, Long-distance dispersal by planktonic larvae of shoal-water benthic invertebrates among central Pacific islands, Bull. Mar. Sci., 39:241-256.
Schneirla, T. C., 1972, The concept of levels in the study of social phenomena, in: "Selected writings of T. C. Schneirla", L. R. Aronson ed., Freeman, San Francisco.
Scully, E. P., 1979, The effects of gastropod shell availability and habitat characteristics on shell utilization by the intertidal hermit crab Pagurus longicarpus Say, J. Exp. Mar. Biol. Ecol., 37:139-152.
Scully, E. P., 1983, The behavioral ecology of competition and resource utilization among hermit crabs, in: "Studies in Adaptation, The Behavior of Higher Crustacea", S. Rebach and D. W. Dunham eds., Wiley Interscience, N. Y.
Scully, E. P., 1986, Shell investigation behavior of the intertidal hermit crab Pagurus longicarpus Say, J. Crust. Biol., 6:749-756.
Snyder-Conn, E., 1980, Tidal clustering and dispersal of the hermit crab Clibanarius digueti, Mar. Behav. Physiol., 7:135-154.
Snyder-Conn, E., 1981, The adaptive significance of clustering in the hermit

crab Clibanarius digueti, Mar. Behav. Physiol., 8:43-54.
Thomas, D. R., 1969, The use of operant conditioning techniques to investigate perceptual processes in animals, in: "Animal Discrimination Learning", R. M. Gilbert and N. S. Sutherland eds., Academic Press, N. Y.
Thomas, D. R., 1970, Stimulus selection, attention, and related matters, in: "Current Issues in Animal Learning", J. H. Reynierse ed., U. Nebraska Press, Lincoln.
Wicksten, M. K., 1977, Shells inhabited by Pagurus hirsutiusculus (Dana) at Coyote Point Park, San Francisco Bay, California, Veliger, 19:445-446.
Wigal, T., Kucharski, D., and Spear, N. E., 1984, Familiar contextual odors promote discrimination learning in preweaning but not in older rats, Develop. Psychobiol., 17:555-570.
Woodin, S. A., 1986, Settlement of infauna: Larval choice?, Bull. Mar. Sci., 39:401-407.

Design and Analysis of Field Experiments on Competitive Interactions Affecting Behaviour of Intertidal Animals

Antony J. Underwood

Sydney University, Australia

ABSTRACT

Intertidal animals have often been shown to be susceptible to competitive interactions because of shortages of resources. One behaviouur that has commonly been found to be influenced by the local density of a species is dispersal, i.e. density-dependent migration from areas of large density. The simplest field experiments have revealed that this occurs because introduced animals (those brought into an area to increase the density) are more likely to move away than are the original residents. Here, the design and procedures used in such experiments are examined.

Controls are needed for the various types of disturbance caused to animals when they are placed in strange areas to increase the density. Because animals must be moved around in such experiments, it is necessary to have control plots in which disturbed animals are introduced, but densities are not increased. The design and analysis of such an experiment is illustrated for the intertidal limpet, Cellana tramoserica. There was, in fact, no density-dependent process demonstrable in the experiments. All of the increased migration from areas of large density could be attributed to effects of experimental disturbance acting on the introduced limpets.

There are important consequences of this finding for interpretations of experiments on inter- and intra-specific competitive intractions affecting the behaviour of intertidal animals. These are discussed here.

INTRODUCTION

Competition within and among species has often been proposed (sometimes as an article of faith) as one of the most important processes (or sometimes as the only important process) influencing natural populations. Competitive interactions are widely cited as being responsible for many aspects of the life-histories, size-structures and reproductive outputs or spatial distributions of populations, and as a major process moulding the diversity and composition of natural assemblages of organisms (see reviews by Diamond, 1978; Schoener, 1983; Branch, 1984). There is a long tradition for such opinion (Malthus, 1798; Darwin, 1882; Jackson, 1981) and much theoretical modelling of populations is heavily dependent on the assumption that competition is important (e.g. Lotka-Volterra models and their relatives and descendants in Krebs,1978; Williamson, 1972; May, 1974).

Competition occurs when some resource necessary to survival of organisms of the same or different species is available only in limited (i.e. inadequate) quantities. In consequence, organisms suffer when their individual needs for the resource cannot be met (see the definition provided by Birch,1957). There has been increasing sophistication in the categorization of different types of competitive interactions (Schoener, 1983), but the essential concept is quite straightforward.

The basis for widespread acceptance of the importance of competitive interactions in some ecosystems is, however, apparently the existence of patterns in nature that might be explained by competitive processes, rather than demonstrations that resources are in short supply and that competition for these resources leads to the observed patterns (see reviews by Connell, 1983; Schoener, 1983; and arguments in Strong et al., 1984).

Intertidal organisms on rocky shores are particularly susceptible to competitive processes (Connell, 1972; Underwood, 1979; Branch, 1981; Underwood and Denley, 1984). Their two major resources are food and substratum on the shore on which to attach. Both are inextricably linked. All sessile species absolutely require space on which to live on a shore. In addition, they need access to sunlight for energy, or to the water-column for food (and thus their area - the space they occupy - is an important determinant of ability to photosynthesize or feed). Mobile grazing animals must have somewhere to live, but need space over which to feed and, in many cases that have been studied experimentally, algal food has been in limited supply (e.g. Stimson, 1973; Black, 1977; Choat, 1977; Underwood, 1976a, 1978, 1984; Creese, 1980; Creese and Underwood, 1982). Because of this commonality of usage of resources, competition has been observed within and among many species on seashores, including many competitive interactions between organisms from different phyla and very different modes of life (e.g. competitive interactions between barnacles and limpets; Choat, 1977; Underwood et al., 1983; Sutherland and Ortega, 1986).

Because of the great reproductive output of many intertidal species and the vagaries of dispersal and mortality during the "mystery stage" (Spight, 1975) of planktonic life, there are often large fluctuations in the numbers of recruits entering intertidal populations (e.g. Loosanoff, 1964, 1966; Underwood and Denley, 1984; Caffey, 1985; Connell, 1985; Gaines and Roughgarden, 1985). Even where competition has reduced reproductive output on a particular shore, large numbers of recruits can arrive regardless of the previous shortage of resources - because planktonic larvae can arrive from a breeding population in some other area.

Despite this scenario, many processes can act to reduce the effects or prelavence of competition in natural populations. Predation (or grazing) and disease can limit the numbers of organisms so that they do not staturate their resources (e.g. Connell, 1961; Dayton, 1971; Paine, 1974; Menge, 1976). Disturbances due to physical processes (waves, abrasion by sand, desiccation during low tide, etc.) and due to environmental "hazards" (Andrewartha and Birch, 1954) can kill organisms in density-indpendent ways (e.g. Dayton, 1971; Sousa, 1979). Even where competition among individuals does occur, there may be no effect on the distribution of any of the species involved (e.g. Sale, 1977; Creese and Underwood, 1982).

This is the basis for dissent about the importance of competition, which has recently sparked considerable debate (e.g. Simberloff, 1978, 1984; Diamond and Gilpin, 1982; Gilpin and Diamond, 1982). Consequently, the existence or importance of competition cannot be demonstrated by appealing to the fact that it might occur. Nor can the existence of competitive processes be inferred unambiguously simply by observing patterns of distribution, abundance, size-structure, etc., that vary in a manner correlated with intraspecific density or with the numbers of organisms of other species that are present. Much confusion has been foisted on the development of ecological understanding by some authors' refusal to acknowledge that patterns do not, and can not, identify processes (see Dayton, 1973; Schroder and Rosenzweig, 1973; Connor and Simberloff, 1979; Underwood and Denley, 1984). This debate has, of course, not been enhanced by some authors'inability to demonstrate the validity of their proposed patterns - let alone the processes (see the discussion by Simberloff, 1978; Connor and Simberloff, 1979; Strong et al., 1979).

All of this preamble leads to the requirement that the process of competition can only be demonstrated by properly controlled and replicated falsificationist tests (sensu Popper, 1963) of hypotheses based on arguments and models of competition. Connell (1974, 1983) has previously argued the need for experiments in the field. Manipulative experiments on intertidal organisms are numerous and widespread (see reviews by Connell, 1972, 1974; Paine, 1977; Underwood and Denley, 1984; Underwood, 1985; Denley and Dayton, 1985). Yet, the design and analysis of such experiments are often made difficult by problems of poor logic, confounding, lack of controls, inadequate replication or confused statistical considerations - the whole demonic horde of ills that can befall any serious attempt to make experimental order out of untested ecological chaos (see reviews by Hurlbert, 1984; Underwood, 1981, 1985, 1986; and the detailed examples in Underwood, 1983 and Chapman, 1986).

Most of the manipulative experimental studies have focussed on aspects of mortality (e.g. Connell, 1961; Black, 1977; Choat, 1977; Underwood, 1978, 1984), growth (e.g. Underwood, 1976a) or reproductive output (e.g. Fletcher, 1984). There have been relatively few examples of experimental manipulations of competitive interactions that directly affect the behaviour of intertidal animals. Some notable exceptions concerning gastropods are those on limpets (studies by Aitken, 1962; Breen, 1971; Branch, 1975; Mackay and Underwood, 1977).

The present discussion is intended to address some of the problems that are prevalent in the design of field experiments by considering the designs of experiments on competitive interactions that might affect the behavioural regulation of density, dispersion and dispersal of motile intertidal organisms. Sessile organisms pose some different problems (but have a more restricted repertoire of behaviours once they are sessile).

First, I consider examples of experiments on intraspecific effects of increased density on limpets. I then discuss the problems of interspecific interactions. The focus is on the design of better experiments and some problems of statistical analysis of complex designs. The principal intention is to identify some of the problems for interpretation of experiments that are caused by a lack of adequate controls to unconfound the process of competition from anything else incorporated (by chance or by intent) into the experimental design.

A PRIMITIVE EXPERIMENT ON EFFECTS OF INCREASED DENSITY ON DENSITY, DISPERSION AND DISPERSAL

Experimental Design

The simplest experimental test of the hypothesis that intra-specific increases in density will lead to some reduction in density (i.e. that limpets show self-regulation of density) is to compare the densities of animals in experimental plots where extra individuals have been added with those in control, unmanipulated plots. The same experimental design obviously allows tests of hypotheses about the effects of increased intra-specific density on any aspect of the behaviour of the individuals in the study-sites. Such experiments are widespread in the ecological literature, where they form one half of the requirements of a convergence experiment (for limpets, see Stimson and Black, 1975). Where the experimental plots are not isolated (in contrast to the situation on isolated boulders or pier pilings), the animals are free to move into or out of the control and experimental plots and their behavioural responses to increased density are easily observed.

When this type of simple experiment has been done with intertidal limpets, the result has often been a rapid return to the original density in the experimental plots because limpets quickly dispsered away from areas of increased density (Aitken, 1962; Breen, 1971; Mackay and Underwood, 1977). This clearly allows rejection of the null hypothesis that there is no effect of the experimental manipulation and has usually been interpreted to indicate that the behaviour of limpets is modified in response to increased density, causing the animals to move further or faster than would normally be the case. Where limpets are involved, because of their widespread homing behaviour (e.g. Underwood, 1979), the results are also interpreted to indicate a change in the propensity to home when density is increased.

Problems with Interpretation

There is, however, a problem with such interpretations. The outcome of the experiments is confounded hopelessly with the fact that the increased densities of limpets are brought about by introducing new, but disturbed limpets into the region. Consequently, to discover that they then behave differently from limpets at some smaller density may not be a response to density but to any number of other factors that now differ between control (untouched) and experimental (i.e. at greater density, but disturbed) limpets.

AN IMPROVED DESIGN

Controls for Disturbance

Clearly, some better controls for the experimental manipulations are required. The simplest increase in interpretability of this type of experiment is to disturb the resident limpets in a number of experimental areas, thus causing them to be handled, lifted up and carried about, put down again on the shore, but maintained at the original density. These can then be contrasted with control limpets (i.e. in plots where there has been no disturbance and no increase in density - identical to the control plots of the simpler experimental design discussed previously) and plots where densities are experimentally increased after disturbance of the residents.

An Example with the Limpet Cellana tramoserica

The confounding of experimental disturbance and increased density can be demonstrated by an example involving the intertidal limpet *Cellana tramoserica* in New South Wales. This limpet shows variable homing behaviour which can be modified by the presence of other limpets and by availability of food (Mackay and Underwood, 1977). Previous experiments demonstrated that limpets migrated from areas of experimentally enhanced density and that the original residents tended to migrate less than the experimentally introduced ones (Mackay and Underwood, 1977). These results were, however, based on experiments using the simple design described previously.

Experimental Design and Methods

To investigate the problems of experimental disturbance of limpets, manipulations were done in August-September 1986 at the Cape Banks Scientific Marine Research Area (Botany Bay, New South Wales). Study-sites there have been described in Underwood (1975) and Underwood et al. (1983). The areas chosen were at mid-shore levels (0.6 - 0.8 m above low water with a tidal range of about two metres maximum) on a sandstone shore. The mean density of *C. tramoserica* (about 20 mm shell-length) in the area was 93.4 per square metre (S.E. = 12.0, n = 12 metre square quadrats). Four sets of experimental plots were used: Control (C), Disturbed (D), Control + Additional (CA), Disturbed + Additional (DA). Plots were 1 metre square and there were three replicate plots in each treatment. The twelve plots were scattered at random with several metres between each. Control plots were untouched, except that the limpets present were marked with a dab of non-toxic marine paint. In Disturbed plots, limpets were carefully removed from the substratum, marked with paint and replaced in their original positions. In the other two treatments, 300 additional limpets were added to each experimental plot. These were carefully lifted from the rock in nearby areas (some ten metres away), marked and placed at random positions in the experimental plots. In the Control + Additional plots, the original limpets were simply marked with paint of a different colour from the introduced animals. In the Disturbed + Additional plots, the original limpets were lifted up, marked and replaced; they had thus been handled in the same way as the experimentally introduced animals.

After the experiment started (on 18th August 1986), the numbers of original and introduced limpets were counted in each plot at various times during the next month. The total numbers of limpets in each plot (i.e. including animals that had entered the plots after the start of the experiment) were also recorded.

Results

The results were unambiguous. Control and Disturbed treatments maintained their densities throughout the experiment. After only a few days, the densities of limpets in the two treatments where excess limpets had been introduced had decreased to be not different from their original values (Table 1). Thus, rapid self-regulation of density had occurred, brought about by the movement of limpets out of plots where densities had been increased.

Quite clearly, the original inhabitants of the experimental plots were much less likely to leave than were the introduced limpets (Fig. 1) an a mean of only 8 out of 300 introduced animals were still in plots after 29 days (Table 3). The re-establishment of original densities in plots where

Table 1.

Self-regulation of density following experimental increase by the limpet Cellana tramoserica ($\underline{n}$ = 3 plots)

Experimental Treatment	Mean original density (S.E.)	Mean (S.E.) ratio to original number on day: 0	2	3	4	14	16	29
(C) Control	95.0 (12.0)	1.00 -	0.92 (.06)	0.96 (.03)	1.03 (.06)	1.04 (.02)	0.92 (.03)	0.94 (.06)
(D) Disturbed	113.0 (26.1)	1.00 -	0.88 (.06)	0.88 (.04)	1.13 (.05)	1.08 (.05)	1.00 (.10)	0.90 (.09)
(CA) Control and Additional	76.3 (30.8)	6.54 (2.20)	2.89 (.58)	1.94 (.26)	1.45 (.08)	1.13 (.18)	1.07 (.21)	1.13 (.05)
(DA) Disturbed and Additional	89.3 (20.3)	4.87 (1.10)	2.35 (.44)	1.97 (.38)	1.33 (.04)	1.33 (.16)	1.06 (.05)	1.01 (.12)

Table 2.

Analyses of proportions of original and introduced limpets remaining n experimental plots ($\underline{n}$ = 3 plots). C = Control; D = Disturbed; CA = plots with Control and Introduced limpets; DA = plots with Disturbed and Introduced limpets

a) Analyses of variance after 3 and 29 days. Treatments are C and D plots versus Original animals in CA and Da plots versus Introduced animals in CA and DA plots.

Source of variation	Df	MS	F-ratio	P	MS	F-ratio	P
Control vs. Disturbed	1	0.035	8.8	< .005	0.00002	0.00003	ns
Treatments	2	0.289	72.3	< .001	0.078	11.1	< .001
Interaction	2	0.013	3.3	ns	0.00005	0.00007	ns
Residual	12	0.004	-		0.007	-	
Total	17	-	-		-	-	

b) Mean proportions; inequalities indicate significant differences at $\underline{P}$ = 0.05 in Student-Newman-Keuls tests.

3 days.		Control > Disturbed		
		0.64 > 0.33		
	Originals in C and D	=	Originals in CA and DA	>	Introduced CA and DA
	0.61	=	0.61	>	0.24
29 days.		Control = Disturbed		
		0.16 = 0.16		
	Originals in C and D	=	Originals in CA and DA	>	Introduced CA and DA
	0.23	=	0.22	>	0.03

Table 3.

Mean (S.E.) numbers of limpets in experimental plots at the end of the first experiment ($\underline{n}$ = 3 plots; initial mean densities of limpets are in Table 1)

Treatment	Original limpets	Introduced limpets	Immigrant limpets	Total
(C) Control	18.0 (3.5)	-	73.7 (25.9)	91.7 (28.8)
(D) Disturbed	22.3 (1.7)	-	83.4 (35.8)	105.7 (36.2)
(CA) Control and Introduced	14.3 (3.9)	8.0 (5.0)	61.0 (22.9)	83.3 (31.5)
(DA) Disturbed and Introduced	20.7 (8.2)	8.0 (1.0)	57.7 (10.0)	86.4 (14.8)

these had been increased was mostly due to the departure of the introduced limpets. The proportion of original limpets remaining in the plots was significantly greater than that of introduced limpets at the end of the experiment (Table 2). At no stage were the original limpets affected by the presence of increased numbers of limpets (note the lack of significant difference between the proportion of original limpets in Control and Control + Additional plots and between Disturbed and Disturbed + Additional plots in Table 2).

Original limpets did, however, also move out of their plots (only a mean of 21 % of original limpets were still in the plots after 29 days;

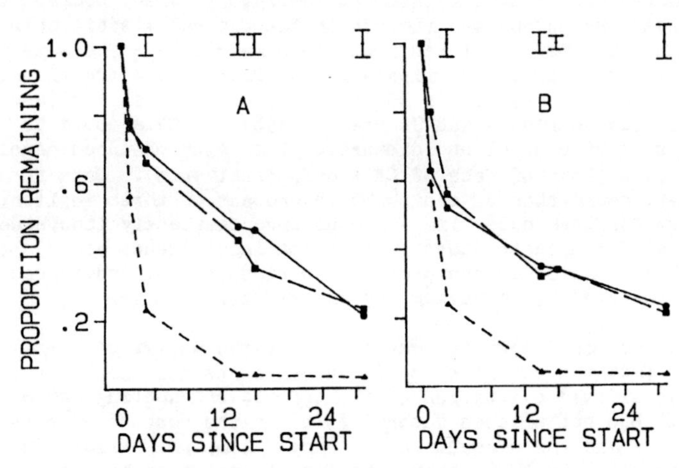

Figure 1. Mean proportion of limpets remaining in experimental plots during the first experiment; pooled Standard Errors for each time of sampling are indicated (above and below 1.0). A is Control (•——) and Control plus additional (■— — are original and ▲---- are introduced limpets); B is Disturbed (•——) and Disturbed plus Additional (■— — are original and ▲---- are introduced limpets).

Fig. 1 and Table 2). Densities were maintained in all the plots by the immigration of new limpets from the surrounding areas (Table 3).

During the first few days of the experiment, there was a greater tendency for disturbed animals (i.e. original limpets that had been handled) to leave the experimental or control plots more quickly than was the case for the undisturbed control limpets (this difference was significant; see Table 2). This difference between untouched and disturbed limpets was no longer detectable by the end of the experiment.

Interpretation of this Experiment

The results of this manipulation, with its more complex, but superior, experimental design were less ambiguous than those obtained in previous experiments. Disturbance of the limpets, *per se*, was important during the first few days of the experiment. If the simpler design had been used, i.e. an experiment without the controls for disturbance, and had been terminated as soon as densities had converged to the control levels (i.e. after about 3 or 4 days in this experiment), the results would have been hopelessly confounded. Such an experiment would only have produced results from the Control and the Control + Additional plots. Introduced limpets in the latter were disturbed, original ones in the former were not.

Disturbance *was* causing limpets to leave the plots faster than undisturbed limpets, even where the densities had not been increased. It was therefore necessary to be able to demonstrate that disturbance *per se* was insufficient to explain the differences between Control and Control + Additional plots. With the improved experimental design, the magnitude of effects of disturbance was successfully identified and shown to be smaller than that of enhanced density and to be insufficient to account for the decline in numbers of limpets in plots where density had been increased.

Further Improvement in Experimental Design is Still Needed

There are still some problems with this improved experimental design. First is a technical difficulty for analysis of the data. The analyses of variance presented here (Table 2) are not completely valid, because the data for Original and Introduced limpets in the Control + Additional plots are not independent (they are taken from the same plots). The same is true for the Original and introduced limpets in the Disturbed + Additional plots. This causes uncertainty in the validity of the analysis (see discussions in Cochran (1947) and Underwood (1981)). This could be avoided by having two sets of control and disturbed plots with enhanced density (i.e. a total of 6 plots of each of CA and TA treatments). Data for the Original limpets could then be monitored in one set of three replicate plots and those for Introduced limpets monitored completely independently in the other set for each treatment. This non-independence would not affect the interpretation of the present experiments, but independence of data should be created by improving the experimental design.

There are much more serious concerns about the design of this experiment, in that the controls for disturbance are not complete. While this design has allowed evaluation of the effects of handling and moving the limpets, it has not allowed determination of the confounded effects of increased density and the presence of "strange" limpets. First, an untouched, original limpet may show behavioural responses to the presence of increased numbers of limpets (part of the original hypothesis being examined) OR responses to the presence of strange limpets regardless of the prevailing density in the region. Thus, some control is needed to determine the effect of introducing new limpets to an area.

In a similar manner, the introduced limpets are disturbed in very different ways from those in the Disturbed treatments of the previous experiment. They have not only been picked up, handled and replaced. Thus, conclusions about the effects of enhanced density are hopelessly confounded with a large number of potential differences between limpets in control and experimental plots. These confounding influences are not eliminated simply by examination of the most obvious effects of experimental handling of the introduced limpets.

AN IMPROVED DESIGN WITH BETTER CONTROLS FOR DISTURBANCES

Types of Disturbance

The next step for improvement of this sort of experiment is to provide controls for those confounding aspects of disturbance that are not functions of density *per se*. When density is experimentally increased in an area of shore, strange animals are introduced (that are "unfamiliar" to the original residents). The introduced limpets are themselves potentially disorientated by being handled (and otherwise disturbed). They have also been moved to new locations. They have never been in these areas before. They are placed in random (or, at the very least, haphazard) sites which may not be suitable and may not be the micro-habitats that undisturbed limpets would be found in, etc., etc. If they are homing individuals, they are not in a previously-occupied home-site.

Support of the notion that intraspecific increases in density have some effect leading to self-regulatory behaviour of intertidal animals requires disproof of the null hypothesis that adding extra limpets causes increased movement out of experimental areas (so that numbers are restored). This, in turn, requires that any increased emigration be solely due to effects of density. Therefore, there must be experimental plots with introduced animals (so that they are disturbed and "unfamiliar"), but with density held constant at control levels. This has been attempted for experiments with *Cellana tramoserica*.

Experimental Treatments and Disturbances

Four experimental treatments were used in the improved design. "Control" plots contained limpets that were not touched (except that a random sample was marked - see later). "Normal" plots contained only limpets that were originally present in the area; some of these were disturbed in various ways (Table 4; see below). Comparison of limpets in Control and Disturbed plots would reveal effects that were simply due to disturbances during establishment of the experiment.

In "Replaced" plots, approximately half of the limpets were removed and were replaced by similarly-sized limpets from elsewhere, to maintain the original density. Again, various of the original limpts were disturbed. Comparison of Replaced plots with Disturbed plots would indicate any differences in behaviour of original or introduced limptes that were due to the presence of strange (i.e. introduced), disturbed limpets at the same density as in Control plots.

Finally, there were "Increased" plots which were treated in exactly the same way as Replaced plots, but then extra limpets were added to increase the density by 50 % over original numbers. Comparison of these plots with the Replaced plots tests the hypothesis that the behaviour of either the original or introduced limpets differs when density is increased. Thus, the replaced plots are the new component added in this design compared with previous experiments.

Table 4.

Experimental treatments in improved design

PLOTS (n = 3)		LIMPETS (n = 10)					
	DENSITY	ORIGINAL				INTRODUCED	
		Undis-turbed	Dist-turbed	Moved (5 cm)	Moved (15 cm)	Replace original	Random position
CONTROL	Normal	X	-	-	-	-	-
NORMAL	Normal	X	X	X	X	-	-
REPLACED	Normal	X	X	X	X	X	X[a]
INCREASED	Enhanced	X	X	X	X	X	X[b]

[a] Each limpet replaced an Original limpet to maintain the initial density

[b] Some limpets replaced Original limpets and then extra animals were added to waCh plot to increase the density

The experimental disturbances were designed to investigate various aspects of the necessary disruptions to normal life of the limpets that were introduced into the Replaced or Increased plots (see Table 4). "Undisturbed" (control) limpets were untouched except for marking. "Disturbed" limpets were lifted from the substratum, handled, marked and replaced in their original positions. Comparison of these with the Untouched animals would reveal any influences on behaviour due solely to the artefacts caused by the necessary intrusion of the experimenter.

"Moved" limpets were treated the same as Disturbed ones, except that they were replaced with random orientation at some distance from their original positions. This treatment was designed to cause these limpets to be in unusual micro-habitats away from their homes (if they were homing) and they would thus be directly comparable to introduced limpets. They would differ from introduced limpets only in that the latter were also in unfamiliar sites at some distance (5 - 10 m) from their original positions and the former had only been moved a short distance within an experimental plot. Experimentally moved limpets were in two groups in each plot; some were moved about 5 cm and others 15 - 30 cm from their original positions. Undisturbed, Disturbed and Moved (both distances) animals were monitored in Normal, Replaced and Increased plots (Table 4).

Introduced animals in Replaced and Increased plots were treated in two ways. Some were placed in the precise location of an original animal that had been removed (Table 4). Each of these limpets was therefore in the exact micro-habitat that another limpet had occupied. Others were placed at random in the plots. Comparisons of these two types of introduced limpets with original, Disturbed and original, Moved limpets, respectively, would indicate any differences in behaviour between original and introdced animals that were directly due to being placed in an unfamiliar micro-habitat and (for homing individuals) not being in a home-site.

Methods and Data Collected

The second experiment was started on 17th October 1986 in an area near

Table 5.

Self-regulation of density in the second experiment

a) Mean (S.E.) numbers of limpets in experimental plots at the start and end of the experiment with the numbers removed and introduced ($\underline{n}$ = 3 plots)

Treatment	Mean original number	Mean number removed	Mean number introduced	Mean final number (Day 3)
(C) Control	123.3 (30.0)	-	-	116.7 (28.4)
(N) Normal	109.0 (20.2)	-	-	109.7 (22.3)
(R) Replaced	133.7 (25.2)	46.7 (6.7)	46.7 (6.7)	116.0 (24.0)
(N) Normal	102.7 (6.3)	40.0 (0)	90.0 (0)	103.7 (8.8)

b) Mean (S.E.) ratio to original number on each day after the start of the experiment and mean (S.E.) Pielou's test statistic for dispersion (see text)

Treatment	Ratio to original number				Pielou's alpha	
	Day since start				Day since start	
	0	1	2	3	0	3
(C) Control	1.00 -	0.97 (.01)	0.95 (.04)	0.93 (.03)	0.91 (.06)	0.83 (.07)
(N) Normal	1.00 -	1.01 (.02)	1.07 (.04)	0.99 (.04)	0.88 (.11)	0.96 (.06)
(R) Replaced	1.00 -	0.91 (.09)	0.89 (.02)	0.86 (.02)	0.92 (.03)	0.83 (.08)
(I) Increased	1.49 (.03)	1.16 (.06)	1.08 (.07)	1.01 (.04)	1.01 (.07)	0.93 (.08)

the first one (see earlier). There were three replicate plots of each of the four experimental treatments (Control, Normal, Replaced and Increased). Mean density of limpets (approximately 20 - 25 mm shell-length) was 117.2 per square metre (S.E. = 10.4, $\underline{n}$ = 12 quadrats). In each plot, samples of ten limpets of each of the disturbances appropriate to that plot were marked individually (see above, Table 4). The original densities in each set of plots and the numbers of animals removed and introduced are summarized in Table 5. After manipulating each plot, the positions of every marked limpet were measured from two fixed points a metre apart at one end of the plot, so that their subsequent positions and movements could be recorded (Underwood, 1977). Positions of every limpet (wherever they were found) were so recorded on each of the next three days (from 18 to 20th October, 1986), with the total number of animals in each plot. There was some imprecision in the determination of distances moved by an individual limpet (see Mackay and Underwood, 1977). Accordingly, limpets that had not moved more than 2 cm from an original position were defined to be homing (in this area the majority of the limpets moved during high tide; unpubl. data). On each day of the experiment, colour photographic slides were taken of each plot to allow calculation of Pielou's (1959) index of dispersion (alpha) by measuring the distance to the nearest limpet from each of twenty points placed randomly within the area of each slide (see Underwood (1976b)for methods). Thus, the daily data consisted of the density and dispersion of limpets in each of three replicate plots of the four treatments and the distances and directions moved by each limpet (n = 10) in a sample representing each of the experimental disturbances (Table 4). A total of 510 individual limpets was monitored daily, but some were lost and a few could not be identified because of loss of marks. All analyses were done with balanced samples of 6 limpets in every treatment.

Asymmetrical Analyses of Data

Because some disturbances were not applicable to limpets in some of the treatments (Table 4), no single analysis could examine all the data. Data representing each plot (i.e. densities, indices of dispersion) could be examined without problems. Data for individual samples of limpets (proportions homing, proportions remaining in plots) were examined in two asymmetrical analyses of variance (modified from designs provided by Winer, 1972; Underwood, 1978, 1984; Fletcher and Underwood, 1987). The first compared data from samples of original limpets (Undisturbed, Disturbed, Moved 5 cm, Moved 15 - 30 cm) in Control, Normal, Replaced and Increased treatments. Except for the Control plots, these two sources of variation were fully orthogonal to each other. There were, however, only Undisturbed limpets in Control plots. These formed an asymmetrical treatment in the analyses (see Results for details). This anlaysis was designed to detect any effects of the various types of disturbance and how these differed according to the type of plot.

Introduced limpets were only present in Replaced and Increased plots (Table 4). These were examined in comparison to the Original limpets (Undisturbed, Disturbed and the two sets of Moved animals) in these two experimental treatments. Again, the undisturbed limpets in the Control plots formed an asymmetrical set of data in the analyses. This analysis was designed to detect any differences between limpets in the plots with greater density (Increased) and those where limpets had been introduced, but at the original density (Replaced). Again, such differences could be examined for the various types of disturbance. These asymmetrical analyses contain data that are not truly independent, given that the various samples representing different disturbances for each treatment (i.e. Normal, Replaced and Increased) were in the same plots. Constraints of available space in the study-site made it impractiacable to have independent plots for each type of disturbance.

The final sets of data were more complex because a replicate sample of distances and directions was available for each type of disturbance in each plot. Thus, Plots formed a nested (hierarchical) component of variation within the different treatments, but were orthogonal to the various experimental disturbances (see Results for clarification). As such, differences among Plots and any variations in effects of, or interactions among, any variable from one plot to another could be detected. These data were strictly independent.

Distances were transformed to natural logarithms to stabilize variances (throughout all analyses, Cochran's tests for homogeneity of variances were non-significant, $P > 0.05$) and because many previous sets of such distances have proven to be approximately exponentially distributed (Mackay and Underwood, 1977; Underwood, 1977; Underwood and Chapman, 1985). Directional data could be examined by the techniques of multi-factorial analysis developed by Underwood and Chapman (1985) and Chapman (1986).

Table 6.

Analyses of proportions of limpets remaining in experimental plots by the third day

a) Original limpets in Control, Normal, Replaced and Increased treatments

Source of variation	SS	Df	MS	F-ratio
All cells	0.131	12	0.011	-
Control vs. Others	0.010	1	0.010	0.35 ns
Other cells	0.121	11	0.011	-
Treatments (T)[a]	0.002	2	0.001	0.03 ns
Disturbance (D)[b]	0.085	3	0.028	1.04 ns
T X D	0.034	6	0.006	0.21 ns
Residual	0.707	26	0.027	-
Total	0.837	38	-	-

[a] Normal vs. Replaced vs. Introduced
[b] Untouched vs. Disturbed vs. Moved 5 cm vs. Moved 15 cm

b) All limpets in Control, Replaced and Increased treatments

Source of variation	SS	Df	MS	F-ratio	P
All cells	0.977	12	0.082	-	
Control vs. Others	0.074	1	0.074	1.82	ns
Other cells	0.903	11	0.082	-	
Treatments (T)[a]	0.0003	1	0.0003	0.001	ns
Disturbance (D)[b]	0.885	5	0.177	4.34	< 0.01
T X D	0.018	5	0.004	0.09	ns
Residual	1.060	26	0.041	-	
Total	2.037	38	-	-	

[a] Replaced vs. Introduced
[b] Untouched vs. Disturbed vs. Moved 5 cm vs. Moved 15 cm vs. Introduced to replace an original limpet vs. Introduced to random position

Results

As found in the previous experiment, the density of limpets where it was increased (Increased plots) quickly returned to the original level (Table 5). After only 3 days, there was no longer any difference in the mean numbers of limpets in any of the four treatments (analysis of variance, $F = 3.7$; 3, 8 df; $P > 0.05$). Thus, the behaviour of limpets in the Increased plots must have been different from that in the other three types of plots.

At no point in the experiment did the manipulations alter the pattern of spatial dispersion. At the start, there was no difference between the four treatments and this did not alter until the experiment ended (Table 5). Generally, limpets were scattered at random over the area of the plots (the individual values ranged from 0.64 to 1.20 over the 12 plots and four days of measurement). This pattern has not been found in other areas at Cape Banks, where dispersion of _Cellana tramoserica_ was previously found to be regular (Underwood, 1976b). No simple explanation for ths difference is offered, but the habitats studied here contained a small cover (< 5 %) of barnacles (_Chamaesipho tasmanica_; previously _C. columna_) which is known to afffect movements and positions of limpets (Underwood et al., 1983). Clumped disperion of _Cellana_ has also been recorded on other shores in New

South Wales (pers.obs.) and dispersion clearly varies from place to place. This warrants further investigation and is not discussed here.

The reduction in density where it had been increased was, as described in the previous experiment, brought about by the emigration of limpets from the experimental plots and their replacement by new immigrants. The rate of departure of introduced limpets was always greater than that of original limpets and the proportion of the former leaving any plot was greater than that of the latter. Thus, density was restored as introduced animals left at a faster rate. The pattern in the data was very clear. After the third day of the experiment, there was no difference in the proportion of marked limpets remaining in Control, Normal, Replaced or Increased plots (note lack of significance of Treatments in analyses in Table 6 and the mean proportions in Fig. 2). In contrast, there were significant differences among the different types of disturbance. There were no differences among the mean proportions remaining of original animals treated in different ways (i.e. Undisturbed, Disturbed and the two types of Moved limpets). Significntly greater proportions of these remained than of the two types of introduced limpets (i.e. placed into an original limpet's position or at random)(Student-Newman-Keuls tests on means in Figure 2; $\underline{P} < 0.05$; see

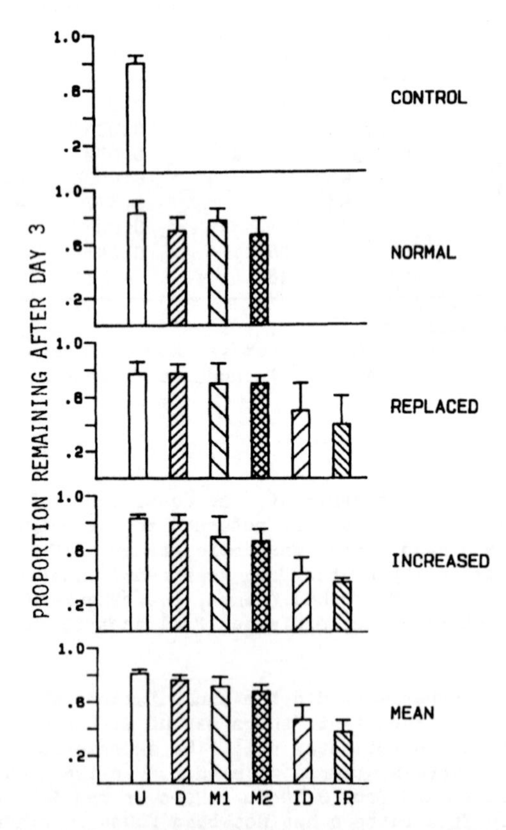

Figure 2. Mean (+ S.E.) proportions of limpets remaining in each type of experimental plot after 3 days of the second experiment, and the mean over all plots. Original limpets: U = Undisturbed, D = Disturbed, M1 = Moved 5 cm, M2 = Moved 15 cm); Introduced limpets: ID = Introduced to position of an original limpet, IR = Introduced to random position.

Table 7.

Analyses of distances moved by limpets by the third day

a) Original limpets in Control, Normal, Replaced and Increased treatments

Source of variation	SS	Df	MS	F-ratio	P
All cells	66.70	12	5.55	-	
Control vs. Others	2.85	1	2.85	1.51	ns
Other cells	63.85	11	5.80	-	
Treatments (T)[a]	3.01	2	1.50	0.49	ns
Disturbance (D)[b]	51.29	3	17.10	5.52	< 0.05
T X D	9.55	6	1.59	0.51	ns
Plots	62.30	26	2.40	-	
Plots (Treatments)	24.94	8	3.11	1.98	ns
D X Plots (T)	37.44	18	2.08	1.32	ns
Residual	306.90	195	1.57	-	
Total	435.89	233	-	-	

[a] Normal vs. Replaced vs. Introduced
[b] Untouched vs. Disturbed vs. Moved 5 cm vs. Moved 15 cm

b) All limpets in Control, Replaced and Increased treatments

Source of variation	SS	Df	MS	F-ratio	P
All cells	100.92	12	8.41	-	
Control vs. Others	11.38	1	11.38	4.37	ns
Other cells	89.54	11	8.14	-	
Treatments (T)[a]	0.51	1	0.51	0.19	ns
Disturbance (D)[b]	91.38	5	18.28	7.02	< 0.05
T X D	5.57	5	1.12	0.43	ns
Plots	53.48	26	2.06	-	
Plots (Treatments)	15.62	6	2.60	2.06	ns
D X Plots (T)	37.86	20	1.99	1.58	ns
Residual	246.04	195	1.26	-	
Total	416.29	233	-	-	

[a] Replaced vs. Introduced
[b] Untouched vs. Disturbed vs. Moved 5 cm vs. Moved 15 cm vs. Introduced to replace an original limpet vs. Introduced to random position

analysis in Table 6b). Thus, being introduced into an unfamiliar area caused limpets to leave faster than was the case for limpets previously present there, regardless of the overall density present.

This result was matched exactly by the distances moved by limpets in different plots. Again, there were no differences among treatments (Control, Normal, Replaced and Introduced; analyses of variance in Table 7, means in Fig. 3). There were, however, differences among the different types of disturbance. Among the original limpets, those that were Moved (5 or 15 - 30 cm) subsequently wandered further than those that were untouched or simply lifted and replaced in the same spot (analysis of variance in Table 7a; SNK test, $\underline{P} < 0.05$). As can be seen in Figure 3, there was no difference between those limpets experimentally moved the longer and those moved the shorter distance.

When the introduced limpets were compared with original animals in Replaced and Inreased plots (Table 7b), this pattern was found again.

Table 8.

Analyses of proportions of limpets homing in experimental plots on the third day

a) Original limpets in Control, Normal, Replaced and Increased treatments

Source of variation	SS	Df	MS	F-ratio	P
All cells	0.617	12	0.051	-	
Control vs. Others	0.023	1	0.023	1.06	ns
Other cells	0.594	11	0.054	-	
Treatments (T)[a]	0.0002	2	0.0001	0.0004	ns
Disturbance (D)[b]	0.546	3	0.182	8.41	< 0.005
T X D	0.048	6	0.008	0.36	ns
Residual	0.563	26	0.022	-	
Total	1.180	38	-	-	

[a] Normal vs. Replaced vs. Introduced
[b] Untouched vs. Disturbed vs. Moved 5 cm vs. Moved 15 cm

b) All limpets in Control, Replaced and Increased treatments

Source of variation	SS	Df	MS	F-ratio	P
All cells	0.617	12	0.051	-	
Control vs. Others	0.048	1	0.048	2.71	ns
Other cells	0.569	11	0.052	-	
Treatments (T)[a]	0.00008	1	0.00008	0.0006	ns
Disturbance (D)[b]	0.554	5	0.111	6.24	< 0.005
T X D	0.016	5	0.003	0.17	ns
Residual	0.462	26	0.018	-	
Total	1.079	38	-	-	

[a] Replaced vs. Introduced
[b] Untouched vs. Disturbed vs. Moved 5 cm vs. Moved 15 cm vs. Introduced to replace an original limpet vs. Introduced to random position

Undisturbed and Disturbed animals moved less distance than those that had been experimentally displaced at the start. The two types of introduced limpets, hwever, moved even further during the period of observations (SNK tests, $\underline{P} < 0.05$; see Fig. 3). Original limpets that had not been disturbed or that had only been lifted and replaced in the same spot moved an average distance of 22.5 cm during the next three days. Those original animals that had been moved around the experimental plots at the start of the experiment moved an average of 36 cm. Introduced animals, in contrast, moved a mean distance of 63 cm throughout the experiment (see the log-transformed means in Fig. 3). It is not surprising, therefore, that a very great majority of the latter had left the 100 x 100 cm square plots after three days.

The directions moved from their initial positions by the samples of limpets in each treatment were analysed for each day. No set of data was ever significanttly directional, i.e. each set was random by Rayleigh's tests ($\underline{P} > 0.05$; Mardia, 1972). There was no evidence that density, treatment or disturbance affected the orientation of movements of these limpets.

The final attribute of behaviour that was examined was the tendency to home (as defined here). The proportions of limpets still within 2 cm of their original positions (after manipulations were completed) also varied with disturbance, but not with the different types of plots. The results were again consistent with the two variables analysed above. The disturbance caused by lifting and replacing the limpets did not alter the tendency to home. Disturbed limpets did not differ from undisturbed ones (Fig. 4 and analyses in Table 8). After three days, these two groups contained a mean of 25 % of limpets that were homing. In contrast, very few Moved or Introduced limpets showed any tendency to home (a mean of 1 % of limpets were homing to the position that they were in at the end of the manipulation). In both analyses in Table 7, significant differences among disturbances were entirely attributable to the significanlty greater proportion homing in Undisturbed and Disturbed treatments than was the case for the Moved or Introduced limpets.

Other possible homing behaviour was also examined (although not reported here). There was no tendency for Moved limpets to return to their original positions. Very few limpets took up residence in a new position (i.e. neither the initial position nor the one in which they were placed at the start of the experiment).

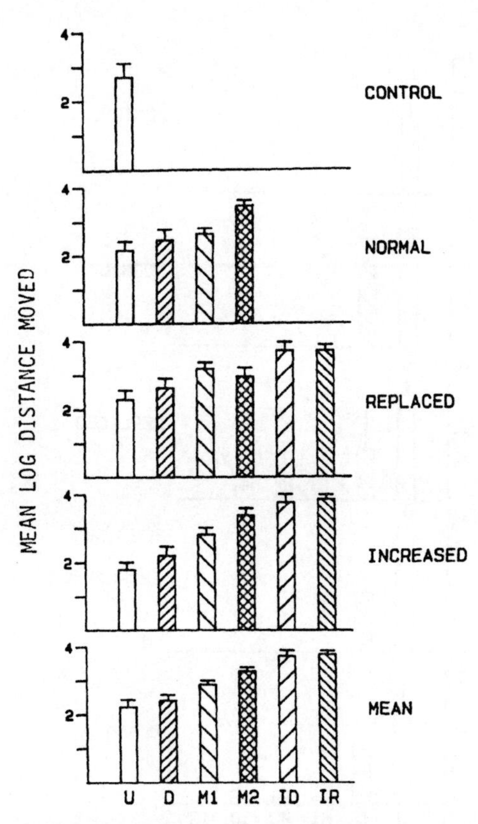

Figure 3. Mean (+ S.E.) distances moved by limpets remaining in each type of experimental plot after 3 days of the second experiment, and the mean over all plots (data trasformed to $\log_e(x + 1)$). All other information as in Figure 2.

Interpretation of This Experiment

The results of this experiment are consistent, straightforward and quite surprising. There is no demonstrable effect of density, as such, on the spatial pattern, homing behaviour, distances travelled, etc., of the limpet. None of the restoration of density after it had been experimentally increased was due to different behaviour caused by increased numbers of limpets in the area (thus increasing the potential rate of encounter of other limpets, etc.). Instead, the results indicate very strongly that regulation of density is brought about by the greater tendency to cease homing and the more rapid rate of movement of disturbed, unfamiliar animals that had been brought into the area. Precisely the same behaviours were shown by these animals when introduced into experimental areas without any increase in density. The original animals in both types of area were also unaffected by the arrival of strange specimens, regardless of density. The only differences discerned in the experiment were due to handling and disturbance of the limpets. The effects of disturbance shown here confirm those described for Cellana by Mackay and Underwood (1977) who found there was no effect on homing caused by picking the animals up and replacing them in their original positions, but that moving them away from their homes tended to decrease the prevalance of homing.

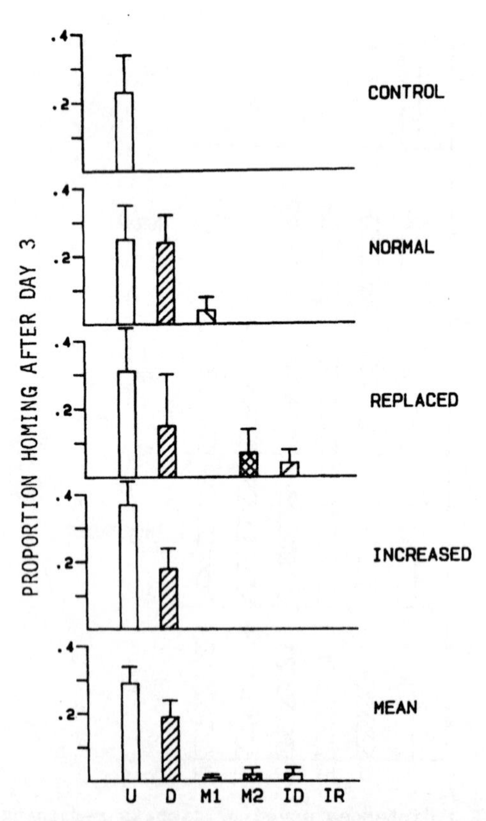

Figure 4. Mean (+ S.E.) proportions of limpets homing in each type of experimental plot after 3 days of the second experiment, and the mean over all plots. All other information as in Figure 2.

Limpets have previously been described to leave or enter areas of different density at different rates (Aitken, 1962; Breen, 1971; Mackay and Underwood, 1977). No previous studies have contained the controls needed to extract such a result from the confounded processes caused by handling and disturbance.

The present results were very consistent among the different types of behaviour and spatial pattern investigated - and totally unexpected. The results make it clear that the investigation of intra-specific competitive interactions affecting behaviour in natural populations in the field is far from an easy task. The consequences of this will be considered below.

SOME EXTRA PROBLEMS IN EXPERIMENTS ON INTERSPECIFIC COMPETITION

Intra- and Inter-specific Processes

As has been summarized elsewhere (Underwood, 1986), there are many reasons why the study of interspecific competition cannot proceed successfully without consideration of the intraspecific processes and interactions affecting each of the species. It has long been realized that requirements for some resource are likely to be much closer among members of a single species than among individuals of different species (e.g. Darwin, 1882; p. 59). There has been frequent discussion of this point (e.g. Pontin, 1969) and it was incorporated into Reynoldson and Bellamy's (1970) thoughtful definitions of necessary, as opposed to sufficient, conditions for competition to occur. Schoener (1983) summarized non-linear interspecific competitive interactions which are, at least in part, a function of intraspecific competition. Thus, when resources are already over-extended because of their use by large numbers of a single species, the addition of a few members of a potentially competing species can have large effects. The arrival of the same numbers of the second species will have less (or no) effect if the numbers of the first species were smaller (or small enough not to be competing among themselves). The magnitude of intraspecific competitive interactions is explicitly incorporated into models of changes in populations and assemblages (e.g. Krebs, 1978; Lawton and Hassell, 1981; Schoener, 1983).

Finally, an understanding of intraspecific competitive processes has been shown to be crucial for correct interpretations of experimental analyses of natural populations (for examples of intertidal organisms, see Creese and Underwood, 1982; Fletcher and Creese, 1985; Ortega, 1986). In these cases, severe intraspecific interactions caused a decline in sustainable abundances of a superior competing species so that its density fell below the numbers necessary to eliminate an inferior competitor from patches of the substratum. Coexistence betweeen the species, even though they competed for food, was made possible by the magnitude of intraspecific interactions within one species.

Clearly, therefore, the previous discussion of experimental analyses of intraspecific competition and its effects on behaviour is very relevant to the more complex problems of interspecific interactions. Two new problems arise when more than one species must be considered together. These are briefly discussed here, as a paraphrase of the more complete treatments in Connell (1983), Underwood (1986) and Chapman (1986). The general design of field experiments between two species, involving inter- and intra-specific interactions at several densities has been discussed elsewhere (see Tables 11.1 and 11.2 in Underwood, 1986). All the points made there apply equally well to studies of behaviour as they do to experiments on abundances, sizes, growth, etc. See also the discussion of indirect methods of determining the intensity of intraspecific interactions in Abrams (1981).

Symmetry and Asymmetry

Numerous reviews of field experiments on competition have drawn attention to the widespread occurrence of interspecific asymmetry (Lawton and Hassell, 1981; Connell, 1983; Schoener, 1983). Thus, the effects of one species (A) on another (B) are not equal to the effects of B on A. Elsewhere (Underwood, 1986), I have demonstrated that this is often going to occur because of differences in the numbers of each species that are used in the experiments. The same will be true where behavioural variables are studied.

As an example, Stimson's (1973) experiments on the owl-limpet, *Lottia gigantea*, demonstrated that the behaviour of large, territorial limpets reduced local abundances of smaller species (of *Acmaea*). Competition was directly for food, but limpets actually "fought" over the space (inside *Lottia*'s territories) on which algal food grew (see also the patterns for *Patella longicosta* described by Branch (1975, 1984)). Clearly, the aggressive behaviour of *Lottia*, in addition to its larger size, caused it to be a more effective individual competitor than any of the smaller limpets in its habitat. To understand the negative correlations in patterns of distribution of *Lottia gigantea* and *Acmaea* species in California, it is crucial to understand the behaviour of the limpets. Conversely, however, to understand the fascinating behaviour of the large, territorial *Lottia*, it is imperative to understand the resources and the patterns of distribution and abundance of competitors and to do experiments to determine the relative strengths of competitive abilities of *L. gigantea* and it competitors.

When more than one species is investigated, experimental designs become more complex. Considerable care must be taken to ensure that interpretations of relative competitive ability (i.e. existence and/or magnitude of asymmetry) are not confounded by use of different numbers (biomasses, sizes, etc.) of the two species. Considerable thought must be given to the logical basis and then to the design of the experiments so that they can properly be interpreted (see the discussion in Underwood, 1986).

Controls for Transplants of Animals

When two (or more) species (or two different size-classes) are to be manipulated in field experiments, there is often a requirement that individuals of either or both (or all) species may need to be removed from one habitat (e.g. one level on a shore, or from an area with one sort of topographical structure, or whatever) to another. This will occur, for example, where one species is consistently found at a higher level on a shore than a second species and competition between the two is proposed as the mechansim determining the boundary of their patterns of distribution (e.g. the discussion in Chapman, 1986 and the analysis in Pielou, 1974). Tests of hypotheses about such competition wil involve removals of each species from some areas (hypothesizing that the other species will then alter its behaviour and move down or up, depending on which is removed). Such experimental tests will, however, also involve transplants of each species into the area occupied by the other (to test hypotheses about direct interference, or about the nature of establishment rather than maintenance of the observed pattern).

Such transplants will need very careful controls. The controls will be more complex than those discussed earlier for intraspecific competition, because the latter only involved transplanting individuals of the same species. These can always be obtained from the same habitat (at the scale of level on a shore, type of topography, wave-exposure, etc.) as that found in the particular sites into which they are transplanted.

Clearly, controls for transplantation from one habitat to another must include similar transplantation within the same habitat. Consider transplantation of individuals of Species A (from higher levels on a shore) to lower levels occupied by Species B to test hypotheses about B's ability to oust A from the lower levels. The behaviour of A's transplanted lower down may be very bizarre because of being moved to a different habitat (e.g. with more or less food, longer periods of submersion, different microflora, etc., etc.), regardless of the presence of Species B. It is necessary to be sure that any subsequent difference in behaviour between Species A when lower down from its behaviour at higher levels is attributable to the presence of Species B. The only way to do this is to provide a whole series of controls. Some A must be transplanted into lower levels where B has been removed and some where B is present. Comparison of these will detect the effect of B - but only on A's that have been transplanted. A second control will be needed to translocate A similar distances, but to new sites at the same level from which they were taken. Comparison of these with similar animals transplanted lower down, but in the absence of B, will detect modifications in behaviour of A that are solely a function of being disturbed, handled and moved to a new habitat (i.e. level on the shore). Finally, disturbed and undisturbed sets of A in their original sites in the original habitat will be needed to detect the effects of handling and translocation to another site.

These aspects of complex experimental designs will not be considered further here - but such controls are missing from most (if not all) accounts of the behaviour of intertidal animals when transplanted to different tidal levels (see the more complete discussions in Underwood, 1986 and Chapman, 1986). Chapman (1986) demonstrated in her experiments that translocation within a habitat had little effect on the subsequent behaviour of the snail *Littorina unifasciata*. Her study, however, has demonstrated the need to investigate - as opposed to the practice of ignoring - such confounding influences in the study of behaviour of natural populations in the field. Such controls will obviously be needed in the design of unambiguous, interpretable tests of hypotheses about competitive interactions between species as influences on their behaviour.

CONCLUSIONS

The investigation of competitive processes is still important for our understanding of distribution, abundance and behaviour of intertidal species. The use of manipulative experiments in the field is increasing and, despite some opinions to the contrary (Diamond, 1986), such experiments are crucial to unravel the complexities of natural phenomena. Results of experiments are, however, often poorly, incorrectly or confusedly interpreted because of problems of inadequate design. Here, the need for better controls (and therefore more complex experimental designs) has been demonstrated for limpets. The same case can be made for any other type of organism. Species that do not show specific consistent homing, as do some limpets, often show very precise patterns of occupancy of micro-habitats. Consequently, disturbances, handling and movement will all potentially affect their behaviour. The situation will be worse where animals are capable of movement over very large distances (e.g. crabs, birds and fish), because our ability to recognize the existence of home-ranges, or requirements for habitat, may be extremely poor.

There has been continuous, recent development of analytical tools for the dissection and interpretation of complex sets of behavioural data (e.g.

the multi-factorial analyses of directional data recently described by Underwood and Chapman, 1985 and Harrison et al., 1986 and the statistical tests for independence of temporal series of observations described by Swihart and Slade, 1985). What is also required, however, is more discussion and thought about, and development and practice of, the sorts of experimental designs that will allow progress to be made more clearly. This discussion demonstrates the need for better designs. All field biologists can contribute to the evolution of experimental procedures because it is the complex biology and natural history and, above all, the behaviour of intertidal organisms in response to their complex and variable environments that simultaneously makes the experiments necessary, but also makes them challenging.

ACKNOWLEDGMENTS

This study was supported by the University of Sydney Research Grant, the Australian Research Grants Committee and funds from the Institute of Marine Ecology. Considerable and expert assistance in the field was provided by K. Astles, S. McNeil, P.R. Scanes and W.J. Steel. This workshop provided the stimulus to overcome my inertia and to do the experiments in response to thinking about the needs for better controls. Thinking about experimental designs has consistently been stimulated by my students (recently and specifically by Drs P.G. Fairweather, W.J. Fletcher, S.J. Kennelly, K.A. McGuinness, M.A. O'Donnell and J.H. Warren; Messrs N.L. Andrew, R.D. Otaiza, N.M. Otway and Ms L.J. Stocker), by my colleague M.G. Chapman and my wife Dr P.A. Underwood. I thank all of these (and my children) for their contributions to, and suffering during, the preparation of this paper.

REFERENCES

Abrams, P., 1981, Alternative methods of measuring competition applied to two Australian hermit crabs, Oecologia, 51:233.

Aitken, J.J., 1962, Experiments with populations of the limpet Patella vulgata L., Irish Nat.J., 14:12.

Andrewartha, H.G., and Birch, L.C., 1954, "The Distribution and Abundance of Animals," University of Chicago Press, Chicago.

Birch, L.C., 1957, The meanings of competition, Amer.Nat., 91:5.

Black, R., 1977, Population regulation in the intertidal limpet Patelloida alticostata (Angas, 1865), Oecologia, 30:9.

Branch, G.M., 1975, Mechanisms reducing intraspecific competition in Patella spp.: migration, differentiation and territorial behaviour, J.Anim.Ecol., 44:575.

Branch, G.M., 1984, Competition between marine organisms: ecological and evolutionary implications, Annu.Rev.Oceanogr.Mar.Biol., 22:429.

Breen, P.A., 1971, Homing behaviour and population regulation in the limpet Acmaea (Colisella) digitalis, Veliger, 14:177.

Caffey, H.M., 1985, Spatial and temporal variation in settlement and recruitment of intertidal barnacles, Ecol.Monogr., 55:313.

Chapman, A.R.O., 1973, A critique of prevailing attitudes towards the control of seaweed zonation on the seashore, Bot.Mar., 41:80.

Chapman, M.G., 1986, Assessment of some controls in experimental transplants of intertidal gastropods, J.Exp.Mar.Biol.Ecol., 103:181.

Choat, J.H., 1977, The influence of sessile organisms on the population biology of three species of acmaeid limpets, J.Exp.Mar.Biol.Ecol., 26:1.

Cochran, W.G., 1947, Some consequences when the assumptions of the analysis of variance are not satisfied, Biometrics, 3:22.

Connell, J.H., 1961, The effect of competition, predation by Thais

lapillus and other factors on natural populations of the barnacle Balanus balanoides, Ecol.Monogr., 31:61.
Connell, J.H., 1972, Community interactions on marine rocky intertidal shores, Ann.Rev.Ecol.Syst., 3:169.
Connell, J.H., 1974, Ecology: field experiments in marine ecology, in: "Experimental Marine Biology," R. Mariscal, ed., Academic Press, New York.
Connell, J.H., 1983, On the prevalence and relative importance of interspecific competition: evidence from field experiments, Am.Nat., 122:661.
Connell, J.H., 1985, The consequences of variation in initial settlement versus post-settlement mortality in rocky intertidal communities, J.Exp.Mar.Biol.Ecol., 93:11.
Connor, E.F., and Simberloff, D.S., 1978, The assembly of species communities: chance or competition? Ecology, 60:1132.
Creese, R.G., 1980, An analysis of distribution and abundance of populations of the high-shore limpet Notoacmea petterdi (Tenison-Woods), Oecologia, 45:252.
Creese, R.G., and Underwood, A.J., 1982, Analysis of inter- and intra-specific competition amongst limpets with different methods of feeding, Oecologia, 53:337.
Darwin C., 1882, "The Origin of Species (6th Edition)," Murray, London.
Dayton, P.K., 1971, Competition, disturbance and community organization: the provision and subsequent utilization of space in a rocky intertidal community, Ecol.Monogr., 41:351.
Dayton, P.K., 1973, Two cases of resource partitioning in an intertidal community: making the right prediction for the wrong reason, Am.Nat., 104:662.
Denley, E.J., and Dayton, P.K., 1985, Competition among macroalgae, in: "Handbook of Phycological Methods. 4. Ecological Field Methods: Macroalgae," M.M. Littler, and Littler, D.S., eds., Cambridge University Press, Cambridge.
Diamond, J.M. 1978, Niche shifts and the rediscovery of interspecific competition, Am.Sci., 66:322.
Diamond, J.M., 1986, Overview: laboratory field experiments, field experiments and natural experiments, in: "Community Ecology," J.M. Diamond, and Case, T.J., eds., Harper and Row, New York.
Diamond, J.M., and Gilpin, M.E., 1982, Examination of the "null" model of Connor and Simberloff for species co-occurrences on islands, Oecologia, 52:64.
Fletcher, W.J., 1984, Intraspecific variation in the population dynamics and growth of the limpet Cellana tramoserica, Oecologia, 63:110.
Fletcher, W.J., and Creese, R.G., 1985, Competitive interactions between co-occurring gastropods, Mar.Biol., 86:183.
Fletcher, W.J., and Underwood, A.J., 1987, Interspecific competition among subtidal limpets: effect of substratum heterogeneity, Ecology, 68:387.
Gaines, S.D., and Roughgarden, J., 1985, Larval settlement rate: a leading determinant of structure in an ecological community of the marine intertidal zone, Proc.Natl.Acad.Sci., 82:3707.
Gilpin, M.E., and Diamond, J.M., 1982, Factors contributing to non-randomness in species co-occurrences on islands, Oecologia, 52:75.
Harrison, D., Kanji, G.K., and Gadsden, R.J., 1986, Analysis of variance for circular data, J.Appl.Stat., 13:123.
Hurlbert, S.H., 1984, Pseudoreplication and the design of ecological field experiments, Ecol.Monogr., 54:187.
Jackson, J.B.C., 1981, Interspecific competition and species' distributions: the ghost of theories and data past, Amer.Zool., 21:889.
Krebs, C.J., 1978, "Ecology: The Experimental Analysis of Distribution and Abundance," Harper and Row, New York.

Lawton, J.H., and Hassell, M.P., 1981, Asymmetrical competition in insects, Nature, 289:793.
Loosanoff, V.L., 1964, Variations in time and intensity of settling of the starfish Asterias forbesi, in Long Island Sound during a twenty-five year period, Bull.Lab.Mar.Biol.Woods Hole, 126:423.
Loosanoff, V.L., 1966, Time and intensity of settling of the oyster, Crassostrea virginica, in Long Island Sound, Bull.Lab.Mar.Biol.Woods Hole, 130:211.
Mackay, D.A., and Underwood, A.J., 1977, Experimental studies on homing in the intertidal patellid limpet Cellana tramoserica (Sowerby), Oecologia, 30:215.
Malthus, T.R., 1798, "An Essay on the Principle of Population," Reprinted by MacMillan, New York.
Mardia, K.V., 1972, "Statisitics of Directional Data," Academic Press, London.
May, R.M., 1974, Biological populations with non-overlapping generations: stable points, stable cycles and chaos, Science, 186:645.
Menge, B.A., 1976, Organization of the New England rocky intertidal community: role of predation, competition and environmental heterogeneity, Ecol.Monogr., 46:355.
Ortega, S., 1985, Competitive interactions among tropical intertidal limpets, J.Exp.Mar.Biol.Ecol., 90:21.
Paine, R.T., 1974, Intertidal community structure. Experimental studies on the relationship between a dominant competitor and its principal predator, Oecologia, 15:93.
Paine, R.T., 1977, Controlled manipulations in the marine intertidal zone, and their contributions to ecological theory, in: "The Changing Scenes in Natural Sciences, 1776-1976," Acad.Nat.Sci. Special Publication, 12:245.
Pielou, E.C., 1959, The use of point-to-plant distances in the study of pattern in plant populations, J.Ecol., 49:271.
Pielou, E.C., 1974, "Population and community ecology: principles and methods," Gordon and Breach, New York.
Pontin, A.J., 1969, Experimental transplantation of nest-mounds of the ant Lasius flavus (F.) in habitat containing also L. niger (L.) and Myrmica scabrinodis Nyl., J.Anim.Ecol., 38:747.
Popper, K.R., 1963, "Conjectures and Refutations: the Growth of Scientific Knowledge," Harper and Row, New York.
Reynoldson, T.B., and Bellamy, L.S., 1970, The establishment of interspecific competition in field populations, with an example of competition in action between Polycelis nigra (Mull.) and P. teguis (Ijima) (Turbellaria, Tricladida), in: "Proceedings of the Advanced Study Institute on Dynamics of Numbers in Populations, Oosterbeck, 1970," P.J. den Boer, and Gradwell, G.R., eds., Centre for Agricultural Publication and Documentation, Wageningen.
Sale, P.F., 1977, Maintenance of high diversity in coral reef fish communities, Am.Nat., 111:337.
Schoener, T.W., 1983, Field experiments on interspecific competition, Am.Nat., 122:240.
Schroder, G.D., and Rosenzweig, M.L., 1975, Perturbation analysis of competition and overlap in habitat utilization between Dipodomys ordii and Dipodomys merriami, Oecologia, 19:9.
Simberloff, D.S., 1978, Using island biogeographic distributions to determine if colonization is stochastic, Am.Nat., 112:713.
Simberloff, D.S., 1984, Properties of coexisting bird species in two archipelagoes, in: "Ecological Communities: Conceptual Issues and the Evidence," D.R. Strong, Simberloff, D.S., Abele, L.G., and Thistle, A.B., eds., Princeton University Press, Princeton.
Sousa, W.P., 1979, Disturbance in marine intertidal boulder fields: the nonequilibrium maintenance of species diversity, Ecology, 60:1225.
Spight, T.M., 1975, Factors extending gastropod embryonic development and

their selective cost, Oecologia, 21:1.
Stimson, J., 1973, The role of the territory in the ecology of the intertidal limpet Lottia gigantea (Gray), Ecology, 54:1020.
Stimson, J., and Black, R., 1975, Field experiments on population regulation in intertidal limpets of the genus Acmaea, Oecologia, 18:111.
Strong, D.R., Simberloff, D.S., Abele, L.G., and Thistle, A.B., 1984, "Ecological Communities: Conceptual Issues and the Evidence," Princeton University Press, Princeton.
Strong, D.R., Szyska, L.A., and Simberloff, D.S., 1979, Tests of community-wide character displacement against null hypotheses, Evolution, 33:897.
Sutherland, J.P., and Ortega, S., 1986, Competition conditional on recruitment and temporary escape from predators on a tropical rocky shore, J.Exp.Mar.Biol.Ecol., 95:155.
Swihart, R.K., and Slade, N.A., 1985, Testing for independence of observations in animal movements, Ecology, 66:1176.
Underwood, A.J., 1976a, Food competition between age-classes in the intertidal neritacean Nerita atramentosa Reeve (Gastropoda: Prosobranchia), J.Exp.Mar.Biol.Ecol., 23:145.
Underwood, A.J., 1976b, Nearest neighbour analyses of spatial dispersion of intertidal prosobranch gastropods within two substrata, Oecologia, 26:257.
Underwood, A.J., 1977, Movements of intertidal gastropods. J.Exp.Mar.Biol.Ecol., 26:191.
Underwood, A.J., 1978, An experimental evaluation of competition between three species of intertidal gastropods, Oecologia, 33:185.
Underwood, A.J., 1979, The ecology of intertidal gastropods, Adv.Mar.Biol., 16:111.
Underwood, A.J., 1981, Techniques of analysis of variance in experimental marine biology and ecology, Annu.Rev.Oceanogr.Mar.Biol., 19:513.
Underwood, A.J., 1983, Spatial and temporal problems in the design of experiments with marine grazers, in: "Proceedings: Inaugural Great Barrier Reef Conference," J.T. Baker, Carter, R.M., Sammarco, P.W., and Stark, K.P., eds., James Cook University Press, Townsville.
Underwood, A.J., 1984, Vertical and seasonal patterns in competition for microalgae between intertidal gastropods, Oecologia, 64:211.
Underwood, A.J., 1985, Physical factors and biological interactions: the necessity and nature of ecological experiments, in: "The Ecology of Rocky Coasts," P.G. Moore, and Seed, R., eds., Hodder and Stoughton, London.
Underwood, A.J., 1986, The analysis of competition by field experiments, in: "Community Ecology: Pattern and Process," D.J. Anderson, and Kikkawa, J., eds., Blackwell Scientific Press, Melbourne.
Underwood, A.J., and Chapman, M.G., 1985, Multifactorial analyses of directions of movements of animals, J.Exp.Mar.Biol.Ecol., 91:17.
Underwood, A.J., and Denley, E.J., 1984, Paradigms, explanations and generalizations in models for the structure of intertidal communities on rocky shores, in: "Ecological Communities: Conceptual Issues and the Evidence," D.R. Strong, Simberloff, D.S., Abele, L.G., and Thistle, A.B., eds., Princeton University Press, Princeton.
Underwood, A.J., Denley, E.J., and Moran, M.J., 1983, Experimental analyses of the structure and dynamics of mid-shore rocky intertidal communities in New South Wales, Oecologia, 56:202.
Williamson, M., 1972, "The Analysis of Biological Populations," Edward Arnold, London.
Winer, B.J., 1971, "Statistical Principles in Experimental Design, 2nd Edition," McGraw-Hill Kogakusha, Tokyo.

Intraspecific Aggression in *Actinia equina*: Behavioural Plasticity in Relation to Over-Winter Survivorship of Adult Anemones

Robin C. Brace

University of Nottingham, U.K.

Series of laboratory-based agonistic encounters were staged using anemones drawn from two densely packed, littoral aggregations, in order to examine behavioural plasticity associated with the juxtapositioning of opponents. Experimental pairing of juxtaposed (on the shore), allogeneic individuals resulted in significantly less aggression than was observed between non-juxtaposed opponents, but the former when paired with the latter, nevertheless, displayed aggression earlier. These results are interpretable in terms of habituation specific to shore-based opponents, and dishabituation (sensitization) when faced by a novel opponent. To explore the postulate that juxtapositioning might bring about a reduction in surface antigenicity, contests involving juxtaposed non-neighbours were conducted. Anemones collected in early and mid-/late winter produced high levels of aggression, a result consistent with specific habituation, but those collected in late autumn displayed relatively little aggression, an outcome in agreement with a change in antigenicity. These results are discussed with respect to the current state of knowledge of self/non-self recognition in anemones and other lower invertebrates.

Affiliated data on the dispersions of contestants on the shore agree with previous investigations, and suggest a scenario of over-winter crowding into relatively protected, shallow surface concavities, accompanied by reduced levels of aggression but preceded by the competitive occupation of such niches. The selective value of this behavioural plasticity is discussed in relation to the known life history characteristics of *Actinia equina*, and the presence of three, ecologically distinct morphs on the British coastline.

INTRODUCTION

The intraspecific aggressive behaviour of *Actinia equina*, which involves the use of acrorhagi, has been fully described in a number of laboratory studies (Bonnin, 1964; Brace, Pavey and Quicke, 1979). To the contrary, information on its functional role on the shore is still surprisingly limited. Recent studies on individuals within aggregations, designed to rectify this omission, commenced with the premise that since *A. equina* broods asexually produced young (Carter and Thorp, 1979; Gashout and Ormond, 1979; Orr, Thorpe and Carter, 1982), most aggregations would have an overtly clonal basis, with clonal boundaries being maintained by aggression. Several recent observations, though, militate against this.

Firstly, biochemical-genetic work has shown that anemones within most aggregations are genetically diverse, and that the clumping of like-genotypes is minimal (Quicke and Brace, 1983; Brace and Quicke, 1985). Secondly, a study of seasonal changes in the dispersion of anemones within an aggregation in Wales (over a 2 year period) revealed a change from gross over-dispersion to randomicity during early winter (Brace and Quicke, 1986b). This was brought about by adults ($\geq$ 15mm pedal disc diameter) congregating within shallow surface concavities, where presumably some protection from environmental inclemencies is afforded. It is during the competitive occupation of such concavities that aggression is likely to be of paramount importance, an assessment substantiated by data on seasonal maxima in aggression provided by Rees (1985) for the same locality. Thirdly, it is now clear that a high proportion of adults survive for many years (Brace and Quicke, 1986a). In view of the aforegoing, it is now concluded that the primary function of intraspecific aggression is to enhance adult survivorship, and hence the long-term production (and widespread dispersal) of syngeneic brooded young, rather than to simply increase the (local) settlement prospects of young in the short-term.

Significantly, graphical analyses of dispersion have revealed that during the late winter period, anemones at the highest densities (i.e. within concavities) are under-dispersed, a feature suggestive of a decline in aggressiveness following juxtapositioning. In this communication, this inferred behavioural habituation has been examined using both juxtaposed (on the shore) and non-juxtaposed contestants. The behavioural data presented, and conclusions reached regarding the mechanisms and degree of specificity involved within the context of self/non-self recognition processes in the lower invertebrates (Klein, 1982), are supplemented by information on the dispersions of contestants on the shore.

MATERIALS AND METHODS

Anemones for behavioural experiments were collected on 3 occasions - late autumn, and early and mid-winter - from parts (each $\sim 1m^2$) of a large ($\sim 10m^2$) aggregation situated on a low shore, vertical rock face on The Worm's Head in South Wales, a locality with which we are already familiar (see above). On each occasion a selection of both juxtaposed anemones (individuals whose pedal discs were in contact with other members of the aggregation, or whose tentacular crowns when expanded, would have obviously made contact with neighbours in a similar state of expansion) and non-juxtaposed individuals (well out of contact with nearest neighbours) was secured. Anemones were returned to the laboratory in individual, self-seal plastic bags, provided with small pieces of slate to which they attach readily, and then allowed at least 7 days acclimation at Nottingham in a non-tidal aquarium (circulating water) prior to use. During this time each anemone was individually housed in a small plastic container.

Each aggressive encounter was initiated by removing the two (expanded) contestants from their respective containers and placing them on the floor of the aquarium. Then they (their substrata) were pushed towards one another until limited tentacular contact was established. Some anemones were used in more than one encounter (maximum of 7 encounters, and means of 3.0, 2.4 and 2.6 for contestants collected in October, December and January, respectively), but intervals of at least 24 h intervened between consecutive encounters. Four contest permutations were examined: A. juxtaposed neighbours; B. non-neighbours; C. juxtaposed non-neighbours; D. juxtaposed non-neighbour versus non-neighbour. Fig. 1 clarifies the usage of this terminology.

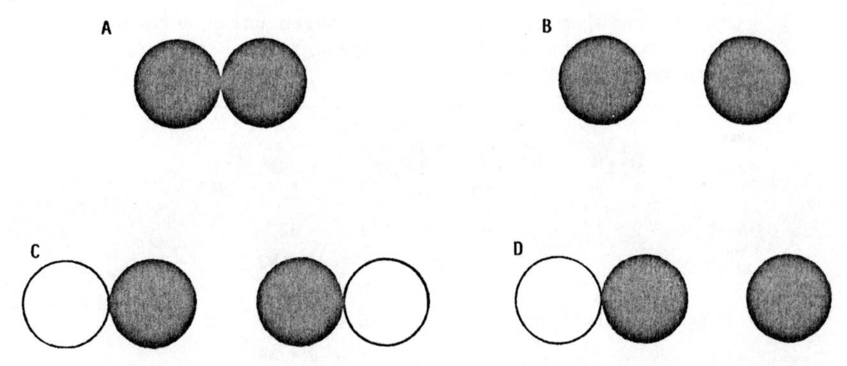

Fig. 1. Diagrams showing the relative positioning of contestants and juxtapositioning to non-contestants on the shore, to explain the terminology used to describe the four contest permutations staged: A. juxtaposed neighbours; B. non-neighbours; C. juxtaposed non-neighbours; D. juxtaposed non-neighbour versus non-neighbour. Contestants are indicated by filled circles.

During each field visit, that part of the aggregation examined, was mapped (see Quicke and Brace, 1983 for details). In the laboratory, the XY coordinates and sizes of anemones were assimilated using a digitizer linked to our computer. The often used measure, R, of Clarke and Evans (1954) was adopted to assess gross dispersion. To examine the dispersion of anemones at different densities within the areas mapped, a graphical analysis of spacing (Weiss, 1981) was employed (see Brace and Quicke, 1985, 1986b for details). A linear plot of the Weiss Index against the ranked, nearest neighbour distances (NND^2) signifies a random distribution, whilst spline convexities and concavities indicate under- and over-dispersion respectively.

For comparative purposes, contests were also conducted using anemones collected from a mid-shore location in South-west England (Trevone, Cornwall), having similar characteristics to that chosen in Wales. Anemones were collected in late winter (30-3-'86), and behavioural experiments conducted in Cornwall following a 2 day acclimation period; few anemones were used in more than one encounter.

RESULTS

Contests Employing Anemones from South Wales

Contest results are shown in Table 1 and permutation comparisons made in Table 2. As predicted, they show that juxtaposed neighbours display significantly lower levels ($P < 0.01$) of aggression towards one another than do non-neighbours; such results are readily explicable in terms of behavioural habituation.

Table 1. Contest Results from Encounters Staged using Anemones taken from Parts of the Large Aggregation Examined in South Wales.

Sampling date	Contest permutation							
	●●		● ●		○●●○		○● ●	
	A	NA	A	NA	A	NA	A	NA
15-10-'86	1	17	11	4	4	13	32	5
							J 23, NJ 9	
2-12-'86	6	15	18	4	13	6	38	7
							J 31, NJ 7	
28-1-'87	3	11	21	2	17	12	31	8
							J 21, NJ 10	

A, aggression displayed by one or both anemones; **NA**, no aggression displayed by either contestant.
J, juxtaposed anemone being first (either earlier or sole) aggressor;
NJ, non-juxtaposed anemone being first (either earlier or sole) aggressor.

Table 2. Pairwise Comparisons of Contest Results from the Four Contest Permutations Staged, using Anemones from South Wales.

Date	Contest permutation comparisons - χ^2 values.					
	●● / ● ●	● ● / ○● ●○	●● / ○● ●○	●● / ○● ●	● ● / ○● ●	○● ●○ / ○● ●
15-10-'86	16.24***	7.94**	NS	33.05***	NS	20.78***
2-12-'86	12.35***	NS	6.35*	20.11***	NS	NS
28-1-'87	18.64***	NS	5.25*	15.10***	NS	NS

Probability levels: *** $P<0.001$, ** $P<0.01$, * $P<0.05$. NS, not significant.

The results of juxtaposed non-neighbour versus non-neighbour contests are not, however, so simply explained. In all cases, a higher level of aggression was noted in comparison to that displayed between juxtaposed neighbours ($P<0.001$); indeed, the levels recorded were not significantly different from those recorded for non-neighbour confrontations. Importantly, in all three sets of contests relating to the surveys conducted in October '86, December '86 and January '87, juxtaposed non-neighbours were more aggressive than their non-neighbour contestants. Such results can again be explained by the occurrence of specific behavioural habituation, which would leave juxtaposed anemones still responsive to non-neighbours, but they are also consistent with a putative reduction in surface antigenicity. In consequence, non-neighbours might fail to recognize and hence not attack juxtaposed non-neighbours due to surface "masking", but the latter, nevertheless, remain aggressive.

To assess these two alternatives further, it is pertinant to examine the results emanating from the contest permutation involving juxtaposed non-neighbours. If surface "masking" is operative, or any habituation induced by juxtapositioning is non-specific, then little aggression should be observed. If, however, any induced habituation is specific to the juxtaposed partner on the shore, then a high level of aggression can be expected. Unfortunately the results obtained are conflicting: contests staged using anemones collected from South Wales in October '86 produced significantly less aggression than either that observed in the non-neighbour permutation ($P < 0.01$) or in juxtaposed non-neighbour versus non-neighbour contests ($P < 0.001$), but the same was not true for the remaining two samples. Juxtaposed non-neighbour versus non-neighbour contests utilizing anemones collected in December '86 and January '87, produced significantly more aggression than those involving juxtaposed neighbours ($P < 0.05$), and a similar level of aggression to that observed between non-neighbours. Direct comparisons of the incidences of aggression between juxtaposed non-neighbours revealed significant differences between October '86 and December '86 ($\chi^2 = 7.25$, $P < 0.01$), and between October '86 and January '87 ($\chi^2 = 5.32$, $P < 0.05$).

Contests Employing Anemones from South-west England

Table 3. Contest Results from Encounters Staged using Anemones taken from the Aggregation Examined in South-West England on 30-3-'87.

Contest permutation							
●●		● ●		○● ●○		○● ●	
A	NA	A	NA	A	NA	A	NA
5	14	13	2	13	6	14	2
						J 11	NJ 3

A, aggression displayed by one or both anemones; NA, no aggression displayed by either contestant.

J, juxtaposed anemone being first (either earlier or sole) aggressor; NJ, non-juxtaposed anemone being first (either earlier or sole) aggressor.

Table 4. Pairwise Comparisons of Contest Results from the Four Contest Permutations Staged, Using Anemones from South-west England.

Contest permutation comparisons - χ^2 values.					
●● / ● ●	● ● / ○● ●○	●● / ○● ●○	●● / ○● ●	● ● / ○● ●	○● ●○ / ○● ●
12.25***	NS	6.76**	13.10***	NS	NS

Probability levels: *** $P < 0.001$, ** $P < 0.01$. NS, not significant.

Similar analyses of contest results arising from the usage of anemones collected in Cornwall (Tables 3 and 4) indicate that they are in complete agreement with the December and January data sets referred to above; juxtaposed non-neighbour contests produced significantly more aggression than those involving two juxtaposed neighbours ($P<0.01$), and in juxtaposed non-neighbour versus non-neighbour contests, the majority of juxtaposed non-neighbours displayed aggression earlier than their non-neighbour opponents.

Dispersions of Anemones in South Wales

Table 5. Size, Density and NND Statistics of Anemones within Parts of the Large Aggregation in South Wales.

Date	Anemone size (PPD) $\pm$ S.D. (mm)‡	Density of anemones (m^{-2})‡‡	NND Statistics		
			Mean inter-centres NND $\pm$ S.D. (mm)	Mean pedal separation NND $\pm$ S.D. (mm)	Clark & Evans', R
15-10-'86	15.2 ± 7.4 (n = 271) (42%)	347 ($0.780m^2$) (7.8%)	29.6 ± 15.8	14.5 ± 15.0	1.11**
2-12-'86	16.4 ± 8.6 (n = 341) (40%)	486 ($0.702m^2$) (13.2%)	26.1 ± 10.4	10.1 ± 9.2	1.15***
28-1-'87	13.0 ± 8.6 (n = 313) (62%)	333 ($0.939m^2$) (6.4%)	27.7 ± 14.4	14.9 ± 13.4	1.01

PDD, pedal disc diameter. ‡ number of anemones in study area and % <15mm in size (i.e. juveniles) given in parentheses. ‡‡ study area and % of that area occupied by pedal discs. *** $P<0.001$, ** $P<0.01$.

Table 5 provides NND statistics relating to anemones comprising the three parts (areas) of the aggregation examined, together with the data on anemone sizes and densities. Note that areas mapped were in the range 0.7-$0.9m^2$ and that the number of anemones considered on each occasion was of the order of 300. Also note that whilst only adult anemones (≥15mm diam.) were used in contests, and that the mean anemones size was >15mm, the mean size in January '87 was only 13.0mm, a figure attributable to the presence of a relatively high proportion of juveniles (62%). The NND data reveals that although pedal occupancy was low (6.4-13.2% of study area) and that the mean inter-centres NNDs were in excess of 25mm, using pedal separation distances (i.e. distances between adjacent disc perimeters) as a measure of NND, mean distances were much shorter (10.1-14.9mm) and thus indicative of the likelihood of a considerable amount of anemone interaction.

Inspection of the R values shows that there was significant, gross over-dispersion of anemones in both October and December '86, but a random distribution in January '87. To look for under-dispersion at high densities (i.e. short NNDs), Weiss plots were constructed for the December and January NND data sets (Fig. 2). With regard to the December data, it is apparent that over-dispersion rather than clumping is indicated in the inter-centres plot (note concavity in the basal spline in Fig. 2A). That this is, however, an artifact brought about by limitations imposed upon spacing by contiguous occupancy (see Brace and Quicke, 1985), is demonstrated by the fact that once pedal disc separation is adopted as the measure of NND, under-dispersion at high densities (indicated by spline convexity in Fig. 2B) is in fact revealed. With regard to the January NND data (Fig. 2C, D), the separations plot (Fig. 2D) not surprisingly, shows similar under-dispersion. (Note that the X axis scales differ between A and C, and between B and D).

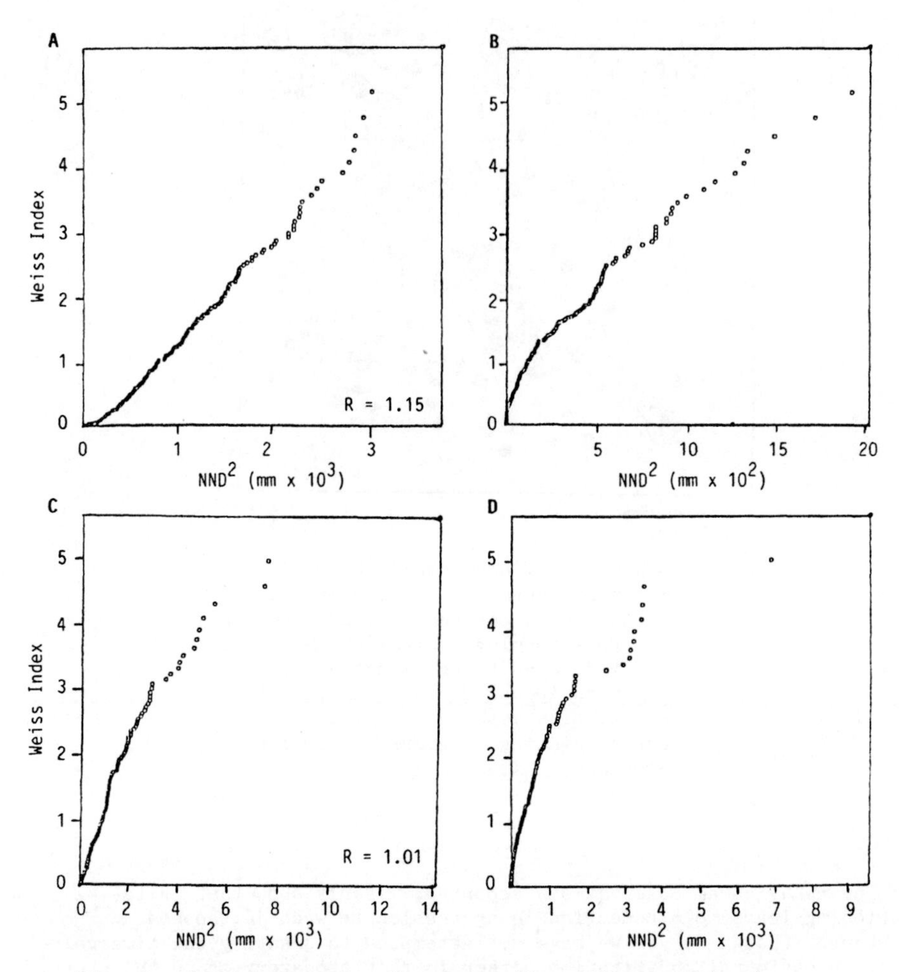

Fig. 2. Graphical tests of anemone dispersions within the parts of the large aggregation examined in South Wales. A,B. Weiss plots for 2-12-'86; C,D. for 28-1-'87. A and C refer to inter-centres NNDs, whilst B and D refer to pedal separation NNDs. Values for Clark and Evans', R, which denote gross dispersions, are provided in A and C. In all graphs, the maximum and minimum values of the Weiss Index and NND^2 are positioned at the extremes of the ordinate and abscissa scales respectively.

Thus not only is there evidence for clumping at high densities in January, when there was gross randomicity in spacing, but also for December when there was gross over-dispersion (a small degree of under-dispersion at high densities was also detected in the October data set). In agreement with previous findings (Brace and Quicke, 1986b), Fig. 3 serves to demonstrate that many of the clusters of anemones were indeed located within shallow depressions in the rock face.

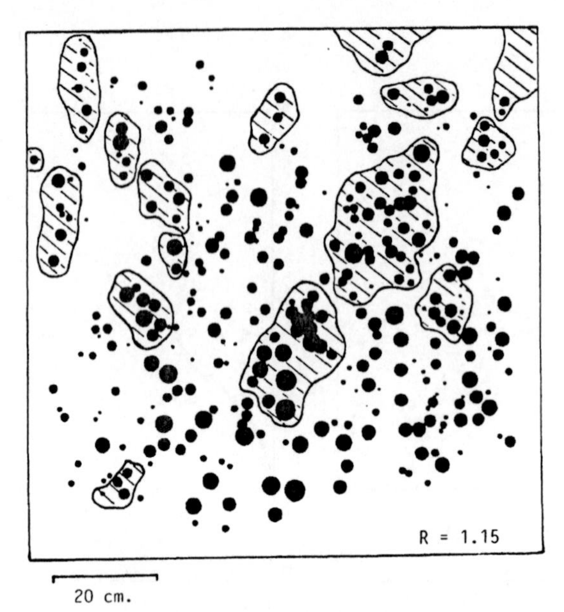

Fig. 3. Map of study area examined in South Wales in December '86 to show the relationship between anemone placements and rock topology. Shallow concavities are indicated by cross-hatching. Anemones (pedal discs) are drawn to scale. The measure, R, of Clark and Evans relating to gross dispersion, is provided.

DISCUSSION

The behavioural observations reported, clearly show that juxta-positioning leads to a reduction in aggression between juxtaposed neighbours, though as yet we have not attempted to quantify the timescale of this behavioural modification either in the laboratory or on the shore. The adaptiveness of this behavioural plasticity is regarded as being interpretable in terms of the reciprocal maximization of over-winter survivorship of adults, following the competitive occupation of relatively protected micro-habitats. Data presented here and elsewhere (Brace and Quicke, 1986b) indicate that the under-dispersion (i.e. clustering) of anemones at high densities is associated with the selective occupation of shallow concavities in rock faces. Since adult *Actinia* can have a life expectancy of several or even tens of years (Ottaway, 1980; Brace and Quicke, 1986a), the potential importance of such areas should be emphasized. That anemones are susceptible to environmental stress is suggested by the marked degrowth, flaccidity and surface necrosis exhibited by many individuals on The Worm's Head during a recent bad winter ('81/'82). At that time, the ratio of storage to structural lipids decreased dramatically (Rees, 1984), a feature which is generally interpreted as signifying a reduction in overall metabolic 'fitness' (Giese, 1960, 1961).

Of course the existence of local variation in habitat quality on most rocky shores is virtually universal, given the usual complex topology of most rock faces, but that this variability is great enough to result in differential survivorship is another matter. Indirect evidence that this is so, however, is forthcoming from two sources: 1. recolonization experiments have shown that previously, sparsely occupied raised areas, effectively remain uncolonized; 2. transects have indicated the presence of strong, differential size-dependent selection of the three morphs (see below) found on British shores, apparently acting through a sharp dichotomy in pedal attachment strength (Quicke et al., 1985) and resulting from the disproportionately greater effects of wave action on larger anemones (see Koehl, 1976). With regard to the presence of several ecologically distinct, multilocus genotypes - which we currently conservatively interpret as being morphs of a single species - on the British coastline (Quicke, Donoghue and Brace, 1983; Quicke and Brace, 1984; Quicke et al., 1985), it is important to note that the two (U and M) which favour vertical faces adhere much more firmly than the third (L), which frequents horizontal surfaces. Moreover, there are affiliated differences in aggressiveness (Brace, in preparation); significantly, U anemones, which predominate on the vertical faces of wave-washed gullies (such as those investigated at The Worm's Head - see Brace and Quicke, 1985, 1986a,b) are far more aggressive than either M or L individuals. Concomitant with this is the feature that the U morph has larger and more numerous acrorhagi than do the others (Donoghue, Quicke and Brace, 1985). Thus the scenario painted by all of these facts is that the selective agencies in force on such vertical faces - which, judging by the relatively high, prevailing anemone densities, are especially advantageous locations - are extreme, and have resulted in both morphological/biochemical and behavioural adaptation.

Turning now to the means by which a reduction in aggression between opposed anemones is achieved, the contest results presented here are clearly consistent in the main with the occurrence of behavioural habituation. But such habituation is evidently specific to the juxtaposed partner, as demonstrated by the results of juxtaposed non-neighbour versus non-neighbour, and juxtaposed non-neighbour contests. Further, the subsequent presentation of a novel stimulus (i.e. a non-neighbouring allogeneic anemone in the former type of contest permutation) leads to an enhanced' (relative to that of opponent) response, which is thus indicative of sensitization (i.e. dishabituation). With regard to the contest permutation involving juxtaposed non-neighbours, whilst results relating to anemones collected in December '86 and January '87 are in agreement with the above interpretation, anemones obtained from The Worm's Head in October '86 to the contrary, displayed little aggression. This lack of aggression is, however, consistent with the alternative supposition put forward in the Introduction that juxtapositioning might result in a reduction in surface antigenicity, which could perhaps be brought about by a simple change in thickness of the external mucous coat.

Behavioural modification or plasticity has in fact been demonstrated previously in Actinia equina in experiments conducted in my laboratory. Each of these involved a number of multiple presentations of an excised tentacle (Brace, Pavey and Quicke, 1978), and indicated not only initial sensitization of the aggressive response, but also subsequent habituation. A similar temporal pattern of response modification was reported by Bigger (1980) for Anthopleura krebsi using the same type of experimental paradigm. Also Purcell and Kitting (1982) have observed habituation in laboratory-housed Metridium senile (which uses 'catch-tentacles' in agonistic encounters), and forwarded evidence implicating that it takes place in the field. The only other published evidence for the occurrence of habituation between allogeneic neighbours on the shore has been provided by Sebens (1984), who found that in crowded situations, where evasive action was impossible, neighbouring A. xanthogrammica displayed little aggression towards one another, but when he brought these habituates into contact with novel opponents by transplantation, they displayed far more aggression. Seben's results are compatible with the data provided here, and with the general finding that aggression in anemones arises following the bringing together of virtually any combination of naive allogeneic contestants (see above and Francis, 1973; Bigger, 1980; Lubbock, 1980; Lubbock and Shelton, 1981; Ayre, 1982). Such an acutely discriminating capacity for allogeneic immunorecognition would, however, appear to be widespread amongst the lower invertebrates (Klein, 1982), and argues for the presence of a large array of histocompatability markers. Specific allogeneic immunocompetence has, for example, been ably demonstrated by tissue transplantation experiments in both corals (Hildemann, Bigger and Johnston, 1979; Hildemann et al., 1980) and sponges (Hildemann, Johnston and Jokiel, 1979; Johnston and Hildemann, 1982; Smith and Hildemann, 1986); in both groups, moreover, a specific inducible memory component has been detected. Progress with anemones has been more tardy, although Sauer, Müller and Weber (1986) have provided evidence to show that the relevant histocompatability markers in Anemonia sulcata are glycoproteins and reside within the body mucus; they have also tentatively espoused the existence of an inducible memory, but their evidence is open to other interpretations. Work by Bigger (1976) and Lubbock (1980) on Anthopleura - concerned solely with acrorhagial nematocyst discharge rather than with the entire (anemone) aggressive sequence - suggests, to the contrary, that the markers are surface bound and non-diffusible. Bonnin (1964) working on Actinia equina showed that an alcohol-soluble fraction of anemone homogenate was a potent elicitor of aggression.

The work of Sauer and co-workers is clearly pertinant in the context of the suggestion raised here that surface antigenicity might be modifiable. Obviously, further experimental work is required in order to assess whether such modification plays a role in mediating behavioural habituation. Finally, it is important to point out that a further research technique is available for investigating self/non-self recognition processes in anemones, that of the extracellular recording of nervous activity. Lubbock and Shelton (1981) have shown that allogeneic stimuli applied to acrorhagi of Anthopleura elegantissima result in intense local activity; syngeneic material is ineffective in this regard. Interestingly, in Actinia equina it has already been shown electrophysiologically that preparations of acrorhagi taken from U morph individuals are generally more responsive than those of L anemones, and that they also respond more vigorously to application of U morph extract than to L morph material (Brace and Storey, in preparation). There is thus clearly great scope for future research using this approach, which, for example, could be used to test the supposition of Bigger (1982) that the response threshold of acrorhagial nematocysts may be dependent upon the state of distension of acrorhagi (and hence overall body posture). However, even if such peripheral agencies influence responsiveness, it is likely, as in control of feeding threshold in Hydra (Koizumi and Maeda, 1981), that "central" mechanisms also play an important role.

ACKNOWLEDGEMENTS

I wish to acknowledge the assistance of the following people with both field work and behavioural experimentation: Vanessa Bullwinkle, Mark Flowers, Richard Lee, Simon Hellyer and Amanda Thorogood.

REFERENCES

Ayre, D.J., 1982, Inter-genotype aggression in the solitary sea anemone Actinia tenebrosa, Mar. Biol, 68:199.

Bigger, C.H., 1976, The acrorhagial response in Anthopleura krebsi: intraspecific and interspecific recognition, in: "Coelenterate Ecology and Behavior," G.O. Mackie, ed., Plenum, New York.

Bigger, C.H., 1980, Interspecific and intraspecific acrorhagial aggressive behavior among sea anemones: a recognition of self and not-self, Biol. Bull. mar. biol. Lab., Woods Hole, 159:117.

Bigger, C.H. 1982, The cellular basis of the aggressive acrorhagial response of sea anemones, J. Morph., 173:259.

Bonnin, J.P., 1964, Recherches sur la 'reaction d'aggression', et sur le functionnement des acrorrhages d'Actinia equina L., Bull. Biol. Fr. Belg., 1:225.

Brace, R.C., Pavey, J., and Quicke, D.L.J., 1979, Intraspecific aggression in the colour morphs of the anemone Actinia equina: the "convention" governing dominance ranking, Anim. Behav., 27:553.

Brace, R.C., and Quicke, D.L.J., 1985, Further analysis of individual spacing within aggregations of the anemone, Actinia equina, J. mar. biol. Ass., U.K., 65:35.

Brace, R.C., and Quicke, D.L.J., 1986a, Dynamics of colonization by the Beadlet anemone, Actinia equina, J. mar. biol. Ass., U.K., 66:21.

Brace, R.C., and Quicke, D.L.J., 1986b, Seasonal changes in dispersion within an aggregation of the anemone, Actinia equina, with a reappraisal of the role of intraspecific aggression, J. mar. biol. Ass., U.K., 66:49.

Carter, M.A., and Thorp, C.H., 1979, The reproduction of Actinia equina L. var. mesembryanthemum, J. mar. biol. Ass., U.K., 59:989.

Clark, P.J., and Evans, F.C., 1954, Distance to nearest neighbour as a measure of spatial relationships in populations, Ecology, 35:445.

Donoghue, A.M., Quicke, D.L.J., and Brace, R.C., 1985, Biochemical-genetic and acrorhagial characteristics of pedal disc colour phenotypes of Actinia equina, J. mar. biol. Ass., U.K., 65:21.

Francis, L., 1973, Intraspecific aggression and its effect on the distribution of Anthopleura elegantissima and some related sea anemones, Biol. Bull. mar. biol. Lab., Woods Hole, 144:73.

Gashout, S.E., and Ormond, R.F.G., 1978, Evidence for parthenogenetic reproduction in the sea anemone Actina equina, J. mar. biol. Ass., U.K., 59:975.

Giese, A.C., 1959, Comparative physiology: annual reproductive cycles of marine invertebrates, A. Rev. Physiol., 21:547.

Giese, A.C., 1960, Lipids in the economy of marine invertebrates, Physiol. Rev., 46:244.

Hildemann, W.H., Bigger, C.H., and Johnston, I.S., 1979, Histoincompatability reactions and allogeneic polymorphism among invertebrates, Transplantn. Proc., 11:1136.

Hildemann, W.H., Jokiel, P.I., Bigger, C.H., and Johnston, I.S., 1980, Allogeneic polymorphism and alloimmune memory in the coral, Montipora verrucosa, Transplantation, 30:297.

Johnston, I.S., and Hildemann, W.H., 1982, Cellular defense systems of the Porifera, in: "The Reticuloendothelial System, Vol. 3", N. Cohen and M.M. Siegel, eds., Plenum, New York.

Klein, J., 1982, "Immunobiology: The Science of Self - Non-self Discrimination," John Wiley, New York.

Koehl, M.A.R., 1976, Mechanical design in sea anemones, in: "Coelenterate Ecology and Behavior", G.O. Mackie, ed., Plenum, New York.

Koizumi, O., and Maeda, N., 1981, Rise of feeding threshold in satiated Hydra, J. Comp. Physiol. A, 142:75.

Lubbock, R., 1980, Clone-specific cellular recognition in a sea anemone, Proc. Natl. Acad. Sci. U.S.A., 77:6667.

Lubbock, R., and Shelton, G.A.B., 1981, Electrical activity following cellular recognition of self and non-self in a sea anemone, Nature, Lond., 289:59.

Orr, J., Thorpe, J.P., and Carter, M.A., 1982, Biochemical genetic confirmation of the asexual reproduction of brooded offspring in the sea anemone Actinia equina, Mar. Ecol. - Prog. Ser., 7:227.

Ottaway, J.R., 1980, Population ecology of the intertidal anemone Actinia tenebrosa. IV. Growth rates and longevities, Aust. J. mar. Freshwat. Res., 31:385.

Purcell, J.E., and Kitting, C.L., 1982, Intraspecific aggression and population distributions of the sea anemone Metridium senile, Biol. Bull. mar. biol. Lab., Woods Hole, 162:345.

Quicke, D.L.J., and Brace, R.C., 1983, Phenotypic and genotypic spacing within an aggregation of the anemone, Actinia equina, J. mar. biol. Ass., U.K., 63:493.

Quicke, D.L.J., and Brace, R.C., 1984, Evidence for the existence of a third, ecologically distinct morph of the anemone, Actinia equina, J. mar. biol. Ass., U.K., 64:531.

Quicke, D.L.J., Donoghue, A.M., and Brace, R.C., 1983, Biochemical-genetic and ecological evidence that red/brown individuals of the anemone Actinia equina comprise two morphs in Britain, Mar. Biol., 77:29.

Quicke, D.L.J., Donoghue, A.M., Keeling, T.F., and Brace, R.C., 1985, Littoral distributions and evidence for post-settlement selection of the morphs of Actinia equina, J. mar. biol. Ass., U.K., 65:1.

Rees, T.D., 1984, The population ecology and behaviour of Actinia equina. Ph.D. Thesis, University of Nottingham.

Sauer, K.P., Müller, M., and Weber, R., 1986, Alloimmune memory for glycoproteid recognition molecules in sea anemones competing for space. Mar. Biol., 92:73.

Sebens, K.P., 1984, Agonistic behaviour in the intertidal sea anemone Anthopleura xanthogrammica, Biol. Bull. mar. biol. Lab., Woods Hole, 166:457.

Smith, L.C., and Hildemann, W.H., 1986, Allogeneic cell interactions during graft rejection in Callyspongia diffusa (Porifera, Demospongia): a study with monoclonal antibodies, Proc. R. Soc. Lond. B, 266:465.

Weiss, P.W., 1981, Spatial distribution and dynamics of populations of the introduced annual Emex australis in south-eastern Australia, J. appl. Ecol., 18:849.

Interspecific Interactions among Selected Intertidal Stomatopods

Roy L. Caldwell

University of California
Berkeley, U.S.A.

INTRODUCTION

Stomatopods, or mantis shrimps as they are commonly known, are predatory crustacea found in most temperate and tropical costal marine habitats. While the group is not large, with fewer than a thousand species in 12 families, it is common to find several relatively abundant species in a given habitat. For example, Dingle et al. (1977) reported eight common species of stomatopods from one intertidal reef flat at Phuket, Thailand. The average density for all species combined was almost four individuals per m2. Reaka and Manning (1980) reported up to six species of *Gonodactylus* occurring on a single reef flat on the Pacific coast of Costa Rica. Such a rich species diversity of stomatopods is not restricted to reef flat habitats. Dingle and Caldwell (1975, 1978) described four species of Squillidae co-occuring on mudflats in the Gulf of Siam and Caldwell (in prep.) studied five squillids and three lysiosquillids on a mudflat in the Bay of Panama. Given that stomatopods are voracious predators and often are quite territorial, why is it that we frequently find more than one species existing in a single habitat?

THE PHUKET STOMATOPOD COMMUNITY

At least part of the answer can be traced to morphological, physiological, and behavioral specializations that allow species to capture and process different prey and to occupy different microhabitats. As an example, let's examine the eight species of stomatopods described from the beach at Phuket (Dingle et al., 1977). That site is characterized by an extensive tidal fluctuation of almost 3 meters. The substrate in the mid-intertidal is mostly shale, but gives way in the lower intertidal to coral rubble, sand and cobble. In the very low intertidal, there are a diversity of live corals which border a subtidal fringing reef.

On a gross functional level, we can divide these species into two groups based on the type of raptorial feeding appendage that they possess: what Caldwell and Dingle (1975, 1976) called spearers and smashers. *Lysiosquilla tredecimdentata* and *Pseudosquilla ciliata* are both spearers. The dactyls of the raptorial appendages are armed with several spines used to impale soft-bodied prey. The other six species, Gonodactylus chiragra, G. viridis, G. smithii, G. mutatus, G. ternatensis, and Haptosquilla glyptocercus, are all smashers and possess raptorial appendages equipped with an inflated dactyl heel and a single terminal spine. This type of appendage is used to crush armored prey such as snails and crabs.

But even between the two spearing species, differences in raptorial appendage morphology and predatory habits are considerable. P. ciliata, which obtains a maximum size of approximately 10 cm and lives in small u-shaped burrows, is a generalist. It takes a wide diversity of bottom-dwelling prey including small fish, shrimp, crabs, annelids, and even other stomatopods. The slender raptorial dactyls are armed with only three spines and are used to seize and hold prey. Large food items are rarely eaten in the field, but are transported back to the safety of the burrow where they can be cut up by the serrated mandibles. This species is strictly diurnal. Individuals hunt away from their burrows only during low tides when water depth is less than 20 cm and large predatory fish have been forced off of the reef flat. When foraging, P. ciliata range widely, using their excellent vision and olfaction to detect prey. The ability to stalk prey is aided by their cryptic coloration. Individuals usually match general substrate color and pattern, although the ability to conform to different backgrounds may take weeks and involves at least one molt (Caldwell, unpublished).

L. tredecimdentata is a much larger species, growing to over 25 cm in length. Males and females form pairs that occupy single, u-shaped burrows that can be up to 10 cm in diameter and 10 m in length. Lysiosquillids are sit-and-wait predators that almost never leave their burrows unless forced to find a new mate. Their bodies are delicate and poorly armored. Their walking legs are short and project laterally, adaptations for moving in a burrow rather than walking on the substrate. The thorax and abdomen are marked by bold, conspicuous yellow and black bands. In contrast to P. ciliata, L. tredecimdentata are extremely vulnerable to large predators when out of their burrows. The elongated raptorial appendages of L. tredecimdentata are specialized for catching fish and squid. The dactyls of males having up to 13 spines, but as is typical for this genus, females have smaller raptorial appendages with fewer dactyl spines. Females also have smaller eyes. This correlates with the fact that males do most of the hunting at the burrow entrance, providing their mates with food. Females usually remain deep inside the burrow, brooding eggs and maintaining the burrow. Male L. tredecimdentata lie in wait for prey at the burrow entrance. During the day, the entrance is usually covered with a thin, flexible cap of sand and mucus with a small hole in the center. The stomatopod sits at the entrance with just its sand-colored antennules and eyes extending through the hole. When a fish swims over the entrance within reach of the raptorial appendages, the stomatopod strikes through the cap and impales its quarry. At night, animals are more likely to wait for prey in an uncapped entrance. Thus, while both species have spearing raptorial appendages, their forms are quite different, specialized for taking very different prey. This difference is reinforced by divergent hunting behaviors which make it very unlikely that even similarly sized individuals are in direct competition.

The six species of gonodactylids have very similar smashing raptorial appendages. All feed primarily on armored prey, such as crabs and gastropods. They are all roughly the same size, all have similar body shapes and armor, are all diurnal, and all live in cavities in hard substrates. In fact, some of these species are so similar that, until our field studies on live animals, preserved specimens could not be distinguished. However, in the field, they vary considerably, not only with respect color, but also with the type of microhabitat occupied and their hunting behavior.

G. chiragra is the largest species, reaching a maximum length of 10 cm. It also is found highest in the intertidal, occuring from the low intertidal up into the mid-intertidal. Individuals occupy cavities in coral rubble or shale. They make extended foraging sorties when the tides are changing and the water depth over the burrow is only a few centimeters.

G. viridis is one of the smaller species in this assemblage, rarely exceeding 5 cm, but it occurs over much of the same tidal range as *G. chiraga*. It usually lives in cavities in rubble. Like *G. chiraga*, it hunts in shallow water to avoid fish predators.

The range of *H. glyptocercus* overlaps extensively with those of *G. chiraga* and *G. viridis*. It is about the same size as *G. viridis*, but is rarely seen foraging far from its cavity. Rather, it occupies tight-fitting, tube-like cavities in rubble or solid bench and lies in ambush for passing prey.

G. smithii occurs in the low intertidal, can achieve a maximum length of 7 cm, and is usually found in association with live corals. It most frequently occupied cavities in the massive coral, *Porites lutea*. Individuals frequently hunted at low tide, climbing through and over the three dimensional coral habitat.

G. mutatus is frequently found in the same general habitat as *G. smithii*, but is smaller, rarely exceeding 4 cm. It is more frequently associated with branching corals and is more likely to hunt from the cavity entrance.

Finally, *G. ternatensis* is the most specialized with respect to microhabitat usage. A large species, reaching almost 9 cm in length, it is usually found living inside heads of the branching coral, *Pocillopora spp*. It forms the cavity by using its raptorial appendages to break away interior branches. These animals rarely ventured far from their home coral colony and feed mostly on the crabs and shrimp associated with it. Thus, at this particular site, these eight species of stomatopods can be separated by morphological and behavioral differences related to microhabitat use and prey specialization.

Pseudosquilla - *Gonodactylus* INTERACTIONS

It is attractive to view apparent cases of resource partitioning as the result of a long process of competitive interaction. However, evidence for competitive exclusion or character displacement operating to structure such assemblages is difficult to obtain. Alternative, more neutral hypotheses, must also be entertained (see Diamond and Case, 1986). Occasionally, nature provides a natural experiment that allows us to glimpse selective processes at work. A recent population explosion of the stomatopod *G. aloha* in Kaneohe Bay on the island of Oahu in Hawaii is one such an example.

As discussed by Kinzie (1968, 1984), prior to World War II, *P. ciliata* was the dominant stomatopod in the sandy back reef areas of Kaneohe Bay. It was frequently collected in association with coral heads and coral rubble. The only gonodactylid found in this area was a rare, dwarf species, *G. hendersoni*. Until this time, Kaneohe Bay was known for its rich and diverse coral reefs. However, following the war, a series of events occurred that greatly changed this habitat (Banner, 1974). A military base was constructed on the bay. A 10 m deep shipping channel was blasted through the barrier reef. Much of the coral rubble tailings were deposited inside the bay as land fill. This was followed by increased sewage disposal into the bay from the military base, and greater siltation caused by erosion of the surrounding hills as a result of housing construction. Over the next several years, there was considerable coral mortality and eutrophic algae began to flourish.

The proliferation of coral rubble in the sandy back reef areas of Kaneohe Bay coincides with the appearence of a gonodactylid previously unknown from Hawaii. This species has clear affinities to the *G. falcatus* species complex which is common throughout the Indo-Pacific. Kinzie (1968, 1984) feels that it was introduced into Hawaii from the Philippines after the war. However, Manning and Reaka (1981) consider it to be an undescribed species endemic to Hawaii and have named it *G. aloha*. Whatever its origin, it is clear that, prior to the 1960's, this gonodactylid was exceedingly rare or absent on Oahu. But as coral rubble accumulated, the population exploded and *G. aloha* became very abundant. Where *P. ciliata* had previously been the dominant stomatopod on the sandy back reefs of Kaneohe Bay, it now shared this habitat with another slightly smaller, but more powerful species of stomatopod.

Hatziolos (1979) studied the behavior and ecology of populations of *P. ciliata* in Thailand, Jamaica, Florida, and Hawaii. In the first three, there has been a long history of co-occurrance with *P. ciliata* and one or more species of *Gonodactylus*. At each location, the *Gonodactylus* were abundant, occupying cavities in live or dead coral. Because of their more powerful smashing raptorial appendage and heavier body armor, gonodactylids invariably dominate aggressive encounters with similarly sized *P. ciliata*.

In Thailand, Jamaica, and Florida, *P. ciliata* live almost exclusively in burrows and, with the exception of juveniles, are rarely found in rock or coral cavities. When Hatziolos (1979) offered Jamaican *P. ciliata* a choice between a burrow and a cavity, all occupied the burrow. In contrast, *P. ciliata* from Hawaii were reported to be associated with coral cavities prior to the spread of *G. aloha* (Edmonson, 1921; Townsley, 1951). While Hatziolos (1979) observed in the mid-1970's that, in the field, most *P. ciliata* were living in burrows, she found in the laboratory that when offered a choice between a burrow and a cavity, almost all took the cavity. She also found that *P. ciliata* from Hawaii were much more aggressive in their defense of burrows or cavities against *Gonodactylus* than were *P. ciliata* from Jamaica.

When Hatziolos (1979) examined *P. ciliata* collected in the field, she found that animals from Hawaii were almost three times more likely to be injured than were those from Jamaica. I found the same difference in 1986 when comparing *P. ciliata* from Kaneohe Bay and *P. ciliata* collected on the Atlantic coast of Panama (Caldwell, unpublished). In this case, I only considered wounds that appeared to have been inflicted by smashing gonodactylids. In Panama, *P. ciliata* co-occurs with the same species of *Gonodactylus* found in Jamaica. This association has probably lasted at least since the closure of the Isthmus of Panama.

It appears that where there has been a long association between *P. ciliata* and gonodactylids, the superior fighting ability of the gonodactylids has allowed them exclusive use of natural cavities, while *P. ciliata* burrowed in soft substrates. The risk of injury fighting with gonodactylids for cavities is apparently so great that *P. ciliata* no longer use cavities even when they are available. In Hawaii, where *P. ciliata* and *G. aloha* have been in conflict for scarcely more than three decades, *P. ciliata* are rarely found living in cavities, but do prefer them to burrows when *G. aloha* are not present.

It is possible that in all four populations, *P. ciliata* have learned to avoid cavities through repeated contacts with resident gonodactylids. In the laboratory, I have shown that a *P. ciliata* will learn to avoid a cavity containing only the odor of a *Gonodactylus* if it has been repelled by a

Gonodactylus occupying a cavity (Caldwell, in prep.). However, it seems more likely that in areas of long association with gonodactylids, populations of P. ciliata have evolved that prefer burrows and avoid cavities where they might encounter Gonodactylus. In Hawaii, there hasn't been sufficient time for selection to effect such a change. The higher rate of wounding in Hawaii suggests that P. ciliata are not avoiding interactions with G. aloha and that they are suffering the consequences of fighting with these more powerful newcomers.

COMPETITION BETWEEN TWO PANAMANIAN GONODACTYLIDS

Occasionally, we find situations where two or more species of stomatopods overlap extensively in habitat use and appear to be in direct competition. G. bredini and G. oerstedii, from the Atlantic coast of central Panama, appear to be such a case. They obtain roughly the same size, occur on the same reef and grass flats, have similar physical tolerances, take the same prey, and occupy the same types and sizes of cavities. Over the past several years, while working primarily on the behavioral ecology of G. bredini, we have collected some data on G. oerstedii. Here, I wish to review some of these observations and discuss why these two species continue to coexist despite their similarity.

Both G. oerstedii and G. bredini are found throughout the Caribbean from Florida to Panama and Venezuela. However, the range of G. bredini extends further north to include the northern Gulf of Mexico, the Carolinas, and Bermuda (Manning, 1969). In most locations where these two species co-occur, there is a fair degree of habitat separation between them, often with another species of Gonodactylus involved. For example, where I studied these species in Curacao, G. bredini occurred almost exclusively in the low intertidal, G. spinulosis was the most abundant species to a depth of 1 m, G. oerstedii was the most common species from 2 to 3 m, and G. curacaoensis, was the most numerous gonodactylid below 4 m. (Fig. 1). The one exception to this pattern was that large breeding G. oerstedii were frequently found as high in the low intertidal as G. bredini.

Natural History

In Panama, at the Galeta Marine Laboratory of the Smithsonian Tropical Research Institute, Gonodactylus are extremely numerous on the turtlegrass (Thalassia testudinum) and rubbly reef flats that surround the laboratory. Densities can reach as high as 25 gonodactylids per m2. While we have collected six species of Gonodactylus at Galeta, only three, G. bredini, G. oerstedii, and G. austrinus, are common. Of the other three, G. curacaoensis occurs on the face of the fringing reef at depths below 3 m and G. torus is a dwarf species occasionally taken at depths exceeding 10 m. G. spinulosis is rare. Of the over 50,000 gonodactylids collected in this area, only three individuals have been G. spinulosis (all male), and it is very unlikely that there is a breeding population on this coast. Considering the three common species, G. bredini is the most abundant, followed by G. oerstedii, and G. austrinus. They occur in a ratio of roughly 7:2:1 (Steger, 1985).

G. austrinus is a habitat specialist, occuring most frequently on the reef crest and fore reef where there is a swift current and/or wave action and where there is rarely a complete exposure at low tide. G. austrinus is also frequently associated with the calcarious algae, Halimeda opuntia, which grows in dense mats over the surface of coral rubble and coralline algae. Juveniles are often found living between the algae and the coral surface, but adults seek out cavities in the rubble. Adult G. austrinus are occasionally collected inshore in the low intertidal, but we now suspect

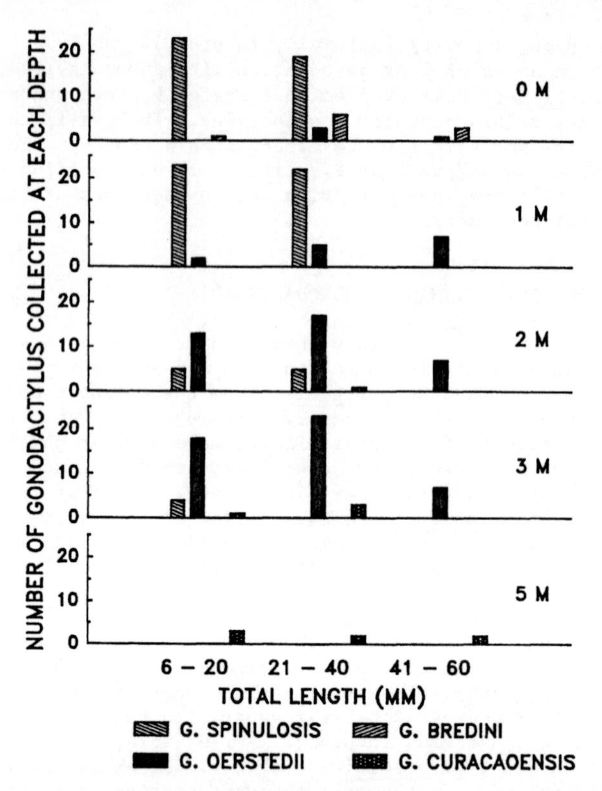

Fig. 1. Number of Gonodactylus collected from 20 pieces of rubble taken at each depth. Collections were from Piscaderabaai, Curacao

that many of these animals were transported there by wave action. For example, in March, 1987, three weeks after several days of high wind and heavy surf, I found several large pieces (20 - 40 cm) of Halimeda-covered coralline algae and coral rubble that had been washed onto exposed Thalassia flats. After several days of exposure, the Halimeda was dying, but inside these pieces of rubble were several G. austrinus. One large block contained 9 G. austrinus including a female with eggs. The closest source for such rubble was over 50 m away on the reef crest where G. austrinus where common. The rubble obviously had been transported inshore along with its resident stomatopods by the recent storm.

G. bredini and G. oerstedii are most abundant at Galeta from the low intertidal down to a subtidal depth of about 0.5 m. While both species occur considerably deeper, they become increasingly rare with depth. There is a tendency for small, non-breeding G. oerstedii (7 - 45 mm) to be more common below low mean water and for large breeding adults (>45 mm) to be found more often in the lower intertidal, but any sized individual can be found from the lower intertidal down. Both adult (>25 mm) and juvenile (7 - 25 mm) G. bredini are more common in the lower intertidal, but they too are found subtidally (Steger, 1985; Caldwell, in prep.).

Physical Tolerances

The upper distributions of *G. bredini* and *G. oerstedii* at Galeta are determined by physical limits. Both species have similar tolerances to the high temperatures, desication, and low salinities that may be encountered during tidal exposures. For example, adults of both species can survive water temperatures up to 39ù for 30 min, but suffer greater than 50% mortality at 40ù (Table 1). With longer exposures at 39°, *G. bredini* is slightly more tolerant than *G. oerstedii*. When animals were held for 90 min in 39ù water, significantly more *G. oerstedii* died or lost their raptorial appendages due to heat stress than did *G. bredini* (Table 2). *G. austrinus*, which normally does not occur in exposed areas, suffered even greater mortality. Tide pool temperatures in excess of 40ù are not uncommon when spring low tides occur during periods of bright sun and light winds.

Table 1. Effect of high temperature exposure on adult *Gonodactylus*. Animals were held in heated water for 30 min. Values are percent healthy after 72 hr. at 25°. Number = ().

Species	38°	39°	40°	41°
G. bredini	100(5)	81(22)	40(5)	0(5)
G. oerstedii	100(5)	83(18)	0(5)	0(5)
G. austrinus	80(5)	90(10)	0(5)	0(5)

When stomatopods are exposed to high temperatures, low salinities, or low oxygen levels, the first indication of severe stress is the permanent extension of the raptorial appendages. Even if the animals survive, the raptorial appendages are useless and are subsequently torn off to regenerate. Occasionally on the reef flats at Galeta, after extreme exposures and high temperatures, we found up to 10 % of the surviving *Gonodactylus* missing one or both raptorial appendages, apparently lost due to heat stress (Caldwell, unpublished).

Both *G. bredini* and *G. oerstedii* females brood their eggs and larvae in the intertidal, but again I found no difference between the two species with respect to the tolerance of eggs or larvae to heat or salinity stress. Surprisingly, the eggs and early larval stages of both species were more tolerant of heat stress than were the adults. For example, when yolked larvae were taken from the cavities of brooding females and were placed in heated water for 30 min, significant mortality did not occur until 41ù (Table 3). Interestingly, when *G. bredini* larvae were tested in heated water after they had completed the fourth stadial molt (non-yolked) and would have left the cavity to enter the plankton (Dingle and Caldwell,

Table 2. Effect of 39° exposure for 90 min on *Gonodactylus* adults. Condition was recorded after 72 hr.

Species	Alive	P.E.	Dead	
G. bredini (N=20)	12	3	5	*
G. oerstedii (N=20)	5	9	6	
G. austrinus (N=20)	1	0	0	

* G = 4.94, p .05

Table 3. Effect of high temperature exposure on larval *Gonodactylus*. Animals were held in heated water for 30 min. Values are percent healthy after 2 hr. Number = ().

Species	39°	40°	41°
G. bredini (yolk)	100(20)	100(40)	30(40)
G. oerstedii (non-yolk)	50(20)	0(20)	0(20)
G. austrinus (yolk)	100(20)	100(20)	0(20)

1972), they were much more susceptible to heat stress. The increased tolerance of early larvae to high temperature and low salinity probably reflects the fact that if a severe exposure occurs while a female is brooding her larvae, she cannot transport them to deeper water to avoid extreme physical conditions.

Even though eggs and early larvae are as tolerant to heat as are adults, I have observed mass mortality during exposures. For example, in May, 1983, the first very low spring tides of the wet season were unusually severe due to light, offshore winds and bright sun. When the first exposures occurred, approximately 40% of the large *G. bredini* and *G. oerstedii* females were brooding eggs and larvae. After five consecutive exposures, I failed to find a single female with eggs or larvae. Most of the females survived by abandoning their cavities and crawling into shallow pools of water under the pieces of rubble that they had occupied or they moved to deeper water. However, most left their eggs or larvae in the cavity where they perished or the eggs were lost when they were removed from the protection of the cavity.

Prey

G. bredini and *G. oerstedii* feed on the same types of prey taken from the same habitats. Both specialize on gastropods and crustaceans. On *Thalassia* flats, snails, crabs, and hermit crabs make up most of their diet (Caldwell et al., 1987). Prey are typically processed in the protection of a cavity, but both species forage actively away from their homes, making several trips a day to find food. The only difference in the animals taken by the two species is that the oldest *G. oerstedii* grow a few millimeters larger than *G. bredini* and therefore can process slightly bigger prey. However, over most of the size range, they feed on the same prey populations.

Cavity Use

While individuals of both species will occupy an empty cavity in almost any hard substrate including nuts, sponges, gastropod shells, sea urchin tests, and even tin cans and bottles, they are most frequently found living in cavities in coral rubble or coralline algae. On *Thalassia* flats, where rubble tends to be less common, we frequently find *G. bredini* and *G. oerstedii* living in hollow coralline algae nodules scattered through the grass. These are formed when gonodactylids find a small piece of rubble and modify it to produce a cavity by pounding pebbles or shell fragments between the branches. Gradually, coralline algae overgrows this structure and cements it together. Most nodules show signs of repeated breakage and repair and are probably occupied by stomatopods for many years. We have tagged some nodules that have persisted in this habitat for up to seven years.

There is little difference in the types of cavities occupied by *G. bredini* and *G. oerstedii*. Individuals of both species, collected from the same habitats, had the same cavity volume to stomatopod length relationships (Steger, 1985, 1987; Caldwell and Steger, in prep.). Either natural or artificial cavities placed in the field were quickly colonized by both species. When the same cavities were placed repeatedly in the same location, they were colonized in proportion to the abundance of the two species in that habitat (Caldwell, in prep.).

There is, however, one subtile difference in cavity use by these two species that may lessen competition between them. *G. bredini* begin to reproduce when they are less than one year of age and about 25 mm in length. Initially, females lay eggs every three or four months, but by the end of their second year (approximately 40 mm), they breed every other month. Few females live longer than three or four years or exceed 50 mm total length. Correlated with early breeding, *G. bredini* grow more slowly and obtain a smaller maximum size than do *G. oerstedii*. Because of a strong positive correlation between body size and eggs number, they also produce relatively few eggs per clutch (100 - 1200) (Caldwell and Steger, in prep.). While our data are less complete for *G. oerstedii*, we found no females breeding at less than 45 mm in total length. The age at first reproduction is probably two years. Females continue to breed over the next few years until they reach a maximum size of approximately 65 mm. *G. oerstedii* delay reproduction and grow rapidly until they are larger than most *G. bredini* in this population. Once breeding commences, because of their larger body size, *G. oerstedii* produce larger egg masses (1000 - 4000 eggs) (Caldwell, in prep.).

Large cavities in the intertidal, which are used by both *G. bredini* and *G. oerstedii* for breeding, are relatively scarce and competition for them is intense (Montgomery and Caldwell, 1984; Steger, 1985, 1987). Competition between *G. oerstedii* and *G. bredini* for breeding cavities is reduced because females of the two species tend to breed at different sizes. Reproductive *G. oerstedii* use large cavities (10 to 60 cc) while *G. bredini* females with eggs are more likely to be found in cavities from 2 to 20 cc. At the same time, *G. oerstedii* less than 45 mm in length are more likely to be found in the subtidal where conditions are apparently less suitable for breeding, but where competition for cavities is less intense. Still, there is considerable overlap in cavity use by these two species and while this partial separation in breeding size may lessen competition between them, it has not eliminated it.

Competition for Cavities

Steger (1985, 1987) has produced evidence from Galeta that there is intense competition for cavities among stomatopods, particularly in the low intertidal where the numbers of *Gonodactylus* are highest. He demonstrated that most cavities in this habitat were occupied by stomatopods; that most gonodactylids occupied a cavity of less than their preferred volume; that when cavities were added to an area, the local density of gonodactylids increased; and that there was a considerable population of gonodactylids looking for empty cavities. This conclusion is supported by laboratory and field observations that both *G. bredini* and *G. oerstedii* will frequently fight to the point of injury to keep or acquire a cavity (Caldwell, 1986 a, 1986 b; Montgomery and Caldwell, 1984; Steger and Caldwell, 1983).

Recently, a man-made disaster occurred on the Panamanian coast that provides additional evidence that *Gonodactylus* in the low intertidal are competing for cavities. In May, 1986, there was an oil spill by a refinery not far from Galeta. Several thousand barrels of oil escaped and a considerable amount of it came ashore on study sites on which my students and I had monitored stomatopod populations since 1979. While we were not present during the actual spill, reports indicate that oil covered the face of the reef and that it was carried at high tide back into the mangroves (many of which are now dead). A sheet of oil was spread on the reef flat. Even a year later, oil can be stirred from these sediments.

In September, 1986, Dr. Rick Steger and I returned to Panama to survey the damage to the stomatopod community. We found that two of our previous study sites had been severely impacted by the oil. Fortunately, two others showed little evidence of damage. When we sampled the gonodactylids on each site, we found that where the oil had come ashore, there were fewer large individuals - particularly females. This was not the case in the unaffected areas (Caldwell and Steger, in prep.). We hypothesize that because large individuals occur most frequently in intertidal, they were more likely to be exposed to the oil. (It appears that the oil floated over the lower back reef and lagoon and produced less damage in these areas.) Also, many of the *Gonodactylus* in the mid-intertidal were probably females brooding eggs and they may have been less willing to abandon their cavities when the oil first hit.

When we examined the volume of cavities occupied by *Gonodactylus* at the site most affected by the oil, we found that individuals occupied cavities of a volume significantly larger than did similar-sized animals at sites where the oil did not come ashore. Furthermore, animals in the oil-impacted area were now living in cavities of the preferred size while animals in the unaffected areas still occupied cavities smaller than the preferred size (see Steger, 1985, 1987 for a description of the cavity volume / size relationship in these species). We also found that *Gonodactylus* from the oil-impacted areas, when compared to animals from unaffected sites, were less likely to bare wounds typical of those inflicted by other stomatopods (Caldwell and Steger, in prep.).

These observations suggest that after the oil spill, competition for larger cavities was considerably reduced in areas where large individuals had disappeared. This allowed the remaining animals to move up into cavities of the preferred volume. It also meant fewer and/or less intense fights for cavities. More recent observations taken in March, 1987, show that as animals grow and/or large individuals move back into the oil-impacted areas, the cavity volume / stomatopod size ratio is decreasing toward pre-spill values and that the incidence of wounds is no longer significantly different from levels recorded from unaffected populations (Caldwell and Steger, in prep.).

Given that G. bredini and G. oerstedii appear to compete directly for cavities and that the possession of an adequate cavity has such a strong effect on fitness, it is perhaps surprising that these two species continue to show such broad overlap in their habitat use at Galeta. One possibility that could explain the persistence of both species is that one or both species are particularly successful in different habitats. Recruits from those other areas could maintain populations at Galeta regardless of any local competitive advantage that one species might enjoy. From our experience, this seems unlikely, since the distributions of breeding G. bredini and G. oerstedii are similar along the central Atlantic coast of Panama. However, as discussed above, there are locations in the Caribbean where these two species exhibit more habitat separation. Almost nothing is known about the dispersal of stomatopod larvae in the Caribbean. We therefore cannot rule out the possibility of gonodactylid larvae arriving at Galeta from distant populations. I feel, however, that a more likely explanation can be found in factors affecting competition for cavities, as well as in the different reproductive strategies discussed above.

Priority Effects

The rock or coral cavities occupied by Gonodactylus offer a strong positional advantage to the resident. The cavity is usually modified so that it has only one entrance. That entrance is only slightly larger than the diameter of the occupant. The resident modifies the cavity by wedging or pounding small pieces of shell and coral into openings or by chipping away material to enlarge the entrance. An intruder attempting to evict a resident must enter through this restrictive opening. Animals much larger than the resident are excluded by the size of the entrance. Animals much smaller in diameter than the entrance usually would be defeated by the resident and in fact generally avoid cavities with large entrances (Caldwell, unpublished).

If contestants for a cavity are equally matched, the resident can use its armored telson to block the entrance. It also has an unobstructed opportunity to strike an intruder attempting entry. It is much more difficult for an intruder to cleanly hit a resident inside a cavity. Also, during a fight for a cavity, the intruder is in the open, creating a disturbance that is likely to attract predators and therefore cannot afford to spend much time contesting a cavity (Caldwell, 1979, 1982, 1985). These asymmetries translate into a clear advantage for the resident. In size and sex-matched contests, the residents usually retain their cavities 70% of the time (Steger and Caldwell, 1983; Caldwell, 1986 a). On the other hand, when intruders gain entry into a cavity, there is about a 50% chance that the resident will be evicted (Caldwell, 1986 a).

Fighting Ability

I can find no significant difference in the ability of G. bredini and G. oerstedii to aggressively compete for cavities. When 22 resident G. bredini greater than 40 mm in total length defended their cavities against size and sex-matched G. oerstedii intruders, 7 were evicted. When the situation was reversed (but using different animals), 9 of 22 G. oerstedii residents were evicted by G. bredini intruders ($G = .38$, $P > .1$). These rates of eviction are similar to those obtained for intraspecific contests for cavities between G. bredini (Caldwell, 1986 a). Even if the fighting abilities of similar sized G. bredini and G. oerstedii were slightly different, the positional advantage conferred to the resident would probably greatly weaken this effect. As a case in point, when I matched G. bredini residents against intruders 15% larger in total length (same sex and species), half successfully defended their cavities (Caldwell, 1986 a).

Effects of Physical Factors on Cavity Availability

But there are also other biological and physical factors that make it unlikely that one species of *Gonodactylus* could exclude the other from the intertidal habitat at Galeta. First, there are seasonal fluctuations in exposures of the reef flat at Galeta. During the dry season from December through April, strong onshore winds prevent prolonged exposures. However, beginning in late April, these winds subside and several consecutive days of severe daytime exposure are possible during the extreme spring tides that occur during this time of the year. When this happens, the *Thalassia* is burned back to the roots, and the gonodactylids living in the lower to mid-intertidal are forced to abandon their cavities and move to deeper water. After the spring tides, animals gradually re-colonize the recently vacated cavities in the intertidal, but the distribution of cavities has been effectively scrambled (Steger, 1985, Caldwell, unpublished).

Occasional storms can also effect the availability of cavities. Intertidal rubble is frequently buried or moved too high in the intertidal for habitation. At some later date, it may be uncovered or carried back into the lower intertidal, providing vacant cavities that can be occupied by either species. For example, in 1981, I tagged 30 pieces of coral rubble and placed them in a low intertidal *Thalassia* bed at Galeta. They were quickly colonized by gonodactylids, but all the tagged rubble had disappeared by 1983. Following a storm in February, 1987, 8 tagged pieces reappeared at the same location and within days all were colonized by stomatopods. Since there is a strong priority effect, with the first animal to discover and occupy a cavity having a good chance of holding it, physical forces that are continually producing vacant cavities reduce the probability that individuals of one species can monopolize this resource.

Effects of Biological Factors on Cavity Availability

Biological processes can also produce frequent cavity turnovers (Caldwell, 1987). Gonodactylids that die or are taken by predators leave behind empty cavities. Animals that grow too large must leave their cavities and find new ones. Also, when animals are away from their cavities searching for food or mates, their is a chance that another stomatopod will discover their vacant cavity and occupy it. We have found that *G. bredini* make up to 12 or more foraging excursions per day, each lasting several minutes (Caldwell et al., 1987). Also, both reproductive males and females are more likely to be away from their cavities during the full moon when mating occurs (Caldwell, 1986 b). Because of the positional advantage afforded to a stomatopod inside a cavity, it is unlikely that an animal will be able to reclaim its home when it returns if the cavity has been occupied by another gonodactylid.

When either *G. bredini* or *G. oerstedii* mate, the male and female pair in one cavity for up to a week. After the female lays her eggs, the male leaves and searches for a new cavity (Dingle and Caldwell, 1972; Caldwell, 1986 b). Each time a pair forms, one cavity is vacated. If the male enters the female's cavity, his cavity will be vacant. If the female enters the male's cavity, her original cavity will be unoccupied, although the female will retain the male's cavity after he leaves. Since females of both species reproduce at Galeta every one to three months, this produces a frequent turnover of large cavities.

Changes in a stomatopod's fighting ability also can increase its vulnerability to eviction. *Gonodactylus* frequently are wounded in fights for cavities. Berzins and Caldwell (1983) have shown that these injuries can severely reduce a resident's ability to defend its cavity. Also, stomatopods must molt periodically to grow and repair their exoskeleton. At the time of the molt, animals are incapable of striking and possess ineffective body armor. At this time, they are easily evicted by other stomatopods or even may be cannibalized. Steger and Caldwell (1983) estimated that *G. bredini* adults suffer fighting deficits due to molting about 20 % of the time. While countermeasures such as concealing the cavity entrance or bluffing a defense have evolved to reduce the risk of eviction during a molt, they are no completely successful. Eighty percent of the *G. bredini* that molted the previous day were evicted from their cavities when matched against similar sized intruders (Steger and Caldwell, 1983). Even when newly-molted residents were 15% larger than intruders, 50% were evicted (Caldwell, 1986 a). These frequent and hidden deficits in fighting ability result in the use of a variety of probing behaviors by intruders to determine a resident's strength. The net result is that animals are frequently displaced from their cavities, sometimes by animals that normally would be of lesser fighting ability.

Because of turnovers in cavities due to physical forces, growth, frequent absences of owners, and evictions due to temporary losses of fighting ability, gonodactylids can't hold cavities indefinitely. If *G. bredini* and *G. oerstedii* have roughly the same search and fighting ability, as they appear to, it is unlikely that competition for cavities will lead to competitive exclusion. This situation is somewhat reminiscent of the lottery hypothesis, based on priority effects, proposed by Sale (1976) to explain species coexistence in reef fishes. However, in this case, resources are continually becoming available throughout the life of an individual. Once in possession of the resource, the animal has a good chance of defending it until unpredictable physical disturbances such as exposures or temporary losses of fighting ability cause it to lose the cavity.

CONCLUSIONS

I have tried here to demonstrate that several species of intertidal stomatopod frequently are found in one habitat. While differences in size, functional morphology and micro-habitat use may contribute to their coexistence, behavioral mechanisms often play a major role mediating species interactions.

In competition for cavities, which are often the limiting resource most affecting stomatopod populations, fighting ability may structure habitat use. This was the case for *P. ciliata* and the gonodactylids with which it interacted. Similarly, *H. glyptocercus* may be able to exclude other small gonodactylids from its preferred cavity type by adopting superior cavity eviction tactics (but also due in part to its heavier telson armor) (Dingle et al, 1973; Caldwell and Dingle, 1978).

On the other hand, the priority effects associated with cavity occupation may be so strong that minor species differences are unimportant in determining which species is able to evict the other from a cavity. In this case, both species may continue to coexist despite intense competition for cavities. In such cases, it is not important what species an opponent is, but rather what is its relative fighting strength. One final study perhaps can best summarize this point. *G. zacae* and *G. bahiahondensis* are similar-sized stomatopods that occupy the same types of cavities in the same habitat on the Pacific coast of Panama. I have shown that a *G. zacae* can recognize specific individuals of its own species by odor. It uses information gained during previous encounters to modify its behavior in subsequent interactions with the same animal. Surprisingly, G. zacae can also recognize individual G. bahiahondensis. Yet it does not appear to recognize, or at least use, information about the species of its zzponent. What appears to be important to a stomatopod are individual characteristics, not species characteristics of an opponent (Caldwell, 1982).

REFERENCES

Banner, A. H., 1974, Kaneohe Bay, Hawaii: urban pollution and a coral reef ecosystem. in: "Proceedings of the Second International Coral Reef Symposium 2", Great Barrier Committee, Brisbane, Australia.

Berzins, I. K. and R. L. Caldwell, 1983, The effect of injury on the agonistic behavior of the stomatopod, Gonodactylus bredini (Manning), Mar. Behav. Physiol., 10:83-96.

Caldwell, R. L., 1979, Cavity occupation and defensive behaviour in the stomatopod Gonodactylus festae: Evidence for chemically mediated individual recognition, Anim. Behav., 27:194-201.

Caldwell, R. L., 1982, Interspecific chemically mediated recognition in two competing stomatopods, Mar. Behav. Physiol., 8:189-97.

Caldwell, R. L., 1985, A test of individual recognition in the stomatopod Gonodactylus festae, Anim. Behav., 33:101-106.

Caldwell, R. L., 1986 a, The deceptive use of reputation by stomatopods, in: "Deception: Perspectives on Human and Non-Human Deceit". R. W. Mitchell and N. S. Thompson, eds., State University of New York Press, N. Y.

Caldwell, R. L., 1986 b., Withholding information on sexual condition as a competitive mechanism, in: "Behavior as a Factor in the Dynamics of Animal Populations", L.C. Drickamer, ed., Privat Publisher, Toulouse.

Caldwell, R. L., 1987, Assessment strategies in stomatopods, Bull. Mar. Sci., (in press).

Caldwell, R. L. and H. Dingle, 1975, Ecology and evolution of agonistic behavior in stomatopods, Naturwissenshaften,62:214-22.

Caldwell, R. L. and H. Dingle, 1976, Stomatopods, Scientific American, January: 80-89.

Caldwell, R. L. and H. Dingle, 1977, Variation in agonistic behavior between populations of the stomatopod, Haptosquilla glyptocercus, Evolution, 31:221-224.

Caldwell, R. L., G. Roderick, and S. Shuster, 1987, Studies of predation by Gonodactylus bredini, Bollettino de Zool., (in press).

Diamond, J. and T. Case, eds., 1986, "Community Ecology", Harper and Row Publishers, New York.

Dingle, H. and R. L. Caldwell, 1972, Reproductive and maternal behavior of the mantis shrimp Gonodactylus bredini Manning (Crustacea: Stomatopoda), Biol. Bull., 142:417-426.

Dingle, H. and R. L. Caldwell, 1975, Distribution, abundance, and interspecific agonistic behavior of two mudflat stomatopods, Oecologia, 20:167-178.

Dingle, H. and R. L. Caldwell, 1978, Ecology and morphology of feeding and agonistic behaviour in mudflat stomatopods (Squillidae), Biol. Bull., 155:134-149.

Dingle, H., R. L. Caldwell, and R. B. Manning, 1977, Stomatopods of Phuket Island, Thailand, Phuket marine biological Center Research Bulletin, 20:1-20.

Dingle, H., R. C. Highsmith, K. E. Evans, and R. L. Caldwell, 1973, Interspecific aggressive behavior in tropical reef stomatopods and its possible ecological significance, Oecologia, 13:55-66.

Edmonson, C. H., 1921, Stomatopoda in the Bernice P. Bishop Museum, Bernice

P. Bishop Museum Occas. Papers, 7(13):281-302.
Hatziolos, M. E., 1979, Ecological correlates of aggression and courtship in the stomatopod Pseudosquilla ciliata, Ph.D. Dissertation, University of California, Berkeley (unpubl.).
Kinzie, R. A. III, 1968, The ecology of the replacement of Pseudosquilla ciliata (Fabricius) by Gonodactylus falcatus (Forskal) (Crustacea; Stomatopoda) recently introduced into the Hawaiian Islands, Pac. Sci., 22:464-475.
Kinzie, R. A. III, 1984, aloha also means goodbye: A cryptogenic stomatopod in Hawaii, Pac. Sci., 38:298-311.
Manning, R. B., 1969, "Stomatopod Crustacea of the Western Atlantic", University of Miami Press, Coral Gables.
Manning, R. B. and M. L. Reaka, 1981, Gonodactylus aloha, a new stomatopod crustacean from the Hawaiian Islands, J. Crust. Biol., 1:190-200.
Montgomery, E. L. and R. L. Caldwell, 1984, Aggressive brood defense by females in the stomatopod Gonodactylus bredini, Behav. Ecol. Sociobiol., 14:247-251.
Reaka, M. L. and R. B. Manning, 1980, The distributional ecology and zoogeographical relationships of stomatopod crustacea from pacific Costa Rica, Smithsonian Contributions to mar. Sci., 7:1-29.
Sale, P. F., 1977, Maintenance of high diversity in coral reef fish communities, Amer. Nat., 111:337-359.
Steger. R., 1985, The behavioral biology of a Panamanian population of the stomatopod, Gonodactylus bredini (Manning), Ph.D. Dissertation, University of California, Berkeley (unpubl.).
Steger, R. 1987, Effects of spatial refuges and postlarval recruitment on the abundance of gonodactylid stomatopods, a guild of mobile prey, Ecology, (in press).
Steger, R. and R. L. Caldwell, 1983, Intraspecific deception by bluffing: A defense strategy of newly molted stomatopods (Arthropoda: Crustacea), Science, 221:558-560.
Townsley, S. J., 1953, Adult and larval stomatopod crustaceans occuring in Hawaiian waters, Pac. Sci., 7:399-437.

Interpreting Differences in the Reproductive Behaviour of Fiddler Crabs (Genus *Uca*)

Michael Salmon and Naida Zucker (*)

University of Illinois at Urbana-Champaign, U.S.A.
(*) New Mexico State University, U.S.A.

Dept. of Biology
New Mexico State University
Las Cruces, NM 88003, USA

INTRODUCTION

A basic characteristic of organisms is the tendency to diversify over time, known as branching evolution or cladogenesis. Particularly dramatic examples of rapid speciation depend upon two categories of events. First, there must be innovation of some new characteristic or group of characters which provide its possessors with the potential to enter a new "adaptive zone" (Simpson, 1953). Secondly, there must exist an opportunity to occupy that zone, perhaps as a result of the absence of important competitors. When both conditions are congruous in space and time, adaptive radiation inevitably follows as near-by niches are subdivided by species which share the invention but diversify appropriately in structure, physiology and behaviour. Examples of such phenomena are well known at the level of species (Darwin's Finches; Hawaiian *Drosophila* and Honeycreepers) and higher taxa (evolution of birds; replacement of dinosaurs by mammals).

The intertidal zone within tropical and warm temperate habitats, protected from strong wave action (e.g., tidal creeks within estuaries; mangroves), is organically rich in interstitial benthic algae, bacteria, ciliates and nematodes. The ocypodids, generally, and the fiddler crabs (Genus *Uca*), specifically, may owe their success to the "invention" of an efficient trophic apparatus for harvesting these energy sources (Miller, 1961). There are well over 80 species of fiddlers, especially abundant in tropical and subtropical bays and estuaries throughout the world though a few occupy supra-tidal areas and one, a non-tidal inland environment. Intertidal species remain within sealed burrows during high tide but emerge to feed as the tide recedes.

All species are sexually dimorphic. Females possess two small claws, both of which are used in feeding. Males feed with one small claw. The other is greatly enlarged and is used primarily in defense, threat and courtship. The primary courtship signal employed by males is a rhythmic raising and lowering movement of the enlarged claw ("waving" or "beckoning") which may also served to deter other males from intruding within territories and/or personal space. In addition to these visual signals, males of some species produce acoustic signals which may also function in courtship and social spacing. The production of sound as a "calling" signal is characteristic of those species which exhibit sexual activity during nocturnal, as well as diurnal, low tides.

The sheer abundance of fiddler crabs, their behavioural complexity, their species diversity and their contrasts in ecology and behaviour have made them ideal subjects for comparative ecological and behavioral studies. But understanding how such contrasts, which are beginning to reveal major patterns, relate to the phylogeny of the group has thus far been difficult, largely because we know so little about species affinities or past evolutionary trends. Morphologically, fiddler crabs fall into two general groups: those with narrow "fronts"(the spaces between the eyestalks) and those in which the front is relatively broad. Bott (1954) divided the American species into two subgenera, Uca and Minuca, respectively, on the basis of these frontal width differences. Crane (1957), after completing a world-wide study of the group, noted contrasts in ecology, courtship, mating location and zoogeographic distribution between the broad and narrow fronts. In a more detailed contribution (Crane, 1975), she further subdivided the fiddlers into five narrow front and four broad front subgenera (Fig. 1). According to this scheme, each subgeneric grouping could be viewed as reflecting varying degrees of advance from "primitive" (most like their presumed totally marine ancestor) to "advanced" evolutionary directions. Central to Crane's proposal was the idea that behavioural, structural and ecological characters within subgenera were concordant in revealing these trends.

The primitive species according to Crane were narrow-fronted, essentially lower intertidal inhabitants of the Indo-West Pacific, the presumed center from which the group arose and radiated to the rest of the world. These species were poorly adapted to prolonged exposure at low tide. Male waving display was briefly exhibited during any one low tide and consisted of a simple (vertical lifting and lowering) waving gesture. Copulations occurred exclusively on the surface, generally at burrows maintained and defended by females.

The "apex" of evolution in the group was presumably shown by those broad-fronted species (subgenus Celuca) in which burrow preferences were in the upper intertidal zone. Such species showed conspicuous male waving displays containing both vertical and lateral components. The most advanced Celuca also exhibited special changes in coloration during periods of courtship and a varied repertoire of male agonistic and courtship signals in the visual, acoustical and tactile domains. Male courtship persisted for prolonged periods, often during most of a single low tide. Mating rarely took place on the surface. Rather, females were attracted, presumably by the vigor and conspicuousness of male display, into the male's burrow. The pair then copulated underground. Finally, Crane proposed that the broad fronts, which were most speciose in the Central American tropics, arose from more advanced narrow front subgenera which dispersed from the Indo-West "center" during the Eocene or late Oligocene to the New World, either as adults by crossing the Bering land bridge (most likely) or as larvae by island hopping across the Pacific basin. She accounted for the presence of broad front species in the Indo-West Pacific by hypothesizing these must have subsequently completed an opposite migration (from the New to the Old World) at a later (unspecified) date (Fig. 1).

Our purpose in this paper is two-fold. First, we critically evaluate Crane's ideas with regard to the zoogeographic origins and evolutionary directions shown by the fiddler crabs. In doing so, we come to very different conclusions based upon what we perceive as more likely alternative hypotheses, given the biology of the animal and historical events (continental drift). Secondly, we review recent studies upon particular broad front species which suggest they overlap consideraly in their behaviour and mating system organization with the narrow fronts. We interprete this situation as suggesting the two groups should be viewed as independent evolutionary "experiments" rather than a "primitive" to "advanced" continuum of species. In our opinion, this view more accurately describes the status of presently extant taxa and is also consistent with the likely origin and early evolution of the entire intertidal ocypodid group.

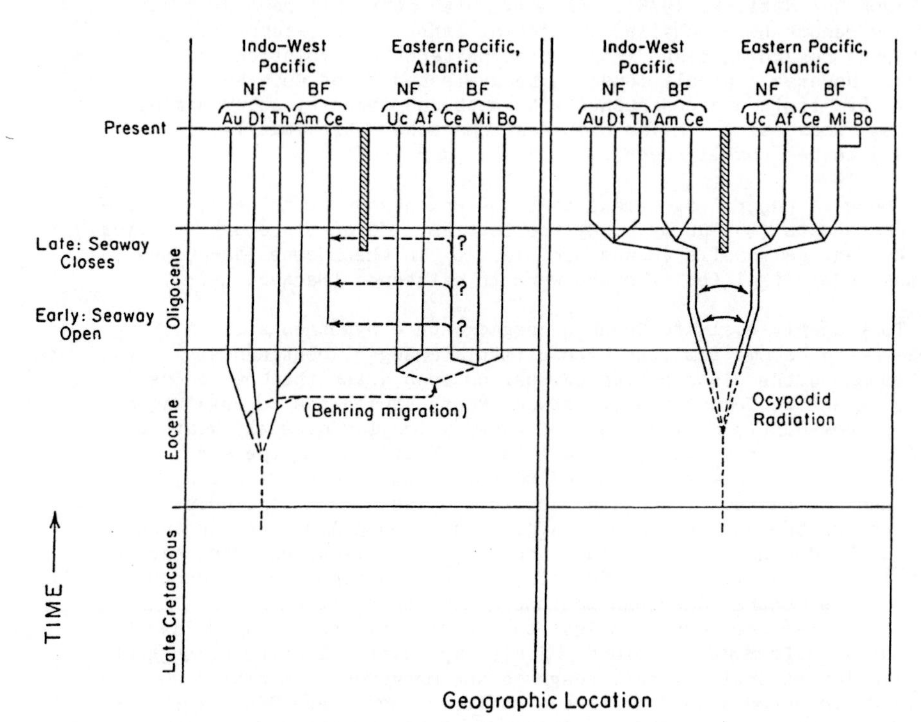

Fig. 1. Phylogeny of fiddler crabs according to Crane's (1975; left) and our (Rt) hypotheses. Crane proposed that the Indo-West Pacific narrow fronts (NF) preceded in origin and gave rise to the broad- (BF) and narrow-fronted subgenera of the New World, probably from migratory Thalassuca (or other) stock during the Eocene-early Oligocene. One BF subgenus (Celuca) then migrated westward at a later (unspecified) date. We postulate an earlier divergence between the broad and narrow fronts and their continuous distribution (double pointed arrows) as a uniform fauna of the Tethyan sea until the late Oligocene. With the closure of the Tethys (vertical hatched bar), the NF and BF subgenèra in each geographic locale had separate radiations. The Boboruca consisting of one species (U. thayeri), and the Minuca, consisting of several species, are probably closely related (Albrecht and von Hagen, 1981). The two phylogenies yield different predictions, discussed in the text.

Abbreviations: AF, Afruca; Am, Amphiuca; Au, Australuca; Bo, Boboruca; Ce, Celuca; Dt, Deltuca; Mi, Minuca; Th, Thalassuca; Uc, Uca.

ZOOGEOGRAPHY AND THE EVOLUTION OF FIDDLER CRABS

Background: Drift and Climate

Discussions of the paleo-zoogeography and origins of fiddlers are inherently dissatisfying because of the virtual lack of any useful fossil record (Crane, 1975). The very fragility of the fiddler crab exoskeleton, coupled with their small size and occupancy of beaches subject to wave action, may have contributed to the paucity of material when compared to larger, more robust and strictly marine crabs (i.e., the Cancrids; Nations, 1975). Clearly, with no fossil record, direct evidence cannot be marshalled to distinguish between alternative hypotheses regarding past distribution patterns or evolutionary trends. However, speculations can be evaluated based upon their simplicity, logic and congruence with patterns shown by other marine organisms possessing similar dispersal characteristics but leaving behind a better fossil record.

There is general agreement that the Brachyrhyncha arose during the late Cretaceous (65-100 my). By the Tertiary, 40 of 51 recent families were evident as fossils (Glaessner, 1969). In the Eocene (40-60 mya), all the modern families of crabs were in existence (Warner, 1977).

To properly evaluate Crane's zoogeographic hypotheses it is necessary to become familiar with climatic changes, continental positions and the distribution of shallow seas after the Cretaceous period, when the Tethys Seaway between North Africa and Eurasia was a vast, predominantly shallow body of water which led directly west into the Atlantic basin (Fig. 2). The Atlantic basin during the early Cretaceous was narrow and bordered to the north by North America, Greenland and Eurasia (Laurasia) and to the south by a continuous shoreline of the yet unseparated African and South American continents (part of Gondawana). By the late Cretaceous, shallow seas were not only extensive between Africa and Eurasia, but also around most of North America, the Central American peninsula and northwestern South America. These seas were tropical to subtropical and characterized by relatively uniform marine biota. For example, fossil evidence suggests that similar extensive coral, seagrass and mangrove communities were widely distributed within all these shallows (McCoy and Heck, 1976; Fig. 2). Furthermore the existence of certain fossil gastropod, bivalve (such as giant clams) and crab (tentatively identified as _Scylla_ and _Macrophthalmus_) genera in the New World, which are now confined to the Indo-West Pacific, speak for a broader distribution of many marine species in the past than at the present time. This was a result of past climatic shifts and alterations of sea level, more severe in the Atlantic than the Indo-West Pacific. Thus, Tertiary extinctions were probably more common in the New World and many taxa with a circumtropical distribution in the Early Tertiary are now limited to the Indo-West Pacific (Vermeij, 1980).

In the late Oligocene (30 mya), a barrier subdivided the Tethys; it was the land corridor between Arabia and Asia (Fig. 2; Hallam, 1981). In theory, interchange of tropical, short lived dispersal stages between the two halves of the Tethyan tropical biota, as Crane proposed, might have occurred by oceanic transport at southern (around southern Africa) or northern (from eastern Asia across the Pacific, to western North America) latitudes. But southerly transport is considered unlikely as with the separation of South America from Antarctica and the establishment of the circum-Antarctic Ocean current during the upper Eocene, there was a correlated significant oceanic cooling in that region (Kennett, 1977). Furthermore such a current would, as it does today, circulate to the East across the vast Pacific basin. Only larval stages capable of prolonged pelagic existence in colder waters could have survived such a trip.

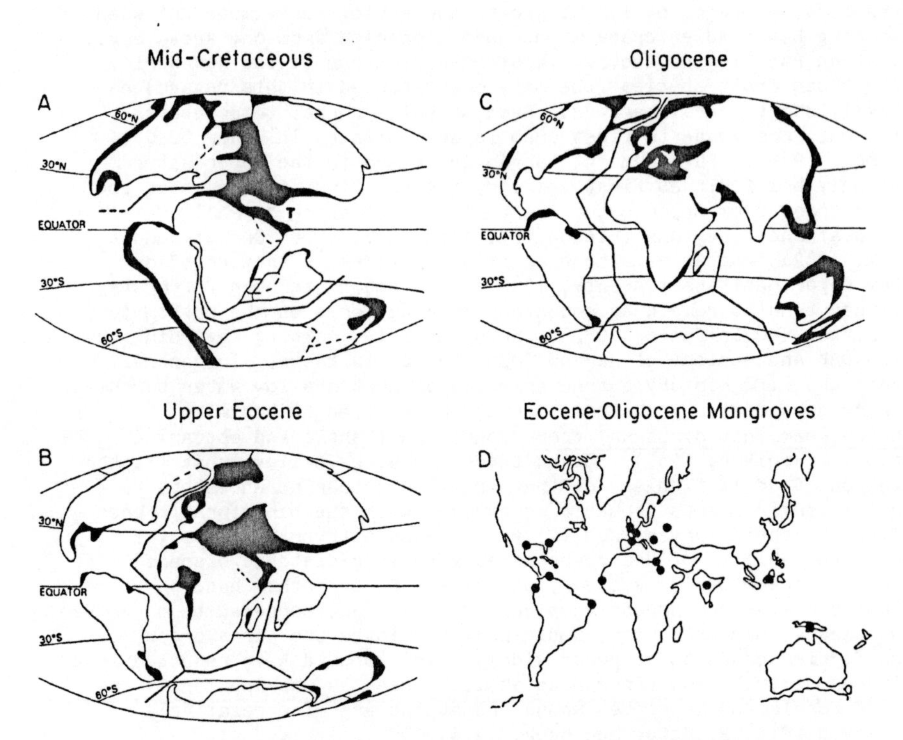

Fig. 2. A-C, continental positions and distribution of shallow seas (darkened) between the Cretaceous and the Oligocene. The Tethys seaway (T) was bissected during the late Oligocene. Thick lines between continents indicate spreading axes and fracture zones (Modified from Cox, Healey and Moore [1973], Smith and Briden [1977], and Hallam [1981]). D, distribution of fossil mangroves, Eocene-Miocene (From McCoy and Heck [1976] based upon several authors).

Transport of propagules at northern latitudes - that is from eastern Asia, across the northern Pacific to North America - also seems unlikely for the same reasons: the great distances involved, the virtual absence of shallow seas (and their biota) along the entire eastern coast of Asia during the late Eocene, and cooler water temperatures. By the Miocene, (23-5 mya) when shallow seas were present along the shores of eastern Asia, further oceanic cooling constituted an effective barrier between tropical shallow water invertebrates of the eastern and western Pacific. The fossil record provides the most convincing evidence that little interchange occurred between tropical or sub-tropical biota of both regions. The closing of Tethys seaway is correlated with abrupt divergence between marine faunas to the east and west of the barrier, probably as a consequence of isolation and, with climatic cooling, of latitudinal zonation (Hallam,

1981 and references therein). These observations are consistent with the hypothesis that a widespread, relatively uniform, tropical Tethyan marine biota was effectively bissected by continental drift some 30 million years ago.

It may, of course, be dangerous to assume that patterns of larval biology today necessarily were the same as those much earlier in time. Nevertheless, patterns of larval growth and ecology are important when considering how fiddler crabs might have dispersed from one area (e.g., the Western Pacific) to another (Western Africa and to the New World). All the broad front species (the only ones studied in this regard) have relatively short larval periods. Even at below normal temperatures, development from zoeae to first crab is accomplished in under 60 days (Vernberg, 1984). Furthermore, larvae are found in the upper water column only briefly: as first zoeae during the first few hours of ebb tide transport from upper estuarine to lower estuarine or shallow continental shelf regions (Pinschmidt, 1963; Goy, 1976; Christy and Stancyk, 1982); and as megalopae on flooding tides as they reinvade shallow water habitats (Christy, 1982a). As second through fifth stage zoeae they tend to remain as sedentary feeding forms on the bottom (Goy, 1976; Dittel and Epifanio, 1982). Unfortunately, nothing regarding the development and ecology of narrow front larvae is known. In general, however, the life history characteristics of most shallow water benthic estuarine decapods suggest their larvae are adapted to maximize <u>retention near</u>, not <u>dispersal from</u>, continental shelf and estuarine locations (Carriker, 1967). Given these facts, we believe that (1) the adaptation of adult fiddlers or their shallow estuarine ancestors to intertidal environments probably co-evolved with the behavioral trends exhibited by their larvae, and (2) given these attributes of larval ecology, the ability of *Uca* zoeae to survive long-distance dispersal across open ocean basins was very unlikely. On the other hand, continuous Cretaceous-Eocene tropical shallow seas, bordered by mangrove communities within estuaries, would have permitted the ancestors of modern fiddler crabs to disperse widely. This should have resulted in a continuous distribution throughout shallow seas, as was apparently the case for corals and mangroves whose propagules are also relatively short-lived (Fig. 2; McCoy and Heck, 1976).

Zoogeography of *Uca*: Present Patterns and Alternative Interpretations

The vast majority of fiddlers are distributed between 35° N and S latitude, with most species concentrated in the Indo-West Pacific, the West coast of Central and South America, and the western Atlantic tropical to subtropical latitudes. Species abundance patterns worldwide are best interpreted with caution for two reasons: (1) Within the past 20 years, thorough surveys have led to descriptions of many new narrow and broad fronted species both in the Indo-Pacific (George and Jones, 1982) and in the Americas (Salmon and Atsaides, 1968; Novak and Salmon, 1974; Thurman, 1981). (2) Present day distributions are not necessarily similar to those of the past (e.g., Carson, 1970). On the other hand, a survey at the level of Crane's subgenera is illuminating, especially in relation to past tectonic and climatic events, and leads to the following generalizations.

1) Two distinct geographical clusters of broad and narrow front species are evident; those of Eastern Africa and the Indo-West Pacific, and those of the New World and Western Africa (Fig. 1).

2) Each geographical cluster contains unique broad- and narrow-fronted subgeneric representatives. An exception is the *Celuca*, found in both areas. However, the Indo-Pacific *Celuca*, a complex of

apparently closely related species (the lactea "complex"), show a number of morphological characters which distinguish them from the Celuca of the New World (Crane, 1975).

3) The narrow fronts of the Indo-Pacific (Australuca, Deltuca, Thalassuca) are distinct in morphology from the narrow fronts found elsewhere. The Afruca of the eastern Atlantic and the Uca of Central America share morphological (Crane, 1975) and biochemical genetic (Albrecht and von Hagen, 1981) characters in common which suggest they are probably more closely related to one another than to the Indo-West Pacific narrow fronts.

Thus, the pattern exhibited is consistent with what one would expect if a widespread fiddler crab fauna, containing then contemporaneous broad and narrow front ancestral "types," was subdivided with the closing of the Tethys into separate species groups. These could then independently evolve into the morphologically parallel forms characteristic of each region.

This hypothesis was considered by Crane (1975), then rejected on the basis of three criteria. First, dispersal of fiddlers across the Atlantic before the closing of the Tethys was impossible because of the distances involved. Secondly, there were no convenient land bridges across the northern Atlantic over which littoral animals could migrate. Thirdly, if significant migration had occurred, more than two genera of ocypodid crabs (Uca, Ocypode) would be found in the Americas as the family consists of at least a dozen, most of which occur in the Indo-West Pacific.

We disagree with these arguments for the following reasons. First, reconstructions of continental positions between the late Cretaceous and Eocene (Fig. 2) suggest that the Pacific Ocean basin was always considerably larger than the Atlantic. Indeed, the continuous, shallow tropical sea between Eurasia and western Africa was only separated by relatively great distances from eastern South America and the Caribbean toward the end of the Eocene. If anything, these observations indicate that larval dispersal between habitats in shallow seas bordering Africa, South America and the Caribbean was far more likely than any eastward displacement. The second argument, based upon the existence of land bridges, is fallacious on two counts. First, land connections between Northern Europe and North America did exist through the late Cretaceous (Hallam, 1981), about the time when the modern Brachyura were evolving. Secondly, it is unlikely that dispersal necessitated any land connection as it is accomplished by pelagic larval, not adult, stages. The third objection, that there is greater generic diversity of ocypodid forms in the Indo-West Pacific than in the New World tropics, is based upon two assumptions which may not be true: (1) that past centers of origin can be reliably identified by present centers of diversity because (2) radiation patterns consist only of a passive spreading of species from their "centers of origin." But what little fossil evidence is available suggests the contrary; that some Brachyura now confined to the Indo-Pacific were once wide-spread. Much the same is true of other marine invertebrate fauna (Ekman, 1953; Vermeij, 1980). The present distribution of species is therefore most likely a consequence of localized ecological factors of the past, such as climatic shifts, local extinctions, differences in habitat complexity, and species-area relationships (e.g., Valentine, 1971; Croizat, Nelson and Rosen, 1974; McCoy and Heck, 1976; Heck and McCoy, 1978; Conner and McCoy, 1979; Abele, 1974; 1982). Of most importance, we see no need to postulate long distance dispersal after the closing of the Tethys by a group which may have already been widely distributed, and whose larvae were probably not capable of surviving such a migration.

Alternative hypotheses are useful to the extent that they generate predictions which can be tested. Crane's phylogeny and ours (Fig. 1) differ most fundamentally in presumed divergence times between, and relationships within, the broad and narrow fronted species. These differences are amenable to analysis by biochemical genetic techniques (Salmon, Ferris, Johnston, Hyatt and Whitt, 1979). For example, Crane hypothesized that the oldest (most "primitive") subgenera were the narrow fronts of the Indo-West Pacific, one or two of which gave rise to the broad and narrow fronts of the New World. If true, the former should show earlier divergence times than the latter. Our hypothesis predicts a simultaneous divergence between the broad and narrow fronts (during their occupancy of habitats throughout the Tethys Sea), followed by their further differentiation into unique subgenera once separated by closure of the seaway. Thus, (1) no one group of narrow fronts should show closer affinities to the broad fronts than any other, and (2) divergence times within both narrow and broad front subgenera of each geographic area may reflect essentially similar events. A key measurement centers around the *Celuca*, found in both the Indo-West Pacific and New World. According to Crane's phylogeny, all species of *Celuca* should be closely related; according to ours, the *Celuca* of the two regions should exhibit genetic distances comparable to those between any other pairs of broad front subgenera, regardless of location.

Evolutionary Directions

Crane (1957, 1975) envisioned the fiddler crabs as a monophyletic group reflecting a uniform evolutionary "direction:" those species most like their presumed marine ancestor were the lower intertidal narrow fronts which never achieved the consistent degree of behavioural, structural or apparently ecological (i.e., upper intertidal) advance of the broad fronts. Implicit in these ideas are several assumptions which we question.

(1) Some fiddler crab species are "primitive." There can be little doubt that the Crustacea, as a group, are predominantly marine organisms and that those species which invaded land did so secondarily. But the marine ancestor of the Ocypodidae (if indeed, the entire family is derived from a single lineage) is unknown, as are its characteristics. Primitive characters are defined as those which exhibit the ancestral, or original, condition (Minkoff, 1983), and are verified by comparison to known ancestors available (generally) as fossils from an historical sequence. No such data exist for any ocypodid crabs.

It is obvious that within the group, some species are better adapted to withstanding exposure than others. Differences of this kind are expected when species are distributed across intertidal gradients; they indicate only differences, not evolutionary "advance." Furthermore, all *Uca* are amphibious and all have a well developed "lung." All show trophic specializations; males show striking allometry of the major chela; and in all, the placement of the legs, alteration of their joints to support the body, and general reduction in body weight suggest animals that are highly derived (i.e., structurally changed and therefore, advanced) from any hypothetical marine ancestor. Put another way, the differences between species of fiddler crabs seem relatively insignificant compared to those which must have accompanied the adaptive shift from a totally marine to a semi-terrestrial existence.

(2) Evolutionary trends in Fiddler Crabs have been unidirectional. Crane (1957, 1975) perceived the fiddler crabs as exhibiting a consistent evolutionary trend of increased "terrestriality." By this, she meant that the most primitive species occupied lower intertidal, permanently damp muddy substrates while those most advanced (i.e., the *Minuca* and *Celuca*) were found in exposed areas where substrates were less muddy, more solid, and subject to periodic (tidal) or prolonged (seasonal) drying. Given that fiddlers are derived from a marine ancestor (certainly reasonable), she assumed that the present wider distribution of species from the lower intertidal to supratidal habitats must have been accomplished orthogenically.

While we agree this is one possibility, we disagree that it is the only one. First of all, decisions regarding trends depend (once again) upon knowledge of the ancestral condition and a fossil record of the intermediates. There are no such data available. Secondly, Crane assumes that the fiddler crabs are direct descendents of a totally marine form. The possibility exists, however, that they evolved from a now extinct intermediate, a generalist adapted to a wide intertidal distribution, and that through time, speciation involved simultaneous evolutionary specialization in different, finer grained, niche directions (toward the lower intertidal zone for some; toward drier, higher intertidal areas for others) within related lineages. This is certainly the pattern within the ocypodid genera as a whole; it is also characteristic of other taxa (Gould and Eldredge, 1977). Thirdly, such trends are apparent for both the broad and narrow front species, suggesting a parallel evolution with both groups.

Within the *Minuca*, for example, there are species which are most common in high salinity, lower intertidal, muddy salt marshes (e.g. *U. pugnax*) and others which, while lower intertidal, are most abundant in brackish areas (*U. minax*) or on lower intertidal rocky beaches (*U. panamensis*). This subgenus also contains the world's most specialized and terrestrial fiddler (*U. subcylindrica*), found so far inland that it completes its life cycle independent of marine habitats (Rabalais and Cameron, 1983).

Within the Australian *Deltuca*, a subgeneric group thought "primitive" by Crane because of its narrow, lower intertidal habitat preferences, similar patterns exists. Most species have in common that they live in shaded environments. But within the group, some prefer lower (*U. dussumieri*), middle (*U. pavo*, *U. capricornis*) or upper (*U. flammula*) intertidal habitats (Warburton, 1978; Williams, 1974). The most terrestrial species known, *U. elegans*, is found in exposed salt flats which are only inundated by the spring tides (Jones and George, ms. in prep.). In terms of intertidal "position," it rivals the most common upper intertidal *Celuca* and *Minuca* in its ecology and perhaps, in its behavioural specializations. For example, preliminary data suggests it courts vigorously during the syzygies, as do many of the Central American broad fronts.

(3) Subgeneric Categories Accurately Reflect Evolutionary Relationships. Perhaps the most difficult task facing systematists is the selection of characters which accurately reflect relationships. The problem is particularly acute when similar adaptive specialization are independently acquired by species representing different groups, as many morphological and behavioral traits then arise due to convergence (Futuyma, 1979; Minkoff, 1983). Crane, to her credit, attempted to minimize these possibilities by considering a broad range of

morphological, ecological and behavioural characters. The extent to which these attempts have been successful await more detailed knowledge of the group as a whole (von Hagen, 1976). However, recent discoveries employing electrophoretic (Albrecht and von Hagen 1981) and behavioural (Salmon, 1987) techniques suggest that the groupings may need extensive revision because of unforeseen complications.

A case in point is the subgenus *Boboruca* erected by Crane (1975) for *U. thayeri*. This species, found on either side of the Panamanian Isthmus and throughout the Caribbean, was placed in its own subgenus because its affinities with other species were enigmatic. In frontal width ("moderately wide"), it resembled the *Celuca* and *Minuca* but in major cheliped armature and display, the most "primitive" of the narrow fronts (*Deltuca*). Albrecht and von Hagen compared this species electrophoretically to nine others representing the *Celuca*, *Minuca*, *Afruca* and *Uca*. The data indicated a close affinity between *U. thayeri* and *U. rapax*, a behaviourally and morphologically "conventional" broad fronted *Minuca*. More surprising, *U. rapax* and *U. thayeri* appeared even more closely related to one another than did *U. rapax* to three other *Minuca* broad front species (*U. burgersi*, *U. mordax*, *U. vocator*).

In terms of its ecology and behavior, Salmon (1987) found his Florida population of *U. thayeri* showed many similarities to those reported for the Indo-West Pacific narrow fronts (see below). He interpreted these behavioral similarities as consequences of evolutionary convergence. It is of significance that convergence also affected waving display form and was correlated with significant alterations in the very morphological characters (frontal width, major cheliped armature) considered by Crane as crucial to teasing apart phylogenetic relationships.

EVOLUTIONARY TRENDS IN THE BEHAVIOUR OF FIDDLER CRABS

The historical scenerio we propose postulates an independent evolution of the broad and narrow front species, beginning in the Cretaceous and accelerated by isolation after the Oligocene. If this proposal is correct, then the two groups should show radiation patterns similar to those witnessed repeatedly within species complexes: intricate patterns of divergence and convergence depending upon the selection pressures encountered by species during the process of evolutionary divergence.

Fiddler crab species differ in trophic specializations (Miller, 1961), temperature and humidity tolerance (Edney, 1961; Wilkens and Fingerman, 1965; Macnae, 1968), substrates into which they burrow (Ringold, 1979; Bertness and Miller, 1984), local population size and number of sympatric congeners (Teal, 1958; Montague, 1980) and in social behaviour and organization of their mating systems (Christy and Salmon, 1984). From a broad perspective it is appropriate to ask the following two questions. How do broad and narrow front fiddlers differ as evolutionary "experiments?" Which of these contrasts between them are most likely to account for behavioural divergence?

Though Crane (1975) alludes to a number of differences between the two groups we believe only three might affect behaviour. These are differences (1) in visual perception as a consequence of eyestalk (frontal) width and length, (2) in female reproductive cycles and operational sex ratios, and (3) in clutch size.

In what follows, we compare the broad and narrow fronts in optical structure and reproductive physiology, then speculate upon how these affect behaviour. We then describe the data from field studies. We conclude that biases are there and that they may be responsible for distinct broad and narrow front divergence patterns. However, they are insufficient to prevent at least certain broad front species from exhibiting behavioural convergence with narrow fronts when they exist under similar ecological conditions (Christy and Salmon, 1984; Salmon, 1987). None of this is meant to minimize the importance of profound differences where broad front ecology differs from the narrow fronts. But we know almost nothing about the adaptations shown by narrow fronts of the upper intertidal zone. Until we do, it is impossible to assess the degree to which evolutionary trends have also been parallel in some cases.

Structural and Physiological Contrasts

(1) Frontal width and eyestalk length. Within the genus narrow fronts are characteristic of five subgenera while wide fronts are typical of four. Species with the narrowest fronts (e.g., in the subgenus *Uca*) are also those which often possess the longest eyestalks. These tend to occupy open, flat, and therefore optically "simple" lower intertidal habitats (Crane, 1975). In contrast, the *Minuca*, which possess the widest fronts (e.g., *U. panamensis*, *U. vocator*), tend to occupy visually "complex" habitats, such as rocky beaches or supratidal areas where vegetation is dense, obscuring a clear view of the horizon.

In a recent study, Zeil, Nalbach and Nalbach (1986) compared optical resolution in the horizontal and vertical plane between crab genera (e.g., *Mictyrus*, *Uca*, *Ocypode*), all classified as relatively "narrow-fronted," and grapsid, xanthid and portunid genera, all classified as "broad fronted." All the former were found to possess narrow vertical corneal pseudopupils and an enhanced optical resolving power in the vertical plane, centered at the horizon. All the "broad fronted" genera lacked this specialization. Thus, the relatively narrow fronts and longer eyestalks of the ocypodids were interpreted as adaptations to improve depth perception and size resolution, based upon monocular visual mechanisms. Comparisons *between* fiddler species showed similar trends; but the magnitude of difference was small. Specifically, vertical resolving power was somewhat *more* acute in the narrow than in the broad front species. However, given the relatively minor differences shown, there appears no reason to assume that visual capabilities vary greatly within the fiddlers. All are profoundly visual animals, responding to a variety of optical social displays. If anything, those of the broad fronts are considered more varied and "conspicuous" (Crane, 1957; 1975).

(2) Female reproductive cycles and operational sex ratios. In all fiddlers, females periodically extrude a mass of eggs which adhere to the pleopods. Embryonic development requires about 12-15 days. In the Australian narrow front *U. vocans*, during the breeding season all females are ovigerous (Salmon, 1984), as within hours or a few days after larval release, females extruded a new clutch. They are also continuously receptive and available on the surface while they feed for males to court. In all broad front species thus far studied, a portion (1/2 to 1/3) of the female population incubates clutches underground and is unavailable to males. The remainder of the population feeds on the

surface, accumulating reserves for another clutch (Christy, 1980; Salmon and Hyatt, 1983). But even this fraction of the female population is "unavailable" to males as while females feed (2-4 weeks), they are unreceptive. It is only for a few hours, shortly before ovulation, that the opercula of most broad front females become pliable and females are behaviourally, as well as physiologically, receptive (Christy, 1980; Greenspan, 1982).

Different patterns of receptivity may result in alteration of fiddler crab mating systems because of a skew in the operational sex ratio of broad fronts toward males (Emlen and Oring, 1977). But in all fiddler crabs, the ratio of available males to receptive females is high. Furthermore, there are broad front species (*U. thayeri*; *U. lactea*) which, despite these physiological and temporal differences in female receptivity, show mating systems and courtship behaviour patterns which are virtually identical to *U. vocans* (see below).

(3) Clutch size. In Australian *U. vocans*, clutch sizes are relatively small, averaging about 8000 zoeae upon hatching in females of 1.80-1.90 cm in carapace width (Salmon, 1984). These smaller clutches remain hidden and protected under the abdominal flap. Thus in this species, the clutch is sheltered from high temperature and low humidity within an effective "brooding pouch." It may be for these reasons that females can forage outside the burrow during brooding (Christy and Salmon, 1984).

For all broad front crabs thus far studied, (*U. rapax*; Greenspan, 1980; *U. pugilator*; Salmon and Hyatt, 1982; *U. thayeri*; Salmon, 1987), clutch sizes are considerably larger. For example, female *U. pugilator* of 1.20-1.30 cm carapace width averaged about 10,000 zoeae/clutch. The mass of eggs is so large that it protrudes conspicuously at the margins of the abdomen. These large clutches may require most broad front females brood their eggs primarily within burrows, where egg masses are protected from environmental fluctuations. In most broad front species, females remain sequestered underground for the entire incubation period (Christy and Salmon, 1984).

The issue goes beyond merely enhancing reproductive success by protecting eggs and developing embryos. All fiddler crabs release their larvae at precise phases of the lunar (tidal) cycle (Christy and Stancyk, 1982; Salmon, Seiple and Morgan, 1986). Hatching is completed rapidly (within minutes) and larval release involves the entire clutch. Developmental regimes provided for the eggs must therefore be uniform, an obvious impossibility if any peripheral eggs in the mass are exposed to different incubation temperatures than those within the mass. These considerations led Christy and Salmon (1984) to propose that narrow fronts can incubate their protected clutches over a wider range of environments than broad fronts.

The behaviour of incubating females supports this hypothesis. For example, ovigerous *U. vocans* organize their feeding behaviour as brief 1-2 min "sallies," which are always followed by a return to their burrow and an entry lasting 10-20 sec (Salmon, 1984). Burrow environments provide access to water, cooler temperatures and high humidity (Edney, 1961) where both females and their eggs may obtain a temporary "reprieve." Most broad front females remain within sealed burrows from the time they ovulate until they release their larvae; they do not emerge to feed. The one known exception *U. thayeri*, where both ovigerous and non-ovigerous females surface-feed. The sallies of the former are shorter, while their in-burrow times are longer, than those of non-ovigerous conspecifics (Salmon, 1987).

Differences in clutch size then, appear to be correlated with major shifts in female brooding behaviour and requirements for successful reproduction. These, in turn, probably alter competitive options available for males. In *U. vocans*, brooding females aggressively defend their burrows and the feeding areas around their entrance. Females require nothing from males other than sperm; males, then, must compete with one another to obtain matings using tactics which improve their ability to search for, and encounter, as many females as possible/courtship effort. They do this by visiting females, one by one, during courtship "sallies" which fan out in all directions from a central burrow. They avoid prolonged fights with other males or, if physically smaller, avoid fighting at all. At intervals of a few days, they abandon burrows, wander to new locations, establish another burrow, then court the females locally present. Males have been observed to copulate with two females in rapid succession (within 15 min) though typically, they obtain only 1-2 matings/fortnight. Females apparently select males on the basis of their courtship "persistence" (Salmon, 1984). These characteristics yield resource-free, promiscuous mating systems in *U. vocans* and probably, most narrow fronts.

Because broad front females usually require underground sites for successful reproduction, in many species males may successfully obtain mates by providing these resources (Christy, 1980; 1982b; 1983). The "classic" case is the temperate species, *U. pugilator*, in which females occupy non-defended ("temporary") burrows while feeding. Because local populations are large, food supplies are quickly depleted. Females must therefore forage widely in search of food. Once sufficient reserves for a new clutch are accumulated, females become receptive and "sample" frantically waving males and their breeding burrows as potential breeding sites. Mate choice is independent of male size, age or any associated differences in courtship behaviour. Rather burrow location and physical "stability" are most critical: the burrow must not collapse as tidal waters rise, especially during the first few hours after ovulation when the eggs are not yet securely attached to the pleopods. Mating takes place within the breeding burrow which males seal ("closing") after a female entry. Males remain with their mates until they ovulate (within 24 hrs), then enclose them within a terminus of the burrow, excavate a new terminus, and return to the surface to court. Males are aggressive; those which can acquire and defend breeding burrows in areas where females prefer to breed obtain one or more mates sequentially over several consecutive low tides. Those unable to do so obtain none. Thus mating systems in *U. pugilator*, and many broad front species, are resource-based and polygynous.

Factors Promoting Alternative Reproductive Strategies

While the larger clutches so characteristic of broad front fiddlers may ultimately predispose them toward resource-based mating systems, this is not inexorably so. Under what conditions might broad front females deviate from this pattern, that is, gain in fitness by incubating in their own, rather than in a male's, burrow? The following may be important selective factors altering female behaviour patterns.

First, the process of sampling males involves inherent risks, as females travelling from one male to the next have no burrows of their own and move across unfamiliar areas. Grackle predation upon prospecting females is apparently severe in the broad front *U. beebei* (Christy, pers. comm.). In habitats where predation pressures are high, females might increase their lifetime reproductive success by remaining at their own burrows; males might then compete by attempts to establish burrows near clusters of females (Yamaguchi, 1971).

Secondly, where female burrows are located near reliable supplies of food, foraging can be completed locally. Burrows can then serve as refuges from predators, storage sites for excess food collected during low tide, and brooding sites. The generally large population densities achieved by temperate New World broad fronts, however, usually results in local food depletion as masses of individuals move over the habitat in dense "droves" or "herds" (Montague, 1980; Robertson, et al., 1980). However, this is not always the case in some temperate habitats (Salmon, 1987) or in those occupied by Central American species (Christy, 1987c). It is infrequently the case for both broad and narrow front species of the Indo-Pacific (Yamaguchi, 1971; Frith and Brunenmeister, 1980; Salmon, 1984). The correlation between low population densities, locally available supplies of food near female burrows and resource-free mating systems appears strong (Christy and Salmon, 1984).

Thirdly, in certain (benign) habitats even females with large clutches might successfully complete incubation independently of males. Areas shaded by mangroves or tall grasses might constitute environments where physical extremes in temperature and humidity are minimized. Thus, females might not require the special subsurface habitats provided within male burrows to protect their clutches. *Uca thayeri* are typically found within heavily shaded mangrove habitats. Females show strong burrow fidelity and are courted by males at the burrow entrance. They are also the only species of broad front in which females routinely feed outside their burrows while ovigerous (Salmon, 1987). Another broad front, *U. vocator*, is found in tall grass near brackish streams in Trinidad. Surface copulations are common (von Hagen, 1970) and females are therefore likely to brood on their own. In North Carolina *Spartina* marshes, ovigerous female *U. minax* are abundant within open burrows which, like those of *U. thayeri*, have funnel-shaped entrances believed to enhance the ability of females to defend them. Female breeding burrows are located in shaded, lower intertidal areas. The display burrows of males are characteristically situated on higher terrain, exposed to direct sun. Unfortunately, we do not know if sedentary females are visited by other males, if all females mate within male burrows, then depart to ovulate and incubate elsewhere, or if different females at particular times or locations prefer each "strategy."

In some narrow fronts, such as Japanese populations of *U. vocans* (Nakasone, Akamine and Asato, 1983) and Australian *U. longidigita* (Zann, pers. comm.), females mate in male burrows, then depart to incubate clutches elsewhere. If, as in Australian *U. vocans*, most narrow front females mate by burrows they defend, these differences are also examples of alternative behavioural strategies. Unfortunately, the causes of these variations in female behaviour are unknown.

Empirical Studies: Variation in Broad Front Mating Systems and Behaviour

The general patterns of courtship shown by broad fronts, and initially described by Crane (1957; 1975), upon closer examination, show superficial resemblances which mask unanticipated complexities. First of all, not all males simply wave energetically to attract females to breeding burrows. In some (e.g., *U. deichmanni*), males approach and attempt to carry or maneuver females into their burrows (Zucker, 1983). Similar attempts to maneuver females also occur in *U. rapax*, (Salmon, 1967), *U. pugilator* (Christy 1980), and other species (Crane, 1975), though they do not usually involve direct contact between the sexes. Perhaps most surprising, male waving can occur persistently, yet not necessarily be directed toward females. It may, in fact, be far more important as a threat display directed at male competitors when courtship of females involves local "neighborhoods of dominance" (Christy, 1987a).

(1) Variation within the *Celuca*. In *U. lactea*, an Indo-Pacific *Celuca* studied at Amakusa, Japan, Yamaguchi (1971) reported males waved actively during the breeding season. Yet of 86 copulations witnessed, 83 were surface matings involving males who sallied to near-by female "nests" (burrows) and courted them at the burrow entrance. Unlike some narrow fronts, males at his study site rarely left their burrows to court distant females. Rather, most surface courtships (and copulations) tended to occur between a male and females located in the "neighborhood," i.e., no farther than 35 cm away. More recently, Murai, Goshima and Henmi (in press) examined relationships between ovarian development, mating location, incubation sites and behaviour of females at the same site. They also observed a preponderance of surface matings (78%). Some non-ovigerous females, when displaced from their burrows by males, wandered through the crab colony. A minority (35%) of these had ripe ovaries, were attracted to waving males, followed them into their burrows and mated underground. Underground mating females used male burrows as sites for ovulation and incubation. But most females successfully defended their own burrows where they brooded their clutches. When brooding females were displaced from burrows, they took over empty burrows or dug new ones for themselves.

The same species in Taiwan also copulates predominantly on the surface. In a field study, Lin (1986) marked neighboring males and females. She found that both sexes mated multiply, often within the same morning or afternoon. Males waved vigorously, especially during full or new moon low tides when most copulations occurred. Waving displays were primarily "vertical" when directed toward male competitors but "lateral" as males approached females for surface courtships, though courtship approaches did not always involve waving. The most successful males, in terms of number of copulations achieved, were those which (1) spent the greatest amount of time "stroking" females during surface courtships, and (2) returned for repeated courtships when females were initially unreceptive.

A thorough analysis of the mating system of *U. beebei* in Panama is forthcoming (Christy, 1987b; 1987c). As in *U. lactea*, females often show both surface and underground mating associations. Thirty-two of 44 observed matings occurred underground. Females mating underground were more often attracted to displaying males that constructed pillars behind their breeding burrow entrances then to displaying males which did not. Once males attract females to their burrows by waving, they show a "raised carpus" display which exposes the darker ventral surface of the chela while held vertically; then, males precede females into breeding burrows. Christy suggests that pillars act as icons which continue to direct females toward the breeding burrow entrance in the absence of direct visual stimuli from the (underground) courting male. The shape and orientation of the pillar must be a potent visual stimulus when eyes are specialized to detect vertical images on the horizon (Zeil, et al., 1986).

While females are more often attracted to breeding burrows with pillars, they make final choices between mates based upon the physical characteristics of the male breeding burrow, those which one might expect to be most important to regulation of the brooding environment. In order of importance, these were burrow length, depth and water content (Christy, 1987c). Shortly after mating, females ovulate and males abandon the breeding burrow for the exclusive use of the female.

Twelve observed matings in Christy's study colony were surface copulations between females and their male neighbors. These took place after a male approach, followed by courtship at the female's burrow entrance. Females are continuously receptive. Eight of nine females which could be observed up to 11 days after copulation became ovigerous. All but one was larger than her mate. Further results are forthcoming but it appears that females which surface-mate and breed independently of male assistance are generally of larger body size, and can therefore defend their burrow as incubation sites more effectively than smaller females. Their burrows also are located in sites where food is sufficiently abundant to permit foraging before ovulation.

(2) Variation within the *Minuca*. Considerably less is known about the reproductive behaviour of this broad front group. In *U. rapax*, males establish breeding burrows in close proximity; larger males may be preferred as mates (Greenspan, 1980), and mating takes place within male burrows which are plugged by the male after an attracted (nonovigerous) female has been sequestered below. Some females remained within these burrows for days. When Greenspan later excavated burrows containing attracted females, many were ovigerous. These observations suggest that at least in *U. rapax*, mating systems are also resource-based. Underground mating and brooding within male burrows also appears to be the rule in *U. pugnax* (Greenspan, 1982).

The social behaviour of *U. thayeri* (subgenus *Boboruca*) was studied by Salmon (1987) shortly after its status as a close relative of *U. rapax* (a *Minuca*; Albrecht and von Hagen, 1981) became known. Florida populations are characteristically found at low population densities in the lower intertidal zone, usually obscured by mangrove cover. Salmon studied a large colony where the mangroves had been recently cleared. Here, it was possible for the first time to capture, mark, release and observe both sexes over part of the breeding season. Females became briefly receptive just before ovulating large clutches at six week intervals. In this respect, they resembled other broad front species. But unlike most broad fronts, all females returned to, and defended, permanent breeding burrows. Well over 200 courtships were observed. All took place when males sallied to female burrows. Females showed many similarities with *U. vocans* in their foraging and agonistic behaviour. Many of their threat displays and much of their fighting involved virtually identical acts. In terms of its reproductive physiology, then, this crab was a typical broad front; in terms of its female aggressive behaviour and mating system organization, it showed strong convergence with *U. vocans*.

In summary, then, the behavioural adaptations of broad front fiddlers show patterns which in some instances diverge from, while in others converge with those exhibited by the narrow fronts such as *U. vocans*. The key correlates are ecological variables such as population density, the relative abundance and proximity of food to burrows, and available cover (shade). Differences in breeding requirements of females, especially as a result of clutch size, may predispose broad front species toward underground incubation sites and resource-based mating systems (Christy and Salmon, 1984). However, ecological forces appear capable of reversing the dependence of females upon males to provide these sites. When they do, behaviour and social relationships between the sexes are virtually identical to those observed in lower intertidal narrow fronts.

CONCLUDING REMARKS

In this paper, we attempt to describe more accurately the probable origins, evolution history and adaptive trends shown by the fiddler crabs. In doing so, we come to very different conclusions from those proposed by Crane (1959, 1975). We believe our zoogeographic, evolutionary and behavioural ideas more accurately reflect probable past evolutionary events, radiation patterns and present evolutionary trends in behavioural diversity.

Analyses of the relationships between ecology and behaviour are rapidly advancing, largely due to the clear formulation of hypotheses and their testing by Christy (1982a; 1982b; Christy and Salmon, 1984; Christy, 1987a-c). These hypotheses, for the first time, provide a framework for understanding why males and females in this group behave as they do. Unfortunately, most of the field work continues to revolve around the American broad front species while the narrow fronts of the Indo-West Pacific remain largely unknown. The recent studies on these narrow fronts by behavioural ecologists in Japan are a welcome exception.

Finally, we emphasize that while fiddler crabs are interesting, certainly attractive, and surprisingly complex animals, we do not encourage their study simply because they are "there." For every question in biology, there are organisms ideally suited to obtain the answers (the so-called "August Krogh principle;" Krebs, 1975). The convenient size, abundance and accessibility of fiddler crabs, the ease with which they can be observed, marked and followed, and the precision with which their resource requirements can be measured all contribute to their utility as experimental subjects for field studies. They will, undoubtably, allow us to test many hypotheses of central concern to an understanding of how and why organisms, under an array of environmental conditions, behave as they do.

SUMMARY

The fiddler crabs can be divided into two groups (the broad and narrow fronts) based upon morphology and behaviour. In most broad fronts, males show elaborate courtship displays which attract females to male breeding burrows. Mating, ovulation and the incubation of large clutches by females occurs within the breeding burrow. In most narrow fronts, females defend their own burrows which are used as both shelters and as brooding sites. Mating occurs after brief courtship sallies to females. Females brood their clutches both within, and outside of, their burrows.

Differences between the groups were interpreted by Crane as representing a continuum of evolutionary change from "primitive" (narrow front species) to "complex" (broad fronts). The history of both groups was also seen as reflecting this scenerio as the primitive Indo-West Pacific narrow fronts were envisioned as having given rise to the advanced American broad and narrow fronts after a migration over land bridges during the late Eocene.

We critically evaluate these ideas in the light of more recent paleontological, ethological and ecological studies. We conclude the following:

(1) It is far more likely that the broad and narrow front species (or their immediate ancestors) were pan-Tethyan in distribution during the late Eocene. With the isolation of the Atlantic from the Pacific

during the Oligocene, the narrow and broad front species of each geographic area were isolated from one another. Their evolution since that time has been separate, and to some degree parallel.

(2) Reproductive physiological differences (clutch size) between the two groups may form the basis of constraints predisposing each to differ in their evolutionary directions. In broad fronts, which have large, exposed clutches, females may be predisposed to seek underground incubation sites and to select males on the basis of the "quality" of breeding burrows where these are provided. As a result, mating systems are often resource-based. The smaller clutches of narrow front females enable them to incubate within or near to their burrows. As a result, mating systems of most narrow fronts may be independent of resources (other than sperm) provided by males, and mates are selected on the basis of courtship "techniques."

(3) The broad front species show considerable variation in their breeding behaviour. Under certain ecological conditions, females also incubate clutches without male assistance. As a result, they may show evolutionary convergence with the narrow fronts in their behaviour and in the organization of their mating systems. The circumstances under which broad and narrow front species show convergence appear similar, suggesting both groups opt for common adaptations when their ecology is similar.

(4) We conclude that neither group is more primitive or advanced than the other. Major contrasts at the present time probably reflect a much greater understanding of the diversity shown by broad front species. Some narrow fronts also exist in a broad range of habitats. An analysis of their behavioural diversity in these cases should reveal the nature of their similar and/or unique adaptations and therefore, their basic patterns of evolutionary divergence from the broad fronts.

ACKNOWLEDGEMENTS

We are grateful to Jim Hall and Jeanette Wyneken for their comments on earlier drafts. Barbara Wright typed numerous drafts with patience and accuracy.

LITERATURE CITED

Abele, L. G., 1974, Species diversity of decapod crustaceans in marine habitats. _Ecology_ 55:156-161.

Abele, L., 1982, Biogeography. _In_ "The Biology of Crustacea," Vol. 1, Systematics and the Fossil Record, D. E. Bliss and L. G. Abele, eds., pp. 242-304.

Albrecht, H. and H. O. von Hagen, 1981, Differential weighting of electrophoretic data in crayfish and fiddler crabs. _Comp. Biochem. Physiol._ 70B:393-399.

Bertness, M. D. and T. Miller, 1984, The distribution and dynamics of _Uca pugnax_ (Smith) burrows in a New England salt marsh. _J. Exp. Mar. Biol. Ecol._ 83:211-237.

Carriker, M. R., 1967, Ecology of estuarine benthic invertebrates: A perspective. _In_ "Estuaries," G. Lauff, ed., pp. 442-487. American Association for the Advancement of Science, Washington, D. C.

Carson, H., 1970, Chromosome tracers of the origin of species. _Science_ 168:1414-1418.

Christy, J. D., 1980, The mating system of the sand fiddler crab _Uca pugilator_. Ph.D. thesis, Cornell University, Ithaca, N.Y.

Christy, J. D., 1982a, Adaptive significance of semilunar cycles of larval release in fiddler crabs (Genus Uca): Test of an hypothesis. Biol. Bull. 163:251-263.
Christy, J. D., 1982b, Burrow structure and its use in the sand fiddler crab, Uca pugilator (Bosc). Anim. Behav. 30:487-494.
Christy, J. D., 1983, Female choice in the resource-defense mating system of the sand fiddler crab, Uca pugilator. Behav. Ecol. Sociobiol. 12:169-180.
Christy, J. D., 1987a, Competitive mating, mate choice and mating associations of brachyuran crabs. Bull. Mar. Sci. (in press).
Christy, J. D., 1987b, Pillar function in the fiddler crab Uca beebei (II): Competitive courtship signaling. Ethology (in press).
Christy, J. D., 1987c, Female choice and breeding behavior of the fiddler crab Uca beebei. J. Crust. Biol. (in press).
Christy, J. D. and M. Salmon, 1984, Ecology and evolution of mating systems of fiddler crabs (Genus Uca). Biol. Revs. 59:483-509.
Christy, J. D. and S. E. Stancyk, 1982, Movement of larvae from the North Inlet estuary, S. C., with special reference to crab zoeae. In "Estuarine Comparisons," U. S. Kennedy, ed., pp. 489-503. Academic Press, N.Y.
Conner, E. F. and E. D. McCoy, 1979, The statistics and biology of the species-area relationship. Am. Nat. 113:791-833.
Cox, C. B., I. N. Healey and P. D. Moore, 1973, "Biogeography: An Ecological and Evolutionary Approach." John Wiley and Sons, Inc. N.Y.
Crane, J., 1957, Basic patterns of display in fiddler crabs (Ocypodidae, genus Uca). Zoologica, N.Y. 42:69-82.
Crane, J., 1975, "Fiddler Crabs of the World. Ocypodidae: Genus Uca." Princton University Press, New Jersey.
Croizat, L., G. Nelson and D. E. Rosen, 1974, Centers of origin and related concepts. Syst. Zool. 23:265-287.
Dittel, A. I. and C. E. Epifanio, 1982, Seasonal abundance and vertical distribution of crab larvae in Delaware Bay. Estuaries 5:197-202.
Edney, E. B., 1961, The water and heat relationships of fiddler crabs (Uca spp.). Roy. Soc. So. Africa, Trans. 36:71-91.
Ekman, S., 1953, "Zoogeography of the Sea." Sedgwick and Jackson Limited, London.
Emlen, S. T. and L. W. Oring, 1977, Ecology, sexual selection, and the evolution of mating systems, Science 197:215-223.
Frith, D. W. and S. Brunenmeister, 1980, Ecological and population studies of fiddler crabs (Ocypodidae, Genus Uca) on a mangrove shore at Phuket Island, Western Peninsular Thailand. Crustaceana 39:157-184.
Futuyama, D. J., 1979, "Evolutionary Biology." Sinauer Associates, Inc., Sunderland, MA.
George, R. W. and D. S. Jones, 1982, "A Revision of the Fiddler Crabs of Australia (Ocypodinae: Uca)." Records of the Western Australian Museum, Supplement No. 14, Perth.
Glaessner, M. F., 1969, Decapoda. In "Treatise on invertebrate paleontology," R. C. Moore, ed., Part R. Arthropoda 4. Vol. 2, pp. R400-533.
Gould, S. T. and N. Eldredge, 1977, Punctuated equilibria: the tempo and mode of evolution reconsidered. Paleobiol. 3:115-151.
Goy, J. W., 1976, Seasonal distribution and the retention of some decapod crustacean larvae within the Chesapeake Bay, Virginia. Masters thesis, Old Dominion University, Virginia.
Greenspan, B. N., 1980, Male size and reproductive success in the communal courtship system of the fiddler crab Uca rapax. Anim. Behav. 28:387-392.
Greenspan, B. N., 1982, Semi-monthly reproductive cycles in male and female fiddler crabs, Uca pugnax. Anim. Behav. 30:1084-1092.

Hagen, H. O. von, 1970, Die Balz von Uca vocator als okologisches Problem. Forma Functio 2:238-253.
Hagen, H. O. von, 1976, Review (of J. Crane, "Fiddler crabs of the World"). Crustaceana 31:221-224.
Hallam, A., 1981, Relative importance of plate movements, eustasy, and climate in controlling major biogeographical changes since the early mesozoic. In "Vicariance Biogeography: A Critique," G. Nelson and D. E. Rosen, eds., pp. 303-330. Columbia University Press, N.Y.
Heck, K. L. and E. D. McCoy, 1978, Long-distance dispersal and the reef-building corals of the eastern Pacific. Mar. Biol. 48:349-356.
Jones, D. S. and R. W. George, (in preparation), Tidal activity of the Australian backflat fiddler crabs, Uca elegans and Uca signata.
Kennett, J. P., 1977, Cenozoic evolution of Antarctic glaciation, the circum-Antarctic Ocean, and their impact on global paleoceanography. J. Geophy. Res. 82:3843-3860.
Krebs, H. A., 1975, The August Krogh Principle: "For many problems there is an animal on which it can be most conveniently studied." J. Exp. Zool. 194:221-226.
Lin, H.-C., 1986, The reproductive behavior and mate choice of the fiddler crab Uca lactea lactea, in mid-Taiwan. Masters thesis, Tunghai University, Taiwan.
Macnae, W., 1968, A general account of the fauna and flora of mangrove swamps and forests in the Indo-West-Pacific region. In "Advances in Marine Biology," F. S. Russell and M. Yonge, ed., pp. 73-270. Academic Press, N.Y.
McCoy, E. D. and K. L. Heck, 1976, Biogeography of corals, seagrasses and mangroves: an alternative to the center of origin concept. Syst. Zool. 25:201-210.
Miller, D. C., 1961, The feeding mechanisms of fiddler crabs, with ecological considerations of feeding adaptations. Zoologica N.Y. 46:89-100.
Minkoff, E. C., 1983, "Evolutionary Biology." Addison-Wesley Publication Company, Reading, MA.
Montague, C. L., 1980, A natural history of temperate western Atlantic fiddler crabs (Genus Uca) with reference to their impact on the salt marsh. Contrib. Mar. Sci. 23:25-55.
Murai, M., S. Goshima and Y. Hemni, 1987, The mating system of the fiddler crab, Uca lactea and the effects of phylogeny and current distribution. Bull. Mar. Sci. (in press).
Nakasone, Y., 1982, Ecology of the fiddler crab Uca (Thalassuca) vocans vocans (Linnaeus) (Decapoda:Ocypodidae) I. Daily activity in warm and cold seasons. Res. Pop. Ecol. 24:97-109.
Nakasone, Y., H. Okamine and K. Asato, 1983, Ecology of the fiddler crab Uca vocans vocans (Linnaeus) (Decapoda:Ocypodidae). II. Relation between the mating system and the drove. Galaxea 2:119-133.
Nations, D., 1975, The genus Cancer (Crustacea:Brachyura). Systematics, biogeography and fossil record. Natural History Museum of Los Angeles County, Scientific Bulletin No. 23.
Novak, A. and M. Salmon, 1974, Uca panacea, a new species of fiddler crab from the Gulf Coast of the U.S. Proc. Biol. Soc. Wash. 87:313-326.
Pinschmidt, W. C., Jr., 1963, Distribution of crab larvae in relation to some environmental conditions in the Newport river estuary, North Carolina. Ph.D. thesis, Duke University, Durham, North Carolina.
Rabalais, N. N. and J. N. Cameron, 1983, Abbreviated development in Uca subcylindrica (Stimpson, 1859) (Crustacea, Decapoda, Ocypodidae) reared in the laboratory. J. Crust. Biol. 3:519-541.
Ringold, P., 1979, Burrowing, root mat density, and the distribution of fiddler crabs in the eastern United States. J. Exp. Mar. Biol. Ecol. 36:11-21.

Robertson, J. R., K. Bancroft, G. Vermeer and K. Plaisier, 1980, Experimental studies on the foraging behavior of the sand fiddler crab Uca pugilator (Bosc, 1802). J. Exp. Mar. Biol. Ecol. 44:67-83.

Salmon, M., 1967, Coastal distribution, display and sound production by Florida fiddler crabs (Genus Uca). Anim. Behav. 15:449-459.

Salmon, M., 1984, The courtship, aggressive and mating system of a "primitive" fiddler crab (Uca vocans). Zool. Soc. Lond., Trans. 37:1-50.

Salmon, M., 1987, On the reproductive behavior of the fiddler crab Uca thayeri, with comparisons to U. pugilator and U. vocans: evidence for behavioral convergence. J. Crust. Biol. 7:25-44.

Salmon, M. and S. P. Atsaides, 1968, Behavioral, morphological and ecological evidence for two new species of fiddler crabs from the Gulf coast of the United States. Proc. Biol. Soc. Wash. 81:275-290.

Salmon, M., S. Ferris, D. Johnston, G. W. Hyatt and G. Whitt, 1979, Behavioral and biochemical evidence for the specific distinctiveness of the fiddler crabs Uca spinicarpa and U. speciosa. Evolution 33:182-191.

Salmon, M. and G. W. Hyatt, 1983, Spatial and temporal aspects of reproduction in North Carolina fiddler crabs (Uca pugilator). J. Exp. Mar. Biol. Ecol. 70:21-43.

Salmon, M., W. Seiple and S. Morgan, 1986, Hatching rhythms of fiddler crabs and associated species at Beaufort, N.C. J. Crust. Biol. 6:24-36.

Simpson, G. G., 1953, "The Major Features of Evolution." Columbia University Press, N.Y.

Smith, A. G. and J. C. Briden, 1977, "Mesozoic and Cenozoic Paleocontinental Maps." Cambridge University Press, Cambridge, U.K.

Teal, J. M., 1958, Distribution of fiddler crabs in Georgia salt marshes. Ecology 39:185-193.

Thurman, C. L., 1981, Uca marguerita, a new species of fiddler crab (Brachyura:Ocypodidae) from eastern Mexico. Proc. Biol. Soc. Wash. 94:169-180.

Valentine, J. W., 1971, Plate tectonics and shallow marine diversity and endemism, an actualistic model. Syst. Zool. 20:253-264.

Vermeij, G. J., 1980, "Biogeography and Adaptation: Patterns of Marine Life." Harvard University Press, Cambridge, MA.

Vernberg, F. J., 1984, Fiddler crabs: Ecosystems - Organisms - Molecules. Amer. Zool. 24:293-304.

Warburton, N., 1978, Field observations on behavioural isolation in the Uca species of Ross river estuary. Honor thesis, James Cook University, Townsville, Queensland.

Warner, G. F., 1977, "The Biology of Crabs." van Nostrand Reinhold Company, London, U.K.

Wilkins, J. L. and M. fingerman, 1965, Heat tolerance relationships of the fiddler crab, Uca pugilator, with reference to body color. Biol. Bull. 128:133-141.

Williams, M. J., 1974, A study of the distribution of fiddler crabs (Genus Uca) in Rowes Bay, Townsville, and a morphometric examination of some aspects of growth. Honors thesis, James Cook University, Townsville, Queensland.

Yamaguchi, T., 1971, The courtship behavior of a fiddler crab, Uca lactea. Kamamoto J. Sci., Biol. 10:13-37.

Zeil, J., G. Nalbach and H.-O. Nalbach, 1986, Eyes, eye stalks and the visual world of semi-terrestrial crabs. J. Comp. Physiol. A159:801-811.

Zucker, N., 1983, Courtship variation in the neo-tropical fiddler crab Uca deichmanni: another example of female incitation to male competition? Mar. Behav. Physiol. 10:57-79.

Some Aspect of Agonistic Communication in Intertidal Hermit Crabs

David W. Dunham

University of Toronto
Ontario, Canada

INTERTIDAL ENVIRONMENT AND COMMUNICATION

In a workshop for the discussion of intertidal behavioral adaptations, it is reasonable to raise the question of what effects intertidal habitats and microhabitats have on hermit crab agonistic communication. On one level, we can answer this question straight away - we do not know. We have yet to look in detail at microhabitat differences and their possible effects, as revealed through a fine-scaled analysis of behavior. Some advances have been made, as discussed by Reese (1969), and more will have been reported at this meeting. We can, at this stage, at least ask what kinds of effects one might expect to find, when one would expect to find them, and how one might go about detecting them.

Is there anything unique about the intertidal environment that would have predictive or explanatory value for communication systems? The shallow benthic habitat constituting the intertidal zone shows more species diversity than do deep or pelagic habitats. It is also subject to a much wider range of environmental conditions than is generally found in the oceans. It includes the most biologically productive communities known on earth (Thurman and Webber, 1984). There is obviously a great deal of variance in physical factors, such as wave surge or temperature, among the different subhabitats that comprise the intertidal life zone. For example, the differences between a sandy shoreline with regular wave surge and a high tidal bench inundated irregularly and infrequently by high rogue waves are considerable. They could include differences in turbidity, which would affect the efficiency of visual signals, temperature, which might require animals to partially emerge for evaporative cooling, (restricting their other behaviors), water movement, which could affect the usefulness of distant chemical signals, and background noise, influencing the efficiency of acoustic displays.

Permissive variables dictate the possible array of channels available. Species most active when the light is dim, as are some hermit crabs (Hazlett, 1966a), may either make less use of visual displays (as shown in cave crayfishes, and cave fishes), or may enhance background contrast to make their displays more easily perceived (as in fireflies, fishes in turbid waters, deep sea fishes). When light is not availble, crayfish increase their use of tactile display (Bruski and Dunham, in press). It is not yet known whether hermit crabs compensate in this way, but we are now examining the roles of sensory channels and redundancy in the agonistic communication of crayfishes and hermit crabs.

Clearly then, there are environmental constraints in intertidal habitats that can result in suboptimal conditions for communication through given sensory channels. There are also conditions favorable for the use of given channels. Where water is shallow the good transmission of broad spectral light facilitates visual communication. However, even clear water, if in motion, can distort visual images and have an important effect on visual display (Bartnik, 1970). Low turbidity in quiet waters provides favorable conditions for visual, chemical and acoustic communication.

If indeed subhabitat differences are significant to hermit crabs, should we expect them to be reflected in differences between species found in these different subhabitats? This depends on several factors: (1) the degree to which species are restricted to given habitats; (2) the period of time over which this has been the case; (3) the degree to which variance in signaling systems was available for natural selection to operate on; (4) the degree of competition, or of difference in fitness, that would drive the process of adaptation to different environments.

Microclimatic differences have been shown to be important in visual and acoustic signaling in fiddler crabs (*Uca* spp.) (Salmon, 1965; Salmon and Atsaides, 1968). There are some indications that important differences may exist in intertidal hermit crabs (Hazlett, 1974, 1984b), but this remains largely an unstudied area.

What effects of habitat difference on agonistic communication might we expect to find? In a habitat where environmental conditions made the use of a given sensory channel impractical or expensive, in terms of time and energy investment, we would expect to find a compensatory shift to greater use of an alternative channel. In a habitat where no such use of an alternative channel is facilitated, we would expect a high degree of redundancy in communication. One would expect redundancy to be most strongly expressed in contexts where the cost of failure to be detected by a receiver, or of error in interpretation, is great. This could be redundancy within a channel, as in signal repetition, or across channels with signals in more than one channel having the same message (*sensu* Smith, 1977). It may be difficult in practice to distinguish between redundancy and scaled intensities of a signal, which may really convey different messages.

To what extent are information theoretic measures useful in estimating behaviorally relevant redundancy? What I shall call statistical redundancy (R) can be estimated by a measure based on the Shannon-Wiener information theoretic equation. (The notation below follows the usage of Dingle (1969). See Dingle (1969) and Losey (1978) for informative discussions.)

> $R = 1 - H/H_{max}$, where H represents the information in the overall distribution of following acts in a transition matrix of action-reaction dyads ($H = -\Sigma p_i \log_2 p_i$), and H_{max} represents the log to the base 2 of the number of categories of events (m) in the matrix ($H_{max} = \log_2 m$).

Statistical redundancy, R, will reflect signal repetition within a given communication channel in two ways. Because differential repetition of one or more signals will result in a lower H, R will increase.

For example, simple repetition of a given display will result in a greater R. However, if ALL signals in a system were repeated equally frequently, H would increase, and as H/H_{max} approaches 1, R approaches 0. Therefore it is theoretically possible to have high signal redundancy, but low statistical (information) redundancy.

Another restriction concerns the level of analysis. If signal redundancy were based on repeated cycles of behaviors, analysis of a transition matrix based on a lower order Markovian relationship would not detect it. This is because the relevant unit, perhaps a cycle of acts (as for example in bird song), was not discovered and treated as a unitary behavioral event.

Last, R is also insensitive to cross-channel redundancy, if events of equal frequency in different channels, with the same message, were treated in the same transition matrix. This is because R is derived from event relationships that do not measure the message, but rather the meaning of signals (*sensu* Smith, 1977). We shall return to redundancy after considering agonistic displays in hermit crabs.

VISUAL DISPLAYS AND THEIR EFFECTS ON INTERTIDAL HERMIT CRABS

All intertidal hermit crabs look alike, and they also look different. They look alike because they carry a gastropod shell (in almost all species), and exhibit two long antennae, two chelipeds, and four ambulatory legs. They look different because their size, and the size of the chelae, and the patterning and coloration of most visible parts, including eyestalks, antennules and sometimes maxillipeds, varies greatly among species.

Some commonly ascribed functions of color patterns are inconspicuousness to predators, as in cryptic species, and species and sex discrimination ("recognition") in conspicuous species. Facultative crypsis need not result in uniformity, if different species are found against different backgrounds, although in the same "habitat", as Hazlett (1984b) showed for *Calcinus tibicen* and *Pagurus marshi*. Crypsis is also a matter of degree. In one sense, all hermit crabs are "cryptic" when withdrawn into their shells. However, some like *Pagurus marshi* are cryptic specialists, whose own body color blends into the background, even without adherent epibiota and detritus.

Hermit crabs in a given sympatric assemblage contrast both with their background and with each other. The former facilitates the effectiveness of visual displays generally, and the latter presumably functions in species discrimination. Species discrimination would be expected preceding sexual behavior, and may also be used in agonistic interactions. It can be facilitated by redundancy in social communication (Rand and Williams, 1970). Interspecific dominance among sympatric hermit crabs (Hazlett, 1967, 1974) would suggest that species discrimination could be an important variable in the decision to approach or avoid another hermit crab.

The basic visual displays of the chelipeds, ambulatory legs and shell were described by Hazlett (1966a). The antennae, and the brightly colored antennules and eyestalks, are also conspicuous in social interactions, but their use in visual display has not yet been evaluated. Both demonstrated and probable communication patterns are listed in Table 1.

Table 1. Use of Sensory Channels in Agonistic Communication by Intertidal Hermit Crabs (Paguridae and Diogenidae)

Communication	Representative References
Visual	
Immobility	Reese 1962
Low Body Posture	Hazlett 1966d,1968a
High Body Posture/ Shell Raise	Hazlett 1966a,1966d,1968a
Cheliped Extension	Hazlett 1966a,1966c,Hazlett & Bossert 1965,1966, Hazlett & Estabrook 1974
Cheliped Presentation	---------As Above------------
Ambulatory Raise	---------As Above------------
Cheliped Flicking	Hazlett 1970
Antennae (?)	Not Yet Demonstrated
Antennules (?)	Not Yet Demonstrated
Eyestalks (?)	Not Yet Demonstrated
Tactile	
Palpation 3rd Maxillipeds	Hazlett 1968b,1970 (sexual)
Chelae Touch	Hazlett 1967,1968b (sexual), 1970
Chelae Grasp	Hazlett 1968b,1970
Ambulatory Touch	Hazlett 1967,1968b (sexual), 1970
Ambulatory Grasp	Hazlett 1966a,1967,1968b, 1970
Antennae Contact	Elwood & Glass 1981
Antennae Other?	Not Yet Demonstrated
Percussive	
Chela Strike	Hazlett 1966a
Ambulatory Strike	Hazlett 1966a
Shell Rapping	Hazlett 1966a,1967,1970, Dowds & Elwood 1985
Shell Shaking/Rocking	---------As Above-------
Shell Dislodging-Shaking	Hazlett 1966a,1970
Acoustic Displays (?)	Not Yet Demonstrated
Chemical Displays	Inference in Hazlett 1967, Hazlett 1970 (sexual; likely chemo-tactile)

Experimental determination of display value (social response) of markings on hermit crab ambulatory legs (in *Calcinus tibicen* from the Caribbean) and chelae (in *Pagurus bernhardus* from Europe) was pioneered by Hazlett (1966b, 1969, 1972). Dunham (1978b, 1981) showed in a comparison among three Indo-Pacific species, *Calcinus laevimanus*, *Calcinus seurati* and *Clibanarius zebra*, that both enlargement of the major chela and enhanced brightness of the major chela increased the effectiveness of agonistic displays.

It was also discovered (Dunham, 1981) that chela enlargement lowers feeding efficiency in hermit crabs. In male fiddler crabs (*Uca* spp.) the greatly enlarged major chela is of little use in feeding. Of course, chela enlargement does not disadvantage feeding by definition. Lobsters (Homaridae) and crayfish (Astacidae) use enlarged chelae for both social signaling and feeding, and the two large chelae in *Homarus americanus* are in fact specialized for different kinds of feeding use, crushing versus cutting.

The effect of crypsis on communication has been investigated in the very cryptic Caribbean species *Pagurus marshi*. Brian Hazlett (1966a) first demonstrated that this species showed much poorer responsiveness to its mirror image than do non-cryptic species. It was later shown that individuals with less cryptic cover, both naturally and after cleaning by the investigator, showed better responsiveness to a mirror than did individuals with more cover (Dunham et al., 1986). It was also shown that, by time and movement measures, intraspecific agonistic communication in this species is more efficient between animals that have been cleaned than between more cryptic control animals (Dunham and Tierney, 1983). Thus the ecological specialization of crypsis appears to lower communication efficiency, but to a lesser degree than predicted. Cryptic animal may well have an ability to penetrate intraspecific crypsis.

COMMUNICATION THROUGH OTHER SENSORY CHANNELS

Visual, chemical and tactile signals can all be important, and can have interactive effects, in crustacean social behavior (Siefert, 1982). Distant chemical communication has not been demonstrated in intertidal hermit crabs, but could well be operative. These animals have been shown to respond to specific chemical cues emanating from gastropod predation events (Hazlett and Herrnkind, 1980; Rittschof, 1980a, 1980b; Hazlett, 1982), and even to the calcium emanating from the surface of shells (Mesce, 1982). Predation events are of significance to hermit crabs, because a new, empty shell is likely to be available, and also because other hermit crabs converging on a common site offer the possibility of improving one's fitness by acquiring a better shell (see Hazlett and Herrnkind, 1980, for discussion). We do not know how important chemical social communication may be in hermit crabs. Habitats with minimal water disturbance (Hazlett and Herrnkind, 1980) would certainly permit a chemical gradient to form. There are very specific sensilla for transducing social chemicals on the antennules of some decapod crustaceans (Tierney et al., 1984a, 1984b, 1986), and the possibility of their existence in hermit crabs should be investigated.

Tactile communication, possibly chemo-tactile, is assumed to occur during contact between the antennae of interacting hermit crabs (Elwood and Glass, 1981), and also when the antennae contact other parts of the body. Little is known about the role that this contact plays in social communication in hermit crabs. It has been shown to facilitate the resolution of agonistic interactions in lobsters (Solon and Cobb, 1980), and its increased role in the agonistic behavior of crayfish in the dark was mentioned above (Bruski and Dunham, in press). Its enhanced significance in social communication when the visual channel isn't available is also implied by the large increase in neural connections between antennule aesthetascs and the central nervous system in blind cave crayfish (Orconectes inermis) compared with their epigean congeners (Tierney et al., 1984a, 1986). Tactile input is presumably implicated in shell fights, when the initiator touches the non-initiator with its chelae, grasps the non-initiator's shell with its ambulatory legs, and also later, when it extends itself into the opening of the occupied non-initiator's shell (probing).

Strikes delivered to an agonistic opponent by the chelae or the ambulatory legs can be considered percussive communication, as can the rapping of the initiator's and non-initiator's shells together, or shaking of the non-initiator's shell by the initiator. Hazlett (1987) has shown that alteration of the initiator's shell volume can have a decisive effect on the outcome of a shell fight. He suggested that the fundamental frequency of a shell, set into vibration by rapping, may be a non-bluffable index of shell resource quality, and may be the major source of initiator-shell information available to the non-initiator, withdrawn into its shell, during a shell fight.

Acoustic communication is implicit in the sounds produced by the land hermit crabs Birgus latro and Coenobita spp., although its role in communication has not been established. Both these sounds, and the loud hissing sounds of the large Lamington spiny cray (Euastacus sulcatus) of the Australian rainforest, are also elicited by human approach or handling (Dunham, unpublished observations). Even less is known about sound production by marine hermit crabs. Trizopagurus is known to produce sound (P. MacLaughlin, personal communication), but no behavioral studies have as yet been reported.

WHEN DO INTERTIDAL HERMIT CRABS COMMUNICATE?

Hermit crabs respond to each other, both intra- and interspecifically, when their individual movements bring them together. The response may be as simple as approach and/or avoidance. However, it may also become more complex, involving the use of displays (social signals) and of physical force. The immediate outcome of these "close encounters of the complex kind" can be spacing between the individuals, gaining possession of a patch of food, initiation of a shell fight or initiation of courtship.

The course that an encounter follows depends on decisions on the part of each interacting hermit crab. Such decisions in turn depend on information about the other individual. In agonistic interactions and subsequent behavior that can progress all the way to a shell fight, decisions about the continuation of an interaction are made at choice points. It is therefore useful to consider an interaction in terms of stages. Even at a distance greater than that typical of ritualized social interactions, which varies among species (and investigators) from one or two to several shell diameters, some information is available to hermit crabs about others that are visible (Dunham, 1978a). We infer that some recognition occurs at these distances from the pattern of approach and avoidance seen commonly between hermit crabs in the laboratory that have already interacted (in a confined space) and have established a dominant-subordinate relationship. We also know that large size differences can be evaluated at a distance without social display, because of the long range approach-avoidance response pattern in animals differing greatly in size, e.g. as shown by Hazlett (1968c) for *Clibanarius vittatus*, whose probability of interaction is inversely related to size difference between individuals. Distant interaction can be importantly influenced by the topography of the environment and animal density (Hazlett, 1974).

At closer distances displays may be exchanged, which may lead to cessation of the interaction, or its continuance. Some size evaluation of crab or crab-cum-shell may take place here. Indeed, when environmental factors like turbidity or limited visibility in a highly structured environment preclude distant gathering of information, this may be the first opportunity for mutual reconnaissance by the interactants. Intertidal hermit crabs have well developed antennules, and it would be surprising if some use of chemical information were not made at this distance (as has been established in social behavior of crayfish; Tierney and Dunham, 1982, 1984), especially in more placid microhabitats. This possibility remains to be tested. If a shell fight is initiated (*sensu* Hazlett), initial tactile examination of the non-initiator's shell by the initiator may likewise result in either an end to the interaction, or continuation. Next, rapping (striking of the two shells together) may well provide percussive information on shell size to the non-initiator (Hazlett, 1987), and perhaps to the initiator as well. Other tactile and chemical information may be exchanged during probing into the non-initiator's shell. This behavior is not well understood and may be complex. If the interaction continues to completion, and other factors allow, the initiator retains possession of its original shell while manipulating (legs:tactile) and entering (abdomen:tactile) the non-initiator's shell. Even at that advanced stage, the decison to assume the new shell may be reversed.

What we see here with decreasing distance between the interactants is increasing complexity of behavioral responses, an increase in the number of possible behavioral responses, and an apparent increase in the cross-channel redundancy in communication, at each stage (Table 2). The latter is inferred, since we do not know the precise nature of the messages being communicated in each case.

There is also redundancy in the repetition of signals within a given channel, and rapping is the most intriguing of these. Hazlett's (1987) elegant demonstration of the significance of shell volume on non-initiator motivation for shell exchange in *Clibanarius antillensis*, argues for the function of communicating something about shell size, perhaps fundamental frequency as he suggests. But what does the sequential redundancy in rapping tell us? It has been used as an index of motivation on the part of the initiator. If this influences the non-initiator, it could explain why "rapping" of the abdomen of a naked crab against a non-initiator's shell is effective (Elwood and Glass, 1981; Dowds and Elwood, 1985), even though no initiator's shell exists aobut which information could be conveyed!

If a basic function of redundancy is to minimize error (or, more precisely, to minimize the sum of the cost of redundancy plus the cost of making errors plus the cost of error checking; Dancoff and Quastler, 1953), then it might well be expected to increase as interactions approach those vital stages where the possession of space, food, or a change in shell quality will be decided.

Table 2. The Progression of Agonistic Interaction in Intertidal Hermit Crabs

Interaction Stages	Communication Channels	Typical Behaviors
Pre-Social Distance	*Visual	Approach/Avoidance
Display Distance	*Visual Chemical?	Approach/Avoidance Social Displays
Strike Distance	*Visual Chemical? *Tactile *Percussive	Approach/Avoidance Social Displays Withdrawal into Shell Strike
Close Contact	*Visual (initiator) Chemical? *Tactile *Percussive	 Shell Grab Chela Flicking Withdrawal into Shell Move Away Shell Fight may ensue with Rapping; Probing

*) Use of communication channel demonstrated or strongly inferred; see Table 1.

SUMMARY

1. Some information about at least relative size is exchanged by intertidal hermit crabs at a pre-social distance before social displays are executed. We do not know how this is done, or how important it may be in determining the final outcome of interactions.

2. Displays of the chelipeds, ambulatory legs and shell are important social signals in agonistic interactions. Percussive signals through striking together of shells are important determinants of the outcome of "shell fights".

3. Agonistic communication efficiency is enhanced by specialization of limbs (greater size and brightness). It is diminished by decreased conspicuousness of hermit crabs, either naturally, as in cryptic species, or through experimental manipulation.

4. We do not know how important other visual displays and chemical and tactile information may be in agonistic interactions.

5. Agonistic interactions can be seen as a series of stages with decreasing distance between the interactants, and increasing redundancy of behavior. Decisions to progress or to desist may be made at each stage. The complexity of responses, number of possible responses, and probable cross-channel redundancy increase with sequential stages. This increase in overall redundancy may be related to both communication and error checking as the final outcome of the interaction approaches.

ACKNOWLEDGEMENT

This paper was prepared during the tenure of an operating grant from the Natural Sciences and Engineering Research Council of Canada.

REFERENCES

Bartnik, V. G., 1970, Reproductive isolation between two sympatric dance, Rhinichthys atratulus and R. cataractae, in Manitoba, J. Fish. Res. Board Can., 27:2125-2141.

Bruski, C. A., and Dunham, D. W., 1987, The importance of vision in agonistic communication of the crayfish Orconectes rusticus, I. An analysis of bout dynamics, Behaviour, in press.

Dancoff, S. M., and Quastler, H., 1953, The information content and error rate of living things, pp. 263-273, in: "Information Theory in Biology," H. Quastler, ed., Univ. Illinois Press, Urbana, Illinois.

Dingle, H., 1972, Aggressive behavior in stomatopods and the use of information theory in the analysis of animal communication, Chapter 4, pp. 126-156, in: "Behavior of Marine Animals, Vol. 2: Invertebrates," H. E. Winn, and B. L. Olla, eds., Plenum Press, New York.

Dowds, B. M., and Elwood, R. W., 1985, Shell wars II: the influence of relative size on decisions made during hermit crab shell fights, Anim. Behav., 33:649-656.

Dunham, D. W., 1978a, Effect of chela white on agonistic success in a diogenid hermit crab (Calcinus laevimanus), Mar. Behav. Physiol., 5:137-144.

Dunham, D. W., 1978b, On contrast and communication efficiency in hermit crabs, Crustaceana, 35:50-54.

Dunham, D. W., 1981, Chela efficiency in display and feeding by hermit crabs (Decapoda, Paguridea), Crustaceana, 41:40-45.

Dunham, D. W., and Tierney, A. J., 1983, The communicative cost of crypsis in a hermit crab, Anim. Behav., 31:783-787.

Dunham, D. W., Tierney, A. J., and Franks, P. A., 1986, Response to mirrors by a cryptic hermit crab, Pagurus marshi, Biotropica, 18:270-271.

Elwood, R. W., and Glass, C. W., 1981, Negotiation or aggression during shell fights of the hermit crab Pagurus bernhardus? Anim. Behav., 29:1239-1244.

Hazlett, B. A., 1966a, Social behavior of the Paguridae and Diogenidae of Curacao, Stud. Fauna Curacao, 23:1-143.

Hazlett, B. A., 1966b, Factors affecting the aggressive behavior of the hermit crab Calcinus tibicen. Zeit. Tierpsychol., 23:655-671.

Hazlett, B. A., 1966c, Temporary alteration of the behavioral repertoire of a hermit crab, Nature, 210:1169-1170.

Hazlett, B. A., 1966d, The behavior of some deep-water hermit crabs (Decapoda: Paguridea) from the Straits of Florida, Bull. Mar. Sci., 16:76-92.

Hazlett, B. A., 1967, Interspecific shell fighting between Pagurus bernhardus and Pagurus cuanensis (Decapoda, Paguridea), Sarsia, 29:215-220.

Hazlett, B. A., 1968a, Communicatory effect of body position in Pagurus bernhardus (L.) (Decapoda, Anomura), Crustaceana, 14:210-214.

Hazlett, B. A., 1968b, The sexual behavior of some European hermit crabs (Anomura: Paguridae), Publ. Staz. Zool. Napoli, 36:138-139.

Hazlett, B. A., 1968c, Size relationships and aggressive behavior in the hermit crab Clibanarius vittatus, Zeit. Tierpsychol., 25:608-614.

Hazlett, B. A., 1969, Further investigations of the cheliped presentation display in Pagurus bernhardus (Decapoda, Anomura), Crustaceana, 17:31-34.

Hazlett, B. A., 1970, Tactile stimuli in the social behavior of Pagurus bernhardus (Decapoda, Paguridae), Behaviour, 36:20-48.
Hazlett, B. A., 1972, Stimulus characteristics of an agonistic display of a hermit crab (Calcinus tibicen), Anim. Behav., 20:101-107.
Hazlett, B. A., 1974, Field observations on interspecific agonistic behavior in hermit crabs, Crustaceana, 26:133-138.
Hazlett, B. A., 1982, Chemical induction of visual orientation in the hermit crab Clibanarius vittatus, Anim. Behav., 30:1259-1260.
Hazlett, B. A., 1984, Epibionts and shell utilization in two sympatric hermit crabs, Mar. Behav. Physiol., 11:131-138.
Hazlett, B. A., 1987, Information transfer during shell exchange in the hermit crab Clibanarius antillensis, Anim. Behav., 35:218-226.
Hazlett, B. A., and Bossert, W. H., 1965, A statistical analysis of the aggressive communications systems of some hermit crabs, Anim. Behav., 13:357-373.
Hazlett, B. A., and Bossert, W. H., 1966, Additional observations on the communications systems of hermit crabs, Anim. Behav., 14:546-549.
Hazlett, B. A., and Estabrook, G., 1974, Examination of agonistic behavior by character analysis II. Hermit crabs, Behaviour, 49:88-110.
Hazlett, B. A., and Herrnkind, W., 1980, Orientation to shell events by the hermit crab Clibanarius vittatus (Bosc) (Decapoda, Paguridea), Crustaceana, 39:311-314.
Losey, G. S., Jr., 1978, Information theory and communication, Chapter 3, pp. 43-78, in: "Quantitative Ethology," P.W. Colgan, ed., Wiley-Interscience, New York.
Mesce, K. A., 1982, Calcium-bearing objects elicit shell selection behavior in a hermit crab, Science, 215:993-995.
Rand, A. S., and Williams, E. E., 1970, An estimation of redundancy and information content of anole dewlaps, Amer. Nat., 104:99-103.
Reese, E. S., 1962, Submissive posture as an adaptation to aggressive behavior in hermit crabs, Zeit. Tierpsychol., 19:645-651.
Reese, E. S., 1969, Behavioral adaptations of intertidal hermit crabs, Amer. Zool., 9:343-355.
Rittschof, D., 1980a, Chemical attraction of hermit crabs and other attendants to simulated gastropod predation sites, J. Chem. Ecol., 6:103-118.
Rittschoff, D., 1980b, Enzymatic production of small molecules attracting hermit crabs to simulated gastropod predation sites, J. Chem. Ecol., 6:665-675.
Salmon, M., 1965, Waving display and sound production in the courtship behavior of Uca pugilator, with comparisons to U. minax and U. pugnax, Zoologica, 50:123-150.
Salmon, M., and Atsaides, S. P., 1968, Visual and acoustical signalling during courtship by fiddler crabs (Genus Uca), Amer. Zool., 8:623-639.
Seifert, P., 1982, Studies on the pheromone of the shore crab, Carcinus maenus, with special regard to ecdysone excretion, Ophelia, 21:147-158.
Smith, W. J., 1977, "The Behavior of Communicating," Harvard Univ. Press, Cambridge, Mass.
Solon, M. H., and Cobb, J. S., 1980, Antennae-whipping behaviour in the American lobster Homarus americanus (Milne-Edwards), J. Exp. Mar. Biol. Ecol., 48:217-224.
Thurman, H. V., and Webber, H. H., 1984, "Marine Biology," Charles E. Merrill Publ. Co., London.
Tierney, A. J., and Dunham, D. W., 1982, Chemical communication in the reproductive isolation of the crayfishes Orconectes propinquus and Orconectes virilis (Decapoda, Cambaridae), J. Crust. Biol., 2:544-548.

Tierney, A. J., and Dunham, D. W., 1984, Behavioral mechanisms of reproductive isolation in crayfishes of the genus *Orconectes*, *Amer. Midl. Nat.*, 111:304-310.

Tierney, A. J., Thompson, C. S., Marin, L., and Dunham, D. W., 1984a, Fine structure of antennular chemoreceptors in cambarine crayfishes, *Neurosci. Abstr.*, f10:862.

Tierney, A. J., Thompson, C. S., and Dunham, D. W., 1984b, Site of pheromone reception in the crayfish *Orconectes propinquus* (Decapoda, Cambaridae), *J. Crust. Biol.*, 4:554-559.

Tierney, A. J., Thompson, C. S., and Dunham, D. W., 1986, Fine structure of aesthetasc chemoreceptors in the crayfish *Orconectes propinquus*, *Can. J. Zool.*, 64:392-399.

Intraspecific Variations in Reproductive Tactics in Males of The Rocky Intertidal Fish *Blennius sanguinolentus* in the Azores

Ricardo S. Santos and Vitor C. Almanda (*)

University of Azores, Portugal
(*) I.S.P.A., Lisboa, Portugal

INTRODUCTION

Blennies together with gobies comprise the bulk of resident fish species in the rock intertidal habitats of the warm temperate eastern Atlantic. They are among the most sucessfull groups that colonized intertidal habitats around the world. Moreover their mode of reproduction represents a pattern considerably widespread among warm temperate littoral fishes. The forms that colonized rocky intertidal habitats present a great number of morphological, physiological and behavioural peculiarities that have adaptative valvue in these conditions. The ethology of several species of blennies and related families has been studied in detail (Guitel 1893; Breder 1941; Thompson & Bennet 1953; Qazim 1956 ; Wickler 1957,1961,1964; Robins, Phillips & Phillips 1959; Fishelson 1963,1975; Abel 1964,1973; Gibson 1968b; Losey 1968,1976; Phillips 1971 ,1977; Heymer & Ferret 1976; Nursall 1977,1981; Wirtz 1978; Almada et al. 1983,1987; Denoix 1984; Heymer 1985 ; Santos 1985a,b,1986b; Patzner et al. 1986). Gibson (1969 ,1982,1986) reviewed and summarized most of the available literature on this topic.

Although the descriptive literature on the reproductive behaviour is very large, there are only a few quantitative studies of their paterns of activity (Nursall 1977,1981 for *Ophioblennius atlanticus*; Santos 1985a,b for *B. sanguinolentus* and 1986b for *B. ruber*; Almada et al. 1987 for *B. pilicornis*).

In this paper we present data on the reproductive behaviour of *B. sanguinolentus*, a resident fish of the tide pools in the Azores. We also try to relate ethological data to ecological, life-history and energetic aspects of the reproductive strategy of the species. In so doing we hope that such an integrative approach, already developed for fishes of other habitats and taxa (v.e.g. Gross & Charnov 1980; Gross 1982,1984,1985; Lejeune 1985; Warner & Lejeune 1985; Taborsky et al. in press), will contribute in the future, to a better understanding of the adaptiveness of reproductive styles of intertidal fishes and the selective pressures which act upon them.

THE HABITAT AND GENERAL ECOLOGICAL NOTES

B. sanguinolentus in Azores is an intertidal species. It is found in rock pools or among boulders. Breeding occurs preferentially in pools of the lower rocky intertidal zone. It feeds predominantly on green algae

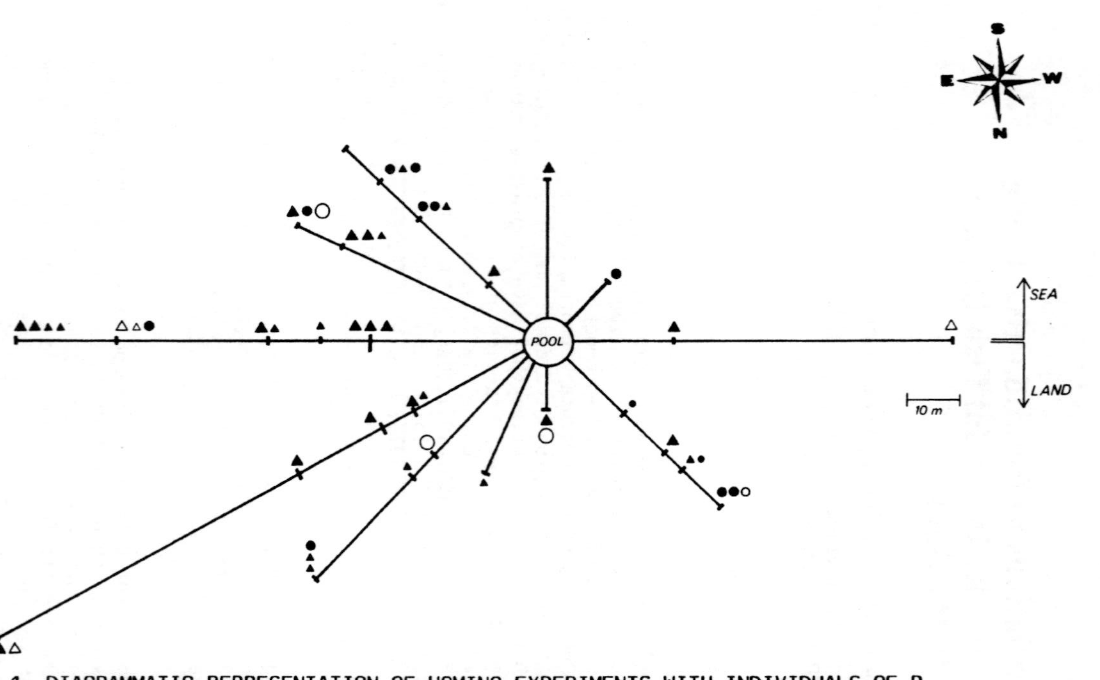

Fig.1. DIAGRAMMATIC REPRESENTATION OF HOMING EXPERIMENTS WITH INDIVIDUALS OF *B. SANGUINOLENTUS*. Big dark triangles represent big mature males who returned to their home pools (or pools of capture) after being released at some distance from them. Small dark triangles represent small males who returned to their home pools. Big and small open triangles, respectively big and small males which were not found again in there home pools. Big circles represent big females, and small circles represent small females for the same conditions referred to in males.

TABLE I

HOMING ABILITY EXPERIMENTS

M+: male presenting developed anal glands; M-: male with undeveloped anal glands; F: females; Y: means that the fish returned to the pool where it was captured (home pool); N: means that it was not found again in home pool.

NO	DATE OF EXPERIMENT	LENGTH (cm)	SEX	DISTANCE (m) AND ORIENTATION FROM HOME POOL		RESULT
1	83/05/18	14.5	M+	11	N	Y
2	83/05/19	-	M+	30	ENE	Y
3	83/05/20	11.9	M+	30	S	Y
4	83/05/20	-	M+	40	ESE	Y
5	83/05/26	-	F	11	N	N
6	83/05/27	11.1	F			N
7	83/06/03	11.9	M+	10	SE	Y
8	83/06/15	13.3	M+	20	W	Y
9	83/06/15	15.9	M+	25	NW	Y
10	83/06/17	15.0	M+	50	ESE	Y
11	83/06/20	12.9	M+	100	E	Y
12	83/06/20	12.8	M+	75	W	N
13	83/06/21	9.4	M-	25	ENE	Y
14	83/06/21	14.2	M+	40	ESE	Y
15	83/06/21	10.9	M+	25	ENE	Y
16	83/06/27	13.4	M+	50	ENE	Y
17	83/06/27	9.6	M-	40	ESE	Y
18	83/06/27	10.9	M+	120	ENE	N
19	83/06/27	13.9	M+	120	ENE	Y
20	83/06/27	10.0	M-	30	NW	Y
21	83/06/27	16.9	M+	120	ENE	Y
22	83/06/28	12.7	M+	50	E	Y
23	83/06/28	10.0	M-	50	E	Y
24	83/06/28	10.0	M-	40	E	Y
25	83/06/28	9.6	M-	25	ENE	Y
26	83/06/30	13.9	M+	30	E	Y
27	83/07/07	12.9	M+	30	E	Y
28	83/07/07	11.3	M+	30	E	Y
29	83/07/07	8.2	M-	40	SE	Y
30	83/07/07	10.7	F	40	SE	Y
31	83/07/07	12.0	F	25	NE	N
32	83/07/07	12.0	M-	30	SE	Y
33	83/07/11	11.6	F	10	SW	Y
34	83/07/12	11.3	M+	80	E	N
35	83/07/12	7.4		30	NW	Y
36	83/07/12	8.3		15	NW	Y
37	83/07/12	8.4		40	NW	N
38	83/07/12	11.4	F	80	E	Y
39	83/07/12	8.9	M-	30	NE	Y
40	83/07/12	10.2	M-	80	E	N
41	83/07/14	-	F	50	ESE	N
42	83/07/14	-	F	50	ESE	Y
43	83/07/27	10.2	F	30	SE	Y
44	83/07/27	10.9	M-	100	E	Y
45	83/07/27	11.4	F	40	SE	Y
46	83/07/27	9.6	M-	100	E	Y
47	83/07/27	10.3	M-	100	E	Y
48	83/07/27	14.8	F	30	SE	Y
49	83/07/27	8.8	M-	60	NE	Y
50	83/07/27	10.5	F	60	NE	Y
51	83/07/27	9.0	M-	60	NE	Y
52	83/07/27	9.1	F	40	NW	Y
53	83/07/27	9.9	F	40	NW	Y

TABLE II

VALUES OF D FOR TAGGED FISHES, WHERE D REPRESENTS THE PATTERN OF PERMANENCE OF FISH IN A GIVEN POOL. It is calculated dividing the number of days of observation by the number of days that the fish was seen in the pool.

NO	1	2	3	4	5	6	7	8	9	10	11	12	13	14	15	16	17	$\bar{X}$	sd
M+	.73	.75	.50	.83	.88	.93	.50	.84	.50	.70	.50	.90	.67	.56	.57	.64	.50	.68	.16
M-	.86	.89	.93	.80	.90	.90	.88	.73	.80	.60	.80	.42	.71	.80	.33	.80	.60	.75	.17
F	.44	.16	.30	.20	.40	.70	.50	.14	.50	.30	-	-	-	-	-	-	-	.36	.18

(Gibson 1968a;Taborsky & Limberger 1980; Goldschmid et al. 1984), but if the opportunity occurs it will scavenge on dead invertebrates (Santos 1985a).

HOME RANGE ASSESSMENT

The fishes have developed hability to return to their home pool when experimentally removed (see Fig.1 and Table I). Observations conducted during the breeding season showed that tagged fishes were found in the pool for many sucessive low tides (Santos 1985b,1986a). Homing ability is developed in parental males, females and small mature males. The results show no significant differences among these sexes and size groups: parental males versus females (Fisher exact probability test: p=0.32, n.s.), parental males versus small mature males (p=0.49,n.s.) small mature males versus females (p=0.21, n.s.). In contrast we found differences (ANOVA: $p > 0.01$) in the regularity of occurence in each pool between territorial males and females (HSD Tukey: $p>0.05$) and between small mature males and females (HSD Tukey: $p>0.05$; Table II and Table III). No significant difference was found between territorial and small mature males (HSD Tukey: n.s.).

These results give support to the idea that, as in other rocky intertidal species, female home ranges are larger than those of the males during the breeding season, and that they continue their feeding excursions. It is not known if they tend to breed in a specific pool or have a breeding scheme with visits to males of different pool. Total collecting of fishes in pools in the breeding season and during low tide, showed that the proportion of males was much higher than in samples performed in other periods (unpublished data). This suggests that the males tend to congregate in nesting areas that are not the most visited by females except for breeding.

The finding that territorial and smaller non-territorial males show strong constancy in a particular pool during the breeding season led us to the hypothesis that these small mature males could be in permanent association with parental territories, acting as sneakers or satellites.

To test this hypothesis we proceeded to a number of behavioural observations of tagged males, and collected data to compare their biology and demography.

TABLE III

Tukey's hsd test for D values

$\bar{D}$	F = .3620	M+ = .6768	M- = .7500
F = .3620	-	.3145*	.3880*
M+ = .6768		-	.7035
M- = .7500			-
.14451	HSD	.18842	* p > 0.05

F: females; M+: males with developed anal glands;
M-: males with small anal glands

BEHAVIOUR OF BREEDING MALES

LOCOMOTION

i. short distance: hopping by simultaneous impulses of pectoral fins.

ii. long distance: sinusoidal anguilliform swimming.

SIGNAL SWIMMING

Swimming-up five to fifteen centimeters off the substratum, followed by pitching and swimming back.

Apparently a self-advertisement action, performed even in the absence of other fishes. Probably functional both in courtship and territorial demarcation.

AGONISTIC BEHAVIOUR

Charging: quick swimming toward an intruder. May or may not lead to bite or butting.
Butting.
Bite.
Chasing.
Fight.
Withdrawal.

Threatening : i. similar to signal swimming, but performed always close to the substratum with less vertical elements. Fish swims towards the intruder and turns back; ii. slow lateral ondulations of the entire body without loss of contact with the substratum.

Side rolling: a submissive posture in which the fish rolls to one side, presenting the dorsal area to other fish.

Fighting: extended agonistic interactions with fishes involved in mutual overt aggression: butting, biting and circle chasing. A fight may include several rounds with immobilization of the fishes at short distances in between.

CLEANING AND MAINTENANCE

Digging: removing detritus with pectoral fins or tail.

Fanning: pectoral or tail fanning. It is assumed to function in aeration and detritus removal. During breeding tail fanning plays the major role.

Mouth transport spitting

Pushing larger objects with head or body

Algal removal around nest entrance: the nests are located in algae covered areas. The male cleans the immediate vicinity of the nest, plucking out the algae and spitting them away from the nest forming an uncovered zone that makes the nest entrance conspicuous.

COURTSHIP

Signal swimming: (see above)

Back and forth movement: rhythmic action in which the male advance and retreat quickly at nest entrance, head facing out. Performed when females are visible.

Others: i. Males also side-roll towards females during courtship. In this case, however, showing the ventral side to the female. ii. During spawning they stay in the nest with female.

For quantitative records the following categories were used:

NA: including swimming, withdrawal and chafing.

AG: including any kind of aggression, and threatening.

AC: fights.

SA: signal swimming.

AL: feeding.

LP: cleaning and maintenance activities when performed out of the nest, but including also excavation, removal of objects from the nest and fanning.

RO: resting out of the nest.

DYNAMICS OF TERRITORIALITY

To study the dynamics of territoriality, we tagged parental males in selected territories. We made animal focal observations for each individual (Altmann 1974). Starting from the nest entrance, we divided the surrounding area in concentric zones whose limits were 20, 40, 60, etc. centimeters from the nest. We grouped activities according to the number of categories listed above. Observation sessions varied from 15 minutes to 2 hours. A total of 35 hours of animal focal observations were conducted during low tide. Occurrences of each activity were recorded as well as the distance from the nest and the nature of other interactions. Examples of three territorial males are given in Figure 2. A map of territory locations in a pool is presented in Figure 3.

If we consider the first zone (A), with a maximal distance from the nest of 20 cm, we can see that 63% of the male's activity occur there. This maximal distance from the nest is less than twice the fish total body length. To all males more than 85% of the activities recorded out of the nest occurred in zones A and B.

If we now consider object removal, including algae, we find that more than 99% take place in zone A. In contrast spitting of the removed materials occur out of the zone A in 69% ($\mp$ 18%) of the times. The contrast between this values shows that the fish invests substantial effort in carrying the removed materials far from the nest and it also confirmes the hypothesis that the fish cleans a restricted area around the nest, that approximately corresponds to the first half of zone A (Fig. 2).

More than 80% of signal swimming is performed in zone A, and never exceeds zone B. These data suggest that fishes use the area they set free from algae around the nests as an arena where displaying is probably most effective (Fig. 2). Feeding corresponds to 10% or less of the total of the activities of the males out of the nests (mean $\pm$ sd = 4.1% $\pm$ 4.3%).

If we now consider that the activities recorded correspond only to the actions performed out of the nest and that in average 90% of the time males were in the nests, we have an idea of how reduced feeding is in parental males during low tide.

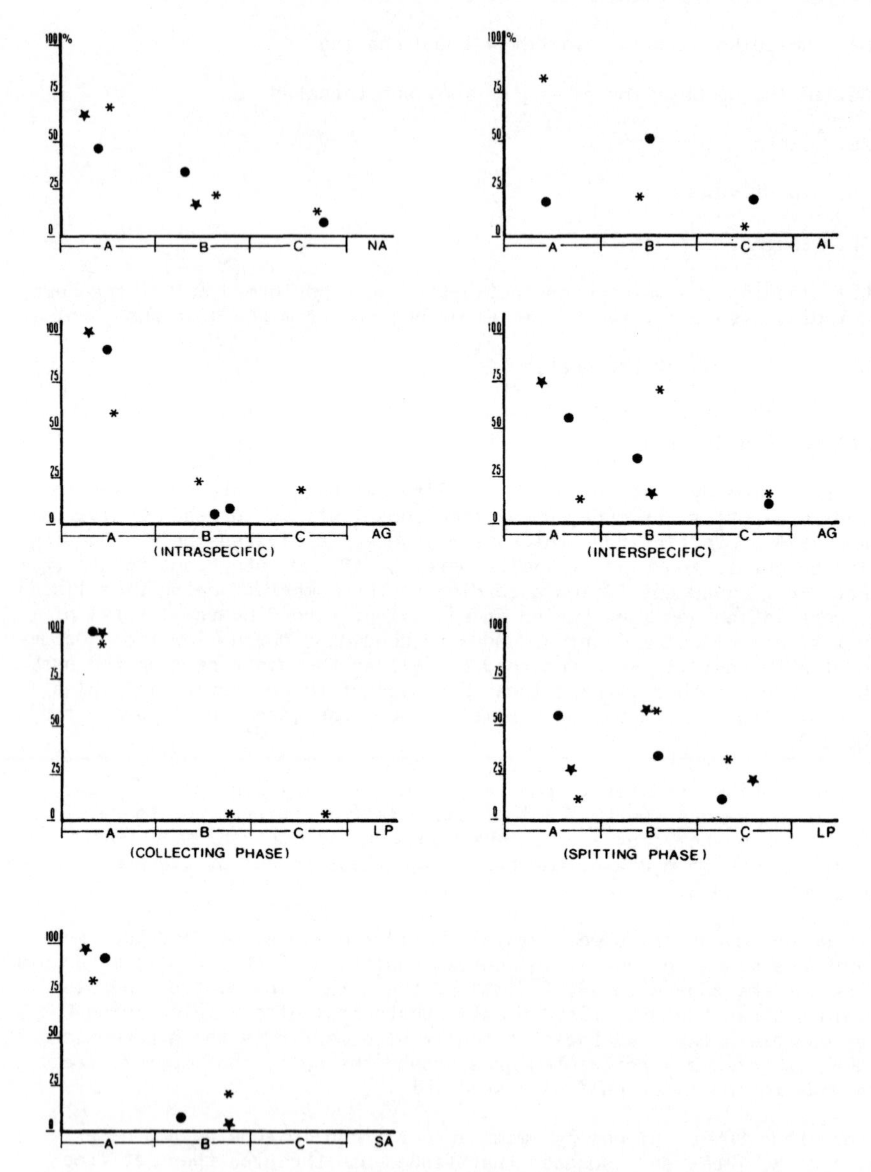

Fig. 2. Proportional distribution (%) of different activities of three males in pool no 1 (see Fig.3) in different areas of their respective territories (see text).

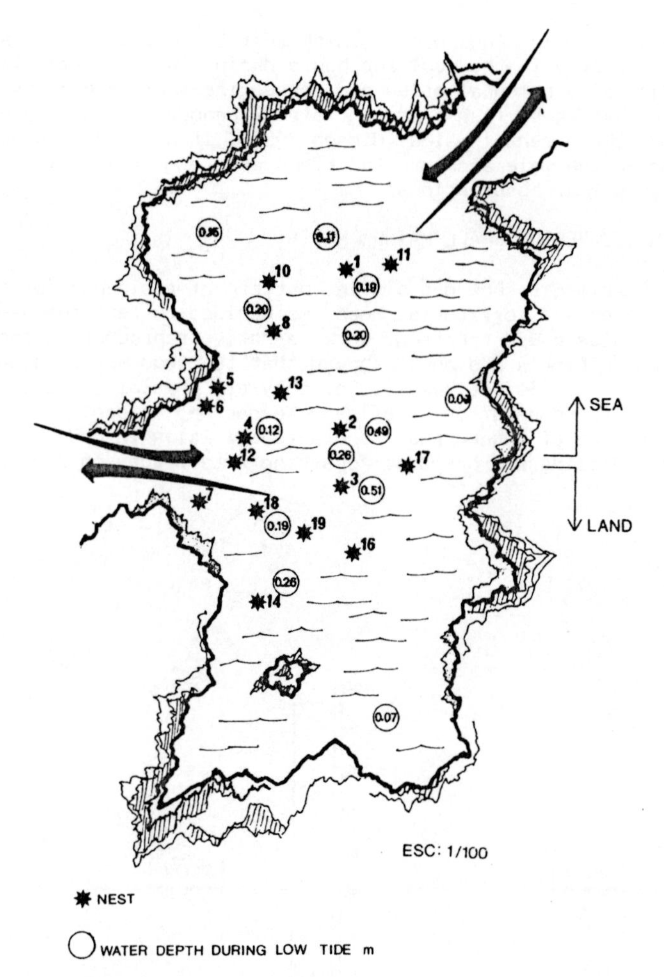

Figure 3. Diagram representing pool no 1, where some behavioural observations were conducted.

When the males are in the nests, they actively fan and clean the eggs. As some of the activities performed out of the nest also have parental value, namely interspecific aggression, we can conclude that parental activities have the dominant role both in time and effort for parental males. If we assume that signal swimming, algae removal and spitting have mainly epigamic or territorial signaling function, we find that a great proportion (mean ± sd = 42.3% ± 17.8%) of the activities of the males in the nests relate with this type of functions.

Finally, aggression makes also a substantial proportion (mean ± sd = 21.5% ± 0.9%) of the activities performed out of the nests. It is interesting to note that a large proportion of the aggression are interspecific (52.1%). Juvenil Liza salien and Diplodus sp., present in the same pools, are by these means effectively driven away. It is a typical situation for fishes nesting in holes, that egg predators are smaller than the parental fish. The simple presence of the male in the nest seems to become effective in driving away these egg predators. It is plausible to think that these transient fishes which are forced to stay in the pool for hours during low tide, may learn to avoid the aggressive parental males and recognize their arenas as unpleasant places. That these fishes present a real danger is shown by experimental removal of the parental males (Gibson 1982). If a spawn is under a stone and we remove the male and turn the stone over, eggs are preyed upon after a few minutes (our observations).

1. TERRITORIAL MALES VERSUS SMALL MATURE MALES

As homing experiments showed a similar pattern of residence for small non-territorial males and territorial ones, we decided to test the hypothesis that these small males could represent an alternative reproductive tactic. Examination of their testis and ducts showed that they had mature gonads and produced sperm (for details see below). So, they are physiological prepared to fertilize eggs if the opportunity arises. We decided to design a second series of observations of tagged small and parental males in order to compare their behaviour. Sampling was based on focussing a territory on each session.

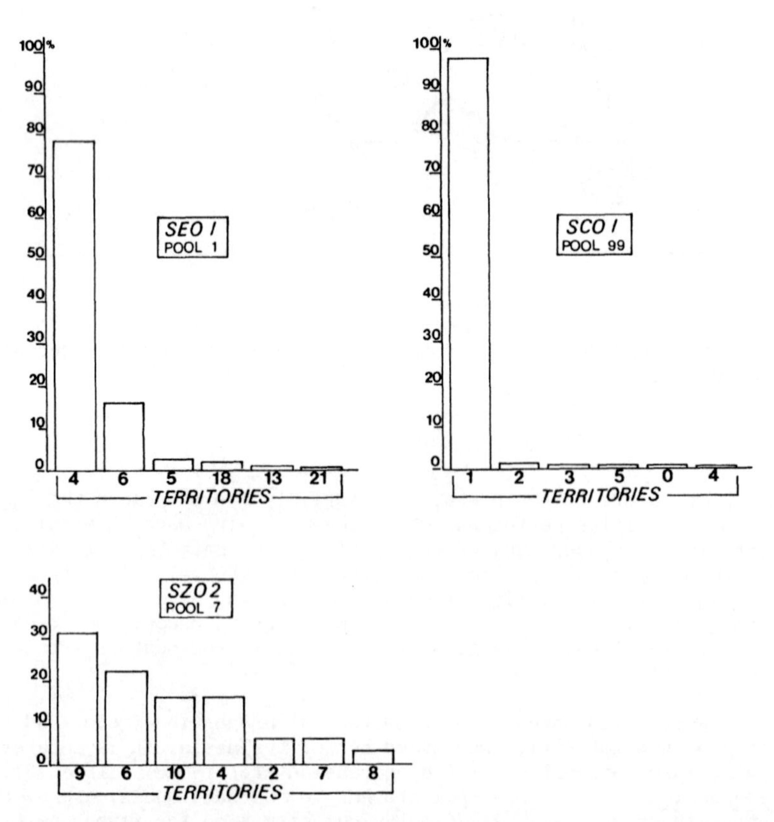

Figure 4. Histograms showing the distribution of activities of three satellites males in different territories.

We also compared their morphology and demography.

1.1. Behaviour of small males

1.1.1. Fidelity to territories

Having mapped a number of territories in pools (Fig. 3), we recorded the distribution of activities of small males, and the territories where they occurred. We adopted as the limits of territory boundaries the distance of 60 cm from the nest. Some results are summarized in Figure 4. They show that these males not only show fidelity to a pool but also fidelity to one or a few territories. Some of them can be classified as typical satellites.

1.1.2. Pattern of activity of small males

The main results can be summarized as follows:

i. All observed fights were between small males (Fig. 5 and Table V).

ii. Chasing was 1.5 times higher for these males, which were involved in 60.57% of all the observed chasing (Fig. 5; Table IV).

iii. Feeding is not significantly different between the two categories of males (Table IV).

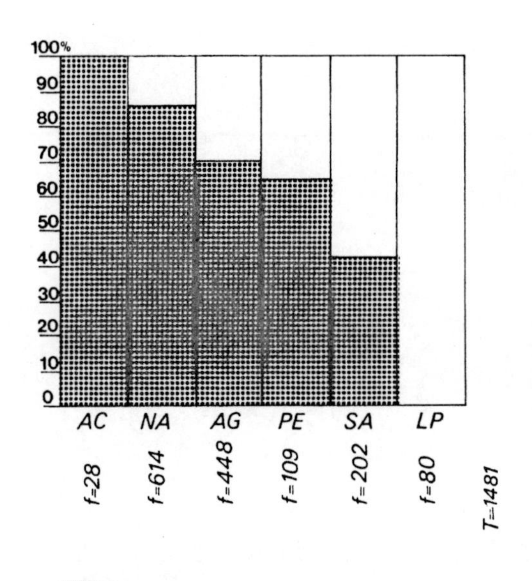

Figure 5. Comparative proportions (%) of different activities of parental males and small males inside parental territories. Global (T) and partial frequencies are given.

TABLE IV. COMPARISON OF PARENTAL AND SATELLITE ACTIVITIES BASED ON TWO-TAILED MANN-WHITNEY U TEST FOR ANALYSE OF THE DIFFERENCES BETWEEN THE MEANS OF THE FREQUENCIES (%) THAT THEY DEDICATED TO EACH ACTIVITY.

ACTIVITY		PARENTALS (n=12)	SATELITES (n=9)	MANN-WHITNEY U	p
NA	mean(%)	17.10	33.19	32	n.s.
	±sd	±12.00	± 7.67		
AG	mean(%)	34.76	23.44	38	n.s.
	±sd	±17.21	±14.17		
AC	mean(%)	0.00	4.72	30	n.s.(0.10)
	±sd		±11.17		
SA	mean(%)	21.65	1.17	4	0.002
	±sd	±18.07	± 3.60		
LP	mean(%)	20.54	0.00	4.5	0.002
	±sd	±14.21			
RO	mean(%)	2.59	36.18	0	0.002
	±sd	± 2.38	±15.10		
AL	mean(%)	3.58	0.72	37	n.s.
	±sd	± 5.09	± 1.57		

iv. Locomotion were insignificant in parentals when compared with satellites (Table IV).

v. *Signal swimming was also exhibited by small males although in a lower proportion (40%). Algae removal was absent (Fig. 5).

vi. Taken together (SA + LP), which we assume to have strong epigamic territorial functions, are strikingly lower, in contrast the most frequent actions of parental males in the area of the nest (Fig. 5; Table IV).

vii. Parental males are dark brown. Small males only show dark colouration when they are not in visual contact with parental males.

1.1.3. Interactions between small and territorial males

i. Fifty percent of parental males aggression were directed satellite males.

ii. It is worthnoting that satellites with a specific territory never attacked the parental male but attacked neighbour parental males (1% of the aggressions performed by satellites) if they approached to much, and in effect drove them back, although they were much larger.

iii. Small males intruded the nests even with the parental male present. At high tide it seems probable that mating has a much higher frequency and intrusions would be expected to increase. As water began to flood the pool the activity of the fishes increased markedly, but observations soon became impossible.

LIFE HISTORY AND REPRODUCTIVE BIOLOGY

1. DEMOGRAPHY

Breeding season of this species, in our study area, lasts from the end of May, beginning of June, until the first days of August. A sample made with Quinaldine in which all fishes of 6 pools were collected in two days of May 86, gave the results presented in Figure 6. From analyssis of modes in the distribution it can be inferred that last year cohort is represented by individuals ranging from 40 mm TL to 100/105 mm. The other three modes must indicate cohorts of preceding years. First cohort represents 65% of the total sample. In this cohort females represent 62% of the total number of females, and males 68% of the total number of males (Fig.6).

An interesting feature of the distribution is the distribution is the broad range of this year class, which must reflect multiple spawning and extended breeding season.

This same pattern is also clear in Figure 7. It also shows that juveniles first begin to settle in July, one and a half to two months after the beginning of the breeding season, and continue to settle until October. Settling size is about 20 mm. In December the larger fishes of the cohort had attained a total length of 80 mm, while the smaller still had 30 mm. This cohort is still clearly separate from cohorts of previous years (Fig.7).

It seems clear that the fishes live three or more years. Otolith studies are in progress, and seem to confirm this view.

2. REPRODUCTIVE BIOLOGY

2.1. Gonadosomatic index

From sample taken during the breeding season, the following data were noted: total fish length total weight, gonad weight, eviscerated weight,and length of males anal glands.

Gonadosomatic indexes (GSI = gonad weight / total weight x 1,000) show that one year fishes are already mature. GSIs of one year males are significantly higher than GSIs of older males (Table V and Fig.8).

Dividing the fish samples in three size classes for each sex (class A individuals less than 90 mm TL; class B individuals between 91 mm and 120 mm TL, class C individuals bigger than 120 mm TL) we found the following

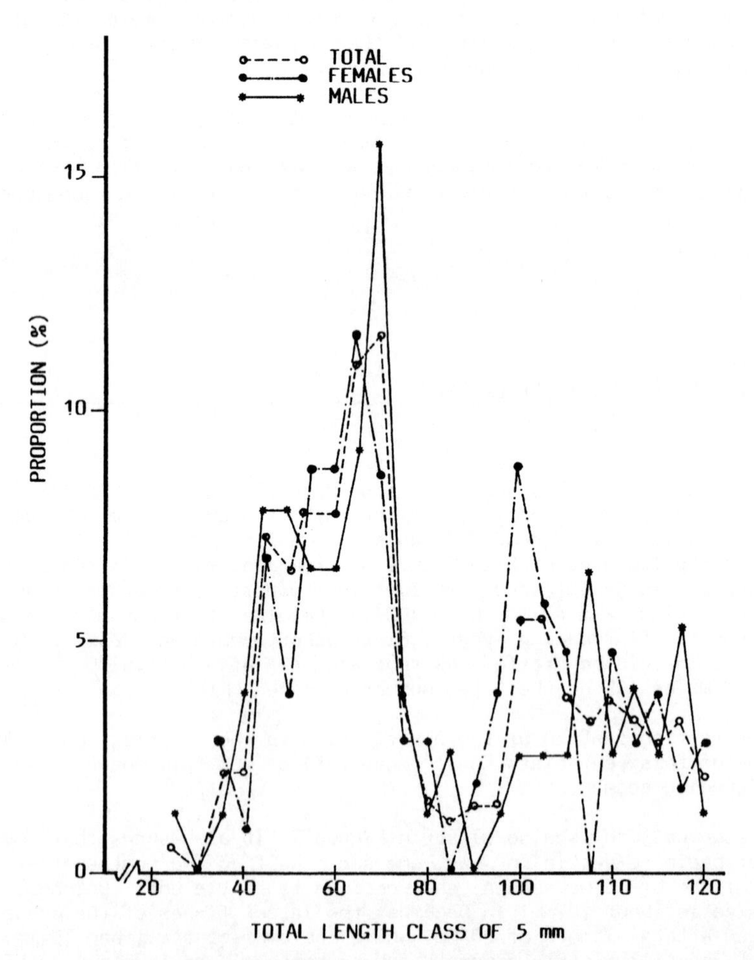

Figure 6. Length class distributions of individuals of the species B. sanguinolentus from a sample collected with Quinaldine in rocky intertidal pools of the Azores.

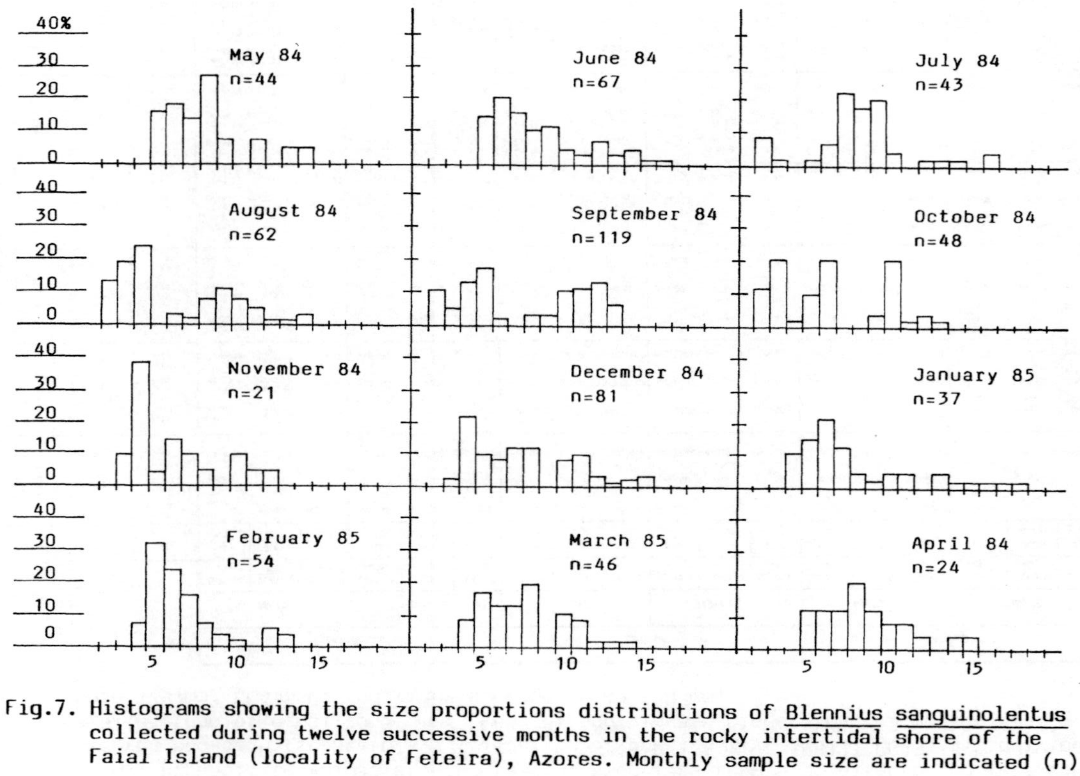

Fig.7. Histograms showing the size proportions distributions of *Blennius sanguinolentus* collected during twelve successive months in the rocky intertidal shore of the Faial Island (locality of Feteira), Azores. Monthly sample size are indicated (n)

TABLE V. CROSS COMPARISONS OF THE SIGNIFICANCE OF THE DIFFERENCE BETWEEN THE MEANS OF GSI = = GONAD WEIGHT/TOTAL WEIGHT x 1000 OF DIFFERENT CLASSES OF INDIVIDUALS OF *BLENNIUS SANGUINOLENTUS*. Student t test was used when variances could be considered homogenes. In other cases we used Mann-Whitney U test (for small samples) or Z (for bigger samples). Probability values are two-tailed. When less than 0.05 differences were considered non-significant (n.s.) (Siegel 1956, Sokal & Rohlf 1981). CLASS A: individuals smaller than or equal to 95 mm; CLASS B: individuals bigger than 95 mm and smaller than 120 mm; CLASS C: individuals bigger than 120 mm.

				MAY 86						JUNE 86						JULY 86					
				MALES			FEMALES			MALES			FEMALES			MALES			FEMALES		
				CLASS A n=33	CLASS B n=10	CLASS C n=21	CLASS A n=37	CLASS B n=21	CLASS C n=28	CLASS A n=5	CLASS B n=11	CLASS C n=19	CLASS A n=3	CLASS B n=3	CLASS C n=2	CLASS A n=6	CLASS B n=5	CLASS C n=12	CLASS A n=5	CLASS B n=6	CLASS C n=3
			GSI mean ± sd	4.61 ±2.12	2.87 ±2.33	1.97 ±0.96	1.33 ±1.01	7.12 ±3.06	9.11 ±3.43	1.48 ±2.75	5.95 ±1.23	2.87 ±1.66	0.46 ±0.06	4.74 ±1.35	5.57 ±1.97	1.15 ±1.47	1.75 ±2.69	1.16 ±0.24	1.05 ±1.31	3.29 ±2.49	3.90 ±1.63
MAY 86	MALES	CLASS A	4.61 ±2.12	-	p< 0.05	p<0.00003	p<0.001	p<0.001	p<0.00003	p<0.01	n.s.	p<0.01	p=0.0047	n.s.	n.s.	p<0.001	p<0.02	p<0.00007	p<0.01	n.s.	n.s.
		CLASS B	2.87 ±2.33	t=2.17 df=41	-	n.s.	p=0.0122	p<0.001	p<0.001	n.s.	p=0.01	n.s.	p<0.02	n.s.	n.s.	n.s.	n.s.	p=0.0256	n.s.	n.s.	n.s.
		CLASS C	1.97 ±0.96	z=4.25	z=0.36	-	p=0.0007	p=0.00003	p<0.00003	p=0.0239	p<0.001	p=0.0418	p=0.0064	p<0.001	p<0.001	n.s.	n.s.	p<0.00023	n.s.	n.s.	p<0.05
	FEMALES	CLASS A	1.33 ±1.01	t=8.28 df=68	z=2.25	Z=3.25	-	p=0.00003	p<0.00003	p=0.0233	p=0.001	p<0.001	p=0.0129	p<0.001	p<0.001	n.s.	n.s.	n.s.	n.s.	n.s.	p<0.001
		CLASS B	7.12 ±3.06	t=3.54 df=52	t=3.84 df=29	z=4.48	z=5.32	-	p<0.05	p<0.001	p=0.0329	p<0.00005	p=0.0030	n.s.	n.s.	p<0.001	p<0.01	p<0.00003	p<0.001	p<0.02	n.s.
		CLASS C	9.11 ±3.43	z=4.62	t=5.18 df=36	z=5.84	z=6.80	t=2.08 df=47	-	p=0.001	p=0.0113	p<0.00003	p=0.0026	p<0.05	n.s.	p<0.001	p<0.001	p<0.00003	p<0.001	p<0.001	p<0.02
JUNE 86	MALES	CLASS A	1.48 ±2.75	t=2.93 df=36	t=1.00 df=13	z=1.98	z=1.96	t=3.77 df=24	t=4.61 df=31	-	p=0.001	n.s.	n.s.	n.s.	n.s.	n.s.	n.s.	n.s.	n.s.	n.s.	n.s.
		CLASS B	5.75 ±1.23	t=1.94 df=42	t=3.65 df=19	t=12.25 df=30	t=12.35 df=46	z=1.84	z=2.28	t=4.53 df=14	-	p<0.001	p<0.02	n.s.	n.s.	p<0.001	p<0.001	p<0.002	p<0.001	p<0.02	p<0.05
		CLASS C	2.87 ±1.66	t=3.02 df=50	t=0.00461 df=27	z=1.73	t=4.22 df=54	z=3.87	z=5.20	t=1.42 df=14	t=5.17 df=28	-	p=0.002	n.s.	p<0.05	p<0.05	n.s.	p<0.002	n.s.	n.s.	n.s.
	FEMALES	CLASS A	0.46 ±0.06	z=2.60	u=0	z=2.49	z=2.23	z=2.75	z=2.81	u=6	u=0	u=0	-	p=0.05	n.s.	n.s.	n.s.	p<0.02	n.s.	n.s.	p=0.05
		CLASS B	4.74 ±1.35	t=0.11 df=34	t=1.24 df=11	t=4.41 df=22	t=5.43 df=38	t=1.31 df=22	t=2.12 df=29	t=1.88 df=6	t=1.43 df=12	t=1.80 df=20	u=0	-	n.s.	p<0.01	n.s.	p<0.02	p<0.02	n.s.	n.s.
		CLASS C	5.57 ±1.97	t=0.62 df=33	t=1.45 df=10	t=4.71 df=21	t=5.49 df=37	t=0.69 df=21	t=1.40 df=28	t=1.87 df=5	t=0.36 df=11	t=2.11 df=19	u=0	t=0.57 df=3	-	p<0.02	n.s.	p<0.02	p<0.02	n.s.	n.s.
JULY 86	MALES	CLASS A	1.15 ±1.47	t=3.77 df=37	t=1.55 df=14	t=1.64 df=25	t=0.38 df=41	t=4.59 df=25	t=5.42 df=32	t=0.26 df=9	t=7.01 df=15	t=2.22 df=23	u=9	t=3.55 df=7	t=3.47 df=6	-	n.s.	n.s.	n.s.	n.s.	p<0.05
		CLASS B	1.75 ±2.69	t=2.69 df=36	t=0.81 df=13	z=1.33	z=0.99	t=3.60 df=24	t=4.46 df=31	t=0.16 df=8	t=4.32 df=14	t=1.16 df=22	u=6	t=1.76 df=6	t=1.78 df=5	t=0.77 df=9	-	n.s.	n.s.	n.s.	n.s.
		CLASS C	1.16 ±0.24	z=3.88	z=1.95	z=3.56	z=0.93	z=4.10	z=4.96	u=12	u=0	u=0	u=0	u=0	u=0	u=25	u=24	-	p=0.002	n.s.	p<0.002
	FEMALES	CLASS A	1.05 ±1.31	t=3.55 df=36	t=1.51 df=13	t=1.73 df=24	t=0.54 df=40	t=4.27 df=24	t=5.02 df=31	t=0.31 df=8	t=6.77 df=14	t=1.21 df=22	u=6	t=3.54 df=6	t=3.42 df=5	t=0.11 df=9	t=0.5 df=8	u=2	-	n.s.	n.s.
		CLASS B	3.29 ±2.49	t=1.33 df=37	t=0.32 df=14	z=1.05	z=1.37	t=2.76 df=25	t=3.82 df=32	t=1.01 df=9	t=2.76 df=15	t=0.46 df=23	u=5	t=0.85 df=7	t=1.07 df=6	t=1.69 df=10	t=0.94 df=9	u=24	t=1.64 df=9	-	n.s.
		CLASS C	3.90 ±1.63	t=0.55 df=34	t=0.65 df=11	t=2.84 df=22	t=3.89 df=38	t=1.73 df=22	t=2.51 df=29	t=1.31 df=6	t=2.20 df=12	t=0.95 df=20	u=0	t=0.61 df=4	t=0.92 df=3	t=2.38 df=7	t=1.19 df=6	u=0	t=2.34 df=6	t=0.34 df=7	-

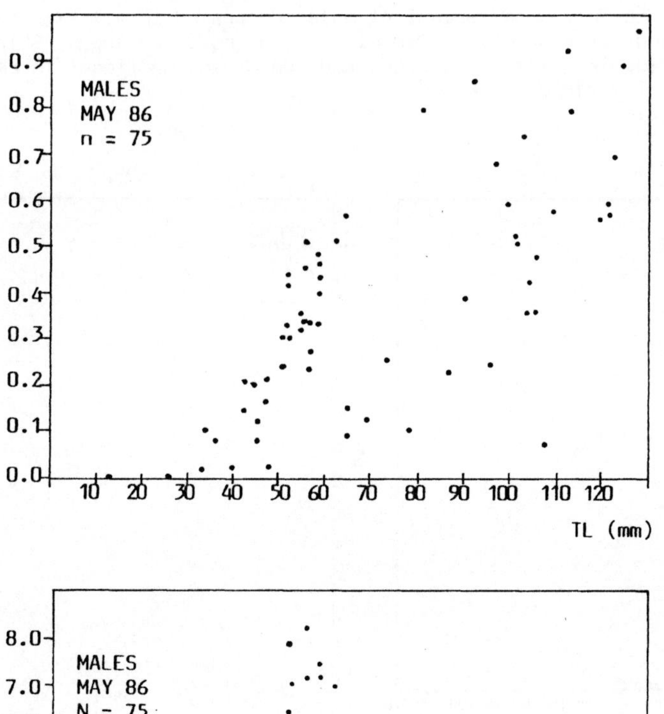

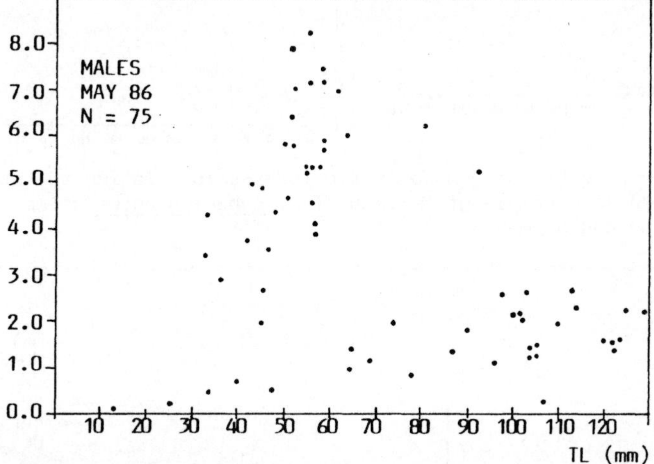

Figure 8. Gonad weights and gonadosomatic indexes in relation to total length in a sample of males of B. sanguinolentus collected in the end of May.

results:

i. In females GSIs are lower in first size class (mean ± sd = 1.33 ± 1.01), intermediate in class B (mean ± sd = 7.12 ± 3.06), and higher in the biggest fishes (mean ± sd = 9.11 ± 3.43). Mean differences are significant (see Table V).

We can conclude that in females gonad weight increases more rapidly than body weight, being strongly accelerated in regard to size (see Fig. 9).

ii. For males, on the contrary, although gonad weight increases almost linearly with size, it represents a decreasing proportion of the total weight (Fig. 8).

From May to July gonad weight of males decreased in all size classes by 77%, 41% and 49% respectively in class A, B and C. In August GSIs are already very low in all size classes, and become insignificant in the following months (Fig.10).

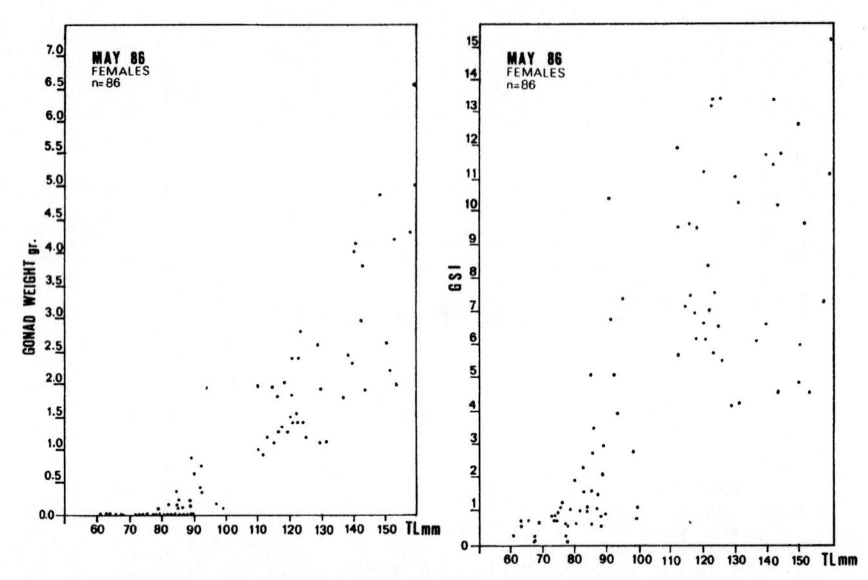

Figure 9. Gonad weights and gonadosomatic indexes in relation to total length in a sample of females of B. sanguinolentus collected in the end of May.

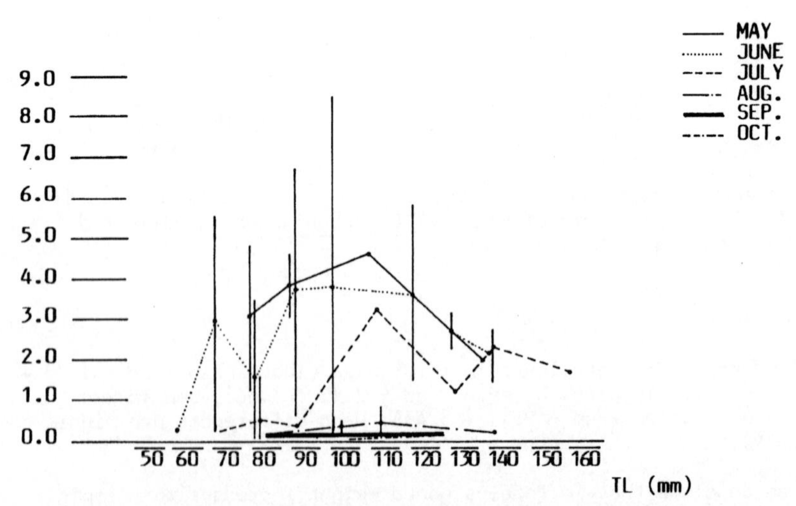

Figure 10. Values of gonadosomatic index of males of B. sanguinolentus during six months, mean and range limits are represented for each total length class.

TABLE VI. CROSS COMPARISONS OF THE SIGNIFICANCE OF THE DIFFERENCE BETWEEN THE MEANS OF KV = WEIGHT-(GONAD WEIGHT+EVISCERATED WEIGHT)/TL^3 x 10,000 OF DIFFERENT CLASSES OF INDIVIDUALS OF *BLENNIUS SANGUINOLENTUS* (see legend Table V).

				MAY 86						JUNE 86						JULY 86					
				MALES			FEMALES			MALES			FEMALES			MALES			FEMALES		
				CLASS A n=27	CLASS B n=7	CLASS C N=19	CLASS A n=30	CLASS B n=15	CLASS C n=23	CLASS A n=3	CLASS B n=3	CLASS C n=11	CLASS A n=2	CLASS B n=0	CLASS C n=2	CLASS A n=6	CLASS B n=5	CLASS C n=12	CLASS A n=5	CLASS B n=6	CLASS C n=3
			KV mean ± sd	17.18 ±3.41	17.11 ±4.96	15.98 ±4.57	19.97 ±7.43	26.84 ±8.08	25.41 ±8.16	10.58 ±2.90	11.30 ±2.00	7.82 ±2.96	25.59 ±14.19	-	13.97 ±1.47	11.63 ±1.30	10.05 ±3.73	6.97 ±1.85	12.90 ±3.37	12.90 ±4.15	9.24 ±0.32
MAY 86	MALES	CLASS A	17.18 ±3.42	-	n.s.	n.s.	p=0.0166	p<0.0003	p<0.0023	p<0.01	p<0.01	p<0.001	n.s.	-	n.s.	p<0.01	p<0.001	p<0.001	p<0.02	p<0.01	p=0.0026
		CLASS B	17.11 ±4.96	t=0.04 df=32	-	n.s.	n.s.	p<0.02	p<0.02	n.s.	n.s.	p<0.002	n.s.	-	n.s.	p=0.026	p<0.05	p<0.002	n.s.	n.s.	p=0.008
		CLASS C	15.98 ±4.57	t=1.58 df=44	t=0.85 df=24	-	p<0.02	p<0.001	p<0.00005	n.s.	n.s.	p<0.001	n.s.	-	n.s.	n.s.	p<0.05	p<0.002	n.s.	n.s.	n.s.
	FEMALES	CLASS A	19.97 ±7.43	z=2.13	t=0.94 df=35	t=2.42 df=47	-	p<0.02	p<0.02	p<0.05	n.s.	p<0.00003	n.s.	-	n.s.	p=0.0099	p<0.001	p<0.00003	p<0.05	p<0.05	p=0.0262
		CLASS B'	26.84 ±8.08	z=6.31	t=2.50 df=20	t=5.11 df=32	t=2.55 df=43	-	n.s.	p<0.01	p<0.01	p<0.002	n.s.	-	n.s.	p<0.002	p<0.001	p<0.002	p<0.01	p<0.001	p<0.02
		CLASS C	25.41 ±8.16	z=3.55	t=2.46 df=28	z=3.98	t=2.48 df=51	t=0.52 df=36	-	p<0.01	p<0.01	p<0.00003	n.s.	-	n.s.	p<0.0005	p<0.01	p<0.00003	p<0.01	p<0.001	p<0.0057
JUNE 86	MALES	CLASS A	10.58 ±2.90	t=3.12 df=28	t=1.90 df=8	t=1.64 df=20	t=2.11 df=31	t=3.25 df=16	t=3.00 df=24	-	n.s.	p<0.02	n.s.	-	n.s.	n.s.	n.s.	p<0.05	n.s.	n.s.	n.s.
		CLASS B	11.30 ±2.00	t=2.83 df=28	t=1.76 df=8	t=1.42 df=20	t=1.95 df=31	t=3.12 df=16	t=2.86 df=24	t=0.29 df=4	-	p<0.01	n.s.	-	n.s.	n.s.	n.s	p<0.01	n.s.	n.s.	n.s.
		CLASS C	7.82 ±2.86	t=9.61 df=36	u=0	t=5.85 df=28	z=4.35	u=2	z=4.62	t=3.04 df=12	t=4.27 df=12	-	p<0.05	-	p<0.001	p<0.001	p<0.05	n.s.	p<0.001	p<0.01	p<0.02
	FEMALES	CLASS A	25.59 ±14.19	z=0.09	u=6	u=14	t=0.93 df=30	t=0.17 df=15	t=0.03 df=23	u=1	u=2	u=0	-	-	n.s.	p<0.05	n.s.	p<0.01	n.s.	n.s.	n.s.
		CLASS B	-	-	-	-	-	-	-	-	-	-	-	-	-	-	-	-	-	-	-
		CLASS C	13.47 ±1.47	t=1.28 df=27	t=0.78 df=7	t=0.38 df=19	t=1.11 df=30	t=2.11 df=15	t=1.90 df=23	t=1.18 df=3	t=1.25 df=3	t=6.25 df=11	t=0.81 df=2	-	-	n.s.	n.s.	p<0.001	n.s.	n.s.	n.s.
JULY 86	MALES	CLASS A	11.63 ±1.30	t=2.80 df=31	u=7	u=28.5	z=2.33	u=6	z=3.39	t=0.66 df=7	t=0.26 df=7	t=6.75 df=15	t=2.06 df=6	-	t=1.84 df=6	-	n.s.	n.s.	n.s.	n.s.	p=0.019
		CLASS B	10.05 ±3.73	t=4.11 df=30	t=2.45 df=10	t=2.66 df=22	t=2.84 df=33	t=4.26 df=18	t=3.96 df=26	t=0.18 df=6	t=0.46 df=6	t=2.44 df=14	t=1.91 df=5	-	t=1.22 df=5	t=0.88 df=9	-	p<0.05	n.s.	n.s.	n.s.
		CLASS C	6.97 ±1.85	t=9.52 df=37	u=0	u=9	z=4.34	u=4	z=4.66	t=2.49 df=13	t=3.34 df=13	t=0.36 df=21	t=4.01 df=12	-	t=4.74 df=12	t=0.64 df=16	t=2.14 df=15	-	p<0.001	p<0.01	p=0.05
	FEMALES	CLASS A	12.90 ±3.37	t=2.150 df=30	t=1.50 df=10	t=1.04 df=22	t=2.03 df=33	t=3.56 df=18	t=3.24 df=26	t=0.86 df=6	t=0.65 df=6	t=4.89 df=14	t=1.58 df=5	-	t=0.36 df=5	t=0.77 df=9	t=1.14 df=8	t=4.60 df=15	-	n.s.	n.s.
		CLASS B	12.90 ±4.15	t=3.08 df=31	t=1.80 df=11	t=1.46 df=23	t=2.45 df=34	t=4.05 df=19	t=3.73 df=27	t=0.50 df=7	t=1.49 df=7	t=3.69 df=15	t=1.80 df=6		t=0.54 df=6	u=16	t=0.77 df=9	t=3.42 df=16	t=0.32 df=9	-	n.s.
		CLASS C	9.24 ±0.32	z=2.80	u=0	u=8	z=1.94	u=3	z=2.53	u=4	u=3	u=1	u=0	-	u=0	u=2	u=6	u=4	u=2	u=3	-

TABLE VII. CROSS COMPARISONS OF THE SIGNIFICANCE OF THE DIFFERENCE BETWEEN THE MEANS OF KS1 = EVISCERATED WEIGHT/TL^3 x 1,000 OF DIFFERENT CLASSES OF INDIVIDUALS OF BLENNIUS SANGUINOLENTUS (see legend Table V).

				MAY 86						JUNE 86						JULY 86					
				MALES			FEMALES			MALES			FEMALES			MALES			FEMALES		
				CLASS A n=33	CLASS B n=7	CLASS C n=15	CLASS A n=38	CLASS B n=15	CLASS C n=23	CLASS A n=3	CLASS B n=3	CLASS C n=11	CLASS A n=2	CLASS B n=0	CLASS C n=2	CLASS A n=17	CLASS B n=5	CLASS C n=12	CLASS A n=5	CLASS B n=6	CLASS C n=3
			KS1 mean ± sd	7.95 ±0.58	8.90 ±0.64	9.72 ±0.55	7.82 ±0.74	8.63 ±0.78	8.84 ±0.73	7.49 ±0.32	8.12 ±0.87	8.71 ±1.03	7.47 ±1.85	-	7.73 ±0.87	8.08 ±0.43	8.58 ±0.35	7.87 ±0.54	7.92 ±0.22	8.06 ±0.40	8.25 ±0.53
MAY 86	MALES	CLASS A	7.95 ±0.58	-	p<0.001	p<0.001	n.s.	p<0.01	p<0.001	n.s.	n.s.	p<0.01	n.s.	-	n.s.	n.s.	n.s.	n.s.	n.s.	n.s.	n.s.
		CLASS B	0.90 ±0.64	t=3.77 df=38	-	p=0.01	p<0.001	n.s.	n.s.	p<0.02	n.s.	n.s.	n.s.	-	n.s.	p<0.01	n.s.	p<0.01	p=0.005	p<0.05	n.s.
		CLASS C	9.72 ±0.55	t=9.73 df=50	t=3.09 df=24	-	p<0.001	p<0.001	p<0.001	p<0.001	p<0.001	p<0.002	p<0.02	-	p<0.001	p<0.001	p<0.001	p<0.001	p<0.001	p<0.001	p<0.001
	FEMALES	CLASS A	7.82 ±0.74	t=0.80 df=69	t=3.57 df=43	t=9.78 df=55	-	p<0.001	p<0.001	n.s.	n.s.	p<0.01	n.s.	-	n.s.	n.s.	p 0.05	n.s.	n.s.	n.s.	n.s.
		CLASS B	8.63 ±0.78	t=3.29 df=46	t=0.77 df=20	t=4.62 df=32	t=3.48 df=51		n.s.	p<0.05	n.s.	n.s.	n.s.	-	n.s.	p<0.002	n.s.	p<0.02	p<0.02	n.s.	n.s.
		CLASS C	8.84 ±0.73	t=4.97 df=54	t=0.21 df=28	t=4.25 df=40	t=5.17 df=59	t=0.81 df=36	-	p<0.01	n.s.	n.s.	p=0.0082	-	n.s.	p<0.00003	n.s.	p<0.001	p=0.0022	p<0.05	n.s.
JUNE 86	MALES	CLASS A	7.49 ±0.32	t=1.31 df=34	t=3.25 df=8	t=6.53 df=20	t=0.75 df=39	t=2.34 df=16	t=3.04 df=24	-	n.s.	n.s.	n.s.	-	n.s.	p<0.05	n.s.	n.s.	n.s.	n.s.	n.s.
		CLASS B	8.12 ±0.87	t=0.45 df=34	t=1.54 df=8	t=4.29 df=20	t=0.66 df=39	t=0.99 df=16	t=1.56 df=24	t=1.14 df=4	-	n.s.	n.s.	-	n.s.	n.s.	n.s.	n.s.	n.s.	n.s.	n.s.
		CLASS C	8.71 ±1.03	t=3.03 df=12	t=0.44 df=16	u=33	t=3.19 df=47	t=0.22 df=24	t=0.41 df=32	t=1.96 df=12	t=0.91 df=12	-	n.s.	-	n.s.	n.s.	n.s.	n.s.	p=0.05	n.s.	n.s.
	FEMALES	CLASS A	7.47 ±1.85	Z=0.07	u=3	u=1	Z=0.06	u=8	z=2.41	u=2	t=0.56 df=3	t=1.43 df=11	-	-	n.s.	n.s.	n.s.	n.s.	n.s.	n.s.	n.s.
		CLASS B	-	-	-	-	-	-	-	-	-	-	-	-	-	-	-	-	-	-	-
		CLASS C	7.73 ±0.87	t=0.50 df=33	t=1.86 df=7	t=4.35 df=19	t=0.17 df=38	t=1.43 df=15	t=1.96 df=23	t=0.33 df=3	t=0.43 df=3	t=1.22 df=11	t=0.17 df=2	-	-	n.s.	n.s.	n.s.	n.s.	n.s	n.s.
JULY 86	MALES	CLASS A	8.08 ±0.43	t=0.70 df=48	t=3.55 df=22	t=9.68 df=34	z=1.14	u=43.5	z=4.20	t=2.16 df=18	t=0.13 df=18	u=58	u=17	-	t=0.91 df=17	-	p<0.05	n.s.	n.s.	n.s.	n.s.
		CLASS B	8.58 ±0.53	t=2.28 df=36	t=0.95 df=10	t=4.25 df=22	t=2.21 df=41	t=0.15 df=18	t=0.76 df=26	t=3.83 df=6	t=1.06 df=6	u=25.5	u=3	-	t=1.57 df=5	t=2.29 df=20	-	p<0.05	p<0.02	n.s	n.s.
		CLASS C	7.87 ±0.54	t=0.41 df=43	t=3.54 df=17	t=8.88 df=29	t=0.21 df=48	t=2.76 df=25	t=3.94 df=33	t=1.08 df=13	t=0.62 df=13	u=33	u=12	-	t=0.29 df=12	t=0.97 df=27	t=2.53 df=15	-	n.s.	n.s.	n.s.
	FEMALES	CLASS A	7.92 ±0.92	t=0.10 df=36	u=2	t=6.86 df=22	z=0.15	u=8	z=2.85	t=1.95 df=6	u=5	u=14	u=5	-	u=5	t=0.74 df=20	t=3.17 df=8	t=0.20 df=15	-	n.s.	n.s.
		CLASS B	8.06 ±0.40	t=0.45 df=37	t=2.56 df=11	t=6.56 df=23	t=0.77 df=42	t=1.42 df=19	t=2.42 df=27	t=1.91 df=7	t=0.13 df=7	u=19.5	u=6	-	t=0.65 df=6	t=0.06 df=21	t=2.04 df=9	t=0.74 df=16	t=0.64 df=9	-	n.s.
		CLASS C	8.25 ±0.53	t=0.86 df=34	t=0.19 df=8	t=4.21 df=20	t=0.98 df=39	t=0.76 df=16	t=1.31 df=24	t=2.00 df=4	t=0.23 df=4	t=0.73 df=12	t=0.75 df=3	-	t=0.70 df=3	t=0.63 df=18	t=1.00 df=6	t=1.06 df=13	t=1.23 df=6	t=0.58 df=7	-

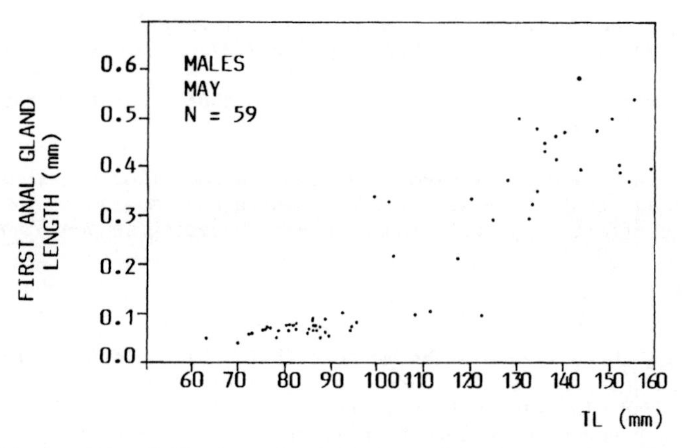

Figure 11. Length of first anal gland in relation to total length from a sample of males of B. sanguinolentus collected in the end of May.

2.2. Growth of anal glands

In contrast with males GSIs, growth of anal glands is only marked in two year old fishes (Fig. 11). For larger sizes it tends to be asymptotic. Out of the breeding season anal glands regress.

2.3. Energy allocation

The breeding modality of B. sanguinolentus led us to the hypothesis that the pattern of investment in reproduction differs between small males, parental males and females.

If females maximize food/egg conversion efficiency, they must continue to feed intensively along the breeding season. Parental males must have reduced feeding rates and rely on reserves previously stored. If this is the case, parental males must show a steady decrease in their reserves. Small males were expected to be intermediate between parental males and females.As data on gut contents, liver weight and oxygen consumption were note available, we proceeded as follows:

i. We compared the eviscerated weight between the three groups (i.e. small males, parental males and females) in the beginning, in the middle and near the end of the breeding season.

ii. For each fish we calculated a coefficient of 'Visceral condition' by subtracting from total weight the sum of eviscerate weight plus gonad weight and dividing this value by total length cube times 1,000. We called this value KV (Table VI). KV reflects a combination of variations in food intake and liver and visceral fat reserves.

Somatic conditions based on eviscerated weight (KS1 = Eviscerated weight/TL^3 x 1,000), is also compared (Table VII).

Analysis of data indicate that:

i. females (class C: May mean GSI = 9.11; June mean GSI = 5.57: n.s.) differ from males (class C: May mean GSI = 1.97; June mean GSI = 2.87: p>0.05) in beginning transfer of energy to gonads well before spawning starts (Table V);

ii. parental males, prior to the breeding season, have converted a substantial amount of energy into reserves, both visceral and somatic, which decreases during breeding (class C: mean KV: May = 15.98; June = 7.82; July = = 6.97 mean KS1: May = 9.72; June = 8.71; July = 7.87) (Table VI).

iii. small males show intermediate pattern, probably due to higher feeding rate than parental males (Table VI);

iv. finally, all fish groups seem to attain the same final level of depletion, suggesting that there are similar constraints imposed in the maximum reproductive effort they are allowed in each breeding season (Table VI and VII).

3. DIANDRY

Comparing the three size classes we note that, while class C corresponds to the males with developed anal glands and relatively low GSI, class A corresponds to one year old males with very reduced anal glands and the highest GSI values. Class B seems to include a mixture of males of both types, or in transition between them (Table V: May Fig. 8).

In our ethological studies parental males always corresponded to class C. Satellites and other non-nesting males which move around the territories, corresponded to class A and B.

Both behavioural data and biological characteristics seem to provide strong evidence for two distinct reproductive tactics in the males of this species. As far as we know, it is the first time such a variation in male tactics is described for a blenny and for rocky intertidal fishes (see also Santos 1985b,1986a). Wirtz (1978) and Jonge (1985) have referred to a similar situation in trypterigids.

DISCUSSION

Blennius sanguinolentus belongs to a species group , that for some authors (Zander 1972,1978,1979; Bath 1976,1981) has generic status (Parablennius), in which the majority of species are subtidal. B. sanguinolentus has preserved in RIH the self-advertisement behaviour of its subtidal relatives.

In the Mediterranean, it is present below the Azores directly swept by wave breaking. Its intertidal occurence in the Azores seems to offer excellent opportunity to the study of the conditions that can lead subtidal benthic fishes to colonize the tidal zone. It also exemplifies to what extent subtidal benthic blennies are pre adapted in many respects, namely in their reproductive style, to colonize the intertidal if opportunity is given.

Azores are 900 miles away from the European continent. The dominant species of the north-eastern Atlantic intertidal blennies, B. pholis, is rare at the Azores. In the subtidal zone, the highly territorial Ophioblennius atlanticus and the Pomacentrid Abudefduf luridus occupy intensely the available rock surface display very aggressively to other small benthic fish (Mapstone & Wood 1975). It is possible that high competition in the subtidal zone, and weak competition in the intertidal were important factors leading the Azorean form of B. sanguinolentus to occupy RIH where it became the dominant species. There, it is out of reach of benthic predators (e.g. Scorpaenidae, Congridae, Muraenidae, and the larger specimens of Gaidropsarus guttatus).

B. sanguinolentus, while presenting the general features of reproductive style of rocky intertidal warm temperate residents, shows a number of peculiarities that illustrates well the evolutionary plasticity that this breeding pattern can display.

Arenas formation and reproductive diethism and dimorphism in males are good examples of this plasticity. It is possible that arena formation and signal-swimming are effective in facilitating the attraction of females in an environment where nests are surrounded by algal clumps.

Concerning the types of males tactics there is an urgent need for work on age determinations. As we did not find immature males overlapping in size with class A, the possibility of these two tactics being alternative life histories as in pacific salmon and sunfish (e.g. Gross 1982,1984) seems improbable. Indeed, if this was the case, we should find males with sizes similar to those of classes A and B that would be sexually immature and would continue to grow, originating class C males at a later age and larger size. The facts presented conform better with the pattern shown by many Labrids in which initial and terminal phase males (e.g. Lejeune 1985, Warner & Lejeune 1985, Taborsky et al. in press) do occur with differences in behaviour and GSI similar to those shown by males of class A and C of B. sanguinolentus. Lejeune (1985) and Taborsky at al. (in press) showed that several species of the European labrid genus Symphodus, where protogynuous hermaphroditism sometimes occur, show this pattern of change from initial to terminal males.

The differences in GSI and energy allocation in soma and gonads in male B. sanguinolentus indicates that class A males maximize the investment in gonads while parental males invest most of their resources in parental and territorial activities. It is possible that beyond a certain threshold an increased investment in testis ceases to be the critical factor to increase reproductive success Fertilization of eggs in the narrow space of a hole must minimize the amount of sperm needed. On the other hand, reproductive success of big males must depend on the efficiency in parental care and ability to control good quality cavities and attract females there. This pattern of investment contrast sharply with heavy allocation of resources into sperm shown by many pelagic species that fertilize eggs out of cavities and do not have parental care duties (Raitt 1932; Iles 1974; Woodhead 1979). On the other hand it presents interesting similarities with the pattern of investment of other small benthic fishes of comparable eco-ethology (Miller 1984; Patzner 1983,1984,1985; Podroschko et al. 1985). Females, on the contrary, invest very heavily in egg production and, judging from GSI values, their fecundity must have accelerating increases with size.

Both parental male and female pattern of investment conform with conditons defined by Sargent & Gross (1986) for the occurrence of uni-parental male care. This incorporation of male parental investment into the breeding system must increase very strongly the chances of survival of each egg. So, for a given quantity of energy invested by females in eggs the yield in surviving fry must be higher. Of course this increase in energy fry conversion efficiency, can not be invoked as a direct selective force and it can be better described as a beneficial side effect. The possession of this breeding system could endow a species with increased potential to colonize the intertidal zone. Low fecundity and the hazards involved in settlement of juveniles must make energy fry conversion rates prohibitively low for fishes without parental care to be successful in this habitat.

The tactics of small males needs further research. Is it frequent among rocky intertidal species? Are these fishes "making the best of a bad job"?

The evidence now available suggests that, in this case, lack of space for establishing territories may not be the crucial factors. Instead we suggest that: i. mortality is must higher for fishes between one and two years, decreasing in consequent years. So, while the chances of a small male to become a territorial one are small, there is an accumulation of parental males of older age; ii. females actively choose nest sites as in Pseudolabrus celidotus (Jones & Thompson 1980), or prefer to spawn in nests that already contain eggs as in Gasterosteus aculeatus (Ridley & Rechten 1981) and Cottus bairdi (Downhower & Brown 1980). The probability of a small male to attract females may be reduced if it adopts a territorial behaviour. High levels of offer of territories, and the possibility for each male to accommodate spawnings of several females facilitate female choice; iii. fanning and other parental activities retain territorial males in the nests 90% of the time, and must be critical for the eggs in waters which temperature reaches levels of 25°C, sometimes 27°C. Thus their patrolling opportunities are very limited, giving the small males many opportunities to approach the nests. The cover of algal clumps like Padina pavonia and the irregular topography in the pools, facilitate movements of small males. Could this behaviour occur commonly on other intertidal fishes living in similar conditions?; iv. as show above nests occur in dense concentrations in some pools. These nests concentration would make female choice economically far less expensive than if nest were very scattered; v. finally it is interesting to know to what extent the parental male can benefit from agonistic behaviour of satellites. They take an active role in chasing egg predators and also other conspecifics away. It is conceivable that a trade off could have evolved, in which parental males have to pay some costs in lost fertilizations compensated for by lower costs in nest defense.

ACKNOWLEDGEMENTS

We would like to thank José Carlos Silva, Norberto Serpa, Alierta Pereira, Olavo Amaral and Carmelina Leal for their technical assistance. We also thank Drs H. Rost Martins and H. Isidro for their helpfull comments and Professors A. B. Vieira and J. A. Martins for their support. Acknowledgements are also due to the participants of B.A.I.L. for their vivid discussions, their questions and remarks. First author (R.S.S.) also wants to thank and enhance the financial support of Calouste Gulbenkian Foundation (Lisbon) and N.A.T.O. Science Fellowships Programmes which made possible the preparation of this paper and the participation in the workshop.

REFERENCES

ABEL, E.F. 1964. Freiwasserstudien zur Fortpflanzungsethologie Zweier Mittermeerfische, Blennius canevae Vinc. und B. inaequalis C.V. Zeitschrift für Tierpsychologie, 21: 205-222.

ABEL, E.F. 1973. Zur ökoethologie des amphibish lebenden fishes Alticus saliens und von Entomachrodus vemiculatus unter besücksichtigung des fortpflanzmgsverhaltens stizungesber. Östeur. Akad. Wiss. Math Naturw. Kl., Abt 1(181): 137-153.

ALMADA, V.; J. Dores; A. Pinheiro; M. Pinheiro & R.S. Santos 1983. Contribuição para o estudo do comportamento de Coryphoblennius galerita (L.) (Pisces: Blenniidae). Memórias do Museu do Mar - Série Zoológica, 2 (24): 1-163.

ALMADA, V.; G. Garcia & R.S. Santos 1987. Padrões de actividade e estrutura dos territórios dos machos parentais de Parablennius pilicornis (Cuvier) (Pisces: Blenniidae) da costa portuguesa. Análise Psicológica, V (2): 261-280.

ALTMANN, J. 1974. Observational study of behaviour: sampling methods. Behaviour, 49: 227-269.

BATH, H. 1976. Revision der Blenniini. Senckenbergiana Biologica 57: 167-234.

BATH, H. 1981. Beitrag zur revalidation von Parablennius ruber (Valenciennes 1936) mit Kritischen Bemerkungen zur Gültigkeit der Gattung Pictiblennius Whitley 1930. Senkenbergiana Biologica, 62 (4/6): 211-224.

BREDER, C.M. 1941. On the reproductive behaviour of the sponge blenny, Paraclinus marmoratus. Zoologica, 26: 233-236.

DENOIX, M. 1984. Zur Biologie des Schleimfisches Parablennius pilicornis Cuvier 1829 (Blenniidae, Perciformes) unter besonderer Berucksichtigung der sekundären Geschlechtsmerkmale des Männchens nach Hormonbehandlung. Dissertation - Universität Tübigen

DOWNHOWER, J.F. & L. Brown 1980. Mate preferences of female mottled sculpins, Cottus bairdi. Animal Behaviour, 28: 728-734.

FISHELSON, L. 1963. Observations on littoral fishes of Israel: I Behaviour of Blennius pavo Risso (Teleostei: Blenniidae). Israel Journal of Zoology, 12: 67-80.

FISHELSON, L. 1975. Observations on the behaviour of the fish Meiacanthus nigrolineatus Smith-Vaniz (Blenniidae). Antr. J. Mar. Freshw. Res., 26: 329-341.

GIBSON, R.N. 1968a. The food and feeding relationships of the littoral fish in the Banyuls region. Vie et Milieu - A, XIX (1): 447-456.

GIBSON, R.N. 1968b. The agonistic behaviour of juvenile Blennius pholis L. (Teleostei) Behaviour, 30: 192-217.

GIBSON, R.N. 1969. The biology and behaviour of littoral fish. Oceanography and Marine Biology Annual Review, 7: 367-410.

GIBSON, R.N. 1982. Recent studies on the biology of intertidal fishes. Oceanography and Marine Biology Annual Review, 20:363-414.

GIBSON, R.N. 1986. Intertidal teleosts: life in a fluctuating environment. In: Tony J. Pitcher (Ed.), The behaviour of teleost fishes, Croom Helm, London

GOLDSCHMID, A.; K. Kotrschal & P. Wirtz 1984. Food and gut length of 14 Adriatic Blenniid fish (Blenniidae; Percomorphi; Teleostei). Zool. Anz., 213 (3/4): 145-150.

GROSS, M.R. 1982. Sneakers, Satellites and Parentals: Polymorphic mating in north american sunfishes. Zeitschrift für Tierpsychologie, 60: 1-26.

GROSS. M.R. 1984. Sunfish, salmon and the evolution of alternative reproductive strategies and tactics in fishes. In: R.J. Wooton and G. Potts (Ed.) Fish Reproduction: Strategies and tactics, Academic Press, New York

GROSS, M.R. 1985. Disruptive selection for alternative life histories in salmon. Nature, London, 313: 47-48.

GROSS, M.R. & E.L. Charnov 1980. Alternative male life histories in Bluegill sunfish. Proceedings National Academy of Sciences, USA, 77: 6937-6940.

GUITEL, F. 1893. Observations sur les moeurs de trois blenniidés: Clinus argentatus, Blennius montagui et Blennius sphynx. Archives de Zoologie Expérimentale et Général, 3eme sér. (I): 325-384.

HEYMER, A. 1985. Stratégie comportamentale du mâle pour la fécondation des oeufs chez Blennius basilicus (Teleostei, Blenniidae). Revue Française d'Aquariologie, 12 (1): 1-4.

HEYMER, A. & C.A. Ferret 1976. Zur ethologie des Mittelmeer-schleimfischeg Blennius rouxi Cocco 1833. Zeitschrift für Tierpsychologie, 41: 121-141.

ILES, T.D. 1974. The tactics and strategy of growth in fishes. In: F.R.H. Jones (Ed.) Sea fisheries research, John Wiley and Sons, Inc., New York

JONES, G.P. & S.M. Thompson 1980. Social inhibition of maturation in females of the temperate wrasse Pseudolabrus celidotus and a comparison with the Blennoid Trypterygion varium. Marine Biology, 59: 247-256.

de JONGE, J. 1985. Difference in mating strategies of Trypterygion tripteronotus and T. xanthosoma. Abstracts of spoken and poster paper: 19th International Ethological conference (Toulouse), 2: 514.

LEJEUNE, P. 1985. Le comportement social des labridés Méditerranéens. Cahiers d'Éthologie Appliquée, 5 (2): XII + 208 pp.

LOSEY, G.S. 1968. The comparative behaviour of some pacific fishes of the genus Hypsoblennius Ph.D. Thesis Scripts Instit. Oceanogr. University of California.

LOSEY, G.S. (Jr.) 1976. The significance of coloration in fishes of the genus Hypsoblennius Gill. Bulletin Southeastern California Academy of Sciences, 75: 183-198.

MAPSTONE, G.M. & E.M. Wood 1975. The ethology of Abudefduf luridus and Chromis chromis (Pisces: Pomacentridae) from the Azores. Journal of Zoology, London 175: 179-199.
MILLER, P.J. 1984. The tokology of gobioid fishes. In: G.W. Potts and R.J. Wooton (Ed.). Fish reproduction: Strategies and tactics, Academic Press, London
NURSALL, J.R. 1977. Territoriality in redlip blennies (Ophioblennius atlanticus - Pisces:Blenniidae). Journal of Zoology, London, 182:205--223.
NURSALL, J.R. 1981. The activity budget and use of territory by a tropical blenniid fish. Zoological Journal of the Linnean Society, 72 (1): 69-92.
PATZNER, R.A. 1983. The reproduction of Blennius pavo (Teleostei, Blenniidae). I. Ovarial cycle, environmental factors and feeding. Hegoländer Meeresuntersuchungen, 36: 105-114.
PATZNER, R.A. 1984. The reproduction of Blennius pavo (Teleostei, Blenniidae). II. Surface structures of the ripe egg. Zool. Anz., 213: 44-50.
PATZNER, R.A. 1985. The reproduction of Blennius pavo (Teleostei, Blenniidae). III. Fecundity. Zool. Anz., 214 (1/2): 1-6.
PATZNER, R.A., M. Seiwald, M. Adlgasser & G. Kaurin 1986. The reproduction of Blennius pavo. V. Reproductive behaviour in natural environment. Zool. Anz., 216 (5/6): 338-350.
PHILLIPS, R.R. 1971. The relationship between social behavior and the use of space in the benthic fish Chasmodes bosquianus Lacèpede (Teleostei: Blenniidae). I.Ethogram. Zeitschrift für Tierpsychologie, 48: 142-174.
PHILLIPS, R.R. 1977. Behavioral field study of the Hawaiian rock skipper Istiblennius zebra (Teleostei: Blenniidae). I. Ethogram. Zeitschrift für Tierpsychologie, 43: 1-22.
PODROSCHKO, S., R.A. Patzner & H. Adam 1985. The reproduction of Blennius pavo (Teleostei, Blenniidae). IV. Seasonal variation in HSI, the liver glycogen value and histological aspects of the liver. Zool. Anz., 215 (5/6): 265-273.
QAZIM, S.Z. 1956. The spawning habits and embryonic development of the shanny (Blennius pholis L.). Proceedings Zoological Society of London, 127: 79-93.
RAITT, D.S. 1932. The fecundity of the haddock . Scientific investigation Fishery Board of Scotland, 1: 1-42.
RIDLEY, M. & C. Rechten 1981. Female sticklebacks prefer to spawn with male whose nests contain eggs. Behaviour, 76: 152-161.
ROBINS, C.R., C. Phillips & F. Phillips 1959. Some aspects of the behavior of the blennioid fish Chaenopsis ocellata Poey. Zoologica, 44: 77-83.
SANTOS, R.S. 1985a. Estrutura e função dos territórios em machos parentais de Blennius sanguinolentus Pallas (Pisces: Blenniidae). Memórias do Museu do Mar - Série Zoologica, 3/29: 1-46.
SANTOS, R.S. 1985b. Parentais e satélites: tácticas alternativas de acasalamento nos machos de Blennius sanguinolentus Pallas (Pisces: Blenniidae). Arquipélago - Série Ciências da Natureza VI: 119-146.
SANTOS, R.S. 1986a. Capacidade de retorno à área vital padrão de dispersão e organização social em Blennius sanguinolentus Pallas (Pisces: Blenniidae) durante a época da reprodução. Psicologia, V(1): 121-131.
SANTOS, R.S. 1986b. Estudos sobre a ecologia e comportamento da fauna litoral dos Açores: I. Primeiras observações sobre o comportamento territorial e parental dos machos de Parablennius ruber (Pisces: Blenniidae), com uma pequena nota sobre os embriões. Açoreana, VI (1): 295-320.
SARGENT, R.C. & Gross, M.R. 1986. Williams' Principle: An explanation of Parental care in teleost fishes. In: Tony J. Pitcher (Ed.) The Behaviour of Teleost Fishes, Croom Helm, London
SIEGEL, S. 1956. Nonparametric Statistics for the Behavioral Sciences. McGraW-Hill Book Company, New York.
SOKAL, R.R. & F.J. Rohlf 1981. Biometry. W.H. Freeman and company San Francisco.

TABORSKY, M. & D. Limberger 1980. The activity rhythm of Blennius sanguinolentus Pallas, an adaptation to its food source? Marine Ecology, 1: 143-153.

TABORSKY, M., B. Hudde and P. Wirtz (in press). Reproductive Behaviour and Ecology of Symphodus (Crenilabrus) ocellatus, a European wrasse with four types of male behaviour. Behaviour.

THOMPSON, J.M. & A.E. Bennett 1953 . The oyster blenny, Omobranchus anolius. Australian Journal of Marine and Freshwater Research, 4: 227-233.

WARNER, R.R. and P. Lejeune 1985. Sex change limited by parental care: a test using four Mediterranean Labrid Fishes, Genus Symphodus. Marine Biology, 87: 89-99.

WICKLER, W. 1957. Vergleichende Verhaltensstudien an Grundfischen, I. Beitrage zur Biologie, besonders zur Ethologie von Blennius fluviatilis Asso im Vergleich zu einigen anderen Bodenfishen. Zeitschrift für Tierpsychologie, 14: 393-428.

WICKLER, W. 1961. Über das verhalten der blenniiden Runula und Aspidontus. Zeitschrift für Tierpsychologie, 18:421-444.

WICKLER, W. 1964. Zur Biologie und Ethologie von Ecsenius bicolor (Pisces, Teleostei, Blenniidae). Zeitschrift für Tierpsychologie, 22: 36-49.

WIRTZ, P. 1978. The behaviour of the Mediterranean Tripterygion species (Pisces: Blenniidae). Zeitschrift für Tierpsychologie, 48: 142-174.

WOODHEAD, A.D. 1979. Senescence in fishes. In: P.J. Miller (Ed.) Fish phenology: anabolic adaptiveness in teleosts. Academic Press, London

ZANDER, C.D. 1972. Beiträge zur Okologie und Biologie von Blenniidae (Pisces) des Mittelmeeres. Helgoländer wiss. Meeresunters, 23: 193-231.

ZANDER, C.D. 1978. Kritische Anmerkungen zur "Revision der Blenniini (Pisces: Blenniidae)" von H. Bath 1977. Z. Zool. Syst. Evol., 16: 290-296.

ZANDER, C.D. 1979. Morphologische und ökologische Untersuchung der Schleimfische Parablennius sanguinolentus (Pallas, 1811) und P. parvicornis (Vallenciennes 1836) (perciformes, Blenniidae). Mitt. Hamb. Zool. Mus. Inst., 76: 469-474.

Behavioural Adaptations of Sandy Beach Organisms: an Ecological Perspective

Anton McLachlan

University of Port Elizabeth
South Africa

INTRODUCTION

Few environments have as little stability or biological structure as exposed sandy beaches on open coasts. The hallmark of all inhabitants of these dynamic systems is consequently a very high degree of motility and the ability to burrow rapidly. Thus the ecologist studying such beaches perceives the interaction between swash/backwash processes, tides and movements of the fauna as being the central arena where the battles are fought and adaptations tested. The result may be interpreted as 'community structure', diversity, zonation or some such parameter.

On sheltered beaches the sand body is relatively stable, permanent burrows can be constructed and biotic interactions can develop and structure the community. The sheltered beach thus has some biological structure and modification of the physical environment and the requirements for motility are less stringent. The contrast between these different beach types is important. Indeed, they should be seen as different types of ecosystems, with the sheltered beach more akin to estuarine sandflats. As biological, physical and chemical processes on these different beach types differ markedly in most cases, trends should not too readily be extrapolated from one to the other. In this review I shall be primarily concerned with exposed beaches and will indicate, where necessary, the sheltered beach situation.

PHYSICAL FEATURES

An essential precursor to any examination of the ecology of sandy beaches is some appreciation of the morphology and dynamics of these systems. The slope of a beach face depends on the interaction of the swash/backwash processes planing it and sand particle size. The stronger the wave action, the flatter a beach will be for any fixed particle size and the coarser the sand the steeper the beach for any fixed level of wave action. Thus storms flatten beaches by moving sand offshore to expand the surf zone while calm conditions have the opposite effect. This results in a range of six morphodynamic states for surf zones and beaches (Short and Wright, 1983) (Fig. 1).

The two extremes are the dissipative and the reflective beach/surf zone. In between these are a series of intermediate states. The reflective end of the scale occurs when conditions are very calm and/or the sediment is coarse. Here all the sand is stored on the subaerial beach, there is no surf zone and waves surge directly up the steep beach face. Often in such cases the tidal range is also small. The beach face is characterised by a step on the lower shore, where incoming waves and backwash collide, depositing sediment, and by a berm or platform above the intertidal. Wave energy is reflected off such a beach face.

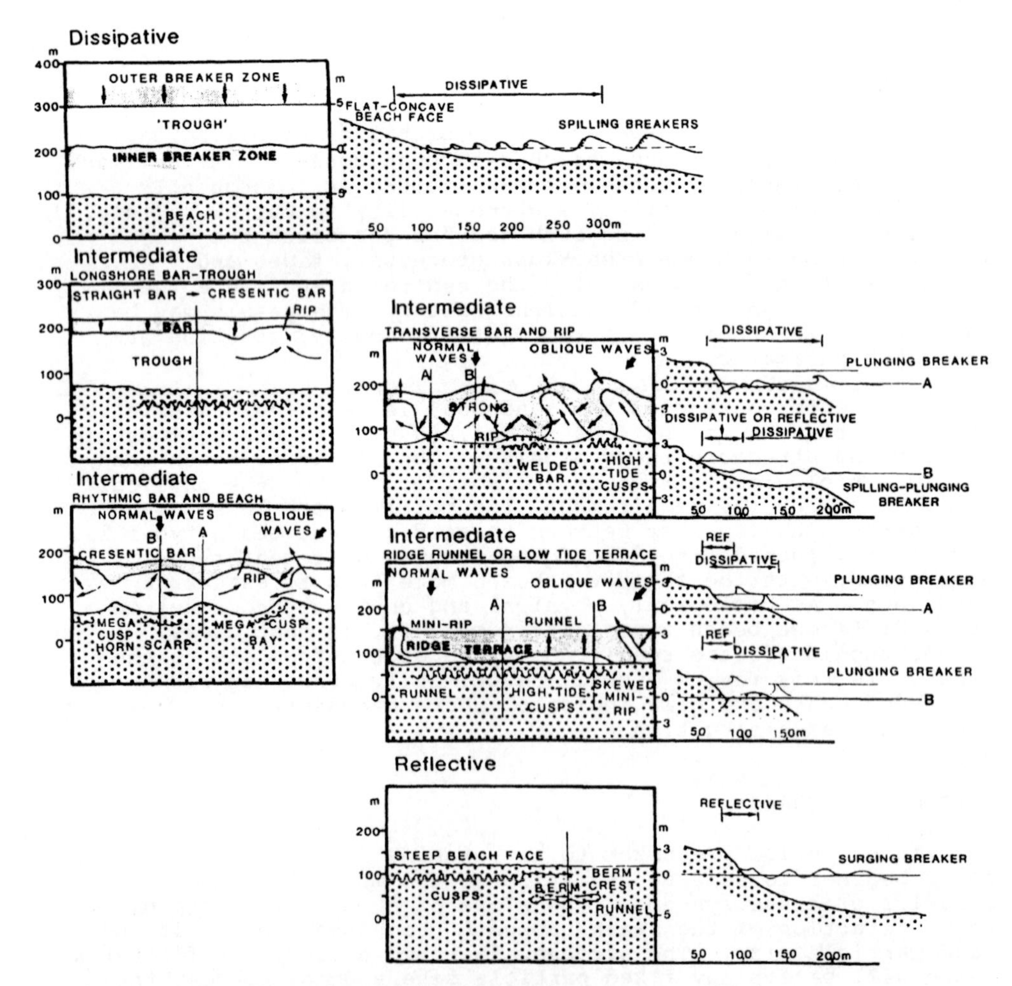

Fig. 1. The six major morphodynamic states of beaches and surf zones (After Short and Wright, 1983).

As bigger waves scour the beach and spread out its sediments to form a surf zone, the reflective beach gives way to a series of intermediate forms. If wave action is sufficiently strong and/or sediment particle size fine enough, the fully dissipative state may ultimately be reached. Here the beach is flat and maximally eroded and the sediment is stored in a broad surf zone which may have multiple bars parallel to the beach. Waves tend to spill and break far from the beach, often reforming and breaking again. In this way most wave energy is consumed in the surf zone before reaching the beach. The swash along the shoreline is usually gentle, although there may be pronounced infragravity waves. Landward water flow takes the form of surface bores while return flow is mainly as undertow, although rip currents may occur. Wave energy is thus dissipated in the surf zone rather than reflected off the beach face.

Between these two extremes are four intermediate states. These beach types are characterised by high temporal variability and sand storage both on the beach and in the surf zone, where bars and troughs demarcate well developed rip currents. Starting with a medium grained beach that has been planed down to a dissipative state after a violent storm, as conditions calm down and sediment starts to move shorewards, the next recognisable state appears as the longshore bar-trough system. Here the outer surf zone bars weld to form a single bar on which the waves break first. If still calmer conditions occur, the bar moves further shorewards and starts to undulate as a result of edge waves, forming the rhythmic bar and beach system. Here on the beach face, a series of rhythmic megacusps emerge. Rip currents are well developed, broaching the bar in between its horns. With prolonged calmness, the horns of the bar migrate shorewards to weld with the beach in places; the rhythmic bar may disappear between the horns. The outcome is a series of transverse bars with well defined rips separating them.

Progressing to yet calmer conditions, all the bars weld with the beach and form a low tide terrace in which ridges and runnels may occur as remnants of the bars and troughs. If the sediment from this terrace moves onto the beach the reflective state is again reached. Complete transition from fully dissipative to reflective states, where sand storage shifts from surf zone to beach, would not normally occur on one beach. Transition towards a dissipative state during storms may occur rapidly while transition towards the reflective extreme during calms is a much slower process.

Dissipative beaches usually occur where waves exceed 2 m and sands are finer than 250 µm, whereas reflective beaches occur where waves are less than 0.5 m high and sands coarser than 400 µm.

Beaches experiencing very large tide ranges (more than 3 m) are different. They are usually of low energy with find sand and have dissipative lower shores but reflective upper beach faces. They generally do not occur on open coasts but in more enclosed areas. Sheltered beaches with very fine sands are essentially dissipative sand flats. Thus dissipative conditions are not dependant on high wave energy, they can be caused by very fine sand. The slope of the beach face is determined by the balance between swash (uprush) and backwash forces. Where swash is strong and there is no backwash, i.e. coarse grained beaches with rapid drainage, a steep beach face is built up. Where backwash is strong, i.e. fine grained beaches where the sediment remains saturated, much scouring occurs and the beach face is planed flat.

The conditions occurring on the beach face are related to the overall morphodynamic state of a beach and are thus determined by the wave climate and sand particle size. Dissipative beaches are characterised by flat slopes traversed by long even swashes (infragravity bores) with periods characteristically 30-50s, whereas reflective beaches have steep faces and short, sharp swash/backwash pulses with periods of 4-10 s (McLachlan and Young, 1982; McLachlan and Hesp, 1984).

The swash climate experienced in the intertidal towards the dissipative extreme is clearly more conducive to macroscopic life, providing near optimum conditions for filter feeders and the general surfing movements of macrobenthos. Steep reflective beaches with rapid pulsing swashes, even where wave energy is relatively low, provide harsh conditions that few organisms can survive. Consequently they have impoverished macrofaunas, often restricted to supralittoral forms.

MACROFAUNA

The most important feature of an intertidal beach is the 'swash climate', which is a function of wave regime and particle size as manifested in a particular beach morphodynamic state. The adaptations of the macrobenthos are primarily to this climate, which in turn determines overall abundance and diversity of the fauna (Fig. 2). Through a variety of conditions, encompassing a large range of wave energy, the macrobenthic fauna increases logarithmically in abundance and linearly in diversity as beaches grade from reflective to dissipative states. Simply stated, this means that the flatter a beach, the wider the surf zone and the longer and more regular the swashes, the more condusive conditions are for macrobenthos. Sand flats, which are simply low energy dissipative beaches, support this generalization by having rich faunas in comparison with most open sea beaches.

The distribution of macrobenthic animals on sandy beaches exhibits patchiness, zonation and fluctuations due to tidal migrations, storms and other disturbances. Patchiness may result from movement and sorting by the swash, localised food concentrations or biological aggregations, possibly for reproduction (Loesch, 1957; Moueza and Chessel, 1976; Hayes,1977; Saloman and Naughton, 1978; Brown, 1982; McLachlan and Hesp, 1984; Sastre, 1985). These three factors are interrelated, for example swash movement may often concentrate both food and benthos in the same parts of the beach.

Patchy distributions have been widely reported (Bally, 1983a). Much work on biological aggregations on sandy beaches has been done on the Californian sand crab *Emerita analoga*. Efford (1965) showed very distinct aggregations which exhibited intraspecific zonation and tidal migrations. He found no evidence for a physical control of aggregation and suggested that it had a biological origin, arising from the behaviour of the animals themselves instead of a physical sorting by the environment. Dillery and Knapp (1969) studied longshore movement of the same species. They showed some longshore movement of individuals following the predominant currents and found the aggregations to be associated with beach cusps. Because individuals moved, but aggregations did not, they concluded that aggregations were made up of 'ever-changing sets of crabs' and that this, plus their association with cusps, suggested a physical basis for aggregation formation. Barnes and Wenner (1968) and Cubit (1969) attributed aggregation to behavioural response to the physical environment. Cubit (1969) showed that changes in the water content of the sand caused emergence and burial and this could explain tidal migrations. Further, he suggested that crabs riding the swashes would be transported laterally into 'convergence areas', such as cusps, faster than they would be transported out of them, and hence aggregations would form. He therefore concluded that aggregation was a result of water movement and not gregarious behaviour.

Fusaro (1980) subsequently described a movement up the beach and a tendency to disperse at night and suggested that not only wave convergence but also some behavioural response was

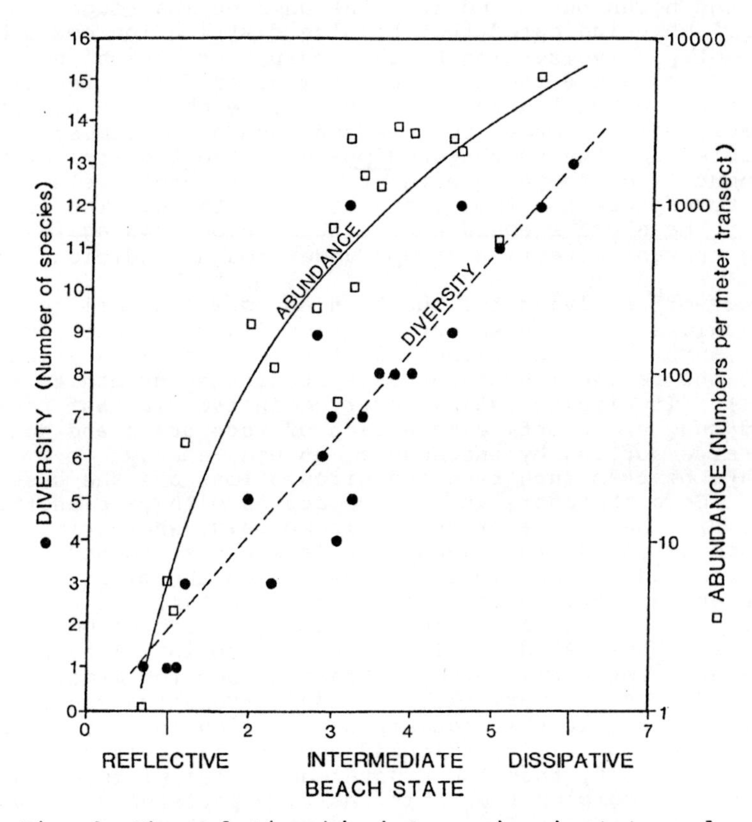

Fig. 2. The relationship between beach state and macrobenthic abundance and diversity (After McLachlan [in prep.]).

responsible for these patterns in *Emerita*. Perry (1980) took this further, indicating that both biotic and abiotic factors were associated with patterns of aggregation. Biotic factors included the avoidance of predation and, more importantly, aggregation for mating during the breeding season; abiotic factors concerned beach cusps. She suggested that higher wave energy and greater fish predation on the lower shore caused crabs to remain clumped in the upper beach. In the case of *Emerita analoga* it therefore seems that aggregations are the consequence of both behavioural responses and sorting by the physical environment in the form of swash patterns.

On beaches near the dissipative extreme, rhythmic features along the shoreline can have wave lengths of hundreds of metres, whereas on reflective beaches cusps have spacings in the order of tens of metres. Cusps on reflective beaches thus provide an ideal microcosm where the response of macrobenthos to some swash processes can be examined. McLachlan and Hesp (1984) did this on some west Australian beaches. The basic features of cusp circulation are illustrated in Fig. 3a. Essentially swash uprush is concentrated on the cusp horns, which are consequently steep, while lateral flow results in most backwash draining down cusp bays as mini-rip currents, the bays thus having flatter slopes. McLachlan and Hesp (1984) found two bivalves concentrated in the cusp bays while a hippid crab aggregated weakly on the horns but moved into the bays during rough conditions. They indicated that bivalve distribution could be the result of passive sorting by the 'cusp circulation' swash patterns. This raised the question of whether filter feeders would actively select flatter parts of the beach, such as cusp bays, because of the longer swashes and inundation times. It is also possible that the scavenger *Hippa* selected the area of maximum input (i.e. the horn with strong swash and little backwash) to capture the food items stranded there. The greater motility of these crabs could enable them to do this against the prevailing current direction except under rough conditions.

One way of resolving the question as to whether filter feeding bivalves actively select areas of flatter slope, would be to find a rhythmic shoreline with the reverse circulation pattern to the reflective beach cusps studied by McLachlan and Hesp (1984). If bivalves still occurred in the flattest region, this would suggest an active selection of such areas and not simply passive sorting by the swash. On high energy intermediate beaches such reversed circulations are the case (Fig. 3b). Here the shoreline has megacusps with wavelengths of 200-300 m. In the centre of the megacusp bays, where rip currents are situated, the beach face is steepest because wave energy is reduced in the rip area, whereas on the megacusp horn, where wave energy is focussed, slopes are flatter. In the eastern Cape we have investigated the distribution of *Donax* in such systems. These studies (T.E. Donn, unpublished) have shown a weak tendency for *Donax* to concentrate around the megacusp horns where slopes are very flat. Swash circulation would tend to concentrate them in the megacusp bays, which indicates some active selection for flatter slopes, even in bivalves which have a much lower motility than the crustaceans. The picture may, however, be more complex than this, as flow patterns can reverse during high tide. Grant (1981) showed that amphipods concentrate on ripple crests in a sheltered sand flat and related this to active preference rather than passive sorting.

This evidence suggests the following scenario: the more motile an animal, the faster it can burrow and the more its behaviour may override the physical forces operating on the beach face. Thus scavenging crabs, being highly motile, can actively concentrate around steep cusp horns. Less motile forms, such as bivalves which surf the swashes, instead of actively swimming, have less control over their movements but can nevertheless exhibit some selection for those parts of beaches with the flattest slopes. Flat slopes and long even swashes are most conducive to filter feeding whereas scavengers are less dependant on slope/swash conditions. It is suggested that these two major feeding groups show some active selection for different swash climates on the beach face. Steep slopes, characterised by strong uprush of the swash but little backwash, because of rapid drainage of the swash into the sand, trap materials on the surface and may provide optimal conditions for scavengers. Flat slopes, composed of finer sands, remain waterlogged and tend not to receive much swash drainage; thus backwashes are strong and swash periods long, providing adequate water coverage for filter feeders. Deposit feeders are only encountered on sheltered shores where the sand is relatively stable.

The dynamics of the beach face are determined by swash period and intensity, beach face slope, sand grain size and the position of the water table. These interact in a complex fashion, some aspects of which are illustrated in Fig. 4 (Duncan, 1964). The upper swash zone, and particularly the area where water infiltrates the sand, is characterised by accretion, whereas the lower swash zone is eroded. The position

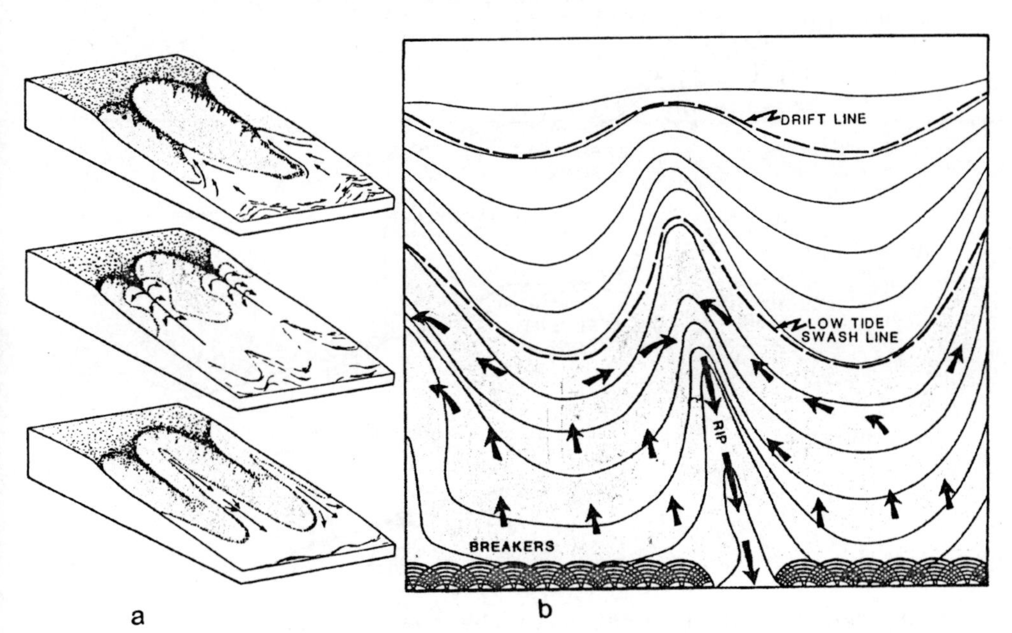

Fig. 3. Circulation patterns on (a) a reflective beach with cusps and (b) a high energy intermediate beach with a rhythmic shoreline.

of the water table influences the position of the erosion and accretion areas. Vilas (1986) coupled the movements of amphipods to the changes in sediment profile (Fig. 5). We need to experimentally separate each of these factors and study the response of different faunal components. Community and population patterns on exposed sandy beaches are the consequence of individual responses to the swash climate, sand movement and liquefaction on the beach face and are not greatly controlled by biological interactions.

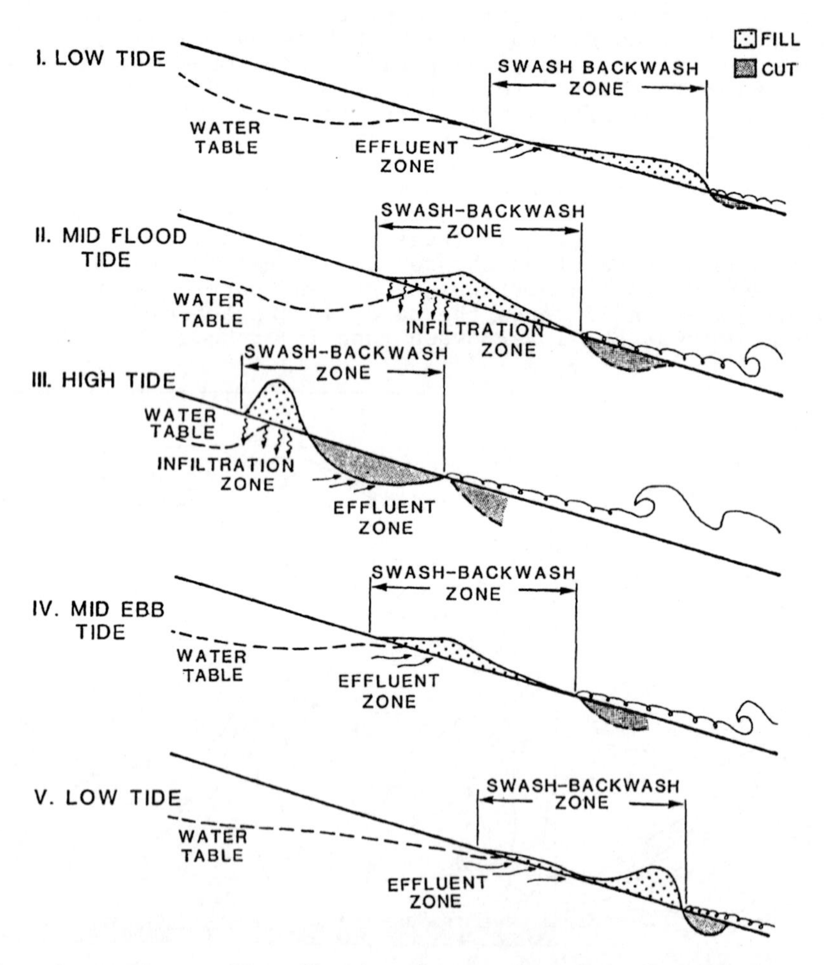

Fig. 4. Sediment distribution in the swash-backwash zone at various stages of a semidiurnal tidal cycle. Successive stages do not indicate composite sediment distribution over the cycle; the initial distribution of sediment at each stage is indicated by a straight profile line (After Duncan, 1964).

Swash period is important in regulating movement and burrowing on the beach face. Where swash period is too short, there is insufficient time for digging and burial, even though Donax for example, only needs to be $^{2}/_{3}$ buried to resist the swash (Trueman, 1975; McLachlan and Young, 1982). Burrowing rate is determined by size, temperature and substrate properties. Most species adapted to exposed beaches can burrow within 30s, many within 5s (Ansell and Trevallion, 1969; Ansell and Trueman, 1973; McLachlan and Young, 1982; Ansell, 1983). The ecophysiology and adaptations of animals to life on sandy beaches have recently been reviewed by Brown (1983) and McLachlan (1983).

Besides motility and the ability to burrow rapidly, other adaptations to these dynamic environments include brood protection (Croker et al., 1975; Wooldridge, 1981; Dexter, 1984) sensitivity to water flow, current direction (Brown and Talbot, 1972; Brown, 1973, 1982) and the degree of thixotropy of the sand as well as the acoustic shock of breaking waves (Turner and Belding, 1957; Wade, 1967; Trueman, 1975; Tiffany, 1972). Most crustaceans show positive rheotaxis and may also sense changes in hydrostatic pressure (Brown, 1973; Enright, 1962). In the presence of turbulence some species burrow while others emerge and swim (Brown, 1973, 1983; Jones and Hobbins, 1985).

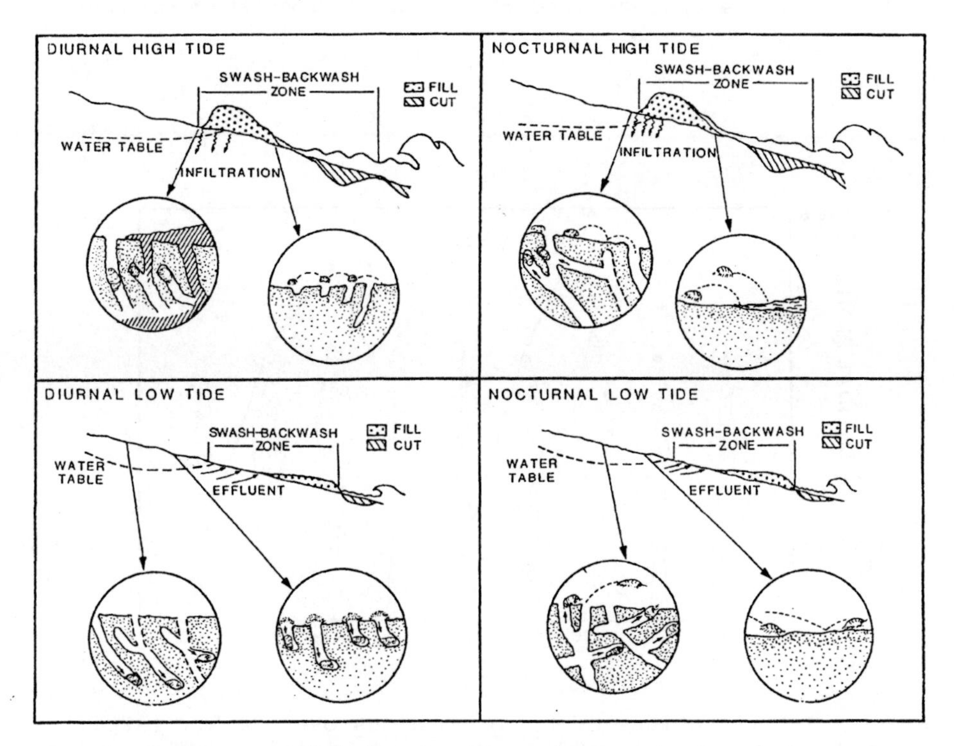

Fig. 5. Responses of Talorchestia saltator to beach face changes (After Vilas, 1986).

The responses of beach macrobenthos to the dynamics of the beach face consist of more than mere reactions to swash/backwash processes. Tides shift the swash zone up and down the shore and most inhabitants of open beaches follow this oscillation by tidal migratory behaviour. In most cases this involves movement on the beach following the swash zone. This can take the form of distinct, almost synchronised movements of whole populations or more gradual shifts of populations as a consequence of scattered individual movements (Fig. 6). In molluscs such as Donax and Bullia it involves no endogenous rhythms, only a series of responses to changing physical conditions such as the thixotropy or dilatancy of the sand and breaking of waves on the beach face (Fig. 7) (Mori, 1938; Turner and Belding, 1957; Ansell and Trevallion, 1969; Trueman, 1971; Tiffany, 1972). The movement of these molluscs is greatly facilitated by the extension of the foot which, presenting a broad surface area to the waves, acts as an underwater sail. This also ensures that the animal is able to dig immediately at the end of an excursion (Brown 1983).

On steep reflective beaches the shock of breaking waves may be an important trigger for emergence (Turner and Belding, 1957; Trueman, 1975; Tiffany, 1972) whereas on more dissipative beaches this is not the case. Liquefaction of the sand alone cannot entirely explain emergence as the sand is always saturated on the lower shore of dissipative beaches. Other triggers are needed, presumably some feature of swash/backwash-erosion/accretion processes occurring on the beach face. Being positioned in the area of erosion (Fig. 4) may, for example trigger emergence into the swash. In crustaceans the

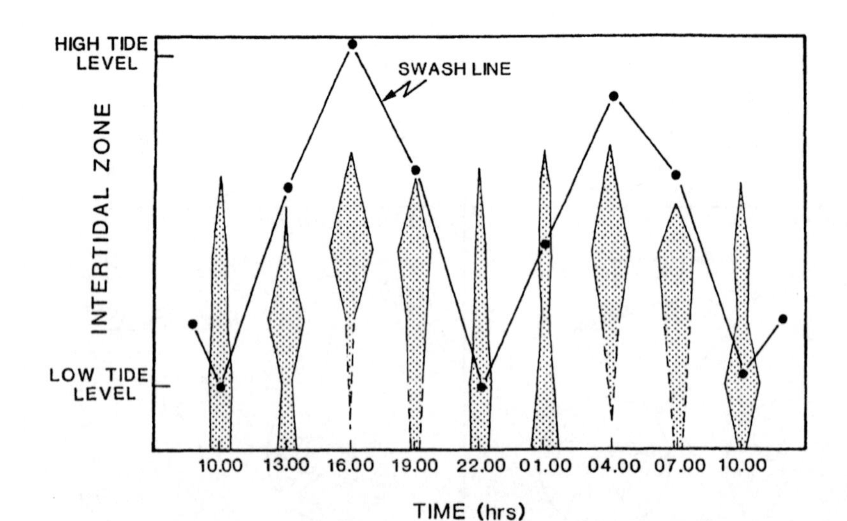

Fig. 6. Tidal migrations of Donax sordidus over 24 h. (After McLachlan et al. 1979). Movement of individuals causes a gradual shift of the population up and downshore.

circatidal rhythms are endogenous and usually superimposed on other rhythms such as circadian rhythms, which include planktonic phases at night, and circasemilunar rhythms, with increased activity during spring tides (Papi and Pardi, 1963; Enright 1963, 1965, 1972; Hamner et al., 1968, 1969; Cubit, 1969; Fincham,1970, 1973; Alheit and Naylor, 1976; MacQuart-Moulin, 1977; McLachlan et al., 1979; Marsh and Branch, 1979; Hager and Croker, 1980; Fusaro, 1980; Jaramillo

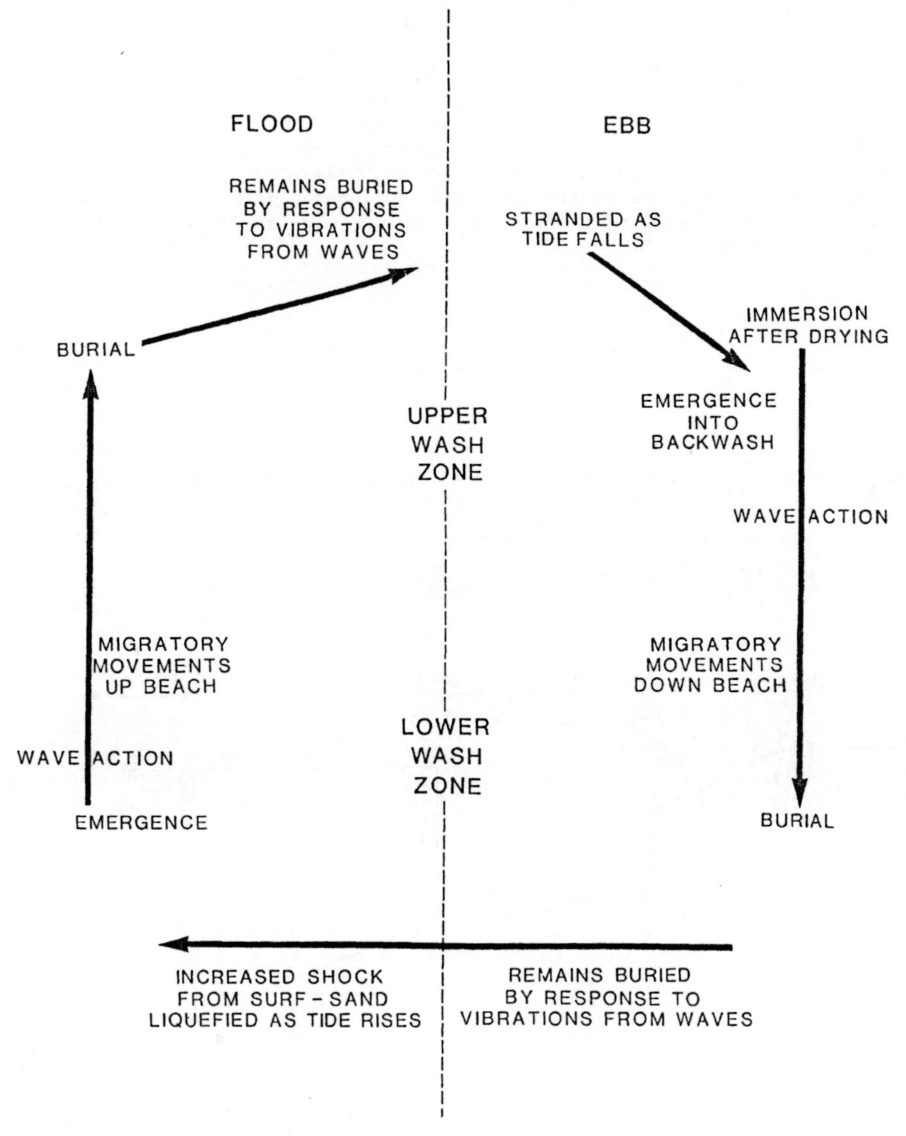

Fig. 7. Main features of the migration cycle of Donax (After Trueman, 1975).

et al., 1980; Scapini and Ugolini, 1981). Swimming behaviour after dark assists reproductive contact of sexes that may be segregated in the sand (Hager and Croker, 1980). The higher motility of crustaceans enables them to migrate more actively than molluscs; they are generally capable of moving about the beach face, even against current flow. Tidal rhythms may be entrained by immersion/exposure cycles, temperature cycles or alternations of hydrostatic pressure (Naylor et al., 1971).

In a review, Jones and Hobbins (1985) (Fig. 8), concluded that rhythmicity in cirolanid isopods was essential for the maintenance of zonation on the lower shore and sublittoral; offshore species showed only circadian and not circatidal rhythms. Endogenous circa-semilunar rhythmicity (Reid and Naylor, 1985) ensures maximum swimming activity on spring tides. In the intertidal, however, zonation is unlikely to be controlled only by rhythmicity, because, should isopods be swept away from their normal habitat, the timing of their activity to the high tide period alone could not return them to their preferred zone (Jones and Hobbins, 1985). Photic reactions, substrate selection and responses to turbulence, pressure and beach slope may all maintain zonation. Jones and Hobbins (1985) recorded differential responses to turbulence and hydrostatic pressure in cirolanids.

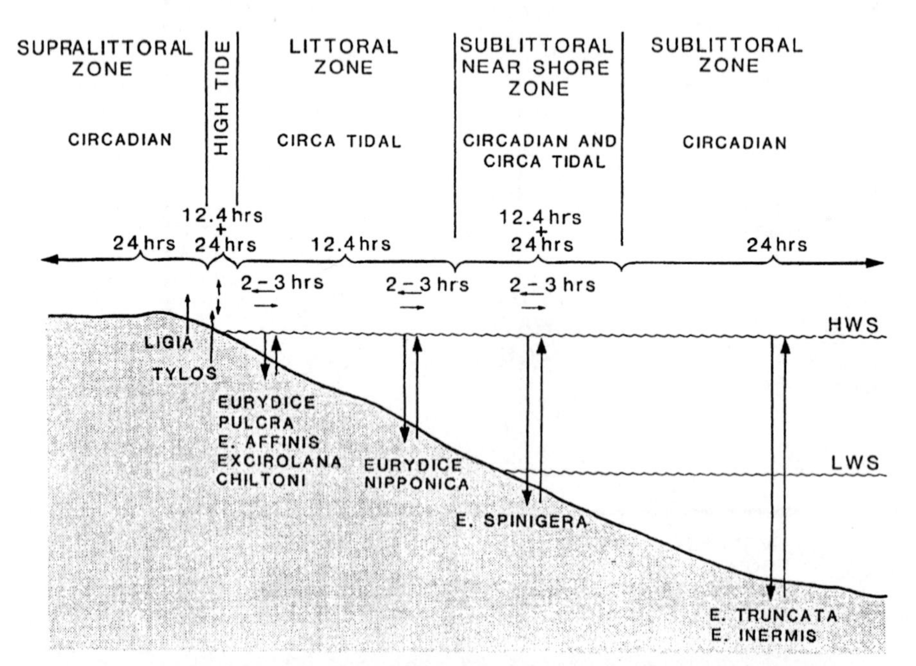

Fig. 8. Zonation and activity rhythms of some sand beach isopods with distributions ranging from supralittoral to offshore (After Jones and Hobbins, 1985).

Circa-semilunar movements have also been noted in a mollusc, Donax serra (McLachlan et al., 1979) and coupled to movements of the groundwater table in the beach face (Donn et al., 1986). As the position of the groundwater table affects the saturation profile and consequently accretion/erosion of the beach face, it would be expected to regulate movements of organisms living in the swash zone which follow cues related to beach face slope.

Ansell and Trueman (1973) estimated the energy costs of migration in Donax and Emerita. They showed that these were small in comparison to energy requirements for maintenance of position and concluded that it was more profitable for the animals to migrate than to maintain their position as the swash zone crossed them. However, in reviewing the genus Donax, Ansell (1983) indicated that energy costs of migration might be higher. These species have a high sensitivity to external stimuli, sensitivity to acoustic and mechanical stimuli in Donax, for example, being effected by the cruciform muscle sense organ (Frenkiel, 1982).

Not all macrofauna migrate tidally. Even in the genus Donax there are species that remain at characteristic levels, some which migrate at times and become stranded, others that move offshore in winter or summer and one where juveniles migrate but adults maintain a fixed position (Ansell, 1983). Nevertheless, tidal migration is still widespread on exposed beaches (Brown, 1983) and the above exceptions mostly refer to sheltered situations.

Tidal migratory behaviour has several advantages for sandy beach animals. It maintains populations in the optimal feeding zone, i.e. the swash, where water coverage is almost continuous but wave action not too severe. The animals are too shallow to be reached by fishes and in a haphazard position for bird predation. Migrating animals are less likely to be stranded as the beach changes with tides, storms and calms as they remain in the zone of sediment reworking. Tidal rhythms enable supralittoral forms to safely move downshore for feeding during the low tide. Tidal migrations have marked effects on zonation: all zones compress towards the high tide period when most of the invertebrate populations are concentrated into a narrow strip. In very sheltered beaches much of the fauna may form fairly permanent burrows and therefore not undergo tidal migrations.

The supralittoral fauna of sandy beaches consists of crustaceans and insects. The crustaceans, ocypodid crabs, oniscid isopods and talitrid amphipods exhibit circadian rhythms and some also follow circatidal rhythms (Jones, 1972; Kensley, 1974; Hamner et al., 1968; Williams, 1980). In most cases maximum activity coincides with the nocturnal low tide. Activity centred around the nocturnal low tide reduces predation, desiccation and the chances of tidal inundation (Marsh and Branch, 1979). Chelazzi and Ferrara (1978) found a transition from nocturnal to diurnal activity coinciding with an increasing distance back from the beach in four isopods. The species furthest from the beach had a diurnal activity pattern and was the most resistant to desiccation while that closest to the beach showed the opposite trend. Jones (1972) found one species of Ocypode to be nocturnal and the other diurnal. The former delayed emergence if the burrow was wet, thereby being less active during spring than neap tides (Barrass, 1983).

In supralittoral forms, where migrations often cover large distances and represent a transition from terrestrial to intertidal conditions, accurate orientation is necessary. This has been best studied in talitrid amphipods. In Talitrus saltator, for example, both the landscape and astronomical solar cues aid navigation, with the latter more important (Pardi and Papi, 1952; Craig, 1973; Hartwick, 1976; Ugolini et al., 1986). These amphipods may turn towards the sea when dehydrated and towards the land when wetted (Ugolini et al., 1986). Other orientation factors include polarised light (Brown, 1983) shore slope (Craig, 1973; Ercolini and Scapini, 1974) and local optical factor (Ercolini et al.,1983). These locomotor activity rhythms of supralittoral amphipods are entrained by photoperiod (Williams, 1980a, b).

All these migrations, responses and orientation mechanisms result in sandy beach animals maintaining characteristic levels on the shore when the tides recede. Thus, despite the dynamicism of the environment, zones may be recognised during low tide. Dahl (1952) first described three zones, based on crustacean distribution patterns and corresponding to the zones of rocky shores. The supralittoral fringe has ocypodid crabs in warm areas, talitrid amphipods in cold areas and oniscid isopods in intermediate areas. The midlittoral has various groups, typically cirolanid isopods, and the sublittoral fringe contains an even higher diversity which includes hippid crabs, haustorid amphipods, portunid crabs and mysids. Subsequently Salvat (1964, 1966, 1967) replaced this biological definition of zones with a physical division into four regions based on moisture content. A zone of dry sand above the shore is equivalent to the supralittoral; below this the zone of retention, reached by all tides, is equivalent to the midlittoral; the zone of resurgence has much water flushing through the sand, drained at low tide it corresponds to the upper boundary of Dahl's sublittoral fringe; the zone of saturation which never dries out and is subject to sluggish interstitial flows, Dahl's sublittoral fringe. The difference between Salvat's latter two zones is small but often distinct (Withers,1977; Bally, 1983b; Wendt and McLachlan, 1985). Boundaries between these zones are not usually sharp (Fig. 9). The ability to burrow into the substrate and so escape desiccation, together with the dynamicism of the environment, make zonation much less marked than on rocky shores. Selection of these zones by the fauna probably results from a complex of interactions including migratory rhythms and responses to beach slope, swash speed, turbulence and sand thixotropy and dilatancy.

Generally crustaceans dominate towards upper tide levels and molluscs towards the lower; crustaceans are often more important on tropical beaches and molluscs on temperate beaches. Crustaceans, because of their greater mobility, are best able to exploit coarse-grained, very dynamic (reflective) beaches, molluscs less so and polychaetes the least, being almost confined to medium to low energy fine-grained shores. Dexter's (1983) studies indicate that crustaceans often dominate the most exposed shores, polychaetes the most sheltered and molluscs intermediate shores.

The depth, slope and orientation of supralittoral burrows together constitute a complex topic which may be related to

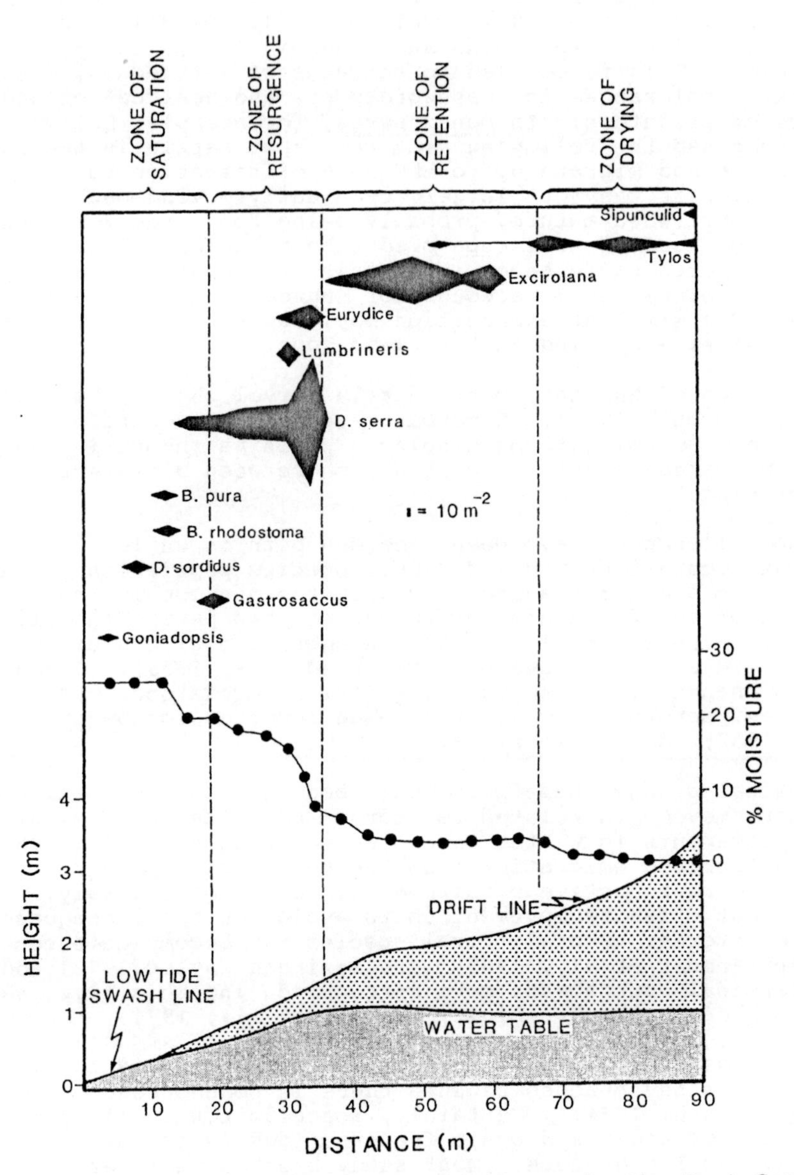

Fig. 9. Zonation of intertidal macrofauna in the East Cape, South Africa, including a moisture profile for the surface sand (After Wendt and McLachlan, 1985).

beach slope, moisture levels in the sand, distance from the sea and temperature (Hill and Hunter, 1973; Williams, 1983; Duncan, 1986).

Intraspecific zonation has been recorded in many sandy beach populations (Wade, 1967; Philip, 1974; De Villiers, 1975; Dexter, 1977; McLachlan and Hanekom, 1979; Haley, 1982; Bally,1983b). This is coupled to patterns of settlement, differential tolerances to swash processes and desiccation and avoidance of predators. In Donax serra, for example, Bally (1983b) proposed the following picture: spat settle in the zone of saturation and migrate up to the zone of retention to avoid fish predation as they get larger, eventually moving back into the sublittoral when mature, probably being too large for fish predators and needing to escape predation by gulls. Edwards (1969) considered differences in osmotic, temperature and desiccation tolerances to account for segregation by size in Olivella. The simplest explanation may, however, be differential size sorting in the swash zone.

Selection of habitat, particularly larval choice, has been reviewed by Brown (1983). Numerous responses by adults also play a rôle. Swimming forms display a positive rheotaxis and may in some cases be able to distinguish between different grades of sand.

Temporal changes have been recorded both in whole macrofauna communities and individual species populations. The most dramatic are those associated with the monsoon in India where much of the fauna disappears during this period (Ansell et al., 1972; McLusky et al., 1975). Longer term fluctuations spanning several years have been recorded (Coe, 1955). Dramatic long term changes may also take the form of occasional mass mortalities, several of which have been reported for Donax spp. (Loesch, 1957; de Villiers, 1975).

Seasonal changes have been described for most species. On/offshore movements related to storms and calms and similar offshore movements in winter have been recorded by Leber (1982 a,b). He ascribed emigration from the beach to temperature declines and reproductive requirements. It is more likely, however, that offshore movement is to avoid, or the consequences of, winter storms. Supralittoral species may become quiescent in winter (Hamner et al., 1968; Bregazzi and Naylor,1972) and some intertidal crustaceans move offshore during this time, so losing much of their rhythmicity (Naylor et al., 1971).

It has generally been assumed that competition is not important on sandy beaches because there is no shortage of food or space (Branch, 1984). Certainly, space is not as limiting as on rocky shores and for deposit feeders, food is not as restrictive as for grazers. Most sandy beach animals are opportunistic and intraspecific competition may be more important than interspecific competition. Opportunists can make use of disturbances. Davis and Von Blaricom (1982) showed the use of feeding pits, created by rays, by motile crustaceans which aggregated in the pits to feed on the organic debris accumulating there. They recorded no competition between potentially competing infaunal species and these opportunists. Thistle (1981) suggested that these opportunists appear because the disturbance concentrates food and that intraspecific depletion of resources is more important than interspecific competition.

Beach animals can partition the habitat both horizontally and vertically. In sheltered situations there are many species which may be more specialised and less opportunistic and there is greater potential for competition. Croker (1967) described spatial partitioning of the habitat between five haustorid amphipods on low energy beaches. Close horizontal and vertical partitioning in the sand suggests that some competition (Croker and Hatfield, 1980) may occur.

On a very sheltered shore, Woodin (1974) showed that tube-building polychaetes inhibit deposit feeders. Wilson (1981), also working on a very sheltered shore, demonstrated negative interactions in a dense assemblage of deposit feeding polychaetes, with large burrowers adversely affecting small surface deposit feeders which in turn ingested the larvae of the former. Large tubes of polychaetes provide protection from predators and disturbance and cause an increase in diversity (Woodin, 1981); sea grass beds have a similar effect. Competitive interactions can thus occur but appear to be limited to low energy beaches.

In all cases where predators have been excluded from soft bottoms, diversity has increased. This is in contrast to the situation on rocky shores where predators are considered to hold potentially competing populations below the level where competitive interactions intrude, thereby increasing diversity (Branch, 1984). This again implies that competition in sediments is less important than on rocky shores. Branch (1984) discussed a number of possible reasons for this which were proposed by C.H. Peterson. These include: the inability of competitors to crush or undercut in sediments, the unlikelihood of overgrowth, the extra depth dimension available, interactions between adults and juveniles that suppress population densities and the improbability of starvation.

Three groups of predators feed on sandy beaches, birds, fishes and invertebrates. The latter includes crabs, gastropods and asteroids. Most work on the effects of such predation has been done in less exposed environments. Reise (1984, 1985) studied the sand flats of the German coast, an environment of very low energy, thus more estuarine than truly open marine. Through caging experiments he showed that large carnivorous fish and birds had little effect on the benthos although they did cause a reduction in mean size by cropping the larger individuals. Birds and fishes limited the numbers of small predators (Fig. 10). Reise (1985) also investigated predation by crabs, shrimps and juvenile fishes and recorded significant effects on meiofauna and small macrofauna as well as juveniles of larger macrofauna; large population increases occurred in the absence of these predators. He concluded that predation is the most important biological interaction on tidal flats and that it decreases upshore. Effects of predation are generally assumed to be most severe in benign habitats and to decrease with increasing harshness (Connell, 1975).

Naticid gastropods are predators of bivalves in sheltered beaches (Ansell, 1982; Berry, 1982). Crabs may be important predators over a wide range of beach types including very exposed beaches. Du Preez (1984) showed the swimming crab Ovalipes to be a significant predator of molluscs on high energy beaches. The importance of crabs as predators is also mentioned by Loesch (1957), Ansell et al. (1972) and Virnstein (1977), although only the latter showed their impact on macrofauna communities in subtidal sand in Chesapeake Bay. Other important predators include juvenile flatfishes (Edwards and Steele, 1969; Poxton et al., 1982) other fishes (Lasiak, 1982; McDermott, 1983; Edwards, 1973) and birds (McLachlan et al., 1980; Myers et al., 1980).

Despite this circumstantial evidence, the importance of competition, disturbance or predation in structuring benthic populations on exposed sandy beaches is still uncertain. The fact that such beaches are so dynamic implies both that caging or other manipulative experiments are incongruous and that significant effects on the benthic populations are unlikely. There seems to be little basis for competition with disturbance a normal part of the environment. Predation certainly occurs and may, in the less exposed situations, have a significant impact on benthic populations.

Underwood and Denley (1984) formulated and criticised some generalisations about biological interactions on rocky shores based on the literature. These generalisations included the ideas that predation is most important in benign environments, competition is more important in harsher environments and physical stress is most important in the harshest environments. The autecological hypothesis of Noy-Meir (1979) is similar, suggesting that in extreme environments biological interactions should be minimal; instead populations are limited by abiotic factors. Exposed sandy beaches clearly fall into the category of physically stressed environments and autecological studies looking at adaptations to abiotic factors are probably the best way of understanding communities and populations in such systems.

INTERSTITIAL FAUNA

The three-dimensional, lacunar interstitial environment is markedly different to that occupied by the macrofauna. It is perpetually dark and dependant on water inputs through the

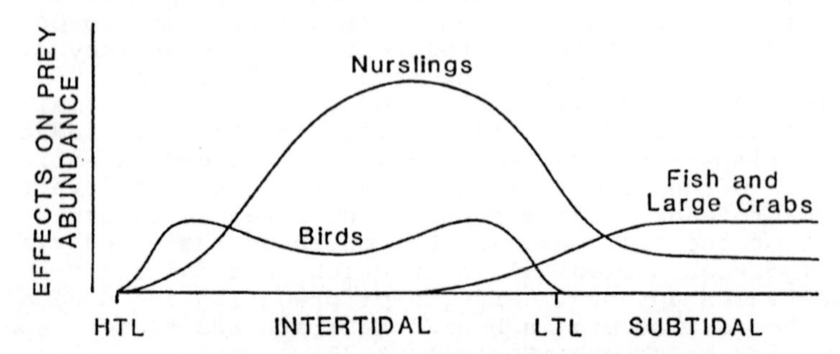

Fig. 10. Distribution of quantitative predator effects on prey populations of North Sea sand flats (After Reise,1985).

surface for a supply of oxygen and food materials. Most physical properties of this system are directly determined by the sediment characteristics and the patterns of water percolation. Particle size and sorting determine porosity and permeability. This limits the body size of burrowing, soft-bodied organisms and on open beaches most are interstitial. Fine sand with limited permeability restricts water percolation and thus tends to be less well oxygenated.

Water filtration, controlling the supply of oxygen and food into the sediment, is the key process controlling the interstitial climate. In the intertidal this is accomplished by swash flushing caused by waves and tides, whereas in the sublittoral it is driven by wave pumping (Riedl, 1971; Riedl and Machan, 1972; Riedl et al., 1972). In the intertidal most input occurs in the upper half of the shore over the high tide period. The volumes of water filtered increase from dissipative to reflective beach states but the residence times of this water in the beach sand decrease. Thus high energy intermediate beaches may filter 10 $m^3.m^{-1}.d^{-1}$ with a residence time around 24 h whereas reflective beaches can filter 50-100 $m^3.m^{-1}.d^{-1}$ with residence times of 1-5 h (McLachlan et al., 1985).

The interstitial system of reflective beaches is subject to extreme hydrodynamic forces and the interstices are very well flushed and oxygenated whereas dissipative beaches tend towards the other extreme. Oxygen availability, the superparameter in interstitial chemistry, is determined by the balance between swash flushing and organic inputs. Stagnant conditions on fine-grained, gently sloping beaches result in poor oxygenation of the sediment and the resultant reduced layers are often close to the surface, especially where there are high organic inputs (Fenchel and Riedl, 1971; McLachlan, 1983). The reflective beach has an interstitial system that is openly three-dimensional, with weak vertical gradients in chemistry (McLachlan et al., 1979) and subject to strong physical forces, whereas the dissipative beach, with its vertically compressed interstitial system, is regulated largely by strong vertical chemical gradients (Fenchel and Riedl, 1971) and faces only weak hydrodynamic forces. Simplified, one can thus perceive the interstitial system of beaches as occurring along a continuum between the physical and chemical extremes, corresponding to the reflective and dissipative beach states, respectively.

From the above description, the interstitial system of the two beach extremes may be zoned as follows (Fig. 11): the chemically controlled system has an upper oxygenated layer, a transition layer, or redox potential discontinuity, and a reduced or black layer where toxic reduced compounds, such as H_2S and NH_4, exclude most metazoans; the physically controlled system has four or five well ogygenated strata similar to those described by Salvat (1964) i.e. zones of: dry sand, retention or moist sand, resurgence around the low tide water table and saturation where oxygen tensions may be low. In the chemically controlled environment metazoan life is focussed towards the surface oxygenated layer whereas physical forces of the other beach extreme concentrate interstitial organisms in the moist layer where there is a combination of maximum water and, therefore, oxygen and food input but minimum hydrodynamic disturbance.

Dominant groups are nematodes, harpacticoid copepods, turbellarians and oligochaetes. Nematodes tend to be more abundant in finer sediments and harpacticoids in coarser sands. Interstitial meiofauna require different adaptive strategies to survive life in each of the two extreme beach states. Chemically controlled sediments bear a greater proportion of large bodied forms, burrowers as opposed to the vermiform interstitial sliders. Furthermore, the physiology and metabolism of the fauna is adapted to lowered oxygen tensions, toxic compounds and reduced organic compounds as food sources. Here, also, the stability of the substrate and concentration of the fauna at the surface facilitate macrofaunal predation on the interstitial fauna.

At the physical extremity the fauna are interstitial forms capable of withstanding strong pulsing currents flowing through the interstices at speeds which may exceed one body length per second. Distribution vertically into the sediment avoids potential macrofaunal predation.

Little is known about adaptations of the interstitial fauna to conditions on physically controlled beaches. These may include vertical migrations coupled to cyclical changes in moisture and temperature in upper beach sands (McLachlan et al., 1977). On beaches where the sands are always saturated migrations are less marked (Harris, 1972). Boaden and Platt (1971) found that nematodes migrated downward when crossed by the tide. Such position changes have been recorded in response to rain, wave disturbance and tidal factors (Bush, 1966; Boaden, 1968; Rieger and Ott, 1971; Meinke and Westheide, 1979). A further adaptation includes the ability to adhere to sand grains to avoid expulsion from the sediment during disturbance.

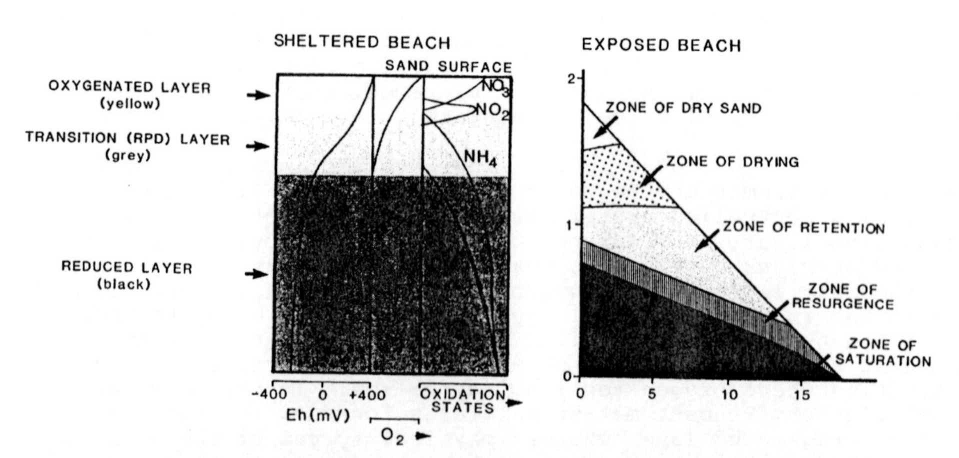

Fig. 11. Schematic representation of the zones or strata of meiofauna distribution in (a) a sheltered beach and (b) an exposed beach (After Fenchel and Riedl, 1971; Pollock and Hummon, 1971).

Most studies on the interstitial system have adopted a 'black box' approach. More work is needed on individual species, their responses to water currents, means of locomotion, feeding habits, etc. Some work on physiological tolerances has been published (Jansson 1967, 1978; Wieser et al., 1974), but unfortunately adaptive strategies may not be reconstructed from this information. The interstitial system is complex and three-dimensional and the navigation and survival strategies of its resident micrometazoans requires further investigation.

CONCLUSION

The behaviour of animals on exposed sandy beaches consists of a series of responses to interactions between swash- backwash processes on the one hand and beach face slope, water content and accretion/erosion on the other. Most of the patterns of animal movement and distribution are the accumulated results of these individual responses. The reaction of macrobenthic fauna to swash processes needs to be investigated in more detail. This can be done by carefully planned field studies and experimental analysis of the effects of isolated factors - such as swash period, swash speed, beach slope, backwash, accretion and erosion - on the burrowing and burial rates and movements of animals in the laboratory. Similarly, responses of interstitial forms to isolated components of interstitial climate need to be studied in the laboratory.

REFERENCES

Alheit, J. and Naylor, E. 1976. Behavioural basis of intertidal zonation in Eurydice pulchra Leach. J. exp. mar. Biol. Ecol. 23: 135-144.

Ansell, A.D. 1982. Experimental studies of a benthic predator-prey relationship. 1. Feeding, growth and egg-collar production in long-term cultures of the gastropod drill Polinices alderi (Forbes) feeding on the bivalve Tellina tenuis (da Costa). J. exp. mar. Biol. Ecol. 56: 235-255.

Ansell, A.D. 1983. The biology of the genus Donax. In McLachlan, A. & T. Erasmus (eds). Sandy beaches as ecosystems. Junk, The Hague, pp. 607-635.

Ansell, A.D. 1983. Species of Donax from Hong Kong: morphology, distribution, behaviour and metabolism. In Proceedings of the second international workshop on the malacofauna of Hong Kong and southern China. B. Morton and D. Dudgeon (eds) Hong Kong Univ. Press.

Ansell, A.D. and Trevallion, A. 1969. Behavioural adaptations of intertidal molluscs from a tropical sandy beach. J. exp. mar. Biol. Ecol. 4: 9-35.

Ansell, A.D., Sivadas, P., Narayanan, B. and Trevallion, A. 1972. The ecology of two sandy beaches in South West India III. Observations on the populations of Donax incarnatus and Donax spiculum. Mar. Biol. 17: 318-332.

Ansell, A.D. and Trueman, E.R. 1973. The energy cost of migration of the bivalve Donax on tropical sandy beaches. Mar. Behav. Physiol. 2: 21-32.

Bally, R. 1983a. Factors affecting the distribution of organisms in the intertidal zones of sandy beaches. In McLachlan, A. and T. Erasmus (eds). Sandy beaches as ecosystems. Junk, The Hague, pp. 391-403.

Bally, R. 1983b. Intertidal zonation on sandy beaches of the west coast of South Africa. Cah. Biol. Mar. 24: 85-103.

Barnes, N.B. and Wenner, A.M. 1968. Seasonal variation in the sand crab Emerita analoga (Decapoda Hippidae) in the Santa Barbara area of California. Limnol. Oceanogr. 13: 465-475.

Berry, A.J. 1982. Predation by Natica maculosa Lamarck (Naticidae : Gastropoda) upon the trochean gastropod Umbonium restiarium (L.) on a Malaysian shore. J. exp. mar. Biol. Ecol. 64: 71-89.

Boaden, P.J.S. 1968. Water movement - a dominant factor in interstitial ecology. Sarsia 34: 125-136.

Boaden, P.J.S. and Platt, H.M. 1971. Daily migration patterns in an intertidal meiobenthic community. Thalass. Jugosl. 7: 1-12.

Branch, G.M. 1984. Competition between marine organisms: Ecological and evolutionary implications. Oceanogr. mar. Biol. Ann. Rev. 22: 429-593.

Bregazzi, P.K. and Naylor, E. 1972. The locomotor activity rhythm of Talitrus saltator (Montagu) (Crustacea: Amphipoda). J. exp. Biol. 57: 375-391.

Brown, A.C. 1973. The ecology of the sandy beaches of the Cape Peninsula, South Africa. Part 4: Observations on two intertidal isopoda, Eurydice longicornis (Studer) and Exosphaesoma truncatitelson Barnard. Trans. roy. Soc. S. Afr. 40: 381-404.

Brown, A.C. 1982. The biology of sandy beach whelks of the genus Bullia (Nassariidae). Oceanogr. Mar. Biol. Ann. Rev. 20: 309-361.

Brown, A.C. 1983. The ecophysiology of sandy beach animals - a partial review. In Sandy beaches as ecosystems (eds) A. McLachlan and T. Erasmus. Junk, The Hague. pp. 575-605.

Brown, A.C. and Talbot, M.S. 1972. The ecology of the sandy beaches of the Cape Peninsula, South Africa. Part 3. A study of Gastrosaccus psammodytes Tattersall (Crustacea: Mysidacea). Trans. roy. Soc. S. Afr. 40: 309-333.

Bush, L.F. 1966. Distribution of sand fauna in beaches at Miami, Florida. Bull. mar. Sci. 16: 58-75.

Chelazzi, G. and Ferrara, F. 1978. Researches on the coast of Somalia. The shore and dune of Sar Uaule. 19. Zonation and activity of terrestrial isopods (Oniscoidea). Monit. Zool. Ital. 8: 189-219.

Coe, W.R. 1955. Ecology of the bean clam Donax gouldi on the coast of southern California. Ecology 36: 512-514.

Connell, J.H. 1975. In Cody, M.C. and J.M. Diamond (eds). Ecology and evolution of communities. Harvard University Press. pp. 480-491.

Craig, P.C. 1973. Behaviour and distribution of the sand beach amphipod Orchestoidea corniculata. Mar. Biol. 23: 101-109.

Croker, R.A. 1967. Niche diversity in five sympatric species of intertidal amphipods (Crustacea : Haustoriidae). Ecol. Monogr. 37: 173-199.

Croker, R.A. and Hatfield, E.B. 1980. Space partitioning and interactions in an intertidal sand-burrowing amphipod guild. Mar. Biol. 61: 79-88.

Croker, R.A., Hager, R.P. and Scott, K.J. 1975. Macroinfauna of northern New England marine sand. II. Amphipod dominated intertidal communities. Can. J. Zool. 53: 42-51.
Cubit, J. 1969. Behaviour and physical factors causing migration and aggregation of the sand crab Emerita analoga (Stimpson). Ecology 50: 118-123.
Dahl, E. 1952. Some aspects of the ecology and zonation of the fauna on sandy beaches. Oikos 4: 1-27.
Davis, N. and Von Blaricom, G.R. 1978. Spatial and temporal heterogeneity in a sand bottom epifaunal community of invertebrates in shallow water. Limnol Oceanogr. 23: 417-417.
De Villiers, G. 1975. Growth, population dynamics, a mass mortality and arrangement of white sand mussels, Donax serra Röding, on beaches in the south-western Cape Province. Invest. Rep. Sea Fish., S. Afr. 109: 1-131.
Dexter, D.M. 1977. Natural history of the Pan-American sand beach isopod, Excirolana braziliensis (Crustacea : Malacostraca). J. Zool., Lond. 183: 103-109.
Dexter, D.M. 1983. Community structure of intertidal sandy beaches in New South Wales, Australia. In McLachlan, A. & T. Erasmus (eds). Sandy beaches as ecosystems. Junk, The Hague. pp. 461-472.
Dexter, D.M. 1984. Temporal and spatial variability in the community structure of the fauna of four sandy beaches in south-eastern New South Wales. Aust. J. mar. Freshw. Res. 35: 663-672.
Dillery, D.G. and Knapp, L.V. 1969. Longshore movements of the sand crab Emerita analoga (Decapoda, Hippidae). Crustaceana 18: 233-240.
Donn, T.E., Clarke, D.J., McLachlan, A. and Du Toit, P. 1986. Distribution and abundance of Donax serra Roding (Bivalvia : Donacidae) as related to beach morphology. I. Semilunar migrations. J. exp. mar. Biol. Ecol. 102: 121-131.
Duncan, G.A. 1986. Burrows of Ocypode quadrata (Fabricius) as related to slopes of substrate surfaces. J. Palaeont. 60: 384-389.
Duncan, J.R. 1964. The effects of water table and tide cycle on swash/backwash sediment distribution and beach profile development. Mar. Geol. 2: 186-197.
Du Preez, H.H. 1984. Molluscan predation by Ovalipes punctatus (De Haan) (Crustacea : Brachyura : Portunidae). J. exp. mar. Biol. Ecol. 84: 55-71.
Edwards, R.R.C. and Steele, J.H. 1969. The ecology of O-group plaice and common dabs at Loch Ewe. I. Population and food. J. exp. mar. Biol. Ecol. 2: 215-238.
Efford, I.E. 1965. Aggregation in the sand crab Emerita analoga (Stimpson). J. Anim. Ecol. 34: 63-75.
Enright, J.T. 1962. Responses of an amphipod to pressure changes. Comp. Biochem. Physiol. 7, 131-145.
Enright, J.T. 1963. The tidal rhythm of activity of a sand beach amphipod. Z. vergl. Physiol. 46: 276-313.
Enright, J.T. 1965. Entrainment of a tidal rhythm. Science 147: 864-867.
Enright, J.T. 1972. A virtuoso isopod : circa-lunar rhythms and their tidal fine structure. J. comp. Physiol. 77: 141-162.
Ercolini, A. and Scapini, F. 1974. Sun compass and shore slope in the orientation of littoral amphipods (Talitrus saltator Montagu). Monit. zool. ital. 8: 85-115.

Ercolini, A., Pardi, L. and Scapini, F. 1983. An optical direction factor in the sky might improve the direction finding of sandhoppers on the seashore. Monit. zool. ital. 17: 313-327.
Fincham, A.A. 1970. Amphipods in the surf plankton. J. mar. biol. Ass. U.K. 50: 177-198.
Fincham, A.A. 1973. Rhythmic swimming behaviour of the New Zealand sand beach isopod Pseudaega punctata. J. exp. mar. Biol. Ecol. 11: 229-237.
Frenkiel, L. 1982. L'organe sensoriel du muscle cruciforme des Tellinacea (mollusques lamellibranches), Structure-ontageneserole. Ph.D. Thesis, Univ. Paris. 149 pp.
Fusaro, C. 1980. Diel distribution differences in the sand crab, Emerita analoga (Simpson) (Decapoda, Hippidea). Crustaceana 39: 287-300.
Grant, J. 1981. Sediment transport and disturbance on an intertidal sandflat: infaunal distribution and recolonisation. Mar. Ecol. Prog. Ser. 6: 249-255.
Hager, R.P. and Croker, R.A. 1980. The sand burrowing amphipod Amphiporeia virginiana Shoemaker 1933 in the tidal plankton. Can. J. Zool. 58: 860-864.
Haley, S.R. 1982. Zonation by size of the Pacific mole crab, Hippa pacifica Dana (Crustacea : Anomura : Hippidae), in Hawaii. J. exp. mar. Biol. Ecol. 58: 221-231.
Hamner, W.M., Smyth, M. and Mulford, E.D. 1968. Orientation of the sand beach isopod Tylos punctatus. Anim. Behav. 16: 405-409.
Harris, R.P. 1972. The distribution and ecology of the interstitial meiofauna of a sandy beach at Whitsand Bay, East Cornwall. J. mar. biol. Ass. U.K. 52: 1-18.
Hartwick, R.F. 1976. Beach orientation in tolitrid amphipods : capacities and strategies. Behav. Ecol. Sociobiol. 1: 447-458.
Hayes, W.B. 1977. Factors affecting the distribution of Tylos punctatus (Isopoda, Oniscoidea) on beaches in southern California and northern Mexico. Pacific Sci. 31: 165-187.
Hill, G.W. and Hunter, R.E. 1973. Burrows of the ghost crab Ocypode quadrata (Fabricius) on the barrier islands, south-central Texas coast. J. sed. Pet. 43: 24-30.
Jansson, B.O. 1967. The significance of grain size and pore water content for the interstitial fauna of sandy beaches. Oikos 18: 311-322.
Jansson, B.O. 1968. B.O. 1968. Quantitative and experimental studies of the interstitial fauna in four Swedish sandy beaches. Ophelia 5: 1-71.
Jaramillo, E., Stotz, W., Bertran, C., Navarro, J., Roman, C. and Varela, C. 1980. Actividad locomotrix de Orchestoidea tuberculata (Amphipoda, Talitridae) sobre la superficie de una playa arenosa de sur de Chile (Mehuin, provincia de Valdivia). Studies on Neotropical fauna and Environment 15: 9-33.
Jones, D.A. 1972. Aspects of the ecology and behaviour of Ocypode ceratophthalmus (Pallas) and O. kuhlii (de Haan) (Crustacea, Ocypodidae). J. exp. mar. Biol. Ecol. 8: 31-43.
Jones, D.A. and Hobbins, C. St. C. 1985. The rôle of biological rhythms in some sand beach cirolanid Isopoda. J. exp. mar. Biol. Ecol. 93: 47-59.
Kensley, B. 1974. Aspects of the biology and ecology of the genus Tylos Latreille. Ann. S. Afr. Mus. 65: 401-471.

Leber, K.M. 1982a. Seasonality of macroinvertebrates on a temperate, high wave energy sandy beach. Bull. mar. Sci. 32: 86-98.
Leber, K.M. 1982b. Bivalves (tellinacea : Donacidae) on a North Carolina beach : contrasting population size structures and tidal migrations. Mar. Ecol. Prog. Ser. 7: 297-301.
Loesch, H.C. 1957. Studies of the ecology of two species of Donax on Mustang Island, Texas. Texas Univ. Inst. mar. Sci., Publ. 4: 201-227.
Macquart-Moulin, C. 1977. Le contrôle de l'emergence et des rages nocturned chez les peracarides des plages de Mediterranee. Eurydice affinis Hansen (Isopoda), Gastrosaccus mediterraneus Bacescu, Gastrosaccus spinifer (Goër) (Mysidacea). J. exp. mar. Biol. Ecol. 27: 61-81.
Marsh, B.A. and Branch, G.M. 1979. Circadian and circatidal rhythms of oxygen consumption in the sandy beach isopod Tylos granulatus Krauss. J. exp. mar. Biol. Ecol. 37: 77-89.
McLachlan, A. 1983. Sandy beach ecology - a review. In McLachlan, A. and T. Erasmus (eds) Sandy beaches as ecosystems. Junk, The Hague, pp. 321-380.
McLachlan, A., Erasmus, T. and Furstenberg, J.P. 1977. Migrations of sandy beach meiofauna. Zool. Afr. 12: 257-277.
McLachlan, A. and Hanekom, N. 1979. Aspects of the biology, ecology and seasonal fluctuations in biochemical composition of Donax serra in the East Cape. S. Afr. J. Zool. 14: 183-193.
McLachlan, A., Wooldridge, T. and Van Der Horst, G. 1979. Tidal movements of the macrofauna on an exposed sandy beach in South Africa. J. Zool., Lond. 188: 433-442.
McLachlan, A. and Young, N. 1982. Effects of low temperature on the burrowing rates of four sandy beach molluscs. J. exp. mar. Biol. Ecol. 65: 275-284.
McLachlan, A. and HESP, P. 1984. Faunal response to morphology and water circulation of a sandy beach with cusps. Mar. Ecol. Prog. Ser. 19: 133-144.
McLachlan, A., Eliot, I.G. and Clarke, D.J. 1985. Water filtration through reflective microtidal beaches and shallow sublittoral sands and its implications for an inshore ecosystem in Western Australia. Estuar. cstl shelf Sci. 21: 91-104.
McLusky, D.S., Nair, S.A., Stirling, A. and Bharoava, R. 1975. The ecology of a central west Indian beach with particular reference to Donax incarnatus. Mar. Biol. 30: 267-270.
Meineke, T. and Westheide, W. 1979. Tide dependant migrations of the interstitial fauna in a sand beach of the Island of Sylt, North Sea, West Germany. Mikrofauna Meeresbodens 75: 203-236.
Mori, S. 1938. Donax semigranosus Dkr. and the experimental analysis of its behaviour on the flood tide. Zool. Mag. Tokyo 50: 1-12.
Moueza, M. and Chessel, D. 1976. Contribution a l'etude de la biologie de Donax trunculus L. (Mollusque : Lassellibranche) dans l'Algero's : analyse statistique de la dispersion le long d'une plage en baie de bon Ismail. J. exp. mar. Biol. Ecol. 21: 211-221.
Naylor, E., Atkinson, R.J.A. and Williams, B.G. 1971. External factors influencing the tidal rhythm of shore crabs. Proc. 2nd interdisc. Conf. Cycle Res. 2: 173-180.

Noy-Meir, I. 1979. Structure and function of desert ecosystems. Israel. J. Bot. 28: 1-19.
Papi, F. and Pardi, L. 1963. On the lunar orientation of sandhoppers. Biol. Bull. 124: 97-105.
Pardi, L. and Papi, F. 1952. Die Sonne als Kompass bei Talitrus saltator Montagu (Amphipoda, Talitidae). Naturwissenschaften 39: 262-263.
Perry, D.M. 1980. Factors influencing aggregation patterns in the sand crab Emerita analoga (Crustacea : Hippidae). Oecologia 45: 379-384.
Philip, K.P. 1974. The intertidal fauna of the sandy beaches of Cochin. Proc. Ind. natr. Sci. Acad. B. 38: 317-328.
Reid, D.G. and Naylor, E. 1985. Free-running, endogenous semilunar rhythmicity in a marine isopod crustacean. J. mar. Biol. Ass. U.K. 65: 85-91.
Reise, K. 1984. Experimental sediment disturbances on a tidal flat : responses of free-living platyelminthes and small polychaeta. Hydrobiologia 118: 73-81.
Reise, K. 1985. Tidal flat ecology, an experimental approach to species interactions. Springer, Berlin. 191 pp.
Riedl, R.J. 1971. How much seawater passes through sandy beaches? Int. Rev. ges. Hydrobiol. 56: 923-946.
Riedl, R.J. and Machan, R. 1972. Hydrodynamic patterns in lotic intertidal sands and their bioclimatological implications. Mar. Biol. 13: 179-209.
Riedl, R.J., Huang, N. and Machan, R. 1972. The subtidal pump : a mechanism of interstitial water exchange by wave action. Mar. Biol. 13: 210-221.
Rieger, R. and Ott, T. 1971. Gezeitenbedingte Wanderungen von Turbellarien und Nematoden eines Nordadriatischen Sandstrandes. Vie Milieu 22: 425-447.
Saloman, C.H. and Naughton, S.P. 1978. Benthic macroinvertebrates inhabiting the swash zone of Panama City Beach, Florida. Northeast Gulf Science 2: 65-72.
Salvat, B. 1964. Les conditions hydrodynamiques interstitielles des sediment meubles intertidaux et la repartition verticale de la jenne endogee. C.R. Acad. Sci. Paris 259: 1576-1579.
Salvat, B. 1966. Eurydice pulchra Leach 1815 et Eurydice affinis Hansen 1905 (Isopodes: Cirolanidae). Taxonomie, ethologie, repartition verticale et cycle reproducteur. Acta Soc. Linn. Bordeaux A 193: 1-77.
Salvat, B. 1967. La macrofauna carcinologique endogee des sediments meubles intertidaux (Tanaidaces, Isopodes et Amphipodes): ethologie, bionomie et cycle biologique. Mem. Mus. natn. Hist. nat. Ser. A 45: 1-275.
Sastre, M.P. 1985. Aggregated patterns of dispersion in Donax denticulatus. Bull. mar. Sci. 36: 220-224.
Scapini, F. and Ugolini, A. 1981. Influence of landscape on the orientation of Talitrus saltator (Crustacea, Amphipoda). Monit. Zool. ital. 15: 324-325.
Short, A.D. and Wright, L.D. 1983. Physical variability of sandy beaches. In McLachlan, A. and T. Erasmus (eds) Sandy beaches as ecosystems. Junk, The Hague, pp. 133-144.
Thistle, D. 1981. The response of a harpacticoid copepod community to a small scale natural disturbance. J. Mar. Res. 38: 381-395.
Tiffany, W.J. 1972. The tidal migration of Donax variabilis Say. Veliger. 14: 82-85.

Trueman, E.R. 1971. The control of burrowing and the migratory behaviour of Donax denticulatus (Bivalvia : Tellinacea). J. Zool., Lond. 165: 453-469.
Trueman, E.R. 1975. The locomotion of soft-bodied animals. Arnold, London. 200 pp.
Turner, J.H. and Belding, D.L. 1957. The tidal migrations of Donax variablilis (Say). Limnol. Oceanogr. 2: 120-125.
Ugolini, A., Scapini, F. and Pardi, L. 1986. Interaction between solar orientation and landscape visibility in Talitrus saltator (Crustacea : Amphipoda). Mar. Biol. 90: 449-460.
Underwood, A.J. and Denley, E.J. 1984. Paradigms, explanations and generalisations in models for the structure of intertidal communities on rocky shores. In Strong, D.R., Simberloff, D., Abele, L.G. and Thistle, A.B. (eds). Ecological communities : conceptual issues and the evidence. Princeton Univ. Press. pp. 151-180.
Vilas, F. 1986. Activity of amphipods in beach sediments and nearshore environments: Playa Ladeira, N.W. Spain. J. cstl Res. 2: 285-295.
Virnstein, R.W. 1977. The importance of predation by crabs and fishes on benthic infauna in Chesapeake Bay. Ecology 58: 1199-1217.
Wade, B.A. 1967. Studies on the biology of the West Indian beach clam Donax denticulatus Linne. I. Ecology. Bull. mar. Sci. 17: 149-174.
Wendt, G. and McLachlan, A. 1985. Zonation and biomass of the intertidal macrofauna along a South Africa sandy beach. Cah. Biol. Mar. 26: 1-14.
Weiser, W., Ott, J., Schiemer, F. and Gnaiger, E. 1974. An eco-physiological study of some meiofauna species inhabiting a sandy beach at Bermuda. Mar. Biol. 26: 235-248.
Williams, J.A. 1980a. The effect of dusk and dawn on the locomotor activity of rhythm of Talitrus saltator (Montagu). J. exp. mar. Biol. Ecol. 42: 285-297.
Williams, J.A. 1980b. Environmental influence on the locomotor activity rhythm of Talitrus saltator (Crustacea: Amphipoda). Mar. Biol. 57: 7-16.
Williams, J.A. 1983. Environmental regulation of the burrow depth distribution of the sand-beach amphipod Talitrus saltator. Estuar. cstl shelf Sci. 16: 291-298.
Wilson, W.H. 1981. Sediment-mediated interactions in a densely populated infaunal assemblage : the effects of the polychaete Abarenicola pacifica. J. mar. Res. 39: 735-748.
Withers, R.G. 1977. Soft-shore macrobenthos along the South west coast of Wales. Estuar. cstl mar. Sci. 5: 467-484.
Woodin, S.A. 1974. Polychaete abundance patterns in a marine soft-sediment environment. The importance of biological interactions. Ecol. Monogr. 44: 171-187.
Woodin, S.A. 1981. Disturbance and community structure in a shallow water sand flat. Ecology. 62: 1052-1066.
Wooldridge, T. 1981. Zonation and distribution of the beach mysid, Gastrosaccus psammodytes. J. Zool., Lond. 193: 183-189.

Eco-Ethology of Mangroves

Richard G. Hartnoll

University of Liverpool
Port Erin, Isle of Man, U.K.

INTRODUCTION

The basic aim of this paper is to review aspects of the behaviour of mangrove animals, in order to see how they relate to the special characteristics of the mangrove environment. To accomplish this it is first necessary to outline some of the more relevant physico-chemical aspects of the mangrove environment, aspects to which the animals must adapt and react in order to exist in a quite stressful environment. It requires a better appreciation of the mangrove than the popular concept of it as a fetid malarial swamp, or as Berry (1972) more graphically depicted the Malaysian mangrove forest as "hot, dirty and infested with snakes and mosquitoes."

The mangrove ecosystem is by definition associated with the mangrove trees, and restricted to such areas. Mangroves are an assemblage of specialised trees and shrubs adapted to the unusual conditions of an inter-tidal saline environment. Over 50 mangrove species can be recognised, from a number of families of flowering plants (Chapman, 1984). The mangrove habit has arisen independently on a number of occasions, but involves in all cases a complex of physical and physiological adaptations. These are designed to cope with the various problems such as support and anchorage in the soft sediment, root respiration in the anoxic sediment, water relations in the saline conditions, and difficulties for seed germination and establishment. Macnae (1968) gives a comprehensive description of the adaptations of the mangrove flora.

There are a number of basic environmental features which characterise the mangrove habitat. It is prone to high temperatures, since mangroves are effectively limited to the tropics (Macnae, 1968, Fig. 1), with only minor extensions beyond that limit. Measurements of field temperatures show the high levels actually encountered (Macintosh, 1978). It is exposed to the air for long periods, since most of the mangrove area is located at mean high water neaps and above (Macnae, 1968; Macintosh, 1978). There can be considerable variation in salinity, due either to evaporation producing hypersaline conditions, or freshwater flow or rain lowering the salinity. There is very little wave action since the mangroves flourish only in shelter and by their presence reduce what limited wave action does occur. Thus there is no splash zone, and the substrate tends to be fine and muddy.

Within this environment the mangrove community exhibits a complex three-dimensional structure as a result of the presence of the mangrove trees (Morton, 1983). Organisms display a pattern of zonation over the substrate from low to high water, but at any point along that slope they also show a vertical zonation up the trees themselves. Resulting from this structural complexity the mangrove fauna can be divided according to basic habit.

There is an epifauna associated with the mangrove trees, comprising a group of organisms which attach to the surface of the mangroves, or which spend most of their time crawling or climbing over the surface. The affinity of this group is with the community of the rocky shore (Morton, 1976). It includes grazing gastropods such as Nerita and Littorina; sessile filter feeders including the barnacles Chthamalus and Balanus, and the rock oyster Crassostrea; and active forms like the grapsid crabs Aratus and Metopograpsus.

The second component of the fauna is associated with the substrate, either burrowing in it, or sheltering under detritus on its surface. The major part of this group is comprised of a diversity of crabs - various fiddler crabs of the genus Uca, other ocypodid crabs including Ocypode and Macrophthalmus, grapsid crabs such as Sesarma, and Cardisoma from amongst the land crabs. The above are all burrowing, whilst various xanthid and portunid crabs shelter under debris, a habitat which they share with the active Metopograpsus. There are various gastropods and bivalve molluscs such as Terebralia and Geloina, and the fish are represented by the highly specialised mud skippers.

The third component is smaller, and comprises species which migrate regularly between the surface of the mud and the surface of the trees. The best known is the gastropod Cerithidea, which rests on the trunks of the mangrove, but descends to feed on the surface of the mud.

The above is not in any way a comprehensive account of the composition of the mangrove fauna - a number of detailed accounts are available (see for example Macnae, 1968; Berry, 1972). It is notable that whilst the mangrove environment does seem to offer quite unusual conditions, very few species are recorded as being restricted entirely to the mangroves (Macnae and Kalk, 1962; Berry, 1975; Morton, 1983). Almost all also occur elsewhere, either on sheltered hard substrates, or on mud or muddy-sand in non-mangrove areas. Thus Berry (1975) concluded that of the various attached species in Malaysian mangroves, only the bivalve Enigmonia is confined to the mangroves, all the others occuring also on rocky shores. Similarly Morton (1983), in a review of mangrove bivalves, could list only a small number specific to that habitat - Enigmonia, Geloina, Laternula and Glauconome. Nevertheless, this is possibly a restricted view, and there are certainly a substantial number of species which occur predominantly within the mangrove, with only scatterred representatives elsewhere - Reid (1985) emphasises this in relation to variuos species of littorinids. The mangrove is universally populated by characteristic assemblages of species. Perhaps a prime interest of the mangrove is that it provides a unique juxtaposition of hard and soft substrate communities, with great potential to interact with each other, and for the organisms to display complex behaviour patterns.

In this review the aim is to examine these behaviour patterns in relation to the environment which the organisms inhabit. The basic format of the paper has been to look at the temporal, or rhythmic aspects of behaviour first, followed by the spatial aspects. Behavioural relations between organisms, whether intraspecific or interspecific, are introduced only where they contribute to understanding the organism-environment interactions. The treatment has been selective, concentrating on those facets of behaviour of particular relevance in the mangrove ecosystem.

RHYTHMIC BEHAVIOUR

Living on the upper regions of the shore in a hot climate makes appropriate patterns of rhythmic behaviour particularly important for mangroves animals. Tidal, diurnal, lunar, and annual cycles all have significant roles to play.

Tidal rhythms

The distinctive feature of tidal rhythms in the mangrove community is that most resident species are active only when out of water. There are exceptions to this of course, with attached filter feeders being the most obvious example - bivalves and barnacles must be active whilst immersed in order to filter suspended food. However, the majority of animals are active when emersed at low tide, and Fig.1 shows a typical example of such a pattern in the fiddler crab *Uca rosea*, a high shore species from Malaysia. The crabs commence activity within half an hour of becoming uncovered, than rapidly cease activity shortly before being reimmersed.

When the tide comes in there are two ways in which immersion may be avoided. One is to enter a burrow and plug the mouth, behaviour characteristics of most crabs - for example *Uca* (Pearse, 1912; Barnwell, 1968; Salmon, 1984, etc), *Sesarma* (Macintosh, 1984) and *Heloecius* (Warren and Underwood, 1986). The other is to climb mangrove trees, as seen in the non-burrowing crabs *Aratus* (Wilson, 1981) and *Metopograpsus*, and even in some species of *Uca* under certain conditions (Von Hagen, 1970). Gastropods like *Cerithidea* also climb trees (Brown, 1971; Berry, 1972). Some mud skippers such as *Periophthalmus chrysospilos* climb trees as the tide rises, others like *P. kalalo* will move to higher ground (Macnae, 1968). There are exceptions to the above generalisations. The sesarmid crab *Neoepisesarma versicolor* leaves its burrow at high tide and climbs the mangrove trunks (Malley, 1977). Some *Uca* populations do not cease activity before immersion, but remain active for some time under shallow water (Teal, 1959; Crane, 1975). Mated pairs of *Periophthalmus* tend to remain in their burrow whilst it is submerged (Macnae, 1968).

What is behind this very strong tendency to avoid immersion on the part of a set of species which are all primarily aquatic in origin? In many this tendency has become so marked that they now live virtually permanently above the water surface, and have became intolerant of any lengthy immersion. The West Indian *Littorina angulifera* experiences 50% mortality after two days immersion (Coomans, 1969). A possible reason is that feeding is more efficient out of water. This could apply to some deposit feeders, though there is no evidence for this, and even so it would not supply a general explanation. A more probable cause is that the risk of predation is substantially greater under water than in air, providing strong selective pressure for terrestrial activity.

There are several studies of predation on *Uca* in air - in Malaysia predators include mudskippers, snakes, spiders, birds and monkeys (Macintosh, 1979). The mudskipper *Periophthalmodon schlosseri* is a particularly effective predator, spotting and chasing crabs from several metres distance - *Uca* may comprise >60% of its diet. However, in water there is potential predation from a variety of fish which migrate into the interdidal at high water (Ong, 1978; Beever et al., 1979; Sasekumar and Thong, 1980), and also by portunid crabs such as *Callinectes* (Odum and Heald, 1972) and *Scylla* (Hill, 1979), which occur in all mangrove areas (Jones, 1984). In combination these could well prove much more formidable than the terrestrial predators.

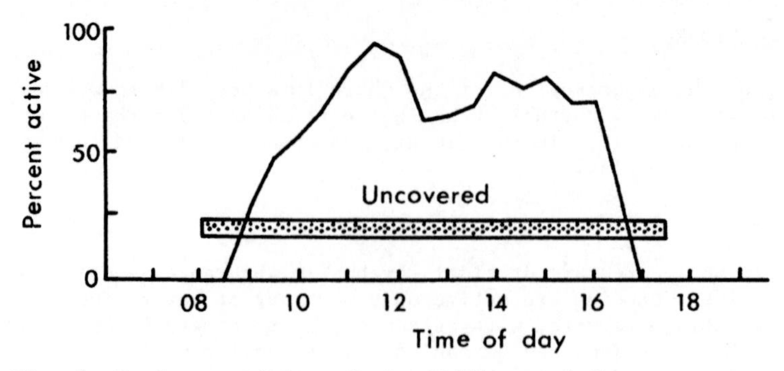

Fig. 1. Surface activity of the fiddler crab Uca rosea on a mangrove shore in Malaysia (after Macintosh, 1982)

Diurnal Rhythms

There is always a measure of interaction between diurnal and tidal rhythms, and it is frequently not easy to completely separate the two. The mangrove fauna contains some species which display predominantly daytime activity, and others which are essentially nocturnal.

In Uca activity is normally restricted to the day. This is well illustrated by Fig.2 which represents activity during a neap tide period when there is no immersion at the shore level to complicate matters. There are many supporting references (e.g. Pearse, 1912; Zucker, 1978; Macintosh, 1978, 1982). In temperate species of Uca there is some nocturnal activity, but there is increasing restriction to the daytime with tropical species (Salmon and Atsaides, 1968; Zucker, 1978), though in the Philippines there may be some activity on bright moonlight nights (Pearse, 1912). Although activity in Uca is essentially diurnal, it can be restricted by very high temperatures during the middle of the day (Macnae and Kalk, 1969; Crane, 1975). Related to this is the diurnal rhythm of colour change in Uca (Abramowitz, 1937; Barnwell, 1963), so that they generally become darker during the middle of the day. However, at high temperatures this rhythm is changed and the colour becomes paler (Brown and Sandeen, 1948). Since pale individuals heat up at a slower rate (Fig.3), this behaviour is suitably adaptive (Wilkens and Fingerman, 1965).

Daytime activity is general in ocypodid crabs (Macintosh, 1984), as Warren and Underwood (1986) noted for Heloecius. However, grapsid crabs are more variable in their response. Many of these are active during the day, and most of these are also active by night (Macnae, 1963; Hartnoll, 1965). Some grapsids are predominantly nocturnal though - Sesarma ricordi and S. curacaoense (Hartnoll, 1965), and Sesarma catenata (Day, 1967). The land crab Cardisoma carnifex and the mud lobster Thalassina anomala are also both mainly nocturnal (Macnae, 1968).

There is little information on diurnal activity patterns in the molluscs. In South Africa the gastropod Cerithidea decollata climbs the shady side of mangrove trunks during the day (Macnae, 1963), presumably as a refuge from the heat. The high zoned bivalve Geloina proxima, during periods of prolonged emersion, gapes during the night to expose the mantle margin and allow some gaseous exchange (Morton, 1975).

It is not clear why the activity of mangrove crustaceans is so predominantly diurnal. Predator avoidance is a possibility, though the risk would seem to appear greater during the day. Perhaps the importance of visual stimuli in these crabs is the main factors.

Lunar Rhythms

The mangrove is restricted to the upper part of the intertidal, which means that much of it will be exposed for long periods during neap tides. The mangrove channels are a very complex and ramifying system, so the higher flushing rates during the spring tides could be important for larval dispersal. Lunar and semi-lunar activity patterns can therefore be important to mangroves species.

High shore species are often uncovered for many days in succession during neap tides, and the substrate becomes hard and dry. Deposit feeders such as *Uca* often remain inactive in their burrows during this period. This applies to the high shore *Uca rosea* and *U. triangularis* in Malaysia (Macintosh, 1978), to high shore *Uca* populations in the Philippines (Pearse, 1912), and to *Uca inversa* in East Africa (Macnae and Kalk, 1962). In Panama *Uca latimanus* is inactive underground for eight-day periods over neap tides (Crane, 1941). This neap tide inactivity may be broken if the substrate becomes damp after rain, or as a result of overnight dew (Macnae and Kalk, 1962).

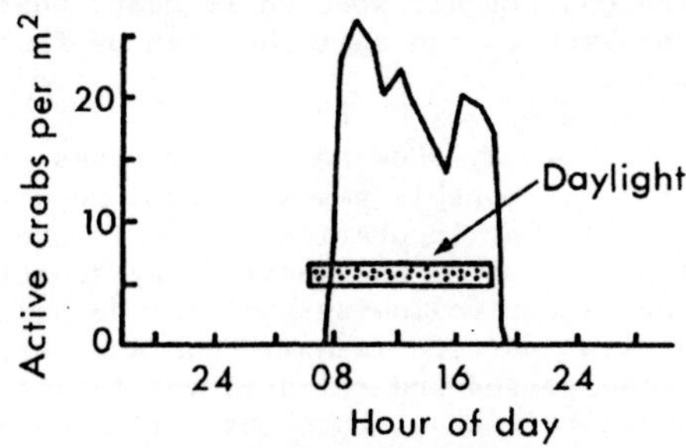

Fig. 2. Surface activity of a high shore population of *Uca* spp. in Malaysia in neap tide period when that shore level is not covered (after Macintosh, 1978)

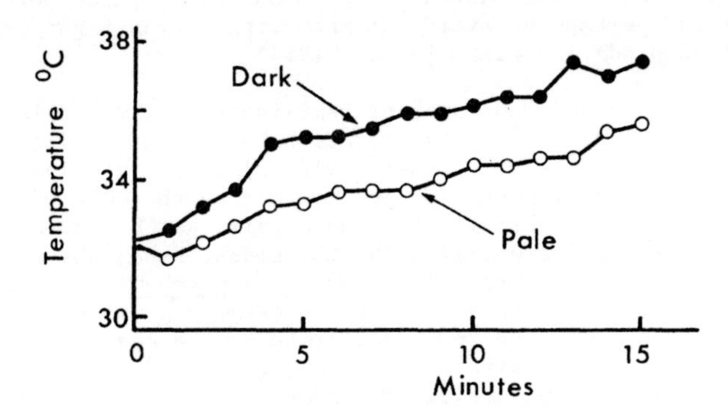

Fig. 3. Rate of heating when exposed to sunshine of pale and dark specimens of *Uca pugilator* (after Wilkens and Fingerman, 1965).

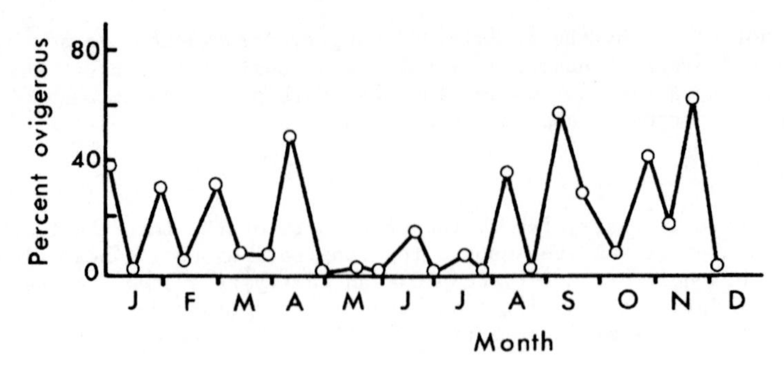

Fig. 4. Percentage of adult females of Uca rosea ovigerous in fortnightly collections (after Macintosh, 1984).

Reproductive activity is also influenced by the spring-neap cycle. For non-migratory high shore species this ensures that there is water available when the larvae are released. For all species it means that larvae can be shed when there is the maximum chance that they can be flushed out to sea for planktonic development.

In Uca courtship activity may show semi-lunar peaks at full moon and at new moon, as in Uca musica, or monthly peaks at full moon in U. latimanus and U. beebei (Zucker, 1978). The incubation period is about 14 days in these tropical conditions, and since the females lay shortly after mating, the larvae will hatch during the following spring tide period. Macintosh (1984) demonstrated a similar pattern in Malaysian species, where crabs also ovulated on one spring tide period and released the larvae on the next. Mid shore species showed larval release at both the full moon and new moon periods - they are immersed at both times. However high shore species such as Uca rosea are covered only by one set of spring tides each month, and show a lunar cycle (Fig. 4). Salmon (1984) and Salmon et al. (1986) have also noted semi-lunar rhythms of larval release in Uca, and have furthermore observed that release is finely timed to coincide with high tide to facilitate the most effective seaward dispersal of the larvae. In Uca vocans larval release occurred at the maxima of new moon springs, but before the maxima of full moon, perhaps to avoid too much light which might expose the larvae to increased predation risk (Salmon, 1984).

Migratory species also show lunar or semi-lunar rhythms of larval release, but these are accompanied by migrations of the berried females. In Aratus pisonii in Jamaica peaks of females with ripe eggs are found at full and new moon, and at this time the females migrate to the edge of the mangrove and enter the water briefly for the eggs to hatch (Warner, 1967). This is an intertidal species entrained by the tides, hence the semi-lunar cycle. It contrasts with the supra-tidal Cardisoma guanhumi which is entrained by the moon, and shows a lunar cycle (Fig. 5). Spawning occurs in peaks at the full moon, when the females migrate to the sea (Gifford, 1962). In both Aratus and Cardisoma larval release occurs by night, presumably to minimise predation on the newly released larvae.

There are also semi-lunar migrations displayed by mangrove molluscs. Various tree-living gastropods such as Littorina and Nerita move higher up the trees during spring tide periods (Barry, 1975). Others are found on the

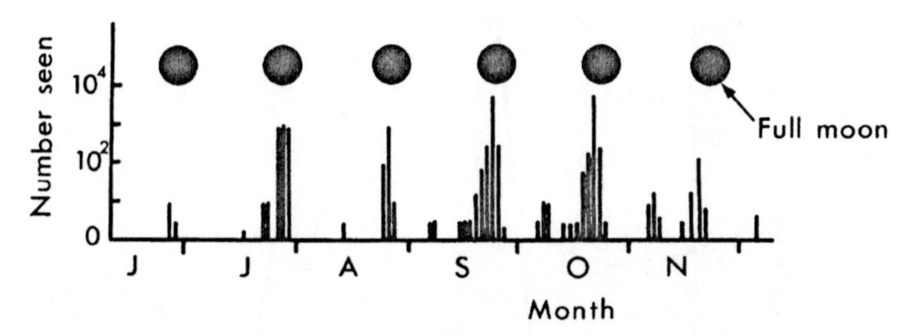

Fig. 5. Number of migrating ovigerous females of Cardisoma guanhumi in relation to the lunar cycle (after Gifford, 1962).

ground when it is emersed during neap tide periods, but climb the trees during spring tides - of these Cerithidea shows the most interesting pattern. The East African Cerithidea decollata spends much of its time clinging to the shaded sides of mangrove trunks. Brown (1971) demonstrated that the numbers aggregated in this way were greatest at spring tides - he suggested that they spread out over the ground after the spring tide period to feed on the newly deposited mud. Further to this Cockcroft and Forbes (1981) have shown that there is an interaction between tidal and semi-lunar cycles. Snails can migrate down onto the mud at low tide, but this activity is inhibited at springs, increasing during the transition to neaps (Fig. 6). They may remain on the mud for several tidal cycles when that level is not covered at neaps. The height of aggregation on the trees also varies over the spring-neap cycle so that the snails there are never submerged at high tide.

Annual Cycles

There are annual cycles of behaviour associated with seasonal weather changes, even in the tropical mangrove areas. They are linked to seasonal rainfall patterns, rather than to temperature cycles. In Singapore the gastropod Cerithidea cingulata migrates up and down the shore correlated with the monson cycle (Vorha, 1970). In various high-shore species of Uca there is a tendency to remain inactive in the burrows for long periods of time, with a marked increase in activity during the rainy season (Crane, 1975; Macintosh, 1984).

In the high level mangrove fauna generally there is a strong tendency to react to terrestrial, as opposed to marine, stimuli. Littorina scabra shows cyclical activity even when continuously above high tide during neap periods, activity occuring as a response to rain or dew (Little and Stirling, 1984). In Malaysia Nerita birmanica has its reproductive activity triggered by heavy rain following periods of drought (Berry et al., 1973).

SPATIAL BEHAVIOUR

Orientation

This applies basically to the positions adopted by sessile species, and there is not much information available, except for the bivalve Enigmonia aenigmatica (Morton, 1976, 1983). This attaches to the prop roots and stems of mangroves, and adopts a consistent relationship to the vertical. It also shows a marked preference for particular sides of the trunks, a preference which varies from site to site. The net result of these preferences is that

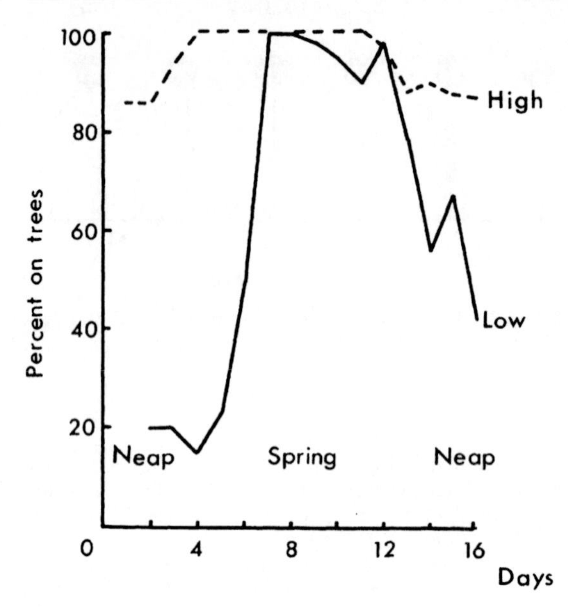

Fig. 6. Percentage of marked Cerithidea decollata on trees at high and low tide over a spring-neap cycle (after Cockcroft and Forbes, 1981)

the bivalve is positioned in such a way that the normal feeding currents are reinforced by the external currents, which will tend to optimise food collection by the species.

Migrations

Generally migrations are rhythmic, and as such they have already been dealt with above. There are however ontogenetic migrations, such as in Scylla serrata. Here small specimens occur far into the mangroves at the head of the creeks, and they migrate seawards as they grow so that the larger ones occur near the mangrove edge (Macnae, 1968). Homing behaviour is an aspect of migration, but has been little studied. In the Philippines fiddler crabs were observed to return to their burrows from up to 12 metres away (Pearse, 1912).

Microhabitat Selection

This is important since the environment of many of the mangrove animals is rigorous, or even marginal, and the selection of the most appropriate microhabitat can go far to minimise stress. It can aid in ensuring an optimal food supply, limiting exposure to high temperature and desiccation, and avoiding predators.

The orientation of Enigmonia, described above, provides one example of microhabitat selection to optimise feeding. The selection of a particular substrate grade by species of Uca offers a further instance. A clear substrate preference is a general phenomenon (Macnae, 1968; Sasekumar, 1974; Hartnoll, 1975; Icely and Jones, 1978), with Table 1 providing a good example. A detailed study of the mouthpart morphology (Miller, 1961; Macnae, 1968; Icely and Jones, 1978) shows a good correlation with substrate grade. The change in preference from sand to fine mud is matched by a shift from a predominance of spoon-tipped setae, to a predominance of plumose

Table 1. Organic content and mean particle size for the preferred substrate of East African *Uca* spp. (after Icely and Jones, 1978)

Species	% organic content	particle size (mm)
U. lactea	3.96	0.185
U. tetragonon	4.16	0.170
U. vocans	4.82	0.164
U. chlorophthalmus	9.12	0.137

setae - each providing most efficient sorting in the chosen substrate. In East Africa this is seen in the sequence *Uca lactea*, *U. vocans*, *U. tetragonon* and *U. chlorophthalmus* (Icely and Jones, 1978); in Malaysia in the sequence *U. lactea*, *U. urvillei* and *U. dussumieri* (Macnae, 1968).

However, it is necessary to employ some caution when assuming that the occupation of a particular substrate or microhabitat is the result of a behavioural choice by the species - species may change the microhabitat by their activity. Dye and Lasiak (1986) demonstrated that the feeding activity of *Uca* altered the sediment composition by affecting the meiobenthos content. Since over 40% of the substrate surface can be sorted on each tide it is not surprising that these effects occur. In Australia *Heloecius cordiformis* consistently burrows in elevated well drained areas. Warren and Underwood (1986) showed that these elevations result from the burrowing of the crab.

The risk of high temperatures and desiccation is always present for animals occupying a high-shore environment in the tropics, and in many mangrove crabs this results in a dependence on the microenvironment of the burrow. *Uca* makes regular visits to the burrow during the course of feeding (Wilkens and Fingerman, 1965; Macintosh, 1978), and these visits are more frequent during higher temperatures. The burrow provides a microenvironment which is cooler and moister than the substrate surface (Crane, 1975; Macintosh, 1978, 1984), and the visit offers a chance to cool down and to replenish the crab's water supply. Macintosh (1978) gives details of these temperature differences (Fig. 7), and at a depth of only 10 cm the burrow can provide a temperature >10°C lower. This dependence on the burrow probably explains why mangrove crabs do not migrate extensively up and down the shore with the tide, as do their more temperate counterparts (Macintosh, 1984).

Mangrove gastropods also show microhabitat selection to avoid heat and desiccation stress. It has been described above how *Cerithidea decollata*, when it climbs the mangrove trunks, selects the shaded side to rest on (Brown, 1971). Various mangrove littorinids and littoral fringe potamonids will hang from shaded surfaces by a dried mucus film during the heat of the day (Vermeij, 1974). The behaviour of the light and dark colour morphs of *Littorina pallescens* offers a particularly instructive example. The light morphs show a preference for the upper surface of leaves, the dark morphs for the undersides, with a greater proportion favouring the undersides at higher temperatures (Cook, 1986). Further study showed that the dark morphs heated up faster than the light morphs (Cook and Freeman, 1986), offering an explanation for the choice of microhabitat.

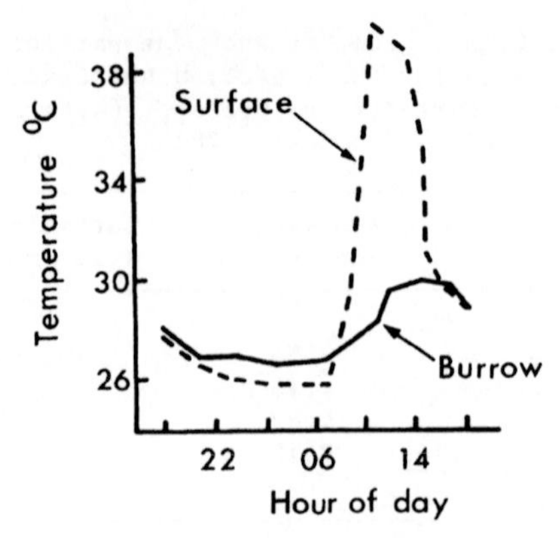

Fig. 7. Diurnal temperature fluctuation on the substrate surface and in the burrow of Uca in an unshaded mangrove area (after Macintosh, 1978)

Mud skippers are at risk from high temperature, and avoid excessively hot locations. Periophthalmus sobrinus avoids situations which cause its body temperature to exceed 35°C (Stebbins and Kalk, 1961).

A further benefit of microhabitat selection comes from selecting a background which matches the animal, an action which should reduce predation. In New Guinea Littorina pallescens prefers to live on mangrove leaves, whereas L. scabra and L. intermedia favour the roots and trunks - in each case the colours match the background (Cook et al., 1985; Reid, 1985). Enigmonia aenigmatica also has colour morphs which match the background, cream on the undersides of leaves, but purple-red on the stems and roots (Morton, 1976, 1983). This is less obviously a result of behavioural selection however, and is possibly an environmental effect.

CONCLUSION

The dominant features of the behaviour of mangrove animals are common to most other upper intertidal communities, representing ways of reducing the stress of the environment for organisms of marine origin. The cycles of behaviour, and the selection of appropriate microhabitats, are clearly adaptive in this respect.

Less obviously adaptive at first sight is the prevailing tendency to concentrate activity in air, rather than in water. Thus most mangrove animals either shelter in refuges at high tide, or migrate ahead of the tide to avoid submersion. It is suggested that this is to avoid high levels of aquatic predation - a hypothesis which is plausible, but by no means proved. If this is the case it emphasizes that biological factors, as well as the physical environment, can be of major importance in determining the patterns of behaviour prevailing in communities.

REFERENCES

Abramowitz, A.A., 1937, The chromatophorotropic hormone of the Crustacea; standardization, properties, and physiology of the eye-stalk glands,

Biol. Bull. mar. biol. Lab. Woods Hole, 72:344-365.
Barnwell, F.H., 1963, Observations on daily and tidal rhythms in some fiddler crabs from equatorial Brazil, Biol. Bull. mar. biol. Lab. Woods Hole, 125:399-415.
Barnwell, F.H., 1968, The role of rhythmic systems in the adaptation of fiddler crabs to the intertidal zone, Am. Zoologist, 8:569-583.
Beever, J.W., D. Simberloff and L.L. King, 1979, Herbivory and predation by the mangrove tree crab Aratus pisonii, Oecologia (Berl.), 43:317-328.
Berry, A.J., 1972, The natural history of west Malaysian mangrove faunas, Malay. Nat. J., 25:135-162.
Berry, A.J., 1975, Molluscs colonizing mangrove trees with observations on Enigmonia rosea (Anomiidae), Proc. malac. Soc. Lond., 41:589-600.
Berry, A.J., R. Lim and A. Sase Kumar, 1973, Reproductive systems and breeding condition in Nerita birmanica (Archaeogastropoda: Neritacea) from Malayan mangrove swamps, J. Zool. Lond., 170:189-200.
Brown, D.S., 1971, Ecology of Gastropoda in a South African mangrove swamp, Proc. malac. Soc. Lond., 39:263-279.
Brown, F.A. and M.I. Sandeen, 1948, Responses of the chromatophores of the fiddler crab, Uca, to light and temperature, Physiol. Zool., 21:361-371.
Chapman, V.J., 1984, Mangrove biogeography, in: "Hydrobiology of the Mangal", F.D.Por and I.Dor, eds., W. Junk Publishers, The Hague.
Cockcroft, V.G. and A.T. Forbes, 1981, Tidal activity rhythms in the mangrove snail Cerithidea decollata (Linn.) (Gastropoda: Prosobranchia: Cerithiidae), S. Afr. J. Zool., 16:5-9.
Cook, L.M., 1986, Site selection in a polymorphic mangrove snail, Biol. J. Linn. Soc., 29:101-113.
Cook, L.M., and P.M. Freeman, 1986, Heating properties of morphs of the mangrove snail Littoraria pallescens, Biol. J. Linn. Soc., 29:295-300.
Cook, L.M., J.D. Currey and V.H. Sarsam, 1985, Differences in morphology in relation to microhabitat in littorinid species from a mangrove in Papua New Guinea (Mollusca: Gastropoda). J. Zool., 206:297-310.
Coomans, H.E., 1969, Biological aspects of mangrove molluscs in the West Indies, Malacologia, 9:79-84.
Crane, J., 1941, Eastern Pacific expeditions of the New York Zoological Society. 26. Crabs of the genus Uca from the west coast of Central America, Zoologica, 26:145-208.
Crane, J., 1975, "Fiddler Crabs of the World (Ocypodidae: genus Uca)", Princeton University Press, Princeton.
Day, J.H., 1967, The biology of the Knysna Estuary, South Africa, in: "Estuaries", G.H. Lauff, ed., A.A.A.S., Washington.
Dye, A.H. and T.A. Lasiak, 1986, Microbenthos, meiobenthos and fiddler crabs: trophic interaction in a tropical mangrove sediment, Mar. Ecol. Prog. Ser., 32:259-264.
Gifford, C.A., 1962, Some observations on the general biology of the land crab, Cardisoma guanhumi (Latreille), in south Florida, Biol. Bull. mar. biol. Lab. Woods Hole, 123:207-223.
Hartnoll, R.G., 1965, Notes on the marine grapsid crabs of Jamaica, Proc. Linn. Soc. Lond., 176:113-147.
Hartnoll, R.G., 1975, The Grapsidae and Ocypodidae (Decapoda: Brachyura) of Tanzania, J. Zool. Lond., 177:305-328.
Hill, B.J., 1979, Aspects of the feeding strategy of the predatory crab Scylla serrata, Mar. Biol., 55:209-214.
Icely, J.D. and D.A. Jones, 1978, Factor affecting the distribution of the genus Uca (Crustacea: Ocypodidae) on an East African shore, Estuar. cstl. mar. Sci., 6:315-325.
Jones, D.A., 1984, Crabs of the mangal ecosystem, in: "Hydrobiology of the Mangal", F.D. Por and I. Dor, eds., W. Junk Publishers, The Hague.
Little, C. and P. Stirling, 1984, Activation of a mangrove snail Littorina scabra scabra (L.) (Gastropoda: Prosobranchia), Aust. J. mar. freshw. Res., 35:607-610.

Macintosh, D.J., 1978, Some responses of tropical mangrove fiddler crabs (Uca spp.) to high environmental temperatures, in: "Physiology and Behaviour of Marine Organisms", D.S. McLusky and A.J. Berry, eds., Pergamon Press, Oxford.

Macintosh, D.J., 1979, Predation of fiddler crabs (Uca spp.) in estuarine mangroves, in: "Proceedings of the Symposium on Mangrove and Estuarine Vegetation in Southeast Asia, 1978, Serdang, Malaysia", Biotrop. Special Publ., 10:101-110.

Macintosh, D.J., 1982, Ecological comparisons of mangrove swamp and salt marsh fiddler crabs, in: "Wetlands: Ecology and Management", B. Gopal, R.E. Turner, R.G. Wetzel and D.F. Whigham, eds., National Institute of Ecology and International Scientific Pubblications, Jaipur.

Macintosh, D.J., 1984, Ecology and productivity of Malaysian mangrove crab populations (Decapoda: Brachyura), in: "Proceedings of the Asian Symposium on Mangrove Environment, Research and Management, Kuala Lumpur, 1984", E. Soepadmo, A.N. Rao and D.J. Macintosh, eds., University of Malaya and UNESCO.

Macnae, W., 1963, Mangrove swamps in South Africa, J. Ecol., 51:1-25.

Macnae, W., 1968, A general account of the fauna and flora of mangrove swamps and forests in the Indo-West-Pacific-region, Adv. mar. Biol., 6:73-270.

Macnae, W. and M. Kalk, 1962, The ecology of the mangrove swamps at Inhaca Island, Mocambique, J. Ecol., 50:19-34.

Macnae, W. and M. Kalk, eds., 1969, "A Natural History of Inhaca Island, Mocambique", Witwatersrand University Press, Johannesburg.

Malley, D.F., 1977, Adaptations of decapod crustaceans to life in mangrove swamps, Mar. Res. Indonesia., 18:63-72.

Miller, D.C., 1961, The feeding mechanism of fiddler crabs, with ecological considerations of feeding adaptations, Zoologica, N.Y., 46:89-100.

Morton, B.S., 1975, The diurnal rhythm and feeding response of the south east Asian mangrove bivalve, Geloina proxima Prime 1864 (Bivalvia: Corbiculacea), Forma Funct., 8:405-418.

Morton, B., 1976, The biology, ecology and functional aspects of feeding and digestion of the S.E. Asian mangrove bivalve, Enigmonia aenigmatica (Mollusca: Anomiacea), J. Zool. Lond., 179:437-466.

Morton, B., 1983, Mangrove bivalves, in: "The Mollusca", vol. 6, W.D. Russel-Hunter, ed., Academic Press, Orlando.

Odum, W.E. and E.J. Heald, 1972, Trophic analyses of an estuarine mangrove community, Bull. mar. Sci., 22:671-738.

Ong, T.L., 1978, Some aspects of trophic relationships of shallow water fishes (Selangor Coast), B. Sc. Honours Thesis, University of Malaya.

Pearse, A.S., 1912, The habits of fiddler crabs, Philippine J. Sci., 7:113-132.

Reid, D.G., 1985, Habitat and zonation patterns of Littoraria species (Gastropoda: Littorinidae) in Indo-Pacific mangrove forests, Biol. J. Linn. Soc., 26:39-68.

Salmon, M., 1984, The courtship, agression and mating system of a "primitive" fiddler crab (Uca vocans: Ocypodidae), Trans. zool. Soc. Lond., 37:1-50.

Salmon, M. and S.P. Atsaides, 1968, Visual and acoustical signalling during courtship by fiddler crabs (genus Uca), Am. Zoologist, 8:623-639.

Salmon, M., W.H. Seiple and S.G.Morgan, 1986, Hatching rhythms of fiddler crabs and associated species at Beaufort, North Carolina, J. crust. Biol., 6:24-36.

Sasekumar, A., 1974, Distribution of macrofauna on a Malayan mangrove shore, J. anim. Ecol., 43:51-69.

Sasekumar, A., T.L. Ong and K.L. Thong, 1980, Predation of mangrove fauna by marine fishes at high tide, in: "Proceedings of Asian Symposium on Mangrove Environment, Research and Management", E. Soepadmo, A.N. Rao and D.J. Macintosh, eds., University of Malaya and UNESCO.

Stebbins, R.C. and M. Kalk, 1961, Observations on the natural history of the mud-skipper, *Periophthalmus sobrinus*, *Copeia*, for 1961: 18-27.

Teal, J.M., 1959, Respiration of crabs in Georgia salt marshes and its relation to their ecology, *Physiol. Zool.*, 40:83-91.

Vermeij, G.J., 1974, Molluscs in mangrove swamps: physiognomy, diversity and regional differences, *Syst. Zool.*, 22:609-624.

Von Hagen, H.-O., 1970, Anpassungen an das spezielle Gezeitenzonen-Niveau bei Ocypodiden (Decapoda, Brachyura), *Forma Funct.*, 2:361-413.

Vorha, F.C., 1970, Some studies on *Cerithidea cingulata* (Gmelin 1790) on a Singapore sandy shore, *Proc. malac. Soc. Lond.*, 39:187-201.

Warner, G.F., 1967, The life history of the mangrove tree crab, *Aratus pisoni*, *J. Zool. Lond.*, 153:321-335.

Warren, J.H. and A.J. Underwood, 1986, Effects of burrowing crabs on the topography of mangrove swamps in New South Wales, *J. exp. mar. Biol. Ecol.*, 102:223-235.

Wilkens, J.L. and M. Fingerman, 1965, Heat tolerance and temperature relationships of the fiddler crab, *Uca pugilator*, with reference to body coloration, *Biol. Bull. mar. biol. Lab. Woods Hole*, 128:133-141.

Wilson, K.A., 1981, Tidal-associated feeding in the mangrove tree crab, *Aratus pisoni*, *Am. Zool.*, 21:1005 (abstract).

Zucker, N., 1978, Monthly reproductive cycles in three sympatric hood-building tropical fiddler crabs (genus *Uca*), *Biol. Bull. mar. biol. Lab. Woods Hole*, 155:410-424.

Physical Processes of the Coastal Sea

Franco Stravisi
Trieste University, Italy

INTRODUCTION

The aim of this lecture is to review some aspects of physical oceanography which can be of interest for people devoted to the study of the coastal environment. Marine or terrestrial life close to the sea is more or less dependent upon seawater. In this way, not only the regular or unregular changes of the sea level and the wave motion may be important; also some physical parameters like seawater temperature, salinity, transparency and the coastal current field and transport mechanisms must play a role at some point of the vital chain. It is therefore useful to keep in mind also these environmental factors for a better understanding of particular problems. In the following, no much more than a list of topics will be, for sake of necessity, presented in a way suitable for a not specialized reader. The wide literature existing in this field will fully cover every further requirement.

PHYSICAL CHARACTERISTICS OF SEAWATER IN THE SURFACE LAYER

Temperature, salinity and density

The maximum variability of the oceanic characteristics is found in the boundary layer, that is near the sea surface and near the coast. The seawater temperature variations are located above the permanent oceanic thermocline, where the sun energy is absorbed and converted into heat. While the mean temperature of the world ocean is something less than 3°C, temperature ranges of about 20°C are found at the sea surface during the year at middle latitudes. The ocean is heated from the surface by light energy absorption, and through the surface heat is lost by long wave radiation, evaporation and contact with the atmosphere. Heat waves propagate downward through the water column; during this propagation the amplitude of the wave is reduced and its phase is delayed. Fig. 1 shows the vertical profile of the mean annual temperature in a coastal seawater column (Gulf of Trieste; the depth is 23 m): a vertical gradient of 0.14°C/m is observed. Fig. 2 shows the temperature variations, at the same place, with the depth and through the year. In this figure, a vertical section gives the temperature profile at a fixed time during the year; a horizontal section describes the annual temperature cycle at a fixed depth.

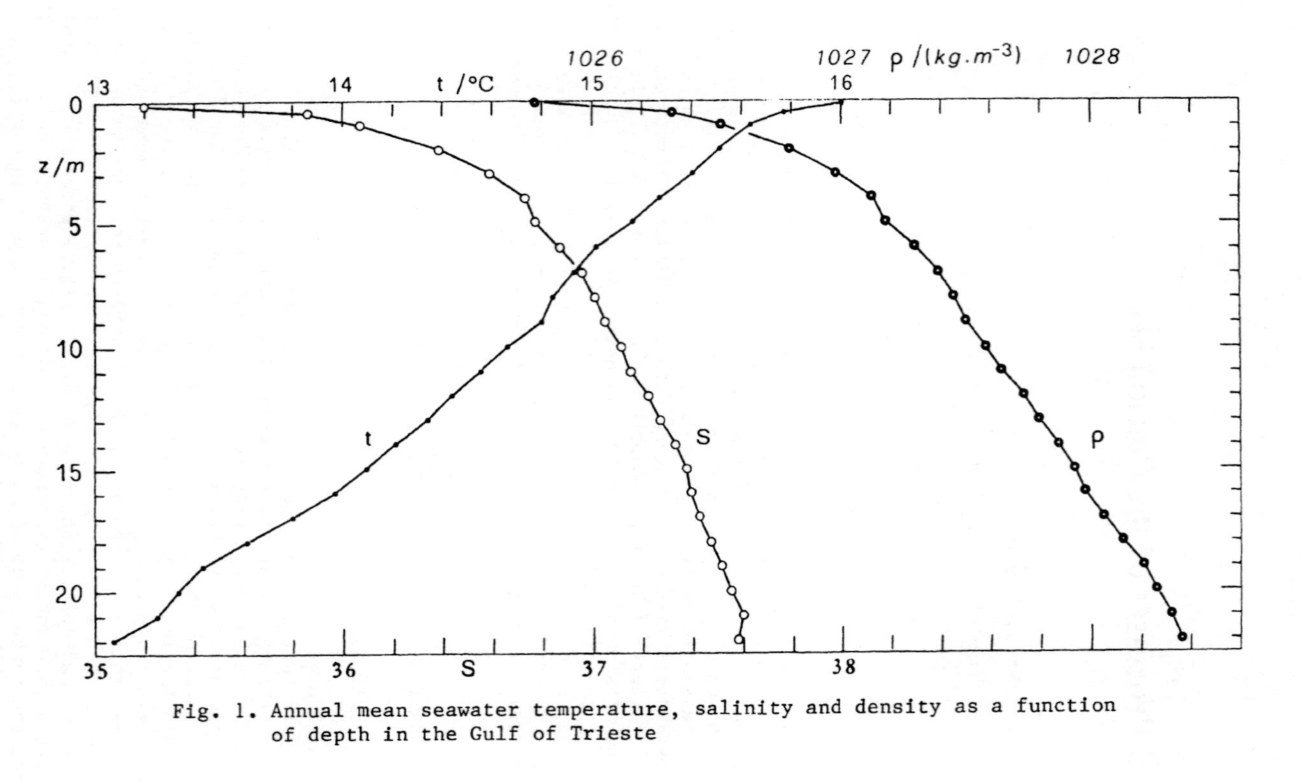

Fig. 1. Annual mean seawater temperature, salinity and density as a function of depth in the Gulf of Trieste

The absolute salinity is defined as the ratio between the mass of salt dissolved in a seawater sample and the mass of the sample. According to international agreements, the practical salinity S has been introduced (UNESCO, 1981a); by definition, S is computed as a function of the in situ measured seawater electrical conductivity, temperature and pressure. S is measured in psu (practical salinity units), adimensional; for example, S = 35 psu means that there arealmost exactly 35 g of salt per 1 kg of seawater: an older notation for this was S = 35 ‰. Figs. 1 and 3 show the vertical variations of salinity as an annual average and through the year; as for the temperature, the maximum variations are found in the surface layer. Daily and annual t,S variations can represent selective conditions for marine life.

The density ρ is defined as the ratio between the mass and the volume of a seawater sample. Units are kilograms per cubic meter. The seawater density exceeding 1000 kg/m^3 is usually called the density anomaly, γ. The seawater density is computed as a function of the in situ salinity, temperature and pressure according to the International Equation of State for Seawater (UNESCO, 1981b). Figs. 1, 4 show the vertical variations of the seawater density as an annual average and through the year: the station is the same of the preceding examples (Gulf of Trieste, 23 m). Vertical density gradients are important because they affect the buoyancy of marine plankton. Horizontal density gradients in the sea are related to horizontal currents.

The thermohaline parameters and the mass field are subject to changes owing both to local conditions (as heating or coastal runoff) and to advection of seawater with different characteristics. Changes by means of advection are particularly important near the shore.

Irradiance and transparency

The underwater radiant field, that is the light distribution as a function of direction and colour, is studied both theoretically and experimentally by physical oceanographers devoted to marine optics. Beyond any physical interest, the underwater light is essential to marine life.

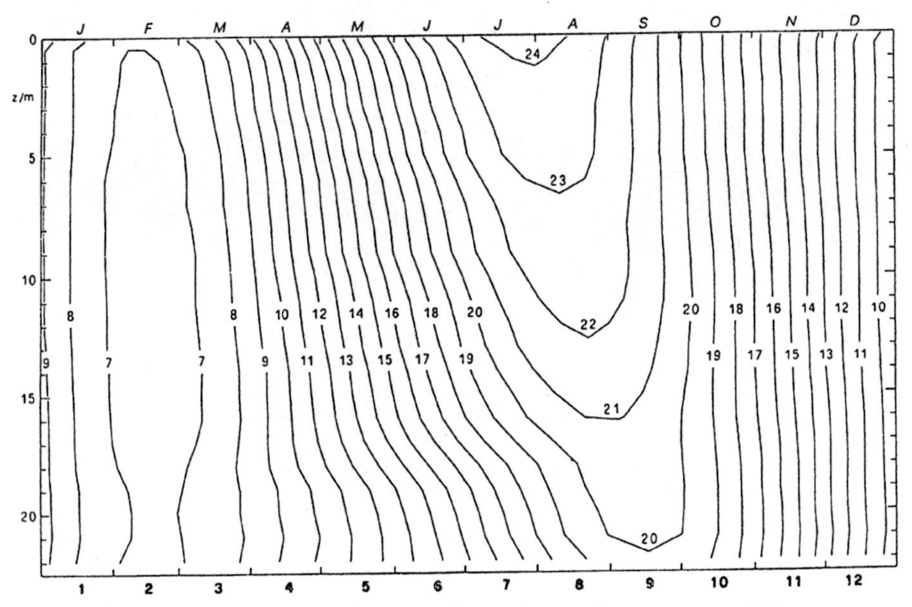

Fig. 2. Mean annual cycle of the seawater temperature /°C as a function of depth (Gulf of Trieste)

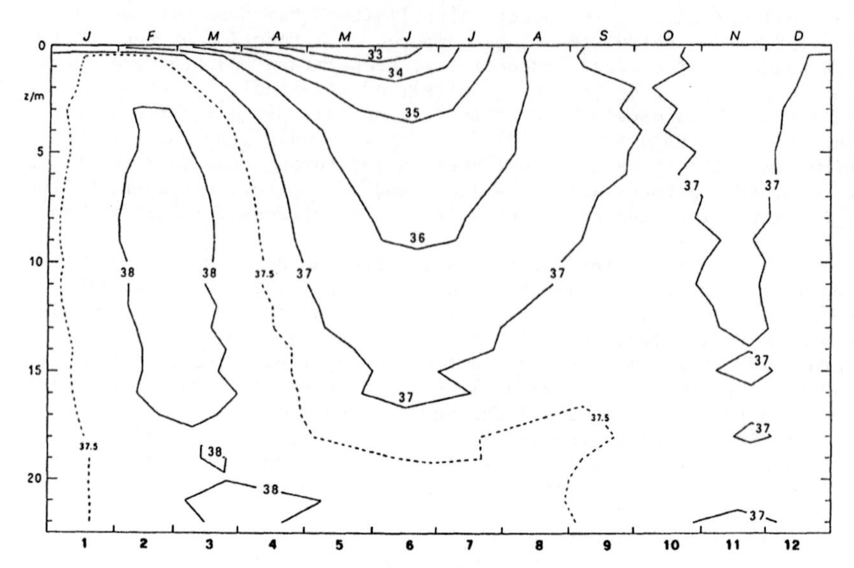

Fig. 3. Mean annual cycle of the seawater salinity /psu as a function of depth (Gulf of Trieste)

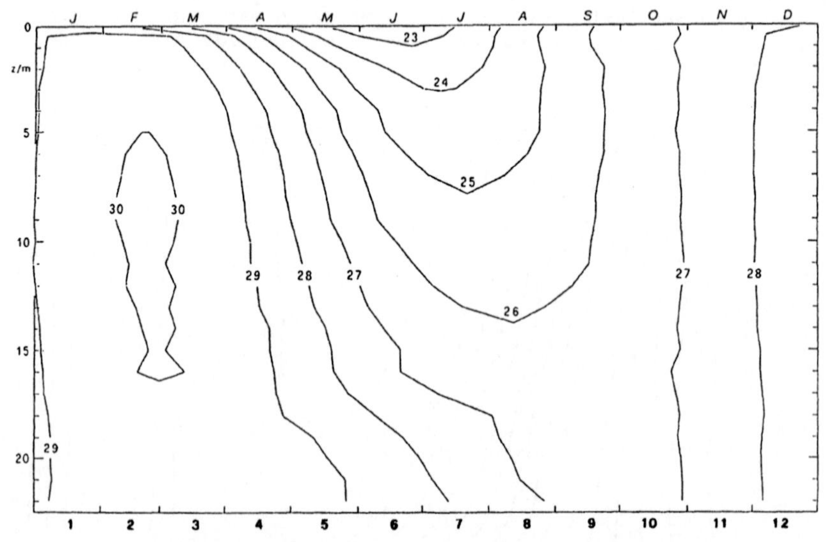

Fig. 4. Mean annual cycle of the seawater density /(kg/m^3) as a function of depth (Gulf of Trieste)

Commonly measured parameters are the underwater downwelling and upwelling irradiance, that is the light power (usually in the visible, or in the active band for the photosynthesis, PAR) impinging on a horizontal unitary surface facing upwards and respectively downwards. The downwelling irradiance is partly direct and partly diffuse light; the upwelling irradiance is diffuse light. Both irradiances vanish almost exponentially with the depth, according to the optical characteristics of seawater. Few centimeters below the sea surface the downwelling irradiance is reduced to about 60 % of the corresponding value above the sea, because of the absorption of the longer wavelengths (red).

The seawater transparency is often indicated by the depth at which a horizontal white disc (the Secchi disc), lowered vertically in the sea, becomes invisible to the observer. Typical Secchi disc depths in clear waters range from 10 m (shallow coastal waters) to 50 m (oligotrophic off-shore waters). At the Secchi disc depth, the downwelling PAR irradiance is about 10 % of the corresponding irradiance above the sea surface.

DYNAMICS OF LONG WAVES IN A BASIN

The laws of motion

Long waves in a basin are those having wavelengths of the order of the horizontal dimensions of the basin itself. Their motion is determined by the departure η of the (smooth) sea surface from the horizontal position at rest and by the average current velocity (the horizontal vector V) of the water column. Examples of long waves are tides, seiches and surges.

The physical laws governing the long wave motion are those describing the conservation of seawater momentum and mass.

The momentum equation (Fig. 5) can be written as follows:

$$\delta V/\delta t = - G + C - F + M + T. \tag{1}$$

This law represents the Newton's second law of dynamics applied to a vertical seawater column moving with velocity **V**. Specific forces at the right hand term are: the restoring force G due to the slope of the sea surface, the Coriolis acceleration C directed at a right angle (to the right in the northern, to the left in the southern emisphere), the bottom friction F, the meteorological force M exerted by the wind and by the atmospheric pressure gradient at the sea surface and the tidal force **T**.

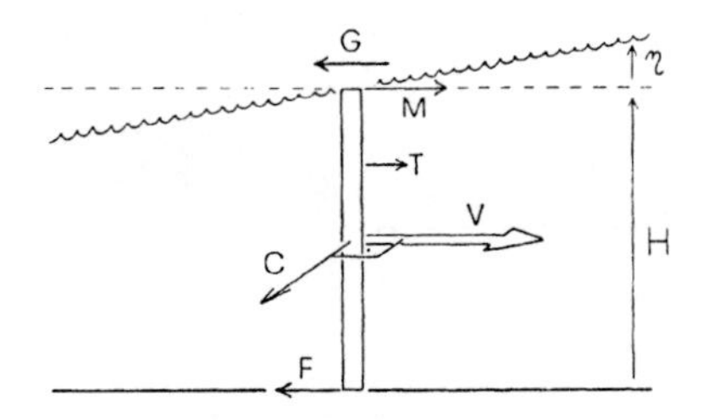

Fig. 5. The momentum equation

The continuity equation

$$\delta\eta/\delta t = - \text{div}(HV), \qquad (2)$$

where the term at the right represent the so called "divergence" of the water flux with a reversed sign, describes the conservation of the seawater volume (or mass) assuming a constant seawater density. This conservation law is illustrated by Fig. 6: a water flux HV positive towards a point is related to a sea level rise, a negative water flux to a sea level depression and vice-versa.

Sea surface elevation and current velocity of a lon wave must satisfy equations (1,2). The examples described in the following sections can be obtained, in the simplest cases, as analytical solutions of the equations of motion; in real cases, adequate solutions can be computed by means of numerical models.

Seiches

An important family of long waves is represented by the so called "seiches". These are the free standing modes of oscillation of a marine basin (Fig. 7) which occur each time its equilibrium has been perturbed by atmospheric events. Waves are possible only with velocity nodes at the opposite coasts, that are venters for the sea level. The first characteristic period for a longitudinal oscillation can be approximated by Merian's formula $T = 2L/\sqrt{gH}$, where $\sqrt{gH}$ (g is gravity) represents the long wave velocity for a mean basin depth H. If the basin is open at one side, 4L is considered instead of twice the basin length L. Let's consider for example the Adriatic Sea: figures can be $H = 200$ m, $L = 800$ km. The basin being open, the period of the main longitudinal seiche given by Merian's formula is 20 h; the observed period is indeed 21.5 h. The first transversal mode, if the width of the Adriatic Sea is 200 km, has a period of 5 h. Higher harmonics, or seiches of higher order, are possible: their periods are $T/2$, $T/3$ and so on.

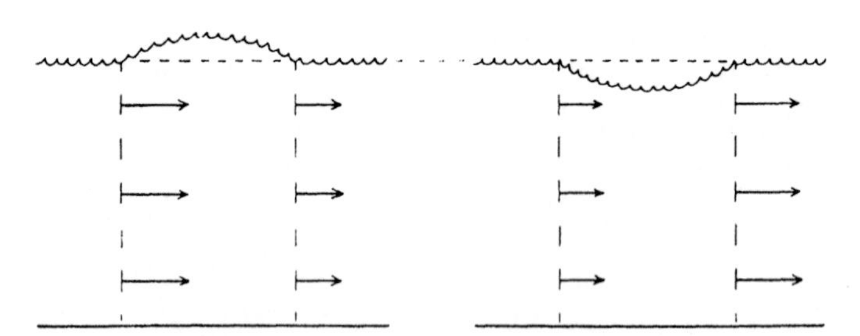

Fig. 6. The continuity equation

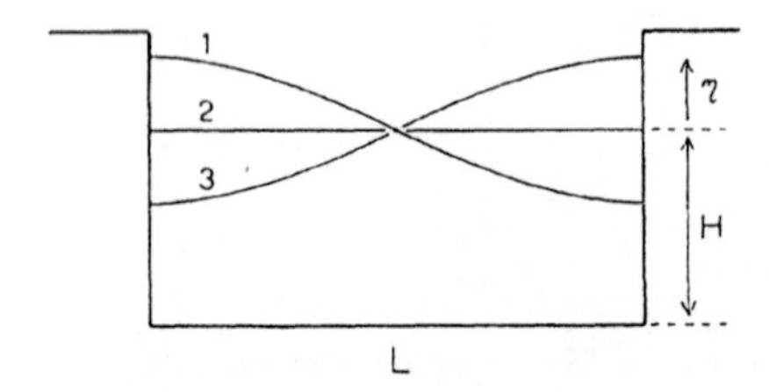

Fig. 7. The first seiche

A seiche can develope in a basin owing to meteorological perturbations; at the coast, maximum sea level elevations are reached at time intervals equal to the period of the seiche, that is of the order of hours. What is interesting is that each basin has a set of characteristic periods os oscillation which depend only on its geometry (shape, horizontal dimensions and depth). The seiches are the solutions of (1,2) in the absence of external forces ($M = T = 0$). Owing to the bottom friction F, the amplitude of a seiche decreases almost exponentially with time: many cycles can be generally observed. The effect of a seiche, as seen by an observer at the coast, is similar to the up and down sea level displacement produced by the astronomical tide.

Storm surges

Surges are aperiodic sea level elevations set up along a downwind coast by the wind and by the horizontal gradient of the atmospheric pressure (that is by a "storm") acting on the sea surface in a basin. If M is the intensity of the meteorological force in equation (1), the slope of the sea surface is M/g and the sea level rise at the coast is $s = ML/g$. During a storm, the seiche elevation is usually added to the surge: if this sea level rise is in phase with the astronomical tide, large coastal floods can occur. The amplitude of these phenomena is increased by the decreasing depth of the basin toward the coast.

THE ASTRONOMICAL TIDE

The tidal force

Astronomical tide is the name reserved to that quasi-periodic component of the sea level change which finds its origin in the presence near the Earth of the moon and the sun, being the effect of other celestial bodies negligible.

The genesis of the tidal force is easily explained by considering the revolution of two celestial bodies (Earth and moon, or Earth and sun) around their center of mass: on the Earth, the centrifugal force C is the same everywhere, while the gravitational force G exerted by the body increases with the inverse square of the distance. The tidal force is the resultant $T = C + G$; it is $T = 0$ at the center of the Earth (since the system is stable), $T \neq 0$ at all other points on the solid earth, ocean or atmosphere. On the earth surface, T is distributed symmetrically around the Earth-body axis (Fig. 10). The tide generating forces of the moon and of the sun, which are in a ratio about 2:1, add together: the resultant lunisolar tidal force has a maximum intensity when the three bodies are aligned, that is during the full and the new moon. At each point of the ocean, the horizontal component of the tidal force acts as a term T in the momentum equation (1), while the vertical component of the tidal force affects the earth gravity by a negligible amount.

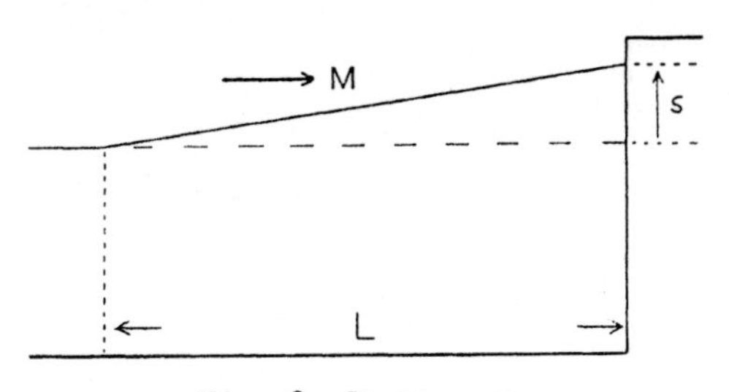

Fig. 8. Storm surge

The tidal force therefore acts on a water column (Fig. 5) in the same way that does the meteorological force; but the tidal force has a permanent cyclic behaviour owing to the regularity of the celestial motions.

The harmonic method of tidal analysis and prediction

The position of the moon and of the sun with respect to the Earth can be described by means of harmonic functions, the periods of which are known: the same periods must therefore characterize the lunisolar tidal force acting on the Earth and also its effects, in particular the astronomical oceanic tides and tidal currents. This is the principle of the harmonic method of tidal analysis and prediction: the astronomical tide is reproduced by means of the sum of a number of "tidal constituents" or "component tides". A component tide is a sinusoidal variation of the sea level: its

Table 1. Periods and relative amplitudes of the main harmonic components of the tidal force

component tide	M_2	S_2	N_2	K_2	K_1	O_1	P_1
period/h	12.42	12.00	12.66	11.97	23.93	25.82	24.07
rel. amplitude	100	47	19	13	58	42	19

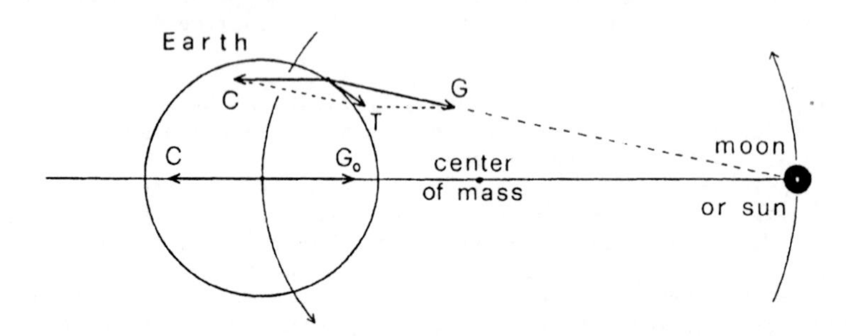

Fig. 9. The tidal force

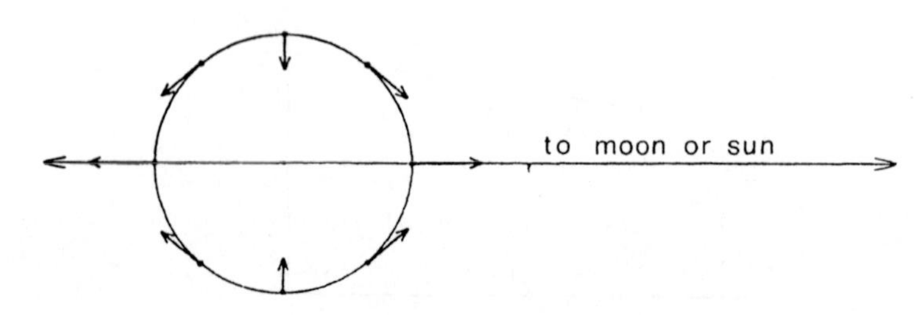

Fig. 10. Distribution of the tidal force

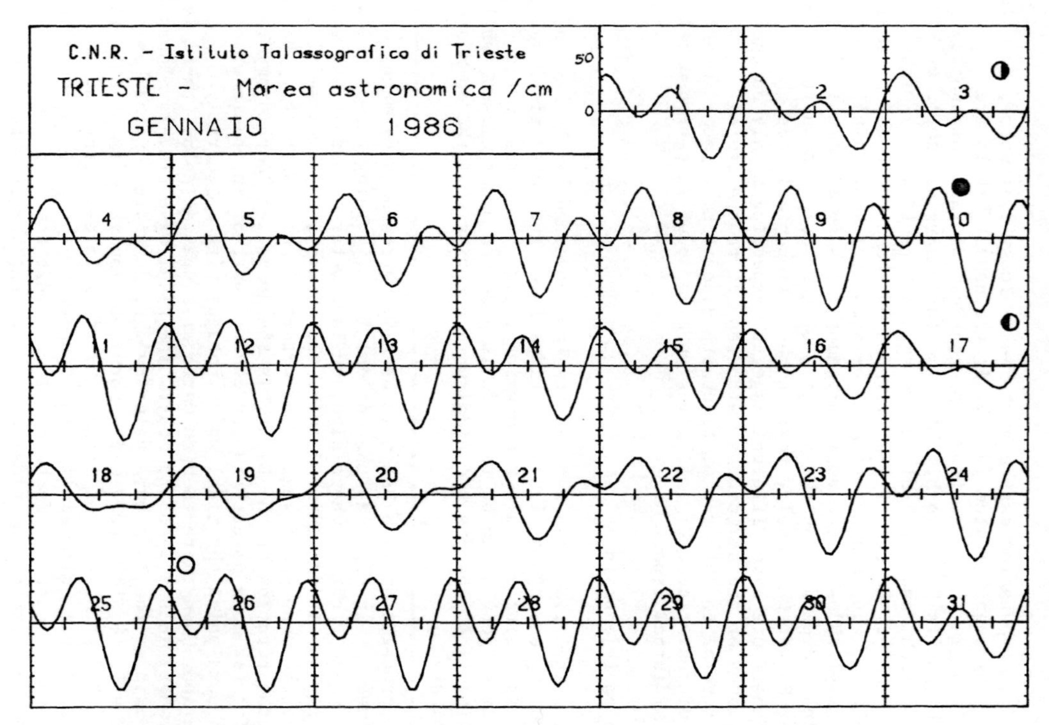

Fig. 11. Astronomical (mainly semidiurnal) tide computed for Trieste

period is exactly known from astronomy. In order to determine a component tide, one has to know also its amplitude and its phase: these two parameters, called "harmonic constants", are characteristic of each point of the ocean. The harmonic constants are computed by performing a suitable analysis on the local sea level records; once that the harmonic constants have been determined, they can be used to compute the local astronomical tide at any desired instant of time, past or future. Annual tide tables are commonly computed and printed in this way for the major harbours in the world; high and low water heights and times are usually derived from the tidal function.

A good tidal computation can be performed, almost everywhere, by using the seven tidal components listed in Tab. 1; the usual symbols are indicated. Periods are in hours; there are four semidiurnal components, indicated by the suffix "2" (two high and two low waters per day), with periods close to 12 h, and three diurnal components (suffix "1") with periods close to 24 h. The relative amplitudes of the corresponding tidal forces are reported, with respect to M , the principal semidiurnal lunar component. The amplitudes of the real component tides in a basin are strongly variable, depending on the geometry of the basin and on its location on the Earth: the result, that is the astronomical tide one observes, can therefore have different aspects, depending mainly on the relative amplitudes of the semidiurnal with respect to the diurnal components. According to this ratio, tides are classified as semidiurnal, mixed type or diurnal tides. Fig. 11 gives an example of an astronomical tide of the semidiurnal type.

Tidal propagation in a basin

The astronomical tide is a wave of the kind governed by the laws of motion described by equations (1,2). The tidal force generates tidal waves in the oceans; in basins of smaller dimensions, like the Adriatic Sea for example, the direct generation of the tide is a second order effect, while the cooscillating tide, that is the tidal wave entering from the open end of the basin, dominates.

The tidal propagation in a basin is usually represented by means of two families of lines, as in Fig. 12: the cotidal lines, joining the points with the same tidal phase (that is where high, or low, water occurs at the same time) and the iso-amplitude lines, joining the points with the same tidal elevation above the mean sea level. A characteristic of almost every basin, deriving from the rotation of the Earth, is the so called "amphidromy": there is a point (or more points) in the central part of the basin with no tide at all, around which the tidal waves rotates with a generally semidiurnal period. When there is high water at a coastal point, there is low water somewhere on the opposite coast. The tidal wave travels around the coast of the basin at the long wave velocity $\sqrt{gH}$, depending on the basin depth H; for H = 1, 10, 100, 1000 m the corresponding velocity of the tide is 11, 36, 113, 356 km/h. This explaines why tides in estuaries and in coastal lagoons advance quite slowly with respect to the propagation along the coast. The above velocity of the tidal wave must not be confused of course with the velocity of the tidal currents, that hardly reach the order of 1·m/s in narrow straits.

The tidal waves can be amplified in marginal seas because of the shallow water effect, coastal morphology and resonance. For example, tidal ranges about 1 m are observed in the northern Adriatic Sea, which has proper periods (the periods of seiches) of 21.5 and 11 h, close to the tidal periods; tidal ranges of the order of 10 m are observed in the English Channel which faces the great volume of the Atlantic Ocean.

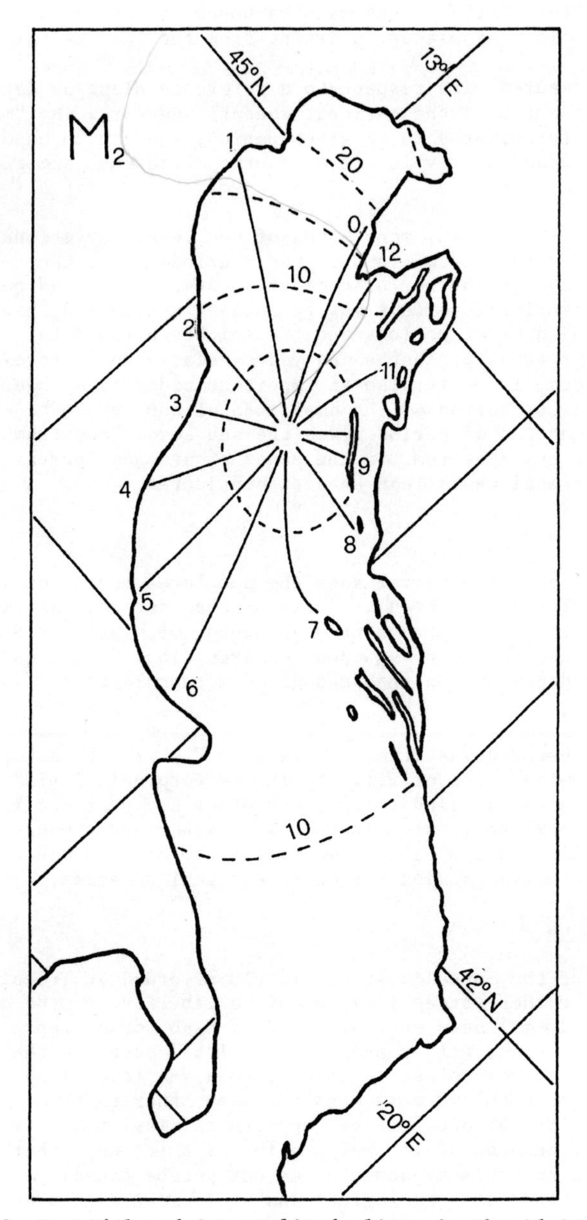

Fig. 12. Co-tidal and iso-amplitude lines in the Adriatic Sea

COASTAL SEA LEVEL VARIATIONS

The spectrum of the sea level variations

The sea level variations are commonly measured by means of suitable tide gauges, generally located in the main harbours of each country. Sea level records exist, in many places, starting from the last century.

Sea level is measured with respect to a reference plane or datum which is arbitrary but connected to the national geodetic network. The "mean sea level" at a station is computed for a given period, and can be used for practical purposes; mean sea levels are of course different, according to the time interval considered.

There are, as we have seen, many kinds of sea level variations: non periodic, as surges or long term trends, quasi-periodic like the astronomical tide, composed by periodic constituents, and damped periodic like seiches. Characteristic periods may be however considered, ranging from tidal semidiurnal and diurnal periods and the seiches periods (of the order of hours), to periods of days, months and years related to meteorological and climatic processes. There is also an important tidal (the nodal) period of 18.6 years. The distribution of the amplitude of the sea level oscillations as a function of period gives the sea level "spectrum". The short period part of the spectrum, in the range of seconds, pertaining to the surface gravitational waves, can be also considered.

Daily variations

At a coastal point, an observer sees the sea level moving on average up and down rather regularly as a result mainly of the astronomical tide and of seiches. While the first is a permanent phenomenon, with greater excursions during the full and the new moon, and smaller excursions during quadratures, seiches are closely dependent on the passage of meteorological perturbations.

An example of observed sea level is given in Fig. 13; by subtracting the computed astronomical tide of Fig. 11, the meteorological tide of Fig. 14 is obtained: there we can find seiches and other non periodic long term variations related to meteoclimatic factors like winds and atmospheric pressure. These variations can be to some extent computed in advance, by means of numerical models, knowing the meteorological forecast.

Interannual variations

A time series of the annual mean sea levels recorded at Trieste is shown in Fig. 15. A rather strong interannual variability, of the order of centimeters, is found. A linear secular trend is also found, representing a relative sea level increase with respect to the local datum at the rate of 13 cm per century. In order to say if we observe a sea level rise, or a coastal subsidence, or both, we must consider many other informations. From world-wide measurements the ocean indeed appears to rise at a rate of 15 cm per century, perhaps because of ice melting in the polar cups (but a reliable explanation for this phenomenon has not yet be found). Around the world, however, one finds many geological and biological evidences of ancient coastlines different from the actual ones.

SURFACE GRAVITY WAVES

The sea surface supports gravitational waves generated by meteorological forces and propagating with wave speeds of the order of

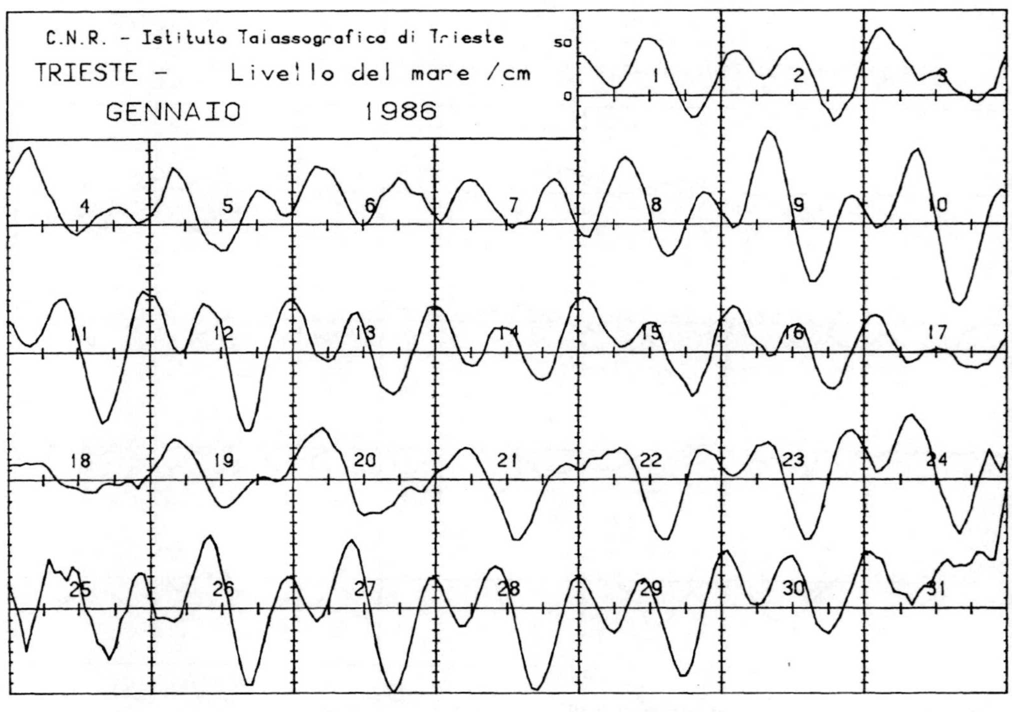

Fig. 13. Observed sea level at Trieste

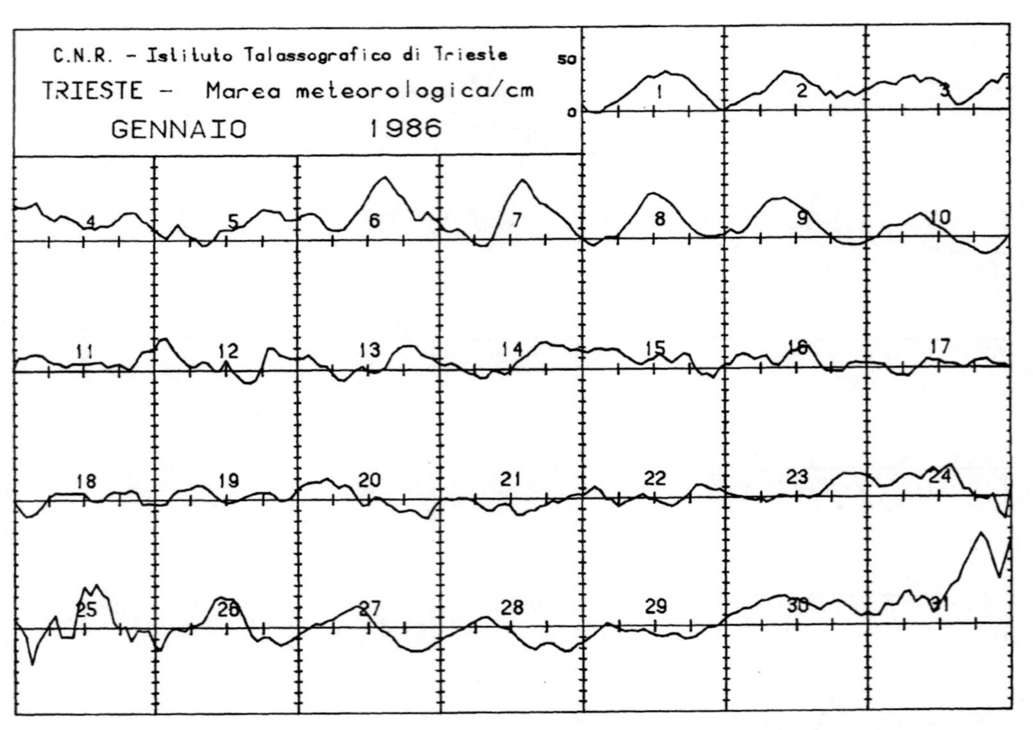

Fig. 14. Meteorological (observed minus astronomical) tide at Trieste

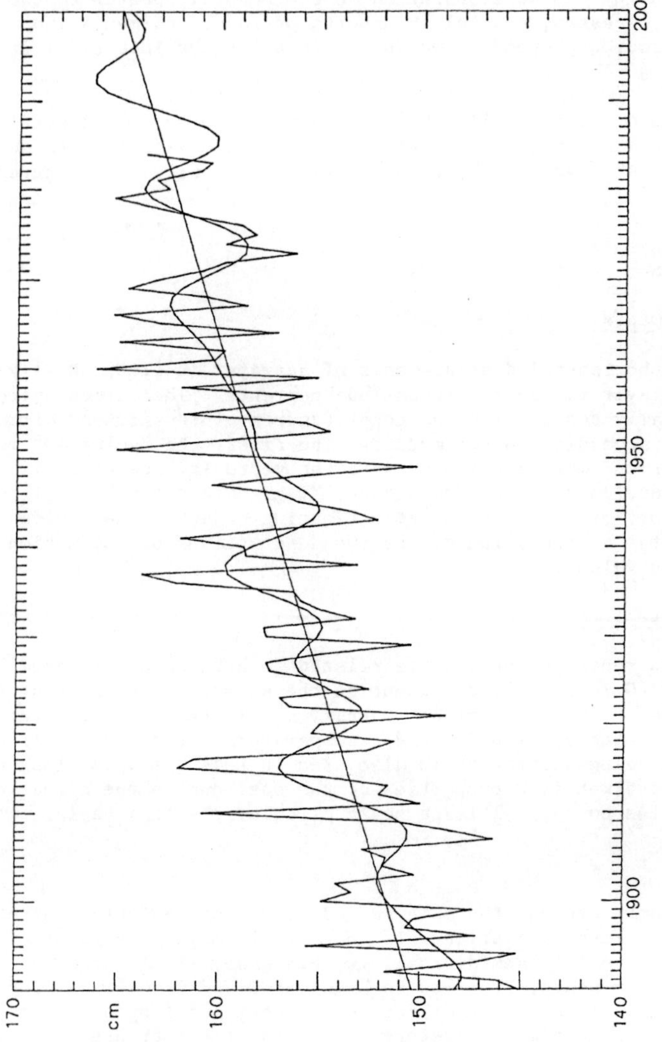

Fig. 15. Annual mean sea levels at Trieste

meters per second, with wavelengths of the order of meters. They affect the ocean to a depth of the order of a wavelength and are therefore called "surface waves", in opposition with long waves that interest the whole water column.

Moving towards the coast in shallow water, the surface waves begin to "feel the bottom": their direction is refracted so that the wave fronts tend to be parallel to the coastline.

The wave energy can be transferred to the coastal zone or to the coast itself: mixing processes, coastal transports of sand, sediments and organisms and erosion phenomena, which can interest the intertidal life, occur.

The absence of waves, following the absence of wind, for a sufficiently long period, can lead in some coastal areas to a calm situation favourable to the formation of "red tides", by permitting local blooms and aggregations of plankton.

COASTAL CURRENTS

Current measurements

The almost horizontal displacements of seawater in time, usually differing from layer to layer, are called "currents". The current velocity, which is physically described by a vector (current intensity and direction), can be measured according to two methods. The first (the "eulerian" method) makes use of current meters which measure the vectorial seawater velocity at a point as a function of time. The second ("lagrangian") method follows the current trajectory by tracking suitable floating objects. The choice of the method is a matter of convenience, the two descriptions of the motion being in principle equivalent.

Gradient currents

Gradient currents in the sea are related to horizontal pressure gradient forces (G in eq. 1), dependent on the seawater density distribution and on the slope of the sea surface. Density gradients arise from the presence of water masses with different temperature and salinity, so that the term "thermohaline currents" is also used in this case. The long term changes of the thermohaline properties of the sea, due to the climatic forcing, give rise to typical large scale circulations in a basin.

Tidal currents

Tidal currents present the same periods and features which characterize the lunisolar sea level variations. If the tide is mainly semidiurnal also tidal currents are mainly semidiurnal. In this case, tidal currents alternate every six hours, from low to high water and vice-versa. The water column moves almost in opposite directions, covering a length of the order of few kilometers each time. Therefore, the total tidal transport during a tidal cycle may be negligible; in the presence of strong tidal velocities, an efficient coastal tidal mixing can be however set up.

Wind currents

Wind or drift currents are driven by the air in motion over the sea surface. Deeper sea layers are moved owing to friction: this motion affects the upper ocean to a depth of the order of one hundred meters (the "Ekman layer").

The circulation of the shallow water coastal seas is strongly affected on the diurnal scale by local sea and land breezes, and at longer time scales by dominant winds. Large scale main gyres are set up in a gulf or bay, together with coastal gyres of smaller dimensions. The velocity of the wind driven current is generally strongly decreasing with the depth. The wind driven currents, which can be very efficient as regards the water transport, add to gradient and to tidal currents, giving a resultant circulation which can be highly complicated and locally unpredictable.

Strong permanent winds almost parallel to a straight coast give rise to important phenomena known as "upwelling" and "downwelling". Owing to the

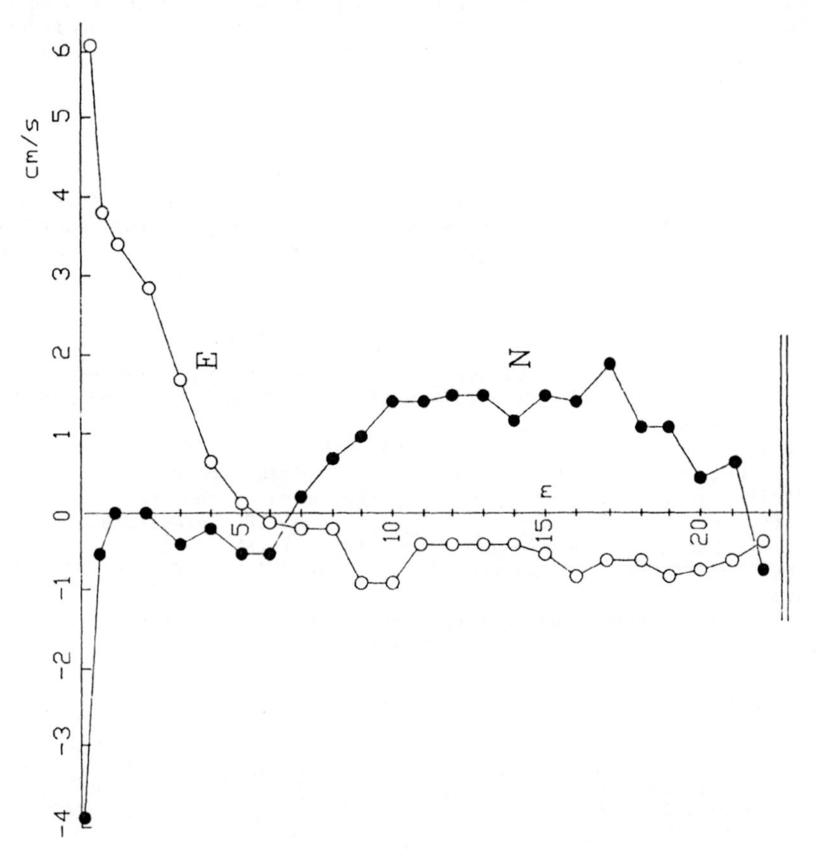

Fig. 16. Average vertical profile of the current velocity N, E components at a point in the Gulf of Trieste

Coriolis force, the wind driven transport in the open sea is at the right (in the northern emisphere) of the direction towards which the wind is blowing. In this way, for example, if a southerly wind is blowing (towards north) along a coast at its left along the same direction, the sea surface transport will be off-shore to the east: a bottom compensation current will move off-shore deep water westwards to the coastal surface layer. This wind produced coastal rise of bottom water is called upwelling; in the same case, a northerly wind would have produced downwelling. These in-shore/off-shore water transports have important effects on the productivity of the coastal sea and on the marine life.

Variability of the current field

The marine circulation in the coastal waters is the result of gradient currents with typical periods of the order of days or more, tidal current and wind currents with mainly a daily period. Turbulence gives rise to gyres of different scales, from meters to kilometers, so that the current velocity at a point can be strongly variable as regards both intensity and direction. A more stable pattern can be obtained by averaging many current measurements, giving the characteristic long term transport in the coastal zone.

Fig. 16 shows a typical average velocity profile from the sea surface to the bottom (23 m) obtained by means of eulerian measurements at a point in the Gulf of Trieste. An important feature which must be stressed is the strong velocity gradient observed at the surface: the current velocity in general rapidly decreases and turns in the first decimeters. A two-layer circulation is distinguished in this case: a surface (6 m) mainly wind driven current flows over a bottom mainly thermohaline current in the opposite direction.

Near the coast, currents tend to flow parallel to the coastline and, in order to save transport, their intensity tend to increase as the water depth decreases.

REFERENCES

Defant, A., 1961, "Physical Oceanography", 2 vols., Pergamon Press.

Neumann, G., and Pierson, W., 1966, "Principles of Physical Oceancgraphy" Prentice Hall.

Sverdrup, H.U., Johnson, M.W., and Fleming, R.H., 1959, "The Oceans: their Physics, Chemistry and General Biology", Prentice-Hall.

UNESCO, 1981, Background papers and supporting data on the Practical Salinity Scale 1978, Technical Papers in Marine Science, 37.

UNESCO, 1981, Background papers and supporting data on the International Equation of State of Seawater 1980, Technical Papers in Marine Science, 38.

UNESCO, 1983, Algorithms for computation of fundamental properties of seawater, Technical Papers in Marine Science, 44.

UNESCO, 1985, The International System of Units (SI) in Oceanography, Technical Papers in Marine Science, 45.

Participants

P. A. Abrams
Department of Ecol. Behav. Biology
University of Minnesota
Minneapolis, MN 55455, U.S.A.

V. C. Almada
I.S.P.A.
Rua Jardim Tabaco 44
1100 Lisboa, Portugal

A. D. Ansell
Dunstaffnage Marine Res. Laboratory
P.O.Box 3, Oban
Argyll - Scotland, U.K.

A. Barkai
Zoology Department
University of Cape Town
Rodenbosch 7700, South Africa

R. S. K. Barnes
University of Cambridge
Department of Zoology
Downing St., Cambridge CB2 3EJ, U.K.

G. Beugnon
Centre de Rech. en Biologie du Comportement
Univ. P. Sabatier, rue de Narbonne 118
Toulouse 31062 France

A. L. Bosman
P. FitzPatrick Inst. Afr. Ornithology
University of Cape Town
Rondebosch 7700, South Africa

E. Bourget
Departement de Biologie
Universite' Laval
Quebec G1l 7P4, Canada

R. C. Brace
Depatment of Zoology
University of Nottingham
University Park - Nottingham, U.K.

G. M. Branch
Zoology Department
University of Cape Town
Rodenbosch 7700, South Africa

R. L. Caldwell
Department of Zoology
University of California
Berkeley, California 94720, U.S.A.

G. Chelazzi
Dipartimento di Biologia
Animale e Genetica
v. Romana 17, 50125 Firenze, Italy

J.-L. Deneubourg
Service Chimie-physique II CP231
Universite' Libre de Bruxelles
Bd du Triomphe, Bruxelles, Belgium

D. W. Dunham
Department of Zoology
University of Toronto
Toronto, Ontario, M5S 1A1 Canada

A. Ercolini
Dipartimento di Biologia
Animale e Genetica
v. Romana 17, 50125 Firenze, Italy

P. R. Evans
Department of Zoology
University of Durham
South Road, Durham, DH1 3LE, U.K.

Ali Said Faqi
National University
Mogadishu
Somalia

S. Focardi
Ist. Naz. Biologia della Selvaggina
v. Stradelli Guelfi 23/a
Ozzano, Emilia, Italy

F. Gherardi
Dipartimento di Biologia
Animale e Genetica
v. Romana 17, 50125 Firenze, Italy

R. N. Gibson
Dunstaffnage Marine Res. Laboratory
P.O.Box 3, Oban
Argyll - Scotland, U.K.

R. G. Hartnoll
Marine Biology Station
Port Erin, Isle of Man
U.K.

B. A. Hazlett
Division of Biological Sciences
University of Michigan
Ann Arbor, Michigan 48109 U.S.A.

W. F. Herrnkind
Department of Biological Science
Florida State University
Tallahassee, Florida 32306 U.S.A.

P. A. R. Hockey
P. FitzPatrick Inst. African Ornithology
University of Cape Town
Rondebosch 7700 South Africa

R. N. Hughes
School of Animal Biology
University College of North Wales
Bangor - Gwynedd, LL57 2UW U.K.

A. MacLachlan
Zoology Dept., Univ. of Port Elizabeth
P.O.Box 1600, Port Elizabeth 6000
South Africa

C. D. McQuaid
Dep. of Zoology & Entomology
Rhodes University, P.O.Box 994
Grahamstown 6140 South Africa

E. Naylor
School of Animal Biology
University College of North Wales
Bangor, Gwynedd LL57 2UW U.K.

D. Neumann
Department of Zoology
Physiol. Ecol. Section
University of Koln, West Germany

L. Pardi
Dipartimento di Biologia
Animale e Genetica
v. Romana 17, 50125 Firenze, Italy

M. Salmon
Dept. of Ecology, Ethology & Evolution
Univ. of Illinois at Urbana-Champaign
Champaign, Illinois 61820 U.S.A.

R. Santos
Dep. of Oceanography & Fish.
University of Azores
Horta 9900 Faial, Azores, Portugal

F. Scapini
Dipartimento di Biologia
Animale e Genetica
v. Romana 17, 50125 Firenze, Italy

P. A. Scherman
Dep. of Zoology & Entomology
Rhodes University, P.O.Box 994
Grahamstown 6140 South Africa

F. Stravisi
Istituto Talassografico di Trieste
v.le R. Gessi 2
34123 Trieste, Italy

A. Ugolini
Dipartimento di Biologia
Animale e Genetica
v. Romana 17, 50125 Firenze, Italy

A. J. Underwood
Dept. of Zoology, School of Biol. Sciences
Sydney University
Sydney, N.S.W. 2006 Australia

M. Vannini
Dipartimento di Biologia
Animale e Genetica
v. Romana 17, 50125 Firenze, Italy

L. West
Hopkins Marine Station
Stanford University
Pacific Grove, California 93050, U.S.A.

D. L. Wolcott
Zoology Department, Box 5577
North Carolina State University
Raleigh, North Carolina 27607, U.S.A.

T. G. Wolcott
Zoology Department, Box 5577
North Carolina State University
Raleigh, North Carolina 27607, U.S.A.

N. Zucker
Department of Biology
New Mexico State University
Las Cruces, New Mexico 88003, U.S.A.

Species Index

Subject Index